48.95
67R

VLSI TECHNOLOGY

This book was set in Times Roman by Information Sciences Corporation.
The editors were T. Michael Slaughter and Madelaine Eichberg;
The production supervisor was Leroy A. Young.
The cover was designed by Joseph Gillians.
The drawings were done by Bell Laboratories, Incorporated.
Halliday Lithograph Corporation was printer and binder.

VLSI TECHNOLOGY

567890HALHAL8987

ISBN 0-07-062686-3

Library of Congress Cataloging in Publication Data
Main entry under title:

VLSI technology.

(McGraw-Hill series in electrical engineering.
Electronics and electronic circuits)
Includes index.
1. Integrated circuits—Very large scale
integration.   I. Sze, S. M., date
II. Series.
TK7874.V566   1983   621.381'73   82-24947
ISBN 0-07-062686-3

McGraw-Hill Series in Electrical Engineering

Consulting Editor
Stephen W. Director, Carnegie-Mellon University

Networks and Systems
Communications and Information Theory
Control Theory
Electronics and Electronic Circuits
Power and Energy
Electromagnetics
Computer Engineering
Introductory and Survey
Radio, Television, Radar, and Antennas

Previous Consulting Editors

Ronald M. Bracewell, Colin Cherry, James F. Gibbons, Willis W. Harman, Hubert Heffner, Edward W. Herold, John G. Linvill, Simon Ramo, Ronald A. Rohrer, Anthony E. Siegman, Charles Susskind, Frederick E. Terman, John G. Truxal, Ernst Weber, and John R. Whinnery

Electronics and Electronic Circuits

Consulting Editor
Stephen W. Director, Carnegie-Mellon University

Gault and Pimmel: *Introduction to Microcomputer-Based Digital Systems*
Grinich and Jackson: *Introduction to Integrated Circuits*
Hamilton and Howard: *Basic Integrated Circuits Engineering*
Hodges and Jackson: *Analysis and Design of Digital Integrated Circuits*
Hubert: *Electric Circuits AC/DC: An Integrated Approach*
Millman: *Microelectronics: Digital and Analog Circuits and Systems*
Millman and Halkias: *Integrated Electronics: Analog, Digital Circuits, and Systems*
Millman and Taub: *Pulse, Digital, and Switching Waveforms*
Peatman: *Microcomputer Based Design*
Pettit and McWhorter: *Electronic Switching, Timing, and Pulse Circuits*
Schilling and Belove: *Electronic Circuits: Discrete and Integrated*
Strauss: *Wave Generation and Shaping*
Sze: *VLSI Technology*
Taub: *Digital Circuits and Microprocessors*
Taub and Schilling: *Digital Integrated Electronics*
Wait, Huelsman, and Korn: *Introduction to Operational and Amplifier Theory Applications*
Wert and Thompson: *Physics of Solids*
Wiatrowski and House: *Logic Circuits and Microcomputer Systems*
Yang: *Fundamentals of Semiconductor Devices*

TECHNOL

Ed
S. M

Bell Laboratories, Incor
Murray Hill, New

McGraw-Hill Book Company

New York   St. Louis   San Francisco   Auckland   Bogotá   Hamburg
London   Madrid   Mexico   Montreal   New Delhi
Panama   Paris   São Paulo   Singapore   Sydney   Tokyo   Toronto

CONTENTS

TK 7874
V566
1983
Chem

v

LIST OF CONTRIBUTORS

A. C. ADAMS
Bell Laboratories
Murray Hill, New Jersey

W. J. BERTRAM
Bell Laboratories
Allentown, Pennsylvania

W. FICHTNER
Bell Laboratories
Murray Hill, New Jersey

D. B. FRASER
Bell Laboratories
Murray Hill, New Jersey

L. E. KATZ
Bell Laboratories
Allentown, Pennsylvania

R. B. MARCUS
Bell Laboratories
Murray Hill, New Jersey

D. A. McGILLIS
Bell Laboratories
Allentown, Pennsylvania

C. J. MOGAB
Bell Laboratories
Murray Hill, New Jersey

L. C. PARRILLO
Bell Laboratories
Murray Hill, New Jersey

C. W. PEARCE
Western Electric
Allentown, Pennsylvania

T. E. SEIDEL
Bell Laboratories
Murray Hill, New Jersey

C. A. STEIDEL
Bell Laboratories
Allentown, Pennsylvania

J. C. C. TSAI
Bell Laboratories
Reading, Pennsylvania

PREFACE

VLSI Technology describes the theoretical and practical aspects of the most advanced state of electronics technology—very-large-scale integration (VLSI). From crystal growth to reliability testing, the reader is presented with all the major steps in the fabrication of VLSI circuits. In addition many broader topics, such as process simulation and diagnostic techniques, are considered in detail. Each chapter describes one aspect of VLSI processing. The chapter's introduction provides a general discussion of the topic, and subsequent sections present the basic science underlying individual process steps, the necessity for particular steps in achieving required parameters, and the trade-offs in optimizing device performance and manufacturability. The problems at the end of each chapter form an integral part of the development of the topic.

The book is intended as a textbook for senior undergraduate or first-year graduate students in electrical engineering, applied physics, and materials science; it assumes that the reader has already acquired an introductory understanding of the physics and technology of semiconductor devices. Because it elaborates on IC processing technology in a detailed and comprehensive manner, it can also serve as a reference for those actively involved in integrated circuit fabrication and process development.

This text began in 1979 as a set of lecture notes prepared by the contributing authors for an in-hours continuing education course at Bell Laboratories. The course, called "Silicon Integrated Circuit Processing," has been given to hundreds of engineers and scientists engaged in research, development, fabrication, and application work of ICs. We have substantially expanded and updated the lecture notes to include the most advanced and important topics in VLSI processing.

In the course of writing *VLSI Technology*, many people have assisted us and offered their support. We would first like to express our appreciation to the management of Bell Laboratories and Western Electric for providing the environment in which we worked on the book. Without their support, this book could not have been written. We have benefited significantly from suggestions made by the reviewers: Drs. L. P. Adda, C. M. Bailey, K. E. Benson, J. E. Berthold, J. B. Bindell, J. H. Bruning, R. E. Caffrey, C. C. Chang, D. L. Flamm, G. K. Herb, R. E. Howard,

E. Kinsbron, P. H. Langer, M. P. Lepselter, J. R. Ligenza, P. S. D. Lin, W. Lin, C. M. Melliar Smith, D. F. Munro, S. P. Murarka, E. H. Nicollian, R. B. Penumalli, J. M. Poate, M. Robinson, D. J. Rose, G. A. Rozgonyi, G. E. Smith, J. W. Stafford, K. M. String, R. K. Watts, and D. S. Yaney.

We are further indebted to Mr. E. Labate and Mr. B. A. Stevens for their literature searches, Ms. D. McGrew, Ms. J. Chee, Ms. E. Doerries, Mr. N. Erdos, Mr. R. Richton, and Mr. N. Timm, with the assistance of Ms. J. Keelan, for technical editing of the manuscript, and Ms. A. W. Talcott for providing more than 3,000 technical papers on IC processing cataloged at the Murray Hill Library of Bell Laboratories. Finally, we wish to thank Ms. J. Maye and the members of the Word-Processing Centers who typed the initial drafts and the final manuscript, Mr. R. T. Anderson and the members of the drafting department who furnished the hundreds of technical illustrations used in the book, and Mrs. T. W. Sze who prepared the Appendixes and Index.

S. M. Sze

VLSI TECHNOLOGY

GROWTH OF THE INDUSTRY

The electronics industry in the United States has grown rapidly in recent years, with factory sales increasing by a factor of 10 since the early 1960s. [See Fig. 1, curve (a).[1, 2]] Electronics sales, which were $114 billion in 1981, are projected to increase at an average annual rate of 15% and finally reach $400 billion by 1990. The integrated circuit (IC) market has increased at an even higher rate than electronic sales [see Fig. 1, curve (b)]. IC sales in the United States were $6.6 billion in 1981 and are expected to grow by 25% annually, reaching $50 billion by 1990. The main impetuses for such phenomenal market growth are the intrinsic pervasiveness of electronic products and the continued technological breakthroughs in integrated circuits. The world market of electronics (about twice the size of the US market) will grow at a comparable rate.[3] In 10 years, it will rival the automobile, chemical, and steel industries in sales volume.

Figure 2 shows the sales of major IC groups and how sales have changed in recent years.[1] In the 1960s the IC market was broadly based on bipolar transistors. Since 1975, however, digital MOS ICs have prevailed. At present, even the intrinsic speed advantage of bipolar transistors is being challenged by MOSFETs. Because of the advantages in device miniaturization, low power dissipation, and high yield, by 1990 digital MOS ICs will dominate the IC market and capture a major market share of all semiconductor devices sold. This book, therefore, emphasizes MOS-related VLSI technology.

DEVICE MINIATURIZATION

Figure 3, curve (a), shows the exponential growth of the number of components per IC chip.[4] Note that IC complexity has advanced from small-scale integration (SSI) to medium-scale integration (MSI), to large-scale integration (LSI), and finally to very-large-scale integration (VLSI), which has 10^5 or more components per chip.

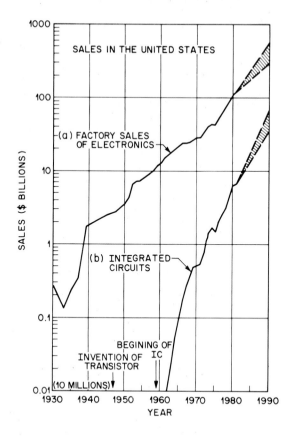

Fig. 1 (a) Factory sales of electronics in the United States for the 52 years between 1930 and 1981 and projected to 1990. (b) Integrated circuit market in the United States for the 20 years between 1962 and 1981 and projected to 1990. *(After Refs. 1 and 2.)*

Although the rate of growth has slowed down in recent years because of difficulties in defining, designing, and processing complicated chips, a complexity of over 1 million devices per chip will be available before 1990.

The most important factor in achieving such complexity is the continued reduction of the minimum device dimension [see Fig. 3, curve (b)]. Since 1960, the annual rate of reduction has been 13%; at that rate, the minimum feature length will shrink from its present length of 2 μm to 0.5 μm in 10 years.

Device miniaturization results in reduced unit cost per function and in improved performance. Figure 4, curve (a), gives an example of the cost reduction. The cost per bit of memory chips has halved every 2 years for successive generations of random-access memories.[5] By 1990 the cost per bit is expected to be as low as ~ 1 millicent for a 1-megabit memory chip. Similar cost reductions are expected for logic ICs.

As device dimension decreases, the intrinsic switching time in MOSFETs decreases linearly. (The intrinsic delay is given approximately by the channel length

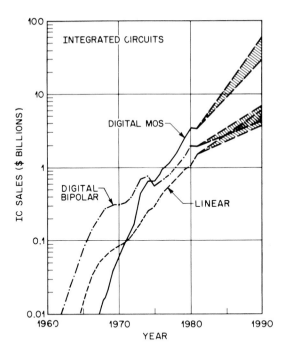

Fig. 2 Sales of major IC groups in the United States. *(After Ref. 1.)*

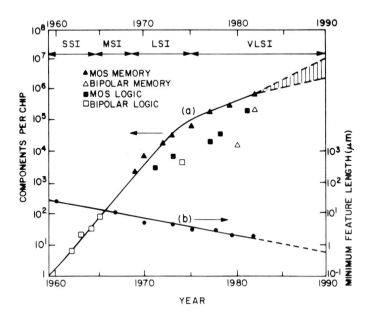

Fig. 3 (a) Exponential growth of the number of components per IC chip. *(After Moore, Ref. 4.)* (b) Exponential decrease of the minimum device dimensions.

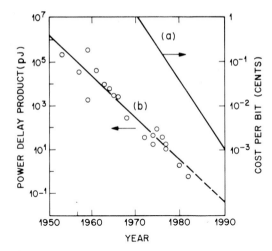

Fig. 4 (a) Reduction of cost per bit of RAM chips. *(After Noyce, Ref. 5.)* (b) Power-delay product per logic gate versus year. *(After Keyes, Ref. 6.)*

divided by the carrier velocity.) The device speed has improved by two orders of magnitude since 1960. Higher speeds lead to expanded IC functional throughput rates. In the future, digital ICs will be able to perform data processing, numerical computation, and signal conditioning at gigabit-per-second rates. Another benefit of miniaturization is the reduction of power consumption. As the device becomes smaller, it consumes less power. Therefore, device miniaturization also reduces the energy used for each switching operation. Figure 4, curve (b), shows the trend of this energy consumption, called the power-delay product.[6] The energy dissipated per logic gate has decreased by over four orders of magnitude since 1960.

INFORMATION AGE

Figure 5 shows four periods of change in the electronics industry in the United States. Each period exhibits normal life-cycle characteristics[7] (i.e., from incubation to rapid growth, to saturation, and finally to decline). The development of the vacuum tube in 1906 and the invention of transistors[8] in 1947 opened the field of electronic circuit designs. The development of integrated circuits[9] in 1959 led to a new generation of logic families. Since 1975, the beginning of VLSI, the frontier has moved to system organization of ICs and the associated software designs.

Many system-oriented VLSI chips, such as speech analysis/recognition and storage circuits, will be built in response to the enormous market demand for sophisti-cated electronic systems to handle the growing complexities of the Information Age.[10, 11] In this age a major portion of our work force can be called "information workers"; they are involved in gathering, creating, processing, disseminating, and using information. Figure 6 shows the changing composition of the work force in the

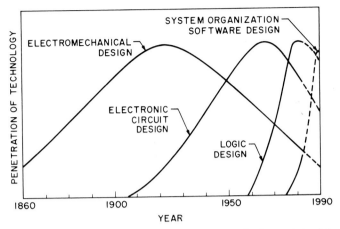

Fig. 5 Penetration of technology into the industrial output versus year for four periods of change in the US electronic industry. *(After Connell, Ref. 7.)*

United States. Prior to 1906, the largest single group was involved in agriculture. In the next period, until the mid 1950s, the predominant group was involved in industry. Currently, the predominant group consists of information workers; about 50% of the total work force is in this category. In Europe and Japan, information workers now constitute about 35 to 40% of the work force, which is also expected to reach 50% before the end of the century.[12] Advances in VLSI will have a profound effect on the world economy, because VLSI is the key technology for the Information Age.

ORGANIZATION OF THE BOOK

Figure 7 shows how the 14 chapters of this book are organized. Chapter 1 considers crystal growth and wafer preparation. VLSI technology is synonymous with *silicon VLSI technology*. The unique combination of silicon's adequate bandgap, stable

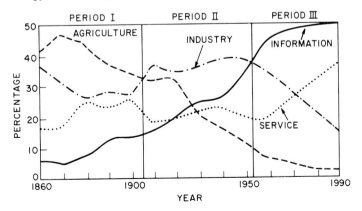

Fig. 6 Changing composition of work force in the United States. *(After Robinson, Ref. 10.)*

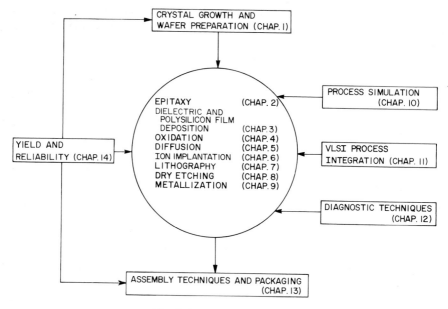

Fig. 7 Organization of this book.

oxide, and abundance in nature ensures that in the foreseeable future, no other semi-conductor will seriously challenge its preeminent position in VLSI applications. (Some important properties of silicon are listed in Appendix A.) Once the silicon wafer is prepared, we enter into the wafer-processing sequence, described in Chapters 2 through 9, and depicted in the wafer-shaped central circle of Fig. 7. Each of these chapters considers a specific processing step. Of course, many processing steps are repeated many times in IC fabrication; for example, lithography and dry etching steps may be repeated 5 to 10 times.

Chapter 10 considers process simulation of all the major steps covered in Chapters 2 through 9. Process simulation is emerging as an elegant aid to process development. This approach is attractive because of its rapid turn-around time and lower cost when compared to the experimental approach. Process simulation coupled with device and circuit simulations can provide a total design system that allows on-line process design and simulation to predict desired device and circuit parametric sensitivities and to facilitate circuit design and layout.

The individual processing steps described in Chapters 2 through 9 are combined in Chapter 11 to form devices and logic circuits. Chapter 11 considers the three most important IC families: the bipolar ICs, the NMOS (n-channel MOSFET) ICs, and the CMOS (complementary MOSFET) ICs. As the device dimension decreases and cir-cuit complexity increases, sophisticated tools are needed for process diagnostics. Chapter 12 covers many advanced diagnostic techniques, such as scanning and transmission electron microscopy for morphology determination, Auger electron spectroscopy for chemical analysis, and x-ray diffraction for structural analysis.

After completely processed wafers are tested, those chips that pass the tests are ready to be packaged. Chapter 13 describes the assembly and packaging of VLSI chips. Chapter 14 describes the yield at every step of the processing and the reliability of the packaged ICs. As device dimensions approach 1 μm, VLSI processing becomes more automated, resulting in tighter control of all processing parameters. At every step of production, from crystal growth to device packaging, numerous refinements are being made to improve the yield and reliability.

To keep the notation simple in this book, we sometimes found it necessary to use a simple symbol more than once, with different meanings. For example, in Chapter 1 S means 4-point probe spacing, in Chapter 7 it means resist sensitivity, while in Chapter 14 it means slope of a failure plot. Within each chapter, however, a symbol has only one meaning and is defined the first time it appears. Many symbols do have the same or similar meanings consistently throughout this book; they are summarized in Appendix B.

At present, VLSI technology is moving at a rapid pace. The number of VLSI publications (i.e., papers with the acronym "VLSI" in the title or abstract) has grown from virtually zero in 1975 to over 1000 in 1981 with an average annual growth rate of over 300%! Note that many topics, such as lithography and process simulation, are still under intensive study. Their ultimate capabilities are still not fully understood. The material presented in this book is intended to serve as a foundation. The references listed at the end of each chapter can supply more information.

REFERENCES

[1] *Electronic Market Data Book 1982*, Electronic Industries Association, Washington, D.C., 1982.

[2] "World Markets Forecast for 1982," *Electronics*, **55**, No. 1, 121 (1982).

[3] "Ten-Year Worldwide Forecast for Electronic Equipment and Components," *Electronic Business*, p. 92 (February 1981).

[4] G. Moore, "VLSI, What Does the Future Hold," *Electron. Aust.*, **42**, 14 (1980).

[5] R. N. Noyce, "Microelectronics," in T. Forester, Ed., *The Microelectronics Revolution*, MIT Press, Cambridge, Mass., 1981, p. 29.

[6] R. W. Keyes, "Limitations of Small Devices and Large Systems," in N. G. Einspruch, Ed., *VLSI Electronics*, Academic, New York, 1981, Vol. 1, p. 186.

[7] J. M. Connell, "Forecasting a New Generation of Electronic Components," *Digest IEEE Spring Compcon.*, **81**, 14 (1981).

[8] W. Shockley, "The Path to the Conception of the Junction Transistor," *IEEE Trans. Electron Devices*, **ED-23**, 597 (1976).

[9] J. S. Kilby, "Invention of the Integrated Circuits," *IEEE Trans. Electron Devices*, **ED-23**, 648 (1976).

[10] A. L. Robinson, "Electronics and Employment: Displacement Effects," in T. Forester, Ed., *The Microelectrons Revolution*, MIT Press, Cambridge, Mass., 1981, p. 318.

[11] J. S. Mayo, "Technology Requirements of the Information Age," *Bell Lab. Rec.*, **60**, 55 (1982).

[12] D. Kimbel, *Microelectronics, Productivity and Employment*, Organization for Economic Cooperational Development, Paris, 1981, p. 15.

CRYSTAL GROWTH AND WAFER PREPARATION

C. W. PEARCE

1.1 INTRODUCTION

Silicon, naturally occurring in the form of silica and silicates, is the most important semiconductor for the electronics industry. At present, silicon-based devices constitute over 98% of all semiconductor devices sold worldwide. Silicon is one of the most studied elements in the periodic table. A literature search on published papers using silicon as a search word yields over 25,000 references. Appendix A is a compilation of some useful constants.[1, 2] Silicon is also a commercially important element for several other major industries, such as glass and gemstones. The commercial value of silicon derives in part from the utility of its mineral forms, which is the way silicon occurs in nature, and from its abundance. Silica is integral to the manufacture of glass and related products, while certain silicates are highly valued as semiprecious gemstones, such as garnet, zircon, and jade. By weight it comprises 25% of the earth's crust and is second only to oxygen in abundance.

Although silicon is generally synonymous with the solid-state era of electronics, as in the use of the term ''silicon chip,'' its mineral forms were used in vacuum-tube electronics. (Silica was used for tube envelopes.) Mica, a silicate, found application as an insulator and capacitor dielectric. Quartz, another silicate, was and still is used as a frequency-determining element and in passive filter applications.

The advent of solid-state electronics dates from the invention of the bipolar transistor effect by Bardeen, Brattain, and Shockley.[3] The technology progressed during the early 1950s, using germanium as the semiconducting material. However, germanium proved unsuitable in certain applications because of its propensity to exhibit high junction leakage currents. These currents result from germanium's relatively narrow bandgap (0.66 eV). For this reason, silicon (1.1 eV) became a practical substitute and has almost fully supplanted germanium as a material for solid-state device fabrication. Silicon devices can operate up to 150°C versus 100°C for germanium.

In retrospect, other reasons could have ultimately led to the same material substitution. Planar processing technology derives its success from the high quality of thermally grown silicon dioxide. Germanium oxide is water soluble and unsuited for device applications. The intrinsic (undoped) resistivity of germanium is 47 Ω-cm, which would have precluded the fabrication of rectifying devices with high breakdown voltages. In contrast, the intrinsic resistivity of silicon is about 230,000 Ω-cm. Thus, high-voltage rectifying devices and certain infrared sensing devices are practical with silicon. Finally, there is an economic consideration—electronic-grade germanium costs 10 times as much as silicon.

Similar problems impeded the widespread use of compound semiconductors. For example, it is difficult to grow a high-quality oxide on GaAs. One element oxidizes more readily than the other, leaving a metallic phase at the interface. Such material is difficult to dope and obtain in large diameters with high crystal perfection. In fact the technology of Group III−V compounds has advanced partly because of the advances in silicon technology.

1.2 ELECTRONIC-GRADE SILICON

Electronic-grade silicon (EGS), a polycrystalline material of high purity, is the raw material for the preparation of single-crystal silicon. EGS is undoubtedly one of the purest materials routinely available. The major impurities of interest are boron, carbon, and residual donors. Pure EGS generally requires that doping elements be in the parts per billion (ppb) range, and carbon be less than 2 parts per million (ppm).[4] These properties are usually evaluated on test ingots rather than measuring on the material itself.[5] In the case of the doping elements, ppb levels are below the capabilities of most laboratory methods, so the doping level is inferred from resistivity measurements on the test ingot.

To obtain EGS requires a multistep process.[4] First, metallurgical-grade silicon is produced in a submerged-electrode arc furnace, as shown in Fig. 1. The furnace is charged with quartzite, a relatively pure form of SiO_2, and carbon in the form of coal, coke, and wood chips. In the furnace a number of reactions take place, the overall reaction being:

$$SiC \text{ (solid)} + SiO_2 \text{ (solid)} => Si \text{ (solid)} + SiO \text{ (gas)} + CO \text{ (gas)} \qquad (1)$$

The process is power intensive, requiring 13 kWh/kg, and metallurgical-grade silicon (MGS) is drawn off at a purity of 98%. Table 1 shows typical purities of various materials used in the arc furnace. The MGS used in the making of metal alloys is not sufficiently pure to use in the manufacture of solid-state devices.

The next process step is to mechanically pulverize the silicon and react it with anhydrous hydrogen chloride to form trichlorosilane ($SiHCl_3$), according to the reaction:

$$Si(\text{solid}) + 3HCl(\text{gas}) => SiHCl_3(\text{gas}) + H_2(\text{gas}) + \text{heat} \qquad (2)$$

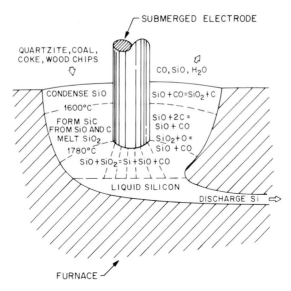

Fig. 1 Schematic of a submerged-electrode arc furnace for the production of metallurgical-grade silicon. *(After Crossman and Baker, Ref. 4.)*

Table 1 Comparison of typical impurity contents in various materials (values in ppm except as noted)

Impurity	Quartzite	Carbon	MGS*	EGS†	Crucible quartz
Al	620	5500	1570	⋯	⋯
B	8	40	44	<1 ppb	⋯
Cu	<5	14	⋯	0.4	0.23
Au	··	⋯	⋯	0.07 ppb	⋯
Fe	75	1700	2070	4	5.9
P	10	140	28	<2 ppb	⋯
Ca	⋯	⋯	⋯	⋯	⋯
Cr	⋯	⋯	137	1	0.02
Co	⋯	⋯	⋯	0.2	0.01
Mn	⋯	⋯	70	0.7	⋯
Sb	⋯	⋯	⋯	0.001	0.003
Ni	⋯	⋯	4	6	0.9
As	⋯	⋯	⋯	0.01	0.005
Ti	⋯	⋯	163	⋯	⋯
La	⋯	⋯	⋯	1 ppb	⋯
V	⋯	⋯	100	⋯	⋯
Mo	⋯	⋯	⋯	1.0	5.1
C	⋯	⋯	80	0.6	⋯
W	⋯	⋯	⋯	0.02	0.048
O	⋯	⋯	⋯	⋯	⋯
Na	⋯	⋯	⋯	0.2	3.7

*Metallurgical-grade silicon.
†Electronic-grade silicon.

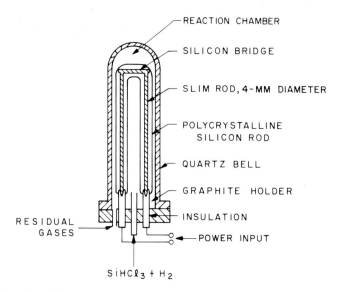

REACTION CHAMBER

SILICON BRIDGE

SLIM ROD, 4-MM DIAMETER

POLYCRYSTALLINE
SILICON ROD

QUARTZ BELL

GRAPHITE HOLDER

INSULATION

RESIDUAL GASES

POWER INPUT

$SiHCl_3 + H_2$

Fig. 2 Schematic of a CVD reactor used for EGS production. *(After Crossman and Baker, Ref. 4.)*

This reaction takes place in a fluidized bed at a nominal temperature of 300°C using a catalyst. Here silicon tetrachloride and the chlorides of impurities are formed. At this point the purification process occurs. Trichlorosilane is a liquid at room temperature (boiling point 32°C), as are many of the unwanted chlorides. Hence purification is done by fractional distillation.

EGS is prepared from the purified $SiHCl_3$ in a chemical vapor deposition (CVD) process similar to the epitaxial CVD processes that is presented in Chapter 2. The chemical reaction is a hydrogen reduction of trichlorosilane.

$$2SiHCl_3(gas) + 3H_2(gas) => 2Si(solid) + 6HCl(gas) \qquad (3)$$

This reaction is conducted in the type of system shown in Fig. 2. A resistance-heated rod of silicon, called a "slim rod," serves as the nucleation point for the deposition of silicon. A complete process cycle takes many hours, and the results in rods, of EGS, which are polycrystalline in structure, up to 20 cm (8 in) in diameter and several meters in length. EGS can be cut from these rods as single chucks or crushed into nugget geometries (Fig. 3). In 1982 the worldwide consumption of electronic-grade polysilicon was approximately 3×10^6 kg.

This CVD process is also used to grow tubes of EGS on carbon mandrils.[6] These tubes are of high purity and strength, and are used as furnace tubes in place of quartz in high-temperature operations (over 1200°C). The tubes are also sectioned and machined to form paddles and wafer carriers for the same high-temperature operations (Fig. 4). In these applications, silicon competes with quartz and silicon-carbide. The choice of silicon as a material for furnace use is advantageous because of its purity and strength. Silicon process tubes show no sagging or similar deformation after several years' use in a furnace process where they are repetitively cycled between 900 and 1250°C. Quartz tubes have limited life in the same process.

Fig. 3 EGS in chunk form loaded into a quartz crucible.

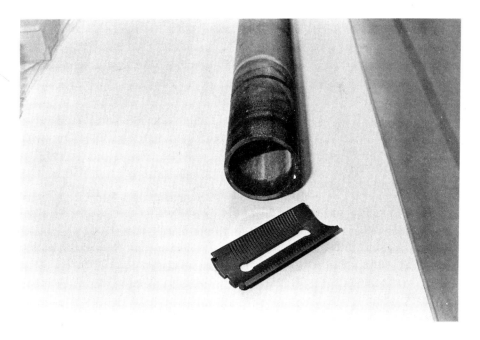

Fig. 4 A polysilicon furnace tube and wafer rack.

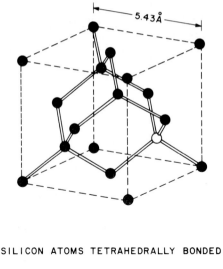

Fig. 5 Schematic of the crystal structure of silicon.

1.3 CZOCHRALSKI CRYSTAL GROWING

A substantial percentage (80 to 90%) of the silicon crystals prepared for semiconductor industry are prepared by the Czochralski (CZ) technique.[7] Virtually all the silicon used for integrated circuit fabrication is prepared by this technique.

1.3.1 Crystal Structure

Silicon has a diamond-lattice crystal structure (Fig. 5), which can be viewed as two interpenetrating face-centered cubic lattices. Each silicon atom has four nearest neighboring atoms to which it is covalently bonded. The lattice constant for silicon is 5.43 Å, and simple geometry reveals that the spacing to the nearest neighbor is 2.35 Å. Dopant atoms (most of Group III and Group V) that substitute for silicon atoms are considered to be occupying substitutional lattice sites. Phosphorous is a substitutional donor, having four of its five valence-band electrons covalently bonded to the four nearest neighbor silicon atoms, leaving the fifth free to support electrical conduction. Similiarly, boron is a substitutional acceptor. Its three valence-band electrons also covalently bond to nearest neighbor silicon atoms. The deficiency of an electron to complete the bonding is the basis for hole conduction. Impurities or dopants that occupy sites not defined by the structure are said to be in interstitial lattice sites.

The principal axes in a crystal can also be used to develop a notation for defining specific directions and planes (Fig. 6). Termed "Miller indices,"[8] they are a series of small integer numbers enclosed in carets, brackets, parentheses, and braces. For

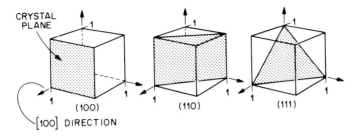

CRYSTAL
PLANE

(100)   (110)   (111)

[100] DIRECTION

Fig. 6 Schematic representation of Miller indices in a cubic lattice system.

example, [111] denotes a specific direction, whereas $\langle 111 \rangle$ denotes the family of all eight directions equivalent to [111]. A (100) notation denotes a particular lattice plane, and {100} denotes all the planes crystallographically equivalent of (100).

The processing characteristics and some material properties of silicon wafers depend on the orientation. The {111} planes have the highest density of atoms on the surface, so crystals grow most easily on these planes. Mechanical properties such as tensile strength are highest for $\langle 111 \rangle$ directions. The moduli of elasticity also show an orientation dependence (Appendix A). Processing characteristics such as oxidation are similarly orientation dependent. For example, {111} planes oxidize faster than {100} planes, because they have more atoms per unit surface area available for the oxidation reaction to occur. The choice of crystal orientation, therefore, is generally not left to the discretion of the crystal grower, but is a device design consideration. Historically, bipolar circuits have preferred $\langle 111 \rangle$ oriented material and MOS devices $\langle 100 \rangle$. There are, of course, exceptions. Growth on other orientations such as $\langle 110 \rangle$ has been demonstrated, but is more difficult to achieve routinely.[9]

A real crystal, as represented by a silicon wafer, differs from the mathematically ideal crystal in several respects. It is finite, not infinite; thus, surface atoms are incompletely bonded. The atoms are displaced from their ideal locations by thermal agitation. Most importantly, real crystals have defects[10, 11] classified as follows: (1) point defect, (2) line defect, (3) area or planar defect, and (4) volume defect. Defects influence the optical, electrical, and mechanical properties of silicon.

Point defects Point defects take several forms as shown in Fig. 7. Any nonsilicon atom incorporated into the lattice at either a substitutional or interstitial site is considered a point defect. This is true whether the atom is an intentional dopant or unintentional impurity. Missing atoms create a vacancy in the lattice called a "Schottky defect," which is also considered a point defect. A silicon atom in an interstitial lattice site with an associated vacancy is called a "Frenkel defect." Vacancies and interstitials have equilibrium concentrations that depend on temperature. From thermodynamic principles the concentration as a function of temperature can be derived and has the following relation:

$$N_d = A \, \exp \, (-E_a / kT) \tag{4}$$

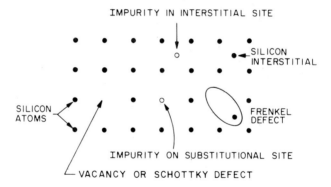

Fig. 7 The location and types of point defects in a simple lattice.

where N_d is the concentration of the point defect, A is a constant, E_a is the activation energy (2.6 eV for vacancies and 4.5 eV for interstitials), T is absolute temperature, and k is Boltzmann's constant.

Point defects are important in the kinetics of diffusion and oxidation. The diffusion of many impurities depends on the vacancy concentration, as does the oxidation rate of silicon. Vacancies and interstitials are also associated with defect formation in processing.[10]

To be electrically active, atoms must usually be located on substitutional sites.[12] When in such sites they introduce an energy level in the bandgap. Shallow levels are characteristic of efficient donor and acceptor dopants. Midgap levels act as centers for the generation and recombination of carriers to and from the conduction and valence bands. Some impurities are entirely substitutional or interstitial in behavior, but others can exist in either lattice position.

Dislocations Dislocations form the second class of defects. Two general categories of dislocations are spiral and line (edge), the terms being aptly descriptive of their shape. Figure 8 is a schematic representation of a line dislocation in a cubic lattice; it can be seen as an extra plane of atoms AB inserted into the lattice. The line of dislocation would be perpendicular to the plane of the page. Dislocations in a lattice are dynamic defects; that is, they can move under applied stress, disassociate into two or more dislocations, or combine with other dislocations. A vector notation developed by Burgers[13] characterizes dislocations in the crystal. The vector notation is also used to describe dislocation interactions.

Crystals for IC usage are generally grown free of edge dislocations,[10] but may contain small dislocation loops from excess point-defect condensation.[14] These defects act as nuclei for precipitation of impurities such as oxygen and are responsible for a swirl pattern seen in wafers.[10] Dislocations (edge type) are also introduced by thermal stress on the wafer during processing[15, 16] or by the introduction of an excessive concentration of an impurity atom. Substitutional impurities with covalent radii larger or smaller than silicon compress or expand the lattice accordingly. The strain S

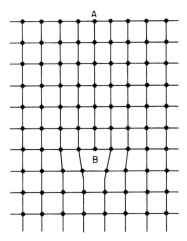

Fig. 8 An edge dislocation in a cubic lattice created by an extra plane of atoms. The line of the dislocation is perpendicular to the page.

(in dynes per cm^2) induced depends on the size of the impurity and its concentration:[10]

$$S = \frac{BCE}{1-V} \tag{5}$$

where B is the lattice contraction constant reflecting the degree of distortion introduced by the impurity ($B = 8 \times 10^{-24}$ cm^3/atom for boron), C is the impurity concentration, E is Young's modulus, and V is Poisson's ratio. Dislocations in devices are generally undesirable, because as they act as sinks for metallic impurities and alter diffusion profiles. Dislocations can be revealed by preferential etching (see Section 1.3.5).

Area (planar) defects Two area defects are twins and grain boundaries. Twinning represents a change in the crystal orientation across a twin plane, such that a certain symmetry (like a mirror image) exists across that plane. In silicon, the twin plane is {111}. A grain boundary represents a transition between crystals having no particular orientation relationship to one another. Grain boundaries are more disordered than twins, and separate grains of single crystals in polycrystalline silicon. Area defects, such as twins or grain boundaries, represent a large area discontinuity in the lattice. The crystal on either side of the discontinuity may be otherwise perfect. These defects appear during crystal growth, but crystals having such defects are not considered usable for IC manufacture and are discarded.

Volume defects Precipitates of impurity or dopant atoms constitute the fourth class of defects. Every impurity introduced into the lattice has a solubility; that is, a concentration that the host lattice can accept in a solid solution of itself and the impurity.

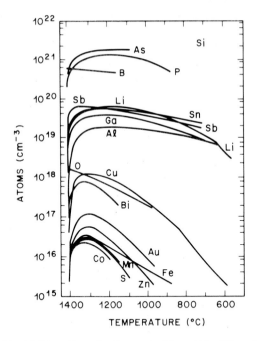

Fig. 9 Solid solubilities of impurity elements in silicon. *(After Milnes, Ref. 12.)*

Figure 9 illustrates the solubility versus temperature behavior for a variety of elements in silicon. Most impurities have a retrograde solubility, which is defined as a solubility that decreases with decreasing temperature. Thus, if an impurity is introduced (at a temperature T_2) at the maximum concentration allowed by its solubility, and the crystal is then cooled to a lower temperature T_1, a supersaturated condition is said to exist (also see Fig. 19). The degree of supersaturation is expressed as the ratio of the concentration introduced at T_2 to the solubility at T_1. The crystal achieves an equilibrium state by precipitating the impurity atoms in excess of the solubility level as a second phase. The kinetics of precipitation depend on the degree of supersaturation, time, and nucleating sites where the precipitates form.

Precipitates are generally undesirable, because they act as sites for dislocation generation. Dislocations result from the volume mismatch between the precipitate and the lattice, inducing a strain that is relieved by dislocation formation. Precipitation in silicon processing has been observed for dopants such as boron, oxygen, and metallic impurities.[17, 18, 19]

1.3.2 Crystal Growing Theory

Growing crystals, in the most general sense, involves a phase change from solid, liquid, or gas phases to a crystalline solid phase. Czochralski growth, named for the inventor, is the process used to grow most of the crystals from which silicon wafers are produced. This process can be characterized, as applied to silicon, as a liquid-

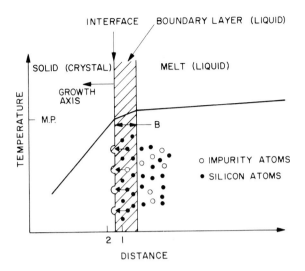

Fig. 10 Temperature gradients, solidification, and transport phenomena involved in Czochralski growth. Positions 1 and 2 represent the location of isotherms associated with Eq. 6 and the crystal solidification at the interface. Impurity atoms are transported across the boundary layer and incorporated into the growing crystal interface. M. P. is the melting point.

solid monocomponent growth system. This section discusses some elements of this process as it relates to the understanding of the properties of the grown crystals. For a more complete treatment of crystal growth, refer to the many excellent books devoted to the subject.[9, 20, 21]

The growth of a CZ crystal involves the solidification of atoms from a liquid phase at an interface. The speed of growth is determined by the number of sites on the face of the crystal and the specifics of heat transfer at the interface. Figure 10 schematically represents the transport process and temperature gradients involved. Macroscopically, the heat transfer conditions about the interface can be modelled by the following equation:[21]

$$L\frac{dm}{dt} + k_l \frac{dT}{dx_1} A_1 = k_s \frac{dT}{dx_2} A_2 \tag{6}$$

where L is the latent heat of fusion, dm/dt is the mass solidification rate, T is the temperature, k_l and k_s are the thermal conductivities of the liquid and solid, respectively, dT/dx_1 and dT/dx_2 are the thermal gradients at points 1 and 2 (near the interface in the liquid and solid, respectively), and A_1 and A_2 are the areas of the isotherms at positions 1 and 2, respectively.

From Eq. 6 the maximum pull rate of a crystal under the condition of zero thermal gradient in the melt can be deduced:[7, 22]

$$V_{max} = \frac{k_s}{Ld} \frac{dT}{dx} \tag{7}$$

where V_{max} is the maximum pull rate (or pull speed) and d is the density of solid silicon. Figure 11 is an experimentally determined temperature variation along a crystal.

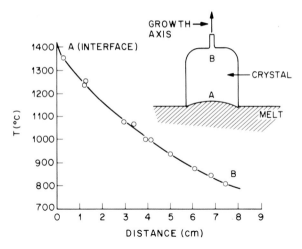

Fig. 11 Experimentally determined temperature gradient in a silicon crystal as referenced to insert showing a growing crystal. *(After deKock and van de Wijgert, Ref. 14.)*

The pull rate influences the incorporation of impurities into the crystal and is a factor in defect generation. Generally, when the temperature gradient in the melt is small, the heat transferred to the crystal is the latent heat of fusion. As a result, the pull rate generally varies inversely with the diameter[7, 22] (Fig. 12). The pull rates obtained in practice are 30 to 50% slower than the maximum values suggested by theoretical considerations.[7]

The growth rate (or growth velocity) of the crystal, actually distinct from the pull rate, is perhaps the most important growth parameter. Pull rate is the macroscopic indication of net solidification rate, whereas growth rate is the instantaneous solidification rate. The two differ because of temperature fluctuations near the interface. The growth rate can exceed the pull rate and even be negative at a given time. When the growth rate is negative, remelting is said to occur. That is, the crystal dissolves back into the melt. The growth rate influences the defect structure and dopant distribution in the crystal on a microscopic scale.

Pull rate affects the defect properties of CZ crystals in the following way. The condensation of thermal point defects in CZ crystals into dislocation loops occurs as the crystal cools from the solidification temperature. This process occurs above 950°C. The number of defects depends on the cooling rate, which is a function of pull rate and diameter, through this temperature range. A pull rate of 2 mm/min eliminates microdefect formation by quenching the point defects in the lattice before they can agglomerate. We find from Fig. 12 that large diameters preclude this pull rate from being achieved for crystal diameters above 75 mm. A related phenomenon is the remelting of the crystal that occurs because of temperature instabilities in the melt caused by thermal convection. This condition can also be suppressed by attaining a pull rate of 2.7 mm/min,[23] which is half the maximum attainable pull rate (Eq. 7). Crystals in which remelt has not been suppressed exhibit impurity striations and defect swirls.[24, 25] Elimination of remelt results in more uniformly doped crystals, dis-

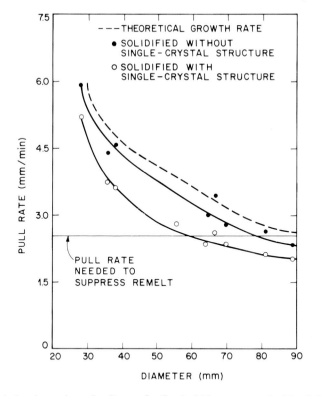

Fig. 12 Theoretical and experimental pull rates for Czochralski-grown crystals. The dashed line is the theoretical growth rate according to Rea (Ref. 7). *(After Digges, Jr. and Shima, Ref. 22.)*

cussed below, but will not necessarily eliminate dopant striation if the growth velocity still varies on a microscopic level (Eq. 10).

As mentioned earlier, every impurity has a solid solubility in silicon. The impurity has a different equilibrium solubility in the melt. For dilute solutions commonly encountered in silicon growth, an equilibrium segregation coefficient k_0 may be defined as:

$$k_0 = \frac{C_s}{C_l} \tag{8}$$

where C_s and C_l are the equilibrium concentrations of the impurity in the solid and liquid near the interface, respectively.

Table 2 lists the equilibrium segregation coefficients for common impurity and dopant atoms. Note that most are below 1, so that during growth, the impurities at the

Table 2 Segregation coefficients for common impurities in silicon

Impurity	Al	As	B	C	Cu	Fe	O	P	Sb
k_0	0.002	0.3	0.8	0.07	4×10^{-4}	8×10^{-6}	1.25	0.35	0.023

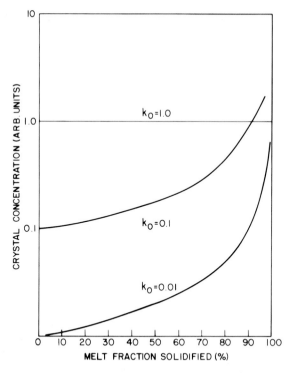

Fig. 13 Impurity concentration profiles for various k_0 with $C_0 = 1$.

interface are left in the liquid (melt). Thus, as the crystal grows, the melt becomes progressively enriched with the impurity.

The distribution of an impurity in the grown crystal can be described mathematically by the normal freezing relation:[21]

$$C_s = k_0 C_0 (1 - X)^{k_0 - 1} \tag{9}$$

where X is the fraction of the melt solidified, C_0 is the initial melt concentration, C_s is the solid concentration, and k_0 is the segregation coefficient.

Figure 13 illustrates the segregation behavior for several segregation coefficients. Experimentally it is found that segregation coefficients differ from equilibrium values and it is necessary to define an effective segregation coefficient k_e:[26]

$$k_e = \frac{k_0}{k_0 + (1 - k_0) \exp(-VB/D)} \tag{10}$$

where V is the growth velocity (or pull rate), D is the diffusion coefficient of dopant in melt, and B is the boundary layer thickness.

The boundary layer thickness is a function of the convection conditions, in the melt. Rotation of a crystal in a melt (forced convection) produces a boundary layer B dimensions defined by:[30]

$$B = 1.8 \, D^{1/3} v^{1/6} W^{-1/2} \tag{11}$$

where W is the rotational velocity.

Our presentation so far represents a first-order approach to the problem. In large melts the convection forced by rotation is often secondary to the thermal convection caused by temperature gradients in the crucible.[30] The effect is to reduce values of B below those of Eq. 11. Since the thermal convection is a random process, the thickness of the boundary layer fluctuates with time, resulting in a variable value for B.

The net result of thermal convection effects is an inhomogeneous distribution of dopant in the crystal on a microscale (Fig. 14). The boundary layer thickness also varies radially across the face of the crystal, resulting in a radial distribution of dopant. Generally, less dopant is incorporated at the edges.

Another effect occuring in heavily doped melts is constitutional supercooling.[28] This effect, particularly prevalent with Sb, occurs when the concentration in the melt

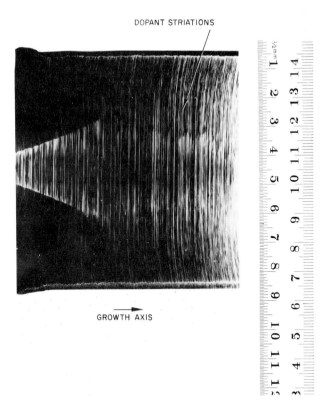

Fig. 14 Dopant striations in a Sb-doped ingot revealed by preferentially etching a longitudinal section from the seed end of an ingot.

ahead of the growing interface is sufficient to depress the solidification temperature (freezing point). When this occurs, the crystal solidifies irregularly and dislocations appear. Constitutional supercooling limits the ultimate dopant incorporation for certain impurities.

The pull speed is also a factor in determining the shape of the growing interface, as are the melt radial temperature gradient and the crystal surface cooling conditions. A proper shape is needed to maintain stability of the growing crystal.[8, 29]

1.3.3 Crystal Growing Practice

A Czochalski crystal growth apparatus, also called a "puller," is shown in Fig. 15. The one pictured weighs 17,600 kg and is 6.5 m tall. This puller can be configured to hold a melt charge of 60 kg of silicon, which can be transformed into a crystal of

Fig. 15 An industrial-sized Czochralski grower. Numbers relate to the four basic parts of the growers.

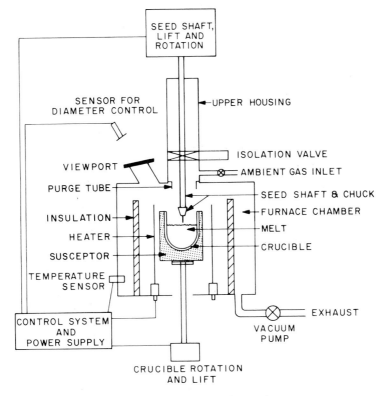

Fig. 16 Schematic representation of a crystal grower.

100-mm diameter and 3.0-m length. The puller has four subsystems as follows (Fig. 16):

1. *Furnace:* crucible, susceptor and rotation mechanism, heating element and power supply, and chamber.
2. *Crystal-pulling mechanism:* seed shaft or chain, rotation mechanism, and seed chuck.
3. *Ambient control:* gas source, flow control, purge tube, and exhaust or vacuum system.
4. *Control system:* microprocessor, sensors, and outputs.

Furnace Perhaps the most important component of the growing system is the crucible (Fig. 3). Since it contains the melt, the crucible material should be chemically unreactive with molten silicon. This is a major consideration, because the electrical properties of silicon are sensitive to even ppb levels of impurities. Other desirable characteristics for crucible material are a high melting point, thermal stability and hardness. Additionally, the crucible should be inexpensive or reusable. Unfortunately, molten silicon can dissolve virtually all commonly used high-temperature materials, such as refractory carbides (TiC or TaC), thus introducing unacceptable levels of the metallic species into the crystal.[30]

Carbon or silicon carbide crucibles are also unsuitable. Although carbon is electrically inactive in silicon, high-quality crystals cannot be grown with carbon-saturated melts.[31] During growth a two-phase solidification occurs, once the solid solubility has been exceeded. The second phase is SiC, which is responsible for dislocation generation and loss of single-crystal structure. The remaining choices for a crucible are either silicon nitride (Si_3N_4) or fused silica (SiO_2), which is in exclusive use today.

Fused silica or quartz does, however, react with silicon, releasing silicon and oxygen into the melt. The dissolution rate is quite substantial, as shown[32] by Fig. 17. The actual rate of erosion is a function of temperature and the convection conditions, either forced or thermal,[27] in the melt. Most of the oxygen in the melt escapes by the formation of gaseous silicon monoxide. The SiO condenses on the inside of the furnace chamber, creating a cleanliness problem in the puller. Crystals grown with these crucibles also contain substantial amounts of interstitial oxygen that can be either beneficial or detrimental, as will be discussed later. The purity of the quartz itself (Table 1) also affects the silicon purity, because the quartz can contain sufficient acceptor impurities to limit the upper values of resistivity that can be grown. The presence of carbon in the melt also accelerates the dissolution rate up to twofold.[33] One possible reaction is:

$$C + SiO_2 \rightarrow SiO + CO \tag{12}$$

Crucibles for large CZ pullers have a diameter-to-height ratio of approximately 1 or slightly greater; common diameters are 25, 30, and 35 cm for charge sizes of 12, 20, and 30 kg, respectively. A 45-cm, 60-kg configuration has even been proposed. Wall thicknesses of 0.25 cm are used, but the silica is sufficiently soft to require the use of

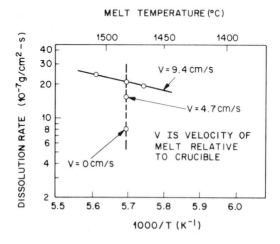

Fig. 17 The dissolution rate of quartz in molten silicon. *(After Hirata and Hoshikawa, Ref. 32.)*

a susceptor for mechanical support. Upon cooling, the thermal mismatch between residual silicon and quartz usually results in the fracture of the crucible.

The feasibility of using silicon nitride as a crucible material has been demonstrated using CVD-deposited nitride.[34] Such an approach is attractive as a means of eliminating oxygen from crucible-grown crystals. However, even the nitride is eroded, resulting in a doping of the crystal with nitrogen, a weak donor. CVD nitride is the only form of nitride with sufficient purity for crucible use. However, this method needs further development before it becomes practical.

The susceptor, as mentioned previously, is used to support the quartz crucible. Graphite, because of its high-temperature properties is the material of choice for the susceptor. A high-purity graphite, such as nuclear grade, is usually specified. This high purity is necessary to prevent contamination of the crystal from impurities that would be volitalized from the graphite at the temperature involved. Besides the susceptor, other graphite parts in the hot zone of the furnace require high purity. The susceptor rests on a pedestal whose shaft is connected to a motor that provides rotation. The whole assembly can usually be raised and lowered to keep the melt level equidistant from a fixed reference point, which is needed for automatic diameter control.

The chamber housing the furnace must meet several criteria. It should provide easy access to the furnace components to facilitate maintenance and cleaning. The furnace structure must be air tight to prevent contamination from the atmosphere, and have a specific design that does not allow any part of the chamber to become so hot that its vapor pressure in the chamber would be a factor in contaminating the crystal. As a rule, the hottest parts of the puller are water cooled. Insulation is usually provided between the heater and the chamber wall.

To melt the charge, radio frequency (induction heating) or resistance heating have been used. Induction heating is useful for small melt sizes, but resistance heating is used exclusively in large pullers. Resistance heaters, at the power levels involved (tens of kilowatts), are generally smaller, cheaper, easier to instrument, and more efficient. Typically, a graphite heater is connected to a dc power supply.

Crystal-pulling mechanism The crystal-pulling mechanism must, with minimal vibration and great precision, control two parameters of the growth process, the pull rate and crystal rotation. Seed crystals, for example, are prepared to precise orientation tolerances, and the seed holder and pulling mechanism must maintain this precision perpendicular to the melt surface. Lead screws are often used to withdraw and rotate the crystal. This method unambiguously centers the crystal relative to the crucible, but may require an excessively high apparatus if the grower is to produce long crystals. Since precise mechanical tolerance is difficult to maintain over a long shaft, pulling with a cable may be necessary. Centering the crystal and crucible is more difficult when using cable. Furthermore, although the cable provides a smooth pulling action, it is prone to pendulum effects. However, since the cable can be wound on a drum, the height of the machine can be smaller than a similar, lead-screw puller. The crystal leaves the furnace through a purge tube, where (if present) ambient gas is directed along the surface of the crystal to affect cooling. From the purge tube, the crystal enters an upper chamber, which is usually separated from the furnace by an isolation valve.

Ambient control Czochralski growth of silicon must be conducted in an inert gas or vacuum, because (1) the hot graphite parts must be protected from oxygen to prevent erosion and (2) the gas around the process should not react with the molten silicon. Growth in vacuum meets these requirements; it also has the advantage of removing silicon monoxide from the system, thus preventing its buildup inside the furnace chamber. Growth in a gaseous atmosphere must use an inert gas, such as helium or argon. The inert gas may be at atmospheric pressure or at reduced pressure. Growing operations on an industrial scale use argon because of its lower cost. A typical consumption is 1500 L/kg of silicon grown. The argon is supplied from a liquid source by evaporation, and must meet requirements of purity relating to moisture, hydrocarbon content, and so on.

Control system The control system can take many forms, and provides control of process parameters such as temperature, crystal diameter, pull rate, and rotation speeds. This control may be closed loop or open loop. Parameters, including pull speed and rotation, with a high response speed are most amenable to closed-loop control. The large thermal mass of the melt generally precludes any short-term control of the process according to temperature. For example, to control the diameter an infrared temperature sensor can be focused on the melt-crystal interface and used to detect changes in the meniscus temperature. The sensor output is linked to the pull mechanism, and controls the diameter by varying the pull rate. The trend in control systems is to digital microprocessor-based systems. These rely less on operator intervention and have many parts of the process preprogrammed.

1.3.4 Impurity and Defect Considerations

Oxygen in silicon is an unintentional impurity arising from the dissolution of the crucible during growth. Typical values[35] range from 5×10^{17} to 10^{18} atoms/cm^3. The reported segregation coefficient for oxygen is 1.25; however, the axial distribution of impurities often reflects the specifics of the puller and process parameters in use, because they influence crucible erosion and evaporation of oxygen from the melt. For example, less dissolution of the crucible occurs as the melt level is lowered in the crucible, and thus less oxygen impurity is available for incorporation.[36] Rotation speeds, ambient partial pressure, and free melt surface area are all factors that determine the level and distribution of oxygen in the crystal.[29] Figure 18 shows a typical diagram of concentration versus fraction solidified. A novel method to reduce crucible erosion is by suppressing thermal convection currents, which can be done by applying a magnetic field to the melt.[37] Such an approach also reduces the thermal fluctuations in the boundary layer, resulting in a more homogeneous distribution of dopant atoms.

As an impurity, oxygen has three effects:[38] donor formation, yield strength improvement, and defect generation by oxygen precipitation. In the crystal as grown, over 95% of the oxygen atoms occupy interstitial lattice sites. Oxygen in this state

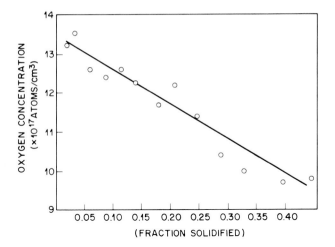

Fig. 18 The axial distribution of oxygen in a Czochralski ingot. *(After Liaw, Ref. 35.)*

can be detected using an infrared absorption line[39] at 1106 cm^{-1}. The remainder of the oxygen polymerizes into complexes, such as SiO_4. T his configuration acts as a donor, thus distorting the resistivity of the crystal caused by intentional doping. These complexes form rapidly in the 400 to 500°C temperature range, with a rate proportional to the oxygen content to the fourth power. The complex formation occurs as the crystal cools. Crystals of larger diameter cool slower and form more complexes. Fortunately, these complexes are unstable above 500°C; so as common practice, crystals or wafers are heated to between 600 and 700°C to dissolve the complexes. During cooling, following the 600 to 700°C treatment, complexes can reform in crystals. Because wafers cool rapidly enough to circumvent this problem, treatment in wafer form is preferred for large diameter material. Common treatment times are tens of minutes for wafers and about an hour for crystals. The longer time for crystals is needed to bring the center of the ingot up to temperature. A typical dissolution rate for these complexes is 5×10^{15} donors/cm^3-h at 700°C.

Oxygen will also combine with acceptor elements to create a second type of donor complex. These complexes form more slowly, 2×10^{14} donors/cm^3-h at 700°C, allowing for a net improvement using the stabilization treatment previously described. The acceptor-element complexes are more resistant to dissolution even with high-temperature processing. From a device viewpoint it is important that material be resistivity stabilized by a suitable heat treatment prior to processing. It is also important to avoid prolonged exposure during processing to the temperature ranges we have discussed. The trend toward low-temperature processing poses a dilemma, because complex formation could occur during the device processing. Complex formation has not been fully researched at the present time.

Oxygen in interstitial lattice sites also acts to increase the yield strength of silicon[40, 41] through the mechanism of solution hardening. Improvements of 25% over oxygen-free silicon have been reported. This beneficial effect increases with concen-

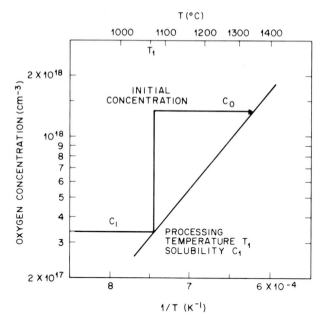

Fig. 19 The solid solubility of oxygen in silicon. *(After Patel, Ref. 38.)*

tration until the oxygen begins to precipitate. Oxygen at the concentration levels mentioned earlier represents a supersaturated condition at most common processing temperatures and will precipitate during processing, given a sufficient supersaturation ratio. Precipitation usually occurs when the oxygen concentration exceeds a threshold value[42] of 6.4×10^{17} atoms/cm^3. The precipitation may proceed homogeneously, but native defects, usually in the crystal from growth, allow heterogeneous precipitation to dominate the kinetics. Figure 19 details the solubility of oxygen in silicon and illustrates the supersaturation effect. A wafer containing an initial concentration of oxygen C_0 when processed at a temperature T_1 results in a supersaturated condition. The supersaturation which is C_0/C_1 results because the oxygen solubility at T_1 is only C_1, and is less than C_0.

The precipitates represent an SiO$_2$ phase. A volume mismatch occurs as the precipitates grow in size, representing a compressive strain on the lattice that is relieved by the punching out of prismatic dislocation loops. Actually, a variety of defects, including stacking faults, are associated with precipitate formation. These defects attract fast diffusing metallic species, which give rise to large junction leakage currents. The ability of defects to capture harmful impurities (called "gettering") can be used beneficially. Defects formed in the interior of a wafer getter impurities from the wafer surface where device junctions are located (Fig. 20). Gettering of defects is explained more fully in Section 1.5.

Carbon is another unintended impurity[43] in the polysilicon, and is transported to the melt from graphite parts in the furnace. Carbon in silicon occupies a substitutional lattice site and is conveniently measured using infrared transmission measurements of

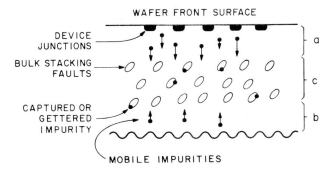

Fig. 20 Schematic of a denuded zone in a wafer cross section and gettering sites. *a* and *b* are zones denuded of defects, *c* represents the region of intrinsic gettering.

an absorption line at 603 cm^{-1}. Because carbon's segregation coefficient is small (0.07), its axial variation is large. Typical seed-end concentrations range from 10^{16} atoms/cm^3 and down. For butt ends at a high percentage of melt solidification, values range[35] up to 5×10^{17} atoms/cm^3. At these levels carbon does not precipitate like oxygen, nor is it electrically active. Carbon has been linked to the precipitation kinetics of oxygen and point defects.[44] In this regard, its presence is undesirable because it aids the formation of defects.

1.3.5 Characteristics and Evaluation of Crystals

Routine evaluation of crystals (also called "ingots" or "boules") involves testing their resistivity, evaluating their crystal perfection, and examining their mechanical properties, such as size and mass. Other less routine evaluations include measuring the crystals' oxygen, carbon, and heavy metal content. The evaluations of heavy metal content are made by minority carrier lifetime measurements or neutron activation analysis.

After growth the crystal is usually weighed and the ingot is then visually inspected. Gross crystalline imperfections like twinning are apparent to the unaided eye. Sections of the ingot containing such defects are cut from the boule, as are sections of the boule that are irregularly shaped or undersized. Total silicon loss can equal 50% at this step. Next the butt (or tang) end of the ingot, or a slice cut from that position, is preferentially etched to reveal defects such as dislocations. A common etchant is Sirtl's etch, which is a 1:1 mixture of HF acid (49%) and five molar chromic acid.[45] This same etch can be used on polished and processed wafers to delineate other types of microdefects or impurity precipitates. Cracks can be detected by a method using ultrasonic waves.[46]

Resistivity measurements are made on the flat ends of the crystal by the four-point probe technique (Fig. 21). Current I (mA) is passed through the outer probes and voltage V (mV) measured between the inner probes. The measured resistance (V/I) is converted to resistivity (Ω-cm) using the formula

$$\rho = (V/I)2\pi S \qquad (13)$$

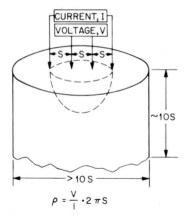

$$\rho = \frac{V}{I} \cdot 2\,\pi\,S$$

Fig. 21 Four-point probe measurement on crystal end. *(Courtesy P. H. Langer.)*

where S is the probe spacing in centimeters. Measurements can be reproduced $\pm 2\%$ if care is taken in selecting instrumentation, probe pressure, and current levels.[47, 48] For example, current levels are raised as the resistivity of the material is lowered, such that the measured voltage is maintained between 2 and 20 mV. The variation of resistivity with ambient temperature is a significant effect—approximately 1%/°C at 23°C for 10 Ω-cm material. The resistivity of the material is related to the doping density through the mobility.[49] Figure 22 shows the relationship for boron- and phosphorus-doped samples.

Boron-doped CZ silicon is available in resistivities from 0.0005 to 50 Ω-cm, with radial uniformities of 5% or better. Arsenic- and phosphorus-doped silicon is available in the range 0.005 to 40 Ω-cm, with arsenic being the preferred dopant in the lower resistivity ranges. Antimony is also used to dope crystals in the 0.01-Ω-cm range. Antimony-doped substrates are preferred as epitaxial substrates because auto-doping effects are minimized (Chapter 2). Radial uniformities for n-doped material range from 10 to 50% depending on diameter, dopant, orientation, and process conditions.[50]

1.4 SILICON SHAPING

Silicon is a hard, brittle material registering 72.6 on the Rockwell "A" hardness scale. The most suitable material for shaping and cutting silicon is industrial-grade diamond, although SiC and Al_2O_3 have also found application. This section highlights major shaping methods, but in some cases alternatives do exist. This section also elucidates the relationship of these operations to the device processing needs required of silicon slices.

Conversion of silicon ingots into polished wafers requires nominally six machining operations, two chemical operations, and one or two polishing operations.[46, 51] Additionally, assorted inspections and evaluations are performed between the major process steps. A finished wafer is subject to a number of dimensional tolerances, dic-

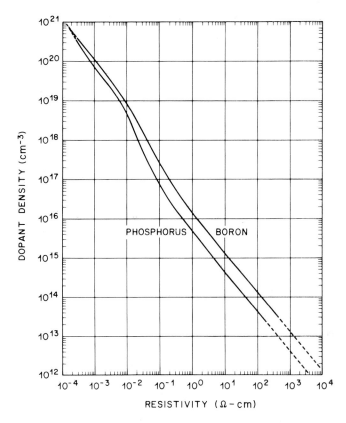

Fig. 22 Conversion between resistivity and dopant density in silicon. *(After Thurber, Mattis, and Liu, Ref. 49).*

tated by the needs of the device fabrication technology. As shown in Table 3, these tolerances are somewhat loose compared to metal machining capability. The existence of organizations to standardize these factors[52] and their measurement[53] proves the maturity of the silicon materials industry. The motivation for standards is twofold. First, it helps to standardize wafer production, resulting in efficiency and cost savings. Secondly, producers of process equipment and fixturing benefit from knowing the wafer dimensions when designing equipment, and so forth.

Table 3 Typical specifications for 100- and 125-mm diameter wafers

Diameter	100 ± 1 mm	125 ± 0.5 mm
Primary flat	30−35 mm	40−45 mm
Secondary flat	16−20 mm	25−30 mm
Thickness	0.50−0.55 mm	0.60−0.65 mm
Bow	60 μm	70 μm
Taper	50 μm	60 μm
Surface orientation	(100) $\pm 1°$	Same
	(111) off orientation	Same

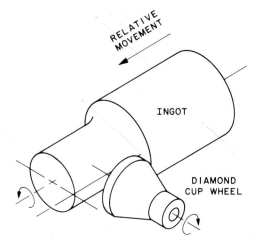

Fig. 23 Schematic of grinding process. *(After Bonora, Ref. 46.)*

1.4.1 Shaping Operations

The first shaping operation removes the seed and tang ends from the ingot. Portions of the ingot that fail the resistivity and perfection evaluations previously mentioned are also cut away. The cuttings are sufficiently pure to be recycled, after cleaning, in the growing operation. The rejected ingot pieces can also be sold as metallurgical-grade silicon. The cutting is conveniently done as a manual operation using a rotary saw.

The next operation is a surface grinding, and is the step that defines the diameter of the material. Silicon ingots are grown slightly oversized because the automatic diameter control in crystal growing cannot maintain the needed diameter tolerance, and crystals cannot be grown perfectly round. Figure 23 shows schematically lathe-like machine tool used to grind the ingot to diameter. A rotating cutting tool makes multiple passes down a rotating ingot until the chosen diameter is attained. Precise diameter control is required for many kinds of processing equipment, and is a consideration in the design of processing and furnace racks.

Following diameter grinding, one or more flats are ground along the length of the ingot. The largest flat, called the "major" or "primary flat," is usually located relative to a specific crystal direction. The location is accomplished by an x-ray technique. The primary flat serves several purposes. It is used as a mechanical locator in automated processing equipment to position the wafer, and also serves to orient the IC device relative to the crystal in a specific manner. Other smaller flats are called "secondary flats," and serve to identify the orientation and conductivity type of the material (Fig. 24). Secondary flats provide a means of quickly sorting and identifying wafers, should mixing occur.

Once the above operations have prepared the ingot it is usually ready to be converted to a wafer geometry. Slicing is important because it determines four wafer parameters: surface orientation, thickness, taper, and bow. The surface orientation is

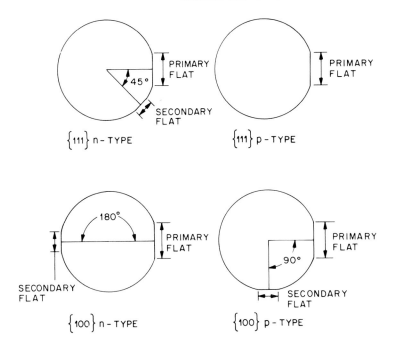

Fig. 24 Identifying flats on silicon wafer. *(SEMI standard, used by permission.)*

determined by cutting several wafers, measuring the orientation by an x-ray method, and then resetting the saw until the correct orientation is achieved. Wafers of $\langle 100 \rangle$ orientation are usually cut "on orientation" (Table 3). The tolerances allowed for orientation do not adversely affect MOS device characteristics such as interface-trap density. The other common orientation, $\langle 111 \rangle$, is usually cut "off orientation," as required for epitaxial processing (Chapter 2). Routine manufacturing tolerances are also acceptable here.

The wafer thickness is essentially fixed by slicing, although the final value depends on subsequent shaping operations. Thicker wafers are better able to withstand the stresses of subsequent thermal processes (epitaxy, oxidation, and diffusion), and as a result exhibit less tendency to plastically or elastically deform in such processing. A major concern in slicing is the blade's continued ability to cut wafers from the crystal in very flat planes. If the blade deflects during slicing, this will not be achieved. By positioning a capacitive sensing device near the blade, blade position and vibration can be monitored, and higher-quality cutting achieved. If a wafer is sliced with excessive curvature (bow), subsequent lapping operations may not be able to correct it, and surface flatness objectives cannot be obtained by polishing.

Inner diameter (ID) slicing is the most common mode of slicing. ID slicing uses a saw blade whose cutting edge is on the interior of an annulus. Figure 25 shows a schematic of the process. The ingot is prepared for slicing by mounting it in wax or epoxy on a support, and then positioning the support on the saw. This procedure ensures that the ingot is held rigid for the slicing process. Some success has been

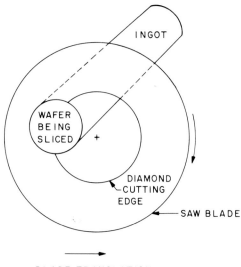

BLADE TRANSLATION

Fig. 25 Schematic of ID slicing process.

obtained mounting ingots in a fixture using hydraulic pressure. The saw blade is a thin sheet of stainless steel (325 μm), with diamond bonded on the inner rim. This blade is tensioned in a collar and then mounted on a drum that rotates at high speed (2000 r/min) on the saw. Saw blades up to 58 cm in diameter with a 20 cm opening are available. Thus, slicing capability up to nearly the ID opening of 20 cm exists. The blade is moved relative to the stationary ingot. The cutting process is water cooled. The kerf loss (loss due to blade width) at slicing is 325 μm, which means that approximately one-third of the crystal is lost as sawdust. Cutting speeds are nominally 0.05 cm/s, which, considering that wafers are sliced sequentially, is a rather slow process. Another shortcoming is the drum's finite depth, which limits the length of the ingot section that can be cut into wafers. Another style ID saw mounts the blade on an air-bearing and provides rotation using a belt drive. This arrangement allows any length of ingot to be sliced; after slicing, individual wafers are recovered opposite the feed side and placed in a cassette. Such a saw, which hydraulically mounts the ingot, represents a highly automatic approach to sawing.

The wafer as cut varies enough in thickness to warrant an additional operation, if the wafers are intended for VLSI application. A mechanical, two-sided lapping operation (Fig. 26), performed under pressure using a mixture of Al_2O_3 and glycerine, produces a wafer with flatness uniform to within 2 μm . This process helps ensure that surface flatness requirements for photolithography can be achieved in the subsequent polishing steps. Approximately 20 μm per side is removed.

A final shaping step is edge contouring, where a radius is ground on the rim of the wafer (Fig. 27). This process is usually done in cassette-fed high-speed equipment. Edge-rounded wafers develop fewer edge chips during device fabrication and aid in controlling the buildup of photoresist (Chapter 7) at the wafer edge. Chip sites

Fig. 26 Double-sided lapping machine.

act as places where dislocations can be introduced during thermal cycles and as places where wafer fracture can be initiated. The silicon chips themselves, if present on the wafer surface, add to the D_0 (defect density) of the IC process reducing yield (Chapter 14).

1.4.2 Etching

The previously described shaping operations leave the surface and edges of the wafer damaged and contaminated, with the depth of work damage depending on the specifics of the machine operations. The damaged and contaminated region is on the order of 10 μm deep and can be removed by chemical etching.

Historically, mixtures of hydrofluoric, nitric, and acetic acids have been used, but alkaline etching, using potassium or sodium hydroxide, has found application.

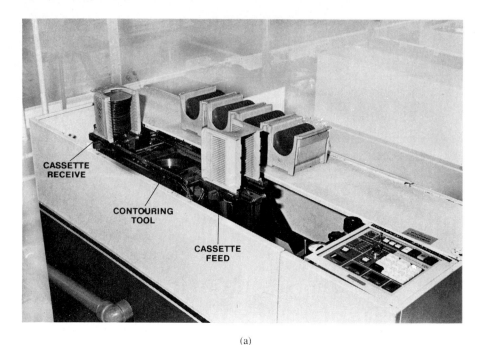

(a)

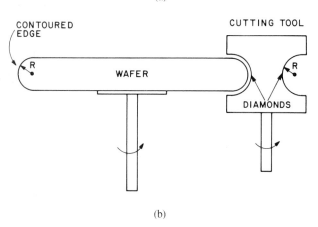

(b)

Fig. 27 (a) Cassette feed edge-contouring tool. (b) Schematic of edge-contouring process.

   The process equipment includes an acid sink, which contains a tank to hold the etching solution, and one or more positions for rinsing the wafers in water. The process is batch in nature, involving tens of wafers. The best process equipment provides a means of rotating the wafer during acid etching to maintain uniformity. Processing is usually performed with a substantial overetch to assure all damage is removed. A removal of 20 μm per side is typical. The etching process is checked frequently by gauging wafers for thickness before and after etching. Etch times are usually on the order of several minutes per batch.

The chemistry of the etching reaction is electrochemical. The dissolution involves oxidation-reduction processes, followed by a dissolution of an oxidation product. In the hydrofluoric, nitric, and acetic acid etching system[46, 51] nitric acid is the oxidant and hydrofluoric acid dissolves the oxidized products according to the reaction:[54, 55]

$$3Si + 4HNO_3 + 18HF => 3H_2 SiF_6 + 4NO + 8H_2O \qquad (14)$$

Acetic acid dilutes the system so that etching can be better controlled. Water can also be used, but acetic acid is preferable because water is a by-product of the reaction. The etching can be isotropic or anisotropic, according to the acid mixture or temperature. In HF-rich solutions, the reaction is limited by the oxidation step. This regime of etching is anisotropic, and the oxidation reaction is sensitive to doping, orientation, and defect structure of the crystal (where the oxidation occurs preferentially). The use of HNO_3-rich mixtures produces a condition of isotropic etching, and the dissolution process is then rate limiting. Over the range 30 to 50°C, the etching kinetics of an HNO_3-rich solution have been found to be diffusion controlled rather than reaction rate limited (Fig. 28). Thus, transport of reactant products to the wafer surface across a diffusion boundary layer is the controlling mechanism. For these reasons, the HNO_3-rich solutions are preferred for removing work damage. Rotating the wafers in solution controls the boundary layer thickness and thereby provides dimensional control of the wafer. The isotropic character of the etch produces a smooth, specular surface. A common etch formulation is 4:1:3; the concentrations are 70% by weight HNO_3, and 49% by weight HF and $HC_2 H_3 O_2$.

Unfortunately, the dimensional uniformity introduced by the lapping step is not maintained across large diameter wafers (>75 mm) to a degree compatible with maintaining surface flatness in polishing. The hydrodynamics of rotating a large diameter wafer in solution do not allow for a uniform boundary layer, so a taper is introduced into the wafer. Projection lithography places demands on surface flatness that necessitates the use of alkaline etching. Alkaline etching is by nature anisotropic, since it depends on the surface orientation. The reaction is apparently dominated by the number of dangling bonds present on the surface. The reaction is generally reaction rate limited, and wafers do not have to be rotated in the solution. Since boundary layer transport is not a factor, excellent unformity can be achieved. As in the acid etching case, the reaction is twofold when a mixture of KOH/H_2O or $NaOH/H_2O$ is used.[56] A typical formulation uses KOH and H_2O in a 45% by weight solution at 90°C to achieve an etch rate of 25 μm/min for {100} surfaces. An occasional problem with the damage removal process is insufficient etching, which can lead to the generation of dislocations in subsequent treatments because of residual damage.

1.4.3 Polishing

Polishing is the final step. Its purpose is to provide a smooth, specular surface where device features can be photoengraved. A main VLSI concern is to produce a surface with a high degree of surface flatness and minimum local slope to meet the requirements of optical projection lithography.[57] Values between 5 and 10 μm are typical

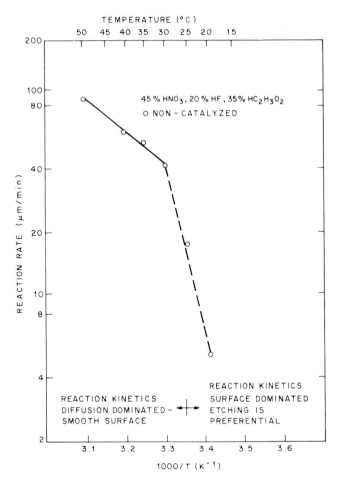

Fig. 28 Typical etch rate versus temperature curve for one mixture of HF, HNO$_3$, and HC$_2$H$_3$O$_2$ acids. *(After Robbins and Schwartz, Ref. 54.)*

surface flatness specifications. The surface is also required to be free from contamination and damage.

Figure 29 shows a typical polishing machine and a schematic of the process. The process requires considerable operator attention for loading and unloading. It can be conducted as a single-wafer or batch-wafer process depending on the equipment. Economics determines the choice of single or batch processing; larger wafers are preferred for single-wafer processing. Single-wafer processing is also felt to offer a better means of achieving surface flatness goals. In both single and batch processing, the process involves a polishing pad made of artificial fabric, such as a polyester felt, polyurethane laminate. Wafers are mounted on a fixture, pressed against the pad under high pressure, and rotated relative to the pad. A mixture of polishing slurry and water is dripped onto the pad to accomplish the polishing (which is both a chemical and mechanical process). The porosity of the pad is a factor in carrying slurry to the

(a)

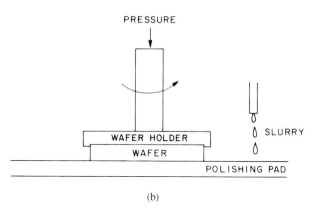

(b)

Fig. 29 (a) Photograph of polishing machine. (b) Schematic of polishing process.

wafer for polishing. The slurry is a colloidal suspension of fine SiO_2 particles (100-Å diameter) in an aqueous solution of sodium hydroxide. Under the heat generated by friction, the sodium hydroxide oxidizes the silicon with the OH^- radical. This is the chemical step. The mechanical step abrades the oxidized silicon away, by the silica particles in the slurry. Polishing rate and surface finish are a complex function of pressure, pad properties, rotation speed, slurry composition, and pH. Typical processes remove 25 μm of silicon. In a batch process involving tens of wafers, silicon removal can take 30 to 60 min; single-wafer processing can be accomplished in 5 min. Single-wafer processes use higher pressures than the batch processes.

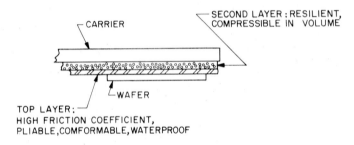

Fig. 30 Schematic of Flex-Mount™ (Flex-Mount is a trademark of Siltec Corp., Menlo Park, California) polishing process. *(After Bonara, Ref. 58.)*

The method of mounting wafers for polishing also deserves attention. Historically, wafers were waxed onto a metal plate. This method is costly and may not give the best surface flatness. An alternative (Fig. 30) is a waxless technique where wafers are applied to a conformal pad, typically a two-layer vinyl.[58] This method is cost effective and eliminates the influence of rear surface particles on front surface flatness. After polishing, wafers are cleaned with acid and/or solvent mixtures to remove slurry residue (and wax), and readied for inspection. Polished wafers are subjected to a number of measurements that are concerned with cosmetic, crystal perfection, mechanical, and electrical attributes.

Figure 31 shows how the industry has used wafers of increasingly larger diameter motivated in part by the trend to larger chip areas.

1.5 PROCESSING CONSIDERATIONS

In the IC processing of silicon wafers it is usually necessary to maintain the purity and perfection of the material.

1.5.1 Gettering Treatments

Many VLSI circuits (dynamic RAMs), require low junction leakage currents. Narrow-base bipolar transistors are sensitive to conductive impurity precipitates, which act like shorts between the emitter and collector (the pipe effect). Metallic impurities, such as transition group elements, are responsible for these effects. These elements are located at interstitial or substitutional lattice sites and are generation-recombination centers for carriers. The precipitated forms of these impurities are usually silicides, which are electrically conductive. To remove impurities from devices, a variety of processing techniques are available, termed "gettering" treatments.[42] "Gettering" is a general term taken to mean a process that removes harmful impurities or defects from the regions in a wafer where devices are fabricated. Among these techniques are ways to pretreat (i.e., pregetter) silicon wafers prior to IC processing. Pregettering provides a wafer with sinks that can absorb impurities as they are introduced during device processing.

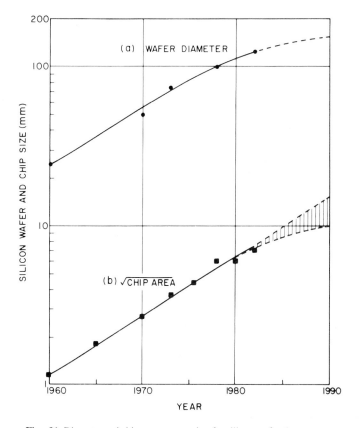

Fig. 31 Diameter and chip area progression for silicon wafers by year.

One technique of removing impurities involves intentionally damaging the back surface of the wafer. Mechanical abrasion methods such as lapping or sand blasting have been used for this purpose. A more controllable process uses damage created by a focused laser beam.[59] This process requires a threshold energy density of 5 J/cm^2. One configuration of this technique involves using a Q-pulsed, Nd:YAG laser. The laser beam is rastered along the back surface to create an array of micromachined spots. Depending on the energy density and proximity of the spots, the silicon lattice is damaged and/or strained by the high-energy pulse. Upon thermal processing, dislocations emanate from the spots. If the stresses placed on the wafer during furnace processing are low, the dislocations remain localized on the back surface. The dislocations represent favorable trapping sites for fast diffusing species. For example, the diffusion length of iron for 30 min at 1000°C is 3000 μm compared to slice thicknesses of 300 to 500 μm. When trapped on the rear surface, these impurities are innocuous.

Another series of methods uses the defects associated with oxygen precipitation for trapping sites. These methods use one or more thermal cycles to produce the desired result.[44] They usually involve a high-temperature cycle (over 1050°C in nitro-

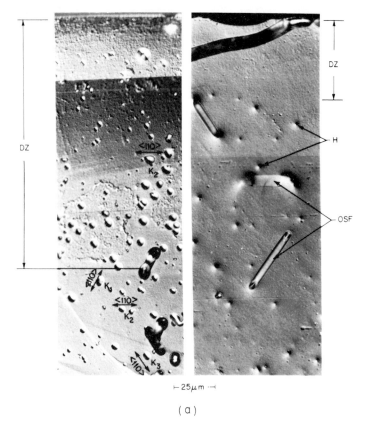

$\vdash 25\mu m \dashv$

(a)

Fig. 32 (a) The left-hand photograph shows a denuded zone (DZ) in a wafer cross section after a preferential etch. K_1, K_2, and K_3 are small stacking fault defects. The right-hand photograph is another wafer cross section showing stacking faults (OSF) and precipitate features (H) below the DZ. *(Courtesy G. A. Rozgonyi.)*

gen), which removes oxygen from the surface of the wafer by evaporation.[60] Denuding can also be accomplished in an oxidizing ambient where substrate oxygen incorporates into the surface oxide layer. Either approach lowers the oxygen content near the surface so that precipitation does not occur, because the supersaturated condition has been removed. The depth of the oxygen-poor region is a function of time and temperature, and depends on the diffusivity of oxygen in silicon (Fig. 32). The region represents a defect-free zone (denuded zone) for device fabrication. Additional thermal cycles can be added to promote the formation of precipitates and defects in the interior of the wafer. This approach is called "intrinsic gettering" because the oxygen is native to the wafer. Intrinsic gettering is attractive because it fills the volume of the wafer with trapping sites. Otherwise, the bulk of the wafer really serves no useful function beyond mechanically supporting the thin layer where the device is formed. Both intrinsic gettering and intentionally damaging the back of the devices have been successfully employed in circuit fabrication processes.[61, 62]

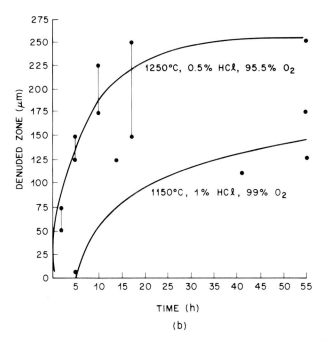

Fig. 32 (b) Denuded zone width for two sets of processing conditions.

1.5.2 Thermal Stress Factors

We generally want to maintain the crystal perfection of wafers through the device fabrication process, and to keep them mechanically undeformed. Wafers are typically processed in furnaces using racks with a high wafer-diameter-to-spacing ratio. Upon removing the wafer from a high-temperature furnace, the wafer edges cool rapidly by radiation to the surroundings, but the wafer centers remain relatively hotter.[15] The resultant temperature gradient creates a thermal stress S that can be estimated as:

$$S = aE \ dT \qquad (15)$$

where a is the coefficient of thermal expansion, E is Young's modulus, and dT is the temperature difference across the slice.

If these stresses exceed the yield strength (the maximum stress the material will accommodate without irreversible deformation) of the material, dislocations will form. Stresses are usually kept to acceptable levels by slowly withdrawing wafers from the furnace to minimize the temperature gradient, or by lowering the furnace temperature[16] prior to removing the wafers to the point where the yield strength at the removal temperature exceeds the stresses imposed (Fig. 33).

Material parameters must also be considered. Oxygen precipitates (useful for gettering) can reduce the yield strength (critical shear stress) up to fivefold (Fig. 33). Wafer thickness and bow must also be considered as bow enhances the thermal

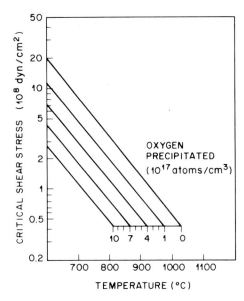

Fig. 33 Yield strength of silicon showing the influence of oxygen precipitates. *(After Leroy and Plougoven, Ref. 63.)*

stress.[63] Thus, the design of furnace cycles must consider the worst case combinations of oxygen precipitation and bow present in the processed wafers. Therefore, because of these effects the thickness of wafers is usually increased as the diameter increases.

1.6 SUMMARY AND FUTURE TRENDS

Silicon wafers have been and will continue to be the predominant material for solid-state device manufacture. In the VLSI device arena, some other material technologies will also become common. The two main contenders are silicon-on-insulator (e.g., Si on SiO_2) and compound semiconductors (notably GaAs). These technologies will be chosen when high-speed circuitry or the need to optimize other circuit parameters are the deciding factors.

The specifications placed on wafers will become more sophisticated for VLSI applications. Unintentional impurities that are now ignored in specifications will need maximum allowable levels placed upon them. This would also be true for carbon and metallic species.[64] Oxygen is already subject to such specifications, but additional control over oxygen precipitation behavior as it relates to the growing process and thermal cycling will probably be forthcoming. Mechanical dimensions will continue to be driven by equipment and processing needs. In particular, lithographic evolution will require flatter wafers. Surface cleanliness and other surface characteristics, because they influence oxide-silicon interface state density, may represent a new class of specification and a new area of study. Laser marking of wafers for identification purposes will also become a new attribute.

Large diameter wafers (>150 mm) are feasible and 200-mm wafers have been produced. Practical implementation is awaiting the need for further productivity improvements and improved circuit fabrication capability, particularly in the lithographic area. Ingots of larger diameters will be grown in big pullers. The slower cooling rates experienced by these such materials may alter the properties of the material, particularly the point-defect kinetics. This topic provides an area for continued research, because properties of the material are related to defects formed in device processing, and thus related to IC yield.

REFERENCES

[1] C. L. Yaws, R. Lutwack, L. Dickens, and G. Hsiu, "Semiconductor Industry Silicon: Physical and Thermodynamic Properties," *Solid State Technol.*, 24, 87 (1981).

[2] J. C. Bailar, Editor, *Comprehensive Inorganic Chemistry*, 1, Pergamon Press, New York, 1973.

[3] W. Shockley, "The Theory of p-n Junctions in Semiconductors and p-n Junction Transistors," *Bell Syst. Tech. J.*, 28, 435 (1949).

[4] L. D. Crossman and J. A. Baker, "Polysilicon Technology," *Semiconductor Silicon 1977*. Electrochem. Soc., Pennington, New Jersey, 1977, p. 18.

[5] Am. Soc. Test. Mater., ASTM Standard, F574, Part 43.

[6] W. Dietze, L. P. Hunt, and D. H. Sawyer, "The Preparation and Properties of CVD-Silicon Tubes and Boats for Semiconductor Device Technology," *J. Electrochem. Soc.*, 121, 1112 (1974).

[7] S. N. Rea, "Czochralski Silicon Pull Rate Limits," *J. Cryst. Growth*, 54, 267 (1981).

[8] R. A. Laudise, *The Growth of Single Crystals*, Prentice Hall, Englewood Cliffs, New Jersey, 1970.

[9] L. D. Dyer, "Dislocation-Free Czochralski Growth of ⟨110⟩ Silicon Crystals," *J. Cryst. Growth*, 47, 533 (1979).

[10] K. V. Ravi, *Imperfections and Impurities in Semiconductor Silicon*, Wiley, New York, 1981.

[11] R. K. Watts, *Point Defects in Crystals*, Wiley and Sons, New York, 1977.

[12] A. G. Milnes, *Deep Levels in Semiconductors*, Wiley, New York, 1973.

[13] J. Friedel, *Dislocations*, Pergamon Press, New York, 1964.

[14] A. J. R. deKock and W. M. van de Wijgert, "The Effect of Doping on the Formation of Swirl Defects in Dislocation-Free Czochralski-Grown Silicon," *J. Cryst. Growth*, 49, 718 (1980).

[15] S. M. Hu, "Temperature Distribution and Stresses in Circular Wafers in a Row during Radiative Cooling," *J. Appl. Phys.*, 40, 4413 (1969).

[16] K. G. Moerschel, C. W. Pearce, and R. E. Reusser, "A Study of the Effects of Oxygen Content, Initial Bow and Furnace Processing on Warpage of Three-Inch Diameter Silicon Wafers," *Semiconductor Silicon 1977*, Electrochem. Soc., Pennington, New Jersey, 1977, p. 170.

[17] S. Kishino, Y. Matsushita, and M. Kanamori, "Carbon and Oxygen Role for Thermally Induced Microdefect Formation in Silicon," *Appl. Phys. Lett.*, 35, 213 (1979).

[18] A. Armigliato, D. Nobili, P. Ostoja, M. Servidori, and S. Solmi, "Solubility and Precipitation of Boron in Silicon," *Semiconductor Silicon 1977*, Electrochem. Soc., Pennington, New Jersey, 1977, p. 638.

[19] W. T. Stacy, D. F. Allison, and T. C. Wu, "The Role of Metallic Impurities in the Formation of Haze Defects," *Semiconductor Silicon 1981*, Electrochem. Soc., Pennington, New Jersey, 1981, p. 344.

[20] B. R. Pamplion, *Crystal Growth*, Pergamon Press, New York, 1983.

[21] W. R. Runyan, *Silicon Semiconductor Technology*, McGraw-Hill, New York, 1965.

[22] T. G. Digges, Jr. and R. Shima, "The Effect of Growth Rate, Diameter and Impurity Concentration on Structure," *J. Cryst. Growth*, 50, 865 (1980).

[23] S. M. J. G. Van Run, "A Critical Pulling Rate for Remelt Suppression in Silicon Crystal Growth," *J. Cryst. Growth*, 53, 441 (1981).

[24] H. Kolker, "The Behavior of Nonrotational Striations in Silicon," *J. Cryst. Growth*, 50, 852 (1980).

[25] J. Chikawa and S. Yoshikawa, "Swirl Defects in Silicon Single Crystals," *Solid State Technol.*, 23, 65 (1980).

[26] J. A. Burton, R. C. Prim, and P. Slichter, *J. Chem. Phys.*, 21, 1987 (1953).

[27] J. R. Carruthers, A. F. Witt, and R. E. Reusser, "Czochralski Growth of Large Diameter Silicon Crystals - Convection and Segregation," *Semiconductor Silicon 1977*, Electrochem. Soc., Pennington, New Jersey, 1977, p. 61.

[28] K. M. Kim, "Interface Morphological Instability in Czochralski Silicon Crystal Growth from Heavily Sb-Doped Melt," *J. Electrochem. Soc.*, 126, 875 (1979).

[29] K. E. Benson, W. Lin, and E. P. Martin, "Fundamental Aspects of Czochralski Silicon-Crystal Growth for VLSI," *Semiconductor Silicon 1981*, Electrochem. Soc., Pennington, New Jersey, 1981, p. 33.

[30] M. H. Liepold, T. P. O'Donnell, and M. A. Hagan, "Materials of Construction for Silicon Crystal Growth," *J. Cryst. Growth*, 40, 366 (1980).

[31] F. A. Voltmer and F. A. Padovani, "The Carbon-Silicon Phase Diagram for Dilute Carbon Concentration," *Semiconductor Silicon 1973*, Electrochem. Soc., Pennington, New Jersey, 1973, p. 75.

[32] H. Hirata and K. Hoshikawa, "The Dissolution Rate of Silica in Molten Silicon," *Jpn. J. Appl. Phys.*, 19, 1573 (1980).

[33] B. Bathey, H. E. Bates, and M. Cretella, "Effect of Carbon on the Dissolution of Fused Silica in Liquid Silicon," *J. Electrochem. Soc.*, 128, 771 (1980).

[34] M. Watanabe, T. Usami, H. Muroaka, S. Matsuo, Y. Imanishi, and H. Nagashima, "Oxygen-Free Silicon Single Crystal from Silicon-Nitride Crucible," *Semiconductor Silicon 1981*, Electrochem. Soc., Pennington, New Jersey, 1981, p. 126.

[35] H. M. Liaw, "Oxygen and Carbon in Czochralski-Grown Silicon," *Semicon. Int.*, 2, 71 (1979).

[36] T. Carlberg, T. B. King, and A. F. Witt, "Dynamic Oxygen Equilibrium in Silicon Melts during Crystal Growth," *J. Electrochem. Soc.*, 127, 189 (1981).

[37] T. Suzuki, N. Isawa, Y. Okubo, and K. Hoshi, "CZ Silicon Growth in a Transverse Magnetic Field," *Semiconductor Silicon 1981*, Electrochem. Soc., Pennington, New Jersey, 1981, p. 90.

[38] J. R. Patel, "Oxygen in Silicon," *Semiconductor Silicon 1977*, Electrochem. Soc., Pennington, New Jersey, 1977, p. 521.

[39] Am. Soc. Test. Mater., ASTM Standard, F121-76, Part 43.

[40] J. Doerschel and F. G. Kirscht, "Differences in Plastic Deformation Behavior of CZ and FZ Grown Silicon Crystals," *Phys. Status Solidi A*, 64, K85 (1981).

[41] K. Sumino et al., "The Origin of the Difference in the Mechanical Strengths of Czochralski Silicon," *Jpn. J. Appl. Phys.*, 19, L49 (1980).

[42] C. W. Pearce, L. E. Katz, and T. E. Seidel, "Considerations Regarding Gettering in Integrated Circuits," *Semiconductor Silicon 1981*, Electrochem. Soc., Pennington, New Jersey, 1981, p. 705.

[43] T. Nozaki, "Concentration and Behavior of Carbon in Semiconductor Silicon," *J. Electrochem. Soc.*, 117, 1566 (1970).

[44] Y. Matsushita, S. Kishino, and M. Kanamori, "A Study of Thermally Induced Microdefects in Czochralski-Grown Silicon Crystals: Dependence on Annealing Temperature and Starting Material," *Jpn. J. Appl. Phys.*, 19, L101 (1980).

[45] D. G. Schimmel, "A Comparison of Chemical Etches for Revealing (100) Silicon Crystal Defects," *J. Electrochem. Soc.*, 123, 734 (1976).

[46] A. C. Bonora, "Silicon Wafer Process Technology: Slicing, Etching, Polishing," *Semiconductor Silicon 1977*, Electrochem. Soc., Pennington, New Jersey, 1977, p. 154.

[47] Am. Soc. Test. Mater., ASTM Standard, F84, Part 43.

[48] Am. Soc. Test. Mater., ASTM Standard, F723, Part 43.

[49] W. R. Thurber, R. L. Mattis, and Y. M. Liu, "Resistivity Dopant Density Relationship for Silicon," *Semiconductor Characterization Techniques*, Electrochem. Soc., Pennington, New Jersey, 1978, p.81.

[50] S. E. Bradshaw and J. Goorissin, "Silicon for Electronic Devices," *J. Cryst. Growth*, 48, 514 (1980).

[51] R. B. Herring, "Silicon Wafer Technology - State of the Art 1976," *Solid State Technol.*, 19, 37 (1976).

[52] Semiconductor Equipment and Materials Institute (SEMI), Mountain View, California.
[53] The American Society for Testing and Materials (ASTM), Committee F-1 on Electronics, Philadelphia, Pennsylvania.
[54] H. Robbins and B. Schwartz, "Chemical Etching of Silicon," *J. Electrochem. Soc.*, 106, 505 (1959); 107, 108 (1960); 108, 365 (1961); and 123, 1909 (1976).
[55] W. Kern, "The Chemical Etching of Semiconductors," *RCA Rev.*, 39, 278 (1978).
[56] I. Barycka, H. Teterycz, and Z. Znamirowski, "Sodium Hydroxide Solution Shows Selective Etching of Boron-Doped Silicon," *J. Electrochem. Soc.*, 126, 345 (1979).
[57] W. A. Baylies, "A Review of Flatness Effects in Microlithographic Technology," *Solid State Technol.*, 24, 132 (1981).
[58] A. C. Bonara, "Flex-Mount Polishing of Silicon Wafers," *Solid State Technol.*, 20, 55 (1977).
[59] C. W. Pearce and V. J. Zaleckas, "A New Approach to Lattice Damage Gettering," *J. Electrochem. Soc.*, 126, 1436 (1979).
[60] K. Yamamoto, S. Kishino, Y. Matsushita, and T. Lizuka, "Lifetime Improvement in Czochralski-Grown Silicon Wafers by the Use of a Two Step Annealing," *Appl. Phys. Lett.*, 36, 195 (1980).
[61] L. E. Katz, C. W. Pearce, and P. F. Schmidt, "Neutron Activation Study of a Gettering Treament for CZ Silicon Substrates," *J. Electrochem. Soc.*, 128, 620 (1981).
[62] M. Ogino, T. Usami, and M. Watanabe, "Microdefects Due to Oxygen Precipitates and Their Application to CMOS LSI and CCD Sensor," *Electrochem. Soc. Extended Abstracts*, 80-2, Abs. 435 (1980).
[63] B. Leroy and C. Plougoven, "Warpage of Silicon Wafers," *J. Electrochem. Soc.*, 127, 961 (1980).
[64] P. F. Schmidt and C. W. Pearce, "A Neutron Activation Analysis Study of the Sources of Transistion Group Metal Contamination in the Silicon Device Manufacturing Process," *J. Electrochem. Soc.*, 128, 630 (1981).

PROBLEMS

1 Iron is an impurity in quartz crucibles. Using a concentration value of 2×10^{18} /cm^3 in the crucible, assume 300 cm^3 of the crucible is dissolved into a 6500-g melt, all at the beginning of the cycle. Calculate the seed (0% solidified) and tang end (90% solidified) iron concentration in the ingot.

2 Using the equation governing crystal growth (Eq. 6), derive an expression relating growth rate inversely to crystal diameter. Assume no temperature gradient in the melt. Since the heat flow down the crystal is small, assume heat flow from the crystal is predominantly from radiation.

3 Calculate the number of gallons of HF and HNO$_3$ acid needed to remove the work damage from 5000 wafers of 100-mm diameter.

4 The seed crystal used in CZ growing is usually "necked down" to a small diameter (3 mm) as a means to initiate dislocation-free growth. Using the yield strength of silicon, calculate the maximum mass of silicon that could be supported by such a seed. Convert this to a length for 100- and 125-mm-diameter crystals.

5 Large growers, such as that pictured in the chapter, require 120 kWh to convert a kilogram of polysilicon into a crystal. Account for this energy in terms of the energy needed to melt the silicon, the radiation loss from the melt surface, and conduction down the crystal. Is all the energy accounted for? Assume that a 10-kg charge of polysilicon is used to grow a 10-cm-diameter crystal from a 25-cm-diameter crucible at a rate of 0.0025 cm/s.

6 Solar cells have been suggested as an alternative energy source. Conduct the following feasibility calculation: How much polysilicon would be required to supply all the United States' electrical needs from 100-mm-diameter Czochralski-grown silicon wafers, and how much land area would this require? Compare the silicon used to current consumption. Use the following data:
1. The average U.S. weekly power consumption is 42×10^9 kWh.
2. Assume that each gram of silicon in the finished cell required 5 g of polysilicon.
3. The average solar energy falling on the earth's surface is 1340 W/m^2; assume 50 h of daylight per week.
4. The cell will convert 8% of all incident energy to electrical power.

7 There are several economic motivations to scale up the melt sizes of industrial crystal growers. Larger melt sizes increase the time a machine is actually growing a crystal, thus making it more productive. Calculate the minimum crucible wall thickness needed under the following conditions: given a charge size of 100 kg, a crucible with a 25% larger volume than that of the silicon volume, a 12.5-cm-diameter crystal growing at a rate of 0.002 cm/s, and a crucible erosion rate of 2×10^{-7} g/cm^2-s. Assume a unity aspect ratio for the crucible. Also calculate the energy loss in kilowatt-hours from the surface of the melt using a temperature of 1450°C.

8 Using the gradient of Fig. 11 and Eq. 7 calculate a maximum pull speed. Assume the latent heat of fusion to be 264 cal/g and the solid thermal conductivity to be 0.05 cal/s-cm-°C. Compare the result to Fig. 12. What do you conclude?

9 At a temperature of 1000°C, calculate the boron concentration in the crystal that would lead to misfit dislocation formation.

TWO

EPITAXY

C. W. PEARCE

2.1 INTRODUCTION

Epitaxy, a transliteration of two Greek words *epi*, meaning "upon," and *taxis*, meaning "ordered," is a term applied to processes used to grow a thin crystalline layer on a crystalline substrate. In the epitaxial process the substrate wafer acts as a seed crystal. Epitaxial processes are differentiated from the Czochralski process described in Chapter 1 in that the crystal can be grown below the melting point. Most epitaxial processes use chemical-vapor deposition (CVD) techniques. A different approach is molecular beam epitaxy (MBE) which uses an evaporation method. These processes will be described in Sections 2.2 and 2.3, respectively. When a material is grown epitaxially on a substrate of the same material, such as silicon grown on silicon, the process is termed homoepitaxy. If the layer and substrate are of different materials, such as $Al_x Ga_{1-x}$ As on GaAs, the process is termed heteroepitaxy. However, in heteroepitaxy the crystal structures of the layer and the substrate should be similiar if crystalline growth is to be obtained.

Silicon epitaxy was developed to enhance the performance of discrete bipolar transistors.[1] These transistors were fabricated in bulk wafers using its resistivity to determine the breakdown voltage of the collector. However, high breakdown voltages necessarily need high-resistivity material. This requirement coupled with the thickness of the wafer results in excessive collector resistance that limits high-frequency response and increases power dissipation. Epitaxial growth of a high-resistivity layer on a low-resistivity substrate solved this problem. Bipolar integrated circuits utilize epitaxial structures in much the same way discrete transistors (Fig. 1) utilize them. The substrate and epitaxial layer have opposite doping types to provide isolation, and a heavily doped diffusion layer serves as a low-resistance collector contact. Unipolar devices such as the junction field-effect transistor (JFET) employ an epitaxial wafer as does the VMOS technology.[2]

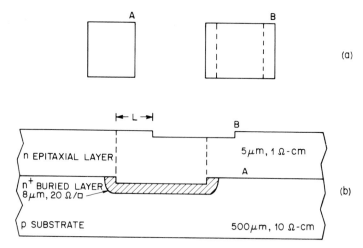

Fig. 1 Cross-sectional schematic (b) of an epitaxial wafer used for integrated circuit fabrication. Part (a) represents a rectangular pattern A present on the substrate prior to epitaxy, whose location is shifted by L and shape distorted to shape B by the epitaxial process.

Epitaxial structures have also been used to improve the performance[3] of dynamic random-access memory devices (RAMs) and CMOS ICs. In JFETs and VMOS circuits the doping profile provided by the epitaxial process is integral to the device structure. In the dynamic RAMs and CMOS circuits, devices could be fabricated in bulk wafers, but certain circuit parameters are optimized using epitaxial material.

The fundamental advantages of epitaxial wafers over bulk wafers are thus two-fold. First, epitaxial layers (one or more) on a substrate, often containing one or more buried layers, offer the device designer a means of controlling the doping profile in a device structure beyond that available with diffusion or ion implantation. Second, the physical properties of the epitaxial layer differ from bulk material. For example, epitaxial layers are generally oxygen and carbon free, a situation not obtained with the melt-grown silicon discussed in Chapter 1.

2.2 VAPOR-PHASE EPITAXY

This section is concerned with several aspects of silicon vapor-phase epitaxy such as: process chemistry, aspects of process hardware, and current capabilities. The CVD of single-crystal silicon is usually performed in a reactor consisting in elemental form of a quartz reaction chamber into which a susceptor is placed. The susceptor provides physical support for the substrate wafers. Deposition occurs at a high temperature where several chemical reactions take place when process gases flow into the chamber.

2.2.1 Basic Transport Processes and Reaction Kinetics

A thorough study of the deposition process involves examining the thermodynamics and kinetics of the chemical reactions and the fluid mechanics of the gas flows in the reactor.[4]

As a starting point for discussing the fluid mechanics of the gas flow let us consider the Reynolds number R_e, a dimensionless parameter that characterizes the type of fluid flow in the reactor.

$$R_e = \frac{D_r \, v \, \rho}{\mu} \tag{1}$$

where D_r is the hydraulic diameter of the reaction tube, v is the gas velocity, ρ is the gas density, and μ is the gas viscosity.

Values of D_r and v are generally several centimeters and tens of cm/s, respectively, for industrial processes. These parameters result in gas flow in the laminar regime,[5] since R_e is less than 2000. Accordingly, a boundary layer of reduced gas velocity will form above the susceptor and at the walls of the reaction chamber. The thickness of the boundary layer y is defined as

$$y = \left[\frac{D_r \, x}{R_e} \right]^{1/2} \tag{2}$$

where x is distance along the reaction chamber.

The carrier gas is usually hydrogen and using its typical values for ρ and μ in Eq. 1 results in values for R_e of about 100.

Figure 2 shows that the boundary layer forms at the inlet to the reaction chamber and increases until the flow is fully established. Although fully established flows are not always encountered in the short lengths of typical reactors, it is across this boundary layer that reactants are transported to the surface. The reaction by-products diffuse back across the boundary layer and are removed by the main gas stream. The fluxes of species going to and coming from the wafer surface are a complex function

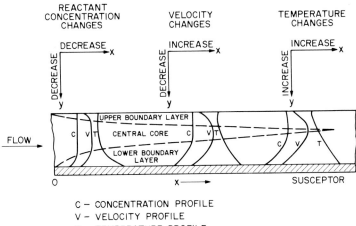

Fig. 2 Boundary layer formation in a horizontal reactor. *(After Ban, Ref. 4.)*

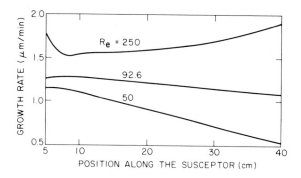

Fig. 3 The influence of R_e number on deposition uniformity. *(After Manke and Donaghey, Ref. 6.)*

of several variables, including temperature, system pressure, reactant concentration, and layer thickness. By convention, the flux is defined as

$$J = D \, dn / dy \tag{3}$$

and approximated as

$$J = \frac{D (n_g - n_s)}{y} \tag{4}$$

where n_g and n_s are the gas stream and surface reactant concentrations, respectively, D is the gas-phase diffusivity which is a function of pressure and temperature, y is the boundary layer thickness, and J is the reactant flux of molecules per unit area per unit time.

The first-order effect of y on the transport process must therefore be taken into consideration when designing the reactor and evaluating the operating conditions. The boundary layer must be adjusted relative to variation in temperature and reactant concentration within the reactor if uniformity of deposition is to be achieved. Figure 3 shows the sensitivity of layer growth rate to the Reynolds number. For a given reactor and set of process conditions R_e is varied by changing the gas flow (velocity). Therefore, y is inversely proportional to gas velocity. Thus, Fig. 3 illustrates how the boundary layer can be varied to achieve growth rate uniformity by varying the gas flow (i.e., R_e) in the reactor. This is a first-order approach to the problem. A rigorous analysis of transport phenomena in a vertical cylinder reactor has been done.[6] This analysis is a numerical solution of the defining equations subject to appropriate boundary conditions. The temperature dependance of the various parameters has been included in this analysis. For example, D is a function of temperature T with a functionality of approximately $T^{3/2}$. Figure 4 shows a substantial temperature gradient normal to the susceptor surface. The steep temperature gradient also complicates the fluid flow, because it creates some turbulence in the gas stream. The importance of this effect relative to the laminar flow is described by the ratio of the Grashof number (Gr) to the square of the Reynolds number.[4, 5] The Grashof number is a dimensionless parameter describing the effect of thermal convection in fluid flow. For Gr/R_e^2 greater than 0.5, the convection effects are found to be significant and can be seen as oscillations in the temperature above the susceptor.

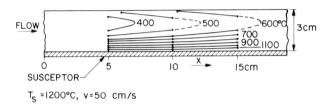

FLOW

3cm

0   5   10   15cm

X

SUSCEPTOR

$T_S = 1200°C$, $v = 50$ cm/s

Fig. 4 Isotherms in a horizontal reactor. *(After Ban, Ref. 4.)*

Reaction kinetics Four silicon sources have been used for growing epitaxial silicon. These are silicon tetrachloride ($SiCl_4$), dichlorosilane (SiH_2Cl_2), trichlorosilane ($SiHCl_3$), and silane[7] (SiH_4). Silicon tetrachloride has been the most studied and seen the widest industrial use. It will be discussed here to exemplify the reaction chemistry. The outline of the discussion is applicable to the other halide compounds.

The overall reaction can be classed as a hydrogen reduction of a gas.

$$SiCl_4(gas) + 2H_2(gas) \Rightarrow Si(solid) + 4HCl(gas) \tag{5}$$

However, a number of intermediate and competing reactions must be considered to understand the reaction fully. A starting point in the analysis is to determine for the Si—Cl—H system the equilibrium constant for each possible reaction and the partial pressure of each gaseous species at the temperature of interest. Equilibrium calculations[8] reveal fourteen species to be in equilibrium with solid silicon. In practice many of the species can be ignored because their partial pressures are less than 10^{-6} atm. Figure 5 shows the important species in the temperature range of interest. The plot is for a particular Cl/H ratio (0.01) which is representative of the ratios that occur in epitaxial deposition. Note that this ratio is constant in the reactor as neither chlorine or hydrogen is incorporated into the layer.

The epitaxial process is not necessarily an equilibrium reaction. Thus, equilibrium thermodynamic calculations may not present the total picture, but relate only to the most probable reactions. In-situ measurements of the reaction process have been made by infrared spectroscopy, mass spectroscopy, and Raman spectroscopy to determine which species are actually present in the reaction. Four species in a $SiCl_4 + H_2$ reaction at 1200°C were detected.[9] Figure 6 illustrates the concentrations of each species at different positions along a horizontal reactor. Notice that the $SiCl_4$ concentration decreases while the other three constituents increase; thus the overall reaction is postulated as

$$SiCl_4 + H_2 \quad \Longleftrightarrow \quad SiHCl_3 + HCl \tag{6}$$

$$SiHCl_3 + H_2 \quad \Longleftrightarrow \quad SiH_2Cl_2 + HCl \tag{7}$$

$$SiH_2Cl_2 \quad \Longleftrightarrow \quad SiCl_2 + H_2 \tag{8}$$

$$SiHCl_3 \quad \Longleftrightarrow \quad SiCl_2 + HCl \tag{9}$$

$$SiCl_2 + H_2 \quad \Longleftrightarrow \quad Si + 2HCl \tag{10}$$

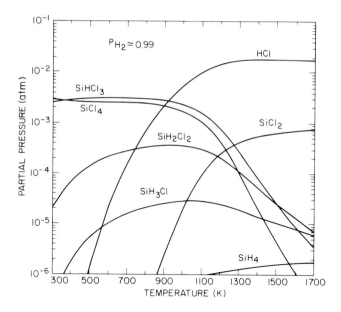

Fig. 5 Temperature variation of the equilibrium gas phase composition at 1 atm and Cl/H = 0.01. *(After Sirtl, Hunt, and Sawyer, Ref. 8.)*

This sequence of reactions is interesting from several viewpoints. The species $SiHCl_3$ and SiH_2Cl_2 are seen as intermediates to the overall reaction. Thus, growth with these halides would start at Eq. 7, 8, or 9. Accordingly, growth with $SiCl_4$ has the highest reported activation energies (1.6 to 1.7 eV) decreasing in turn for $SiHCl_3$ (0.8 to 1.0 eV) and SiH_2Cl_2 (0.3 to 0.6 eV). The reactions are also reversible, and, under the appropriate conditions, the deposition rate can become negative causing the etching process to begin. This observation leads to a more general question about how growth rate varies with temperature. Figure 7 depicts the growth rate variation versus temperature; note the negative deposition rate at low and high temperatures.

Figure 8 shows an Arrhenius plot of growth rate, illustrating the overall reaction process.[10] In region A the process can be characterized as reaction rate or kinetic limited, that is, one of the chemical reactions is the rate-limiting step and is even reversible. Region B represents the situation in which the transport processes are rate limiting, that is, where the growth rate is limited either by the amount of reactant reaching the wafer surface or by the reaction products diffusing away. This regime is termed mass transport or diffusion limited, and the growth rate is linearly related to the partial pressure of the silicon reactant in the carrier gas. The slight increase of the growth rate with temperature in region B is due to the increased diffusivity of the species with temperature in the gas phase. Industrial processes at atmospheric pressure are usually operated in region B to minimize the influence of temperature variations.

2.2.2 Doping and Autodoping

Incorporating dopant atoms into the epitaxial layer involves the same considerations as the growth process requires, for example, mass transport and chemical reactions.[11]

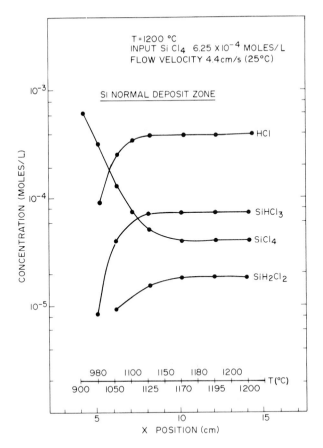

Fig. 6 Species detected by IR spectroscopy in a horizontal reactor using $SiCl_4 + H_2$. *(After Nishizawa and Saito, Ref. 9.)*

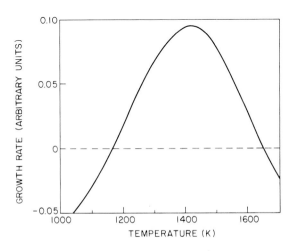

Fig. 7 Growth rate of CVD silicon versus temperature. *(After Sirtl, Hunt, and Sawyer, Ref. 8.)*

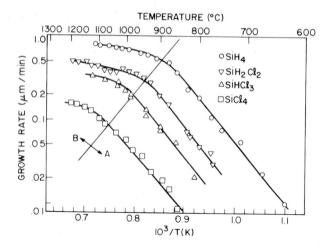

Fig. 8 Temperature dependence of growth rate for assorted silicon sources. *(After Eversteyn, Ref. 10.)*

Typically, hydrides of the impurity atoms are used as the source of dopant. We might expect that these compounds would decompose spontaneously, but they do not. Thermodynamic calculations indicate that the hydrides are relatively stable because of the large volume of hydrogen present in the reaction. Typical of the dopant chemistry is the reaction for arsine, which is depicted with the deposition process in Fig. 9, which shows arsine being absorbed on the surface, decomposing, and being incorporated into the growing layer.

$$2AsH_3 \text{ (solid)} => 2As(gas) + H_2 \text{ (gas)}$$
$$=> 2As(solid) => 2As^+ \text{ (solid)} + 2e \qquad (11)$$

Interactions also take place between the doping process and the growth process. First, in the case of boron and arsenic the formation of chlorides of these species is a competing reaction.[11, 12] Second, the growth rate of the film influences the amount of

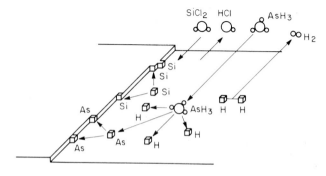

Fig. 9 Schematic representation of arsine doping and growth processes. *(After Reif, Kamins, and Saraswat, Ref. 13.)*

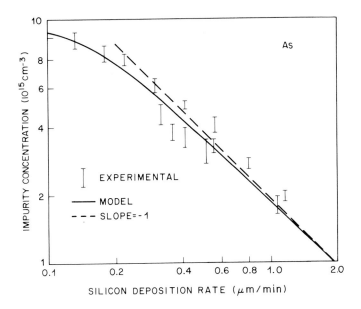

Fig. 10 The influence of growth rate on layer concentration for arsenic doping. *(After Reif, Kamins, and Saraswat, Ref. 13).*

dopant incorporated in the silicon as shown for arsenic in Fig. 10. At low growth rates an equilibrium is established between the solid and the gas phase, which is not achieved at higher growth rates.[13]

In addition to intentional dopants incorporated into the layer, unintentional dopants are introduced from the substrate. The effect, shown in Fig. 11, is termed autodoping.[14] Dopant is released from the substrate through solid-state diffusion and evaporation. This dopant is reincorporated into the growing layer either by diffusion through the interface or through the gas phase. Autodoping is manifested as an enhanced transition region between the layer and the substrate (see Fig. 12). The

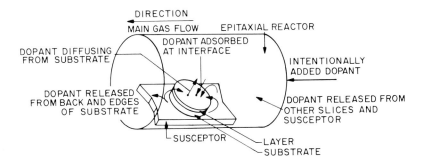

Fig. 11 Sources of dopant for the epitaxial layer, schematically shown in a horizontal reactor. *(After Langer and Goldstein, Ref. 17.)*

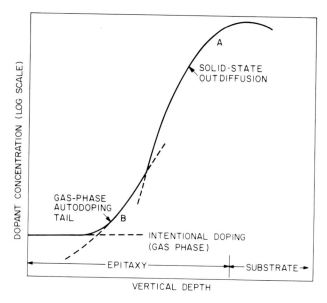

Fig. 12 Generalized doping profile of an epitaxial layer detailing various regions of autodoping. *(After Srinivasan, Ref. 19.)*

shape of the doping profile, close to the substrate, is dominated by solid-state diffusion from the substrate and is a complementary error function[15] if

$$v > 2(D/t)^{1/2} \qquad (12)$$

where v is the growth velocity, D is the substrate dopant diffusion constant, and t is the deposition time. The solid-state outdiffusion aspect of autodoping is easy to visualize; it determines the shape of region A in Fig. 12.

Since the growth velocity easily outpaces the diffusion of dopant, the doping profile in region B is dominated by dopant introduced from the gas phase. If the dopant evaporated from the substrate exceeds the intentional dopant, an autodoping tail develops. Autodoping is a time-dependent phenomena. The dopant evaporating from the wafer surface is supplied from the wafer interior by solid-state diffusion. Thus, the vaporization rate of dopant from an exposed surface is not constant, but decreases with time.

Once the autodoping is diminished the intentional doping predominates, and the profile becomes flat. The extent of the autodoping tail is a function of the substrate dopant species and reaction parameters such as temperature and growth rate. Autodoping limits the minimum layer thickness that can be grown with controlled doping as well as the minimum doping level. Because of the technological importance of autodoping, it has been the subject of many studies.[16-21]

The discussion thus far has centered on equilibrium or at least steady-state reactions. If the dopant flow in the reactor is abruptly altered, it does not result in a rapid change in the doping profile.[22] Molecular beam epitaxy (discussed in Section 2.3) does not have this constraint.

In addition to the chemical cleaning of the substrate, an in-situ vapor-phase etching of the substrate with anhydrous HCl at a nominal temperature of 1200°C usually precedes deposition. The reactions involved are

$$2HCl + Si => SiCl_2 + H_2 \tag{13}$$

$$4HCl + Si => SiCl_4 + 2H_2 \tag{14}$$

Other gases such as HBr and SF_6 have also been proposed for substrate etching.[23] The HCl is supplied as a compressed gas and introduced into the hydrogen mainstream to achieve a concentration of 2 to 3%. Etch rates are on the order of several tenths of a micrometer/min, and total etch depths of up to 5 μm are used on substrates without buried layers. When the sheet resistance of buried layers must be maintained, etch depths are usually kept in the 0.1- to 0.3-μm range. Etching results in a perfectly clean substrate surface, free of native oxides. However, it is not a substitute for poor pre-deposition chemical cleaning. An alternative to in-situ substrate cleaning is to bake the wafers in hydrogen at a high temperature (10 min at 1200°C).

2.2.3 Equipment, Installation, and Safety Considerations

The earliest industrial epitaxial equipment was built by the user. Such equipment generally could handle only a small wafer load and was usually operated manually. In the early 1970s commercial equipment which could handle larger wafer loads and offered process automation became available.[24] Now typical reactors (Fig. 13) weigh 2000 kg and occupy 2 m^2 or more of floor space.

Fig. 13 A radiant-heated barrel reactor.

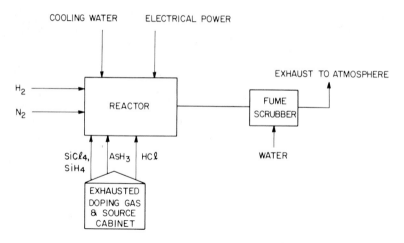

Fig. 14 Schematic of an epitaxial reactor installation.

Several safety considerations must also be addressed in the operation of the reactor. The reactor itself is usually designed with sufficient interlocks to prevent accidents. However the user must safely remove and treat reaction by-products, and arrange for proper delivery of process gases to the reactor. In fact several distinct hazards require consideration: the explosion and fire potential of hydrogen, the corrosive nature of HCl, and the highly toxic nature of the doping gases. The last are particularly dangerous. Arsine, for example, is instantly lethal if a concentration of 250 ppm is inhaled, and exposure at lower levels (35 ppm) poses a health hazard depending on the length of exposure. A complete installation is depicted in Fig. 14. Environmental considerations usually require a water-mist fume scrubber to remove most of the unreacted and reaction products from the carrier-gas stream.

Susceptors in epitaxial reactors are the analogs of crucibles in the crystal growing process. They provide mechanical support for the wafers and are the source of thermal energy for the reaction in induction-heated reactors. The geometric shape or configuration of the susceptor usually provides the name for the reactor. Figure 15 shows three common susceptor shapes—horizontal, pancake, and barrel—which will be discussed in more detail later. Like crucibles the susceptor must be mechanically strong and noncontaminating to the process. Additionally, the susceptor must not react with the process reactants and by-products. Induction-heated reactors require a material that will couple to the rf field. The preferred material has been graphite, although in radiant-heated reactors polysilicon or quartz susceptors are alternatives. Polysilicon susceptors react with HCl, leading to a gradual erosion of the susceptor. This erosion can be prevented with a coating of CVD silicon nitride. Graphite susceptors also requiring a coating because they are relatively impure and soft. A carbon blank is shaped to the required dimensions before the coating is applied. A coating of longstanding use is 50 to 500 μm of silicon carbide applied by a CVD process similar to the silicon CVD process. Other possible coatings include glassy carbon and pyrolytic graphite. The latter forms a dense carbon layer in and on the carbon blank by the

(a)

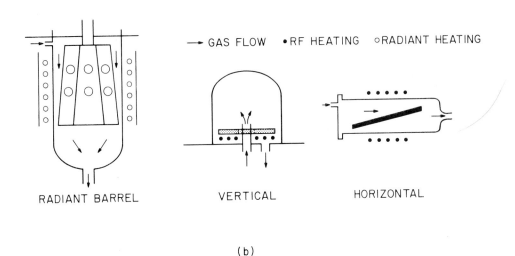

(b)

Fig. 15 (a) Three common susceptor shapes: horizontal, pancake (vertical), and barrel. (b) Schematics of three common reactors.

cracking of methane at elevated temperature. Pinholes and cracks in the coating are persistent susceptor problems caused by the stresses encountered in repetitive thermal cycling and by reactions with metal from tweezers used in wafer loading. These flaws allow impurities in the carbon to escape,[25] which contaminate the epitaxial film and cause defects. Another problem is variation in growth rate and doping caused by temperature nonuniformity due to variable properties of the graphite and coating. The reaction tubes or bell jars are made of high-purity quartz, either clear or opaque, depending on the reactor.

In most reactors, the reaction tube is relatively cool during operation, that is, they are operated "cold wall." Forced-air cooling carries away waste heat. Induction coils and other metal parts are water cooled. Some cold-wall reactors have an outer tube, allowing the reaction tube itself to be water cooled. By way of comparison the usual process for the CVD of polysilicon is a hot-wall operation (Chapter 3) resulting in a coating of silicon on the reactor tube itself.

Historically, energy for the reaction has been supplied by heating the susceptor inductively. The energy is then transported to the wafer by conduction and radiation. Because silicon at room temperature does not heat inductively unless the frequency is above 50 MHz, motor generator sets at 10 kHz or self-excited rf oscillators at 500 kHz are used for heating. In the latter case, plate-input powers up to 100 kW are used by large reactors. A water-cooled coil is placed close to the susceptor so coupling can occur. The coil can be inside or outside the reaction chamber depending on the design of the reactor. Radiant heating, a newer way of supplying energy to the reaction chamber, provides more uniform heating than inductive heating provides[24]. The energy is supplied by banks of quartz halogen lamps.

In most cases, process control involves maintaining gas flows and temperatures at the desired values. In modern equipment the process cycle is generally microprocessor controlled, and the operator only has to load and unload wafers. Sensors monitor the temperature and the microprocessor makes adjustments when they are needed. An optical pyrometer (focused on a wafer inside the reactor) has been used as the temperature-sensing device in rf-heated reactors. Since the temperature is sensed through the quartz reaction tube, the pyrometer actually senses an optically equivalent temperature that is usually 50 to 100°C below the actual temperature due to the emissivity of silicon. This temperature difference should be considered when temperature-dependent curves are studied and compared. Radiant-heated reactors employ sensing elements inside the reaction chamber. Gas flows can be metered using rotometers or mass-flow controllers. The former determine the gas flow by calibrating the position of a stainless-steel or sapphire ball in a glass tube. The calibration is a function of gas viscosity, pressure, temperature, and molecular weight. A mass-flow controller provides a better approach to metering flows. It measures the heat capacity of the material flowing and compares that value to a setpoint. Control is by a solenoid or thermal expansion valve. On-off control is provided by air-operated valves. These valves eliminate any explosion hazard due to sparks, if a hydrogen leak occurs.

Three basic reactor configurations—horizontal, pancake or vertical, and barrel—(Fig. 15) have found widespread use. Each design has its relative merits and disadvantages.[24] Horizontal reactors offer high capacity and throughput; however, controlling the deposition process over the entire susceptor length is a problem. Pancake reactors are capable of very uniform deposition, but suffer from mechanical compexity. Radiant-heated barrel reactors are also capable of uniform deposition, but are not suited for extended operation at temperatures above 1200°C.

A typical process for any configuration includes several steps. First, a hydrogen carrier gas purges the reactor of air. The reactor is then heated to temperature. After thermal equilibrium is established in the chamber, an HCl etch takes place at a temperature between 1150 and 1200°C for 5 min nominally. The temperature is then reduced to the growth temperature with time allowed for stabilization and flushing of HCl as needed. Next, the silicon source and dopant flows are turned on and growth proceeds at a rate of 0.2 to 3.0 μm/min. Once growth is complete, the dopant and silicon flows are eliminated and the temperature reduced, usually by shutting off power. As the reactor cools toward ambient temperature the hydrogen flow is replaced by nitrogen so that the reactor may be opened safely. Depending on wafer diameter and reactor type, capacities range from 10 to 30 wafers per batch. Process cycle times are about 1 h, giving throughputs of nominally 20 wafers per hour.

2.2.4 Process Selection and Capabilities

Epitaxial layers are rarely doped in excess of 10^{17} atoms/cm^3. This concentration is used in a bipolar technology[26] where the epitaxial layer forms the transistor base. The technical feasibility of doping to higher levels, approaching solid solubility, was demonstrated for phosphorus.[27, 28]

The majority of applications require dopings of 10^{14} to 10^{17} atoms/cm^3. Lower doping levels, in the 10^{12} to 10^{14} atoms/cm^3 region, are used for certain types of high voltage and detector devices. These lower values are obtainable[18, 20] if the reactor is clean and the source is pure. Silicon sources with an equivalent purity of less than 10^{13} atoms/cm^3 are commercially available. Rear surface autodoping is often controlled by sealing the rear surface with an oxide or nitride layer. An in-situ sealing can be made in rf-heated reactors by first coating the susceptor with silicon. This layer will be transferred to the wafer during the process. A theoretical lower doping limit of 1.45×10^{10} atoms/cm^3 is the intrinsic doping of silicon at 23°C. Radial uniformities of ±10% are routinely obtained and ±5% are possible in some cases. Variations within a run (batch) and from run to run are on the order of 20% or less, depending on the process and reactor.

The practical upper limit of epitaxial thickness is reached just before the layer overgrows the substrate and the film becomes contiguous with the silicon deposited on the susceptor. If the layer overgrows the substrate, the wafers become hard to separate from the susceptor, and cracking usually occurs. However, film thicknesses of several hundred micrometers, close to the upper limit, are routinely grown for some power device applications. As mentioned previously thin layers are constrained by autodoping considerations, but layers as small as 0.5 μm thickness have been produced.[29] Layers with uniformities of ±5% are routinely produced with variations between runs of ±5% and better.

As in the case of crystal growing, the choice of dopants for epitaxial processes is limited. Boron is used for p-type doping, and arsenic or phosphrous are used for n-type doping. The original method[30] of introducing dopant to the reaction was to mix halides (BCl_3 or PCl_3) of the dopants, which are liquids at room temperature, with the silicon source ($SiCl_4$ or $SiHCl_3$) which is also liquid. Each species was vaporized in a bubbler tank. Such a coupling of the silicon and dopant proved to be inconvenient. For example, a change in doping level required the bubbler tank to be emptied and a new mixture added.

A better approach uses hydrides of the dopants (PH_3, B_2H_6, or AsH_3). These compounds are gases at room temperature and are supplied in compressed-gas cylinders. Since the concentration of dopant in the reactor is in the ppb range, the hydrides are not supplied in pure form but are diluted to between 20 and 200 ppm in hydrogen. Industry practice is to use a system of three flow meters to control the dopant flow.[10] This procedure allows for a three order-of-magnitude range of doping from one cylinder as each flow meter can control over a tenfold range of flows. There is little difference in performance between arsine and phosphine, but most users prefer arsine.

The choice of a silicon source is based on several considerations. Table 1 lists the sources of each presently in use along with characteristic growth rates and tempera-ture ranges.[24] Silane (SiH_4) is usually chosen when a low deposition temperature is needed to minimize boron autodoping and outdiffusion. (Arsenic autodoping increases with lower temperatures.) Silane processes are prone to gas-phase nuclea-tion (the formation of silicon particles in the gas stream above the wafer) which leads to poor film quality. Gas-phase nucleation can be suppressed[7] by adding HCl. Another disadvantage is that silane tends to coat the reactor chamber rapidly, requir-ing frequent cleaning. It also presents a production hazard as it is pyrophoric in con-centrations above about 2%.

Dichlorosilane is a popular choice in many applications.[31] It offers high growth rates at relatively low temperatures. Although a liquid at room temperature, dichloro-silane has a high vapor pressure (>1 atm), so it can be metered directly from a cylinder. No bubbler tank is needed. Trichlorosilane is used for the production of electronic polysilicon as mentioned in Chapter 1. It offers no physical or operational advantage over silicon tetrachloride and is seldom used in epitaxial CVD processes. Silicon tetrachloride is the least expensive and most used of all the silicon sources. It is also a liquid at room temperature, but its low vapor pressure requires a bubbler tank

Table 1 Epitaxial growth of silicon in hydrogen atmosphere

Chemical deposition	Nominal growth rate (μ/min)	Temperature range (°C)	Allowed oxidizer level (ppm)
$SiCl_4$	0.4−1.5	1150−1250	5−10
$SiHCl_3$	0.4−2.0	1100−1200	5−10
SiH_2Cl_2	0.4−3.0	1050−1150	< 5
SiH_4	0.2−0.3	950−1050	< 2

to help vaporization. The high deposition temperatures required of silicon tetra-chloride make it less sensitive to oxidizers in the carrier gas and to the defects they cause.

Epitaxial reactors can generally operate at temperatures between 900 and 1250°C. Selecting the processing temperatures as well as the flow and growth rates is a complex decision based on the film thickness uniformity, doping uniformity, level required, and on the defect levels, pattern shift, and distortion allowed. This chapter explains the process piecemeal, but does not tell how to design a process to meet all the objectives, even though many are contradictory. For example, higher temperatures to reduce pattern shift (see Section 2.2.5) increase autodoping. A systematic approach to process design uses a factorial-design experimental approach to determine the optimum process condition for up to six variables including process temperature.[29, 32] After the factorial design has been used to determine the best operating conditions, a silicon source can be chosen intelligently. Computer programs to simulate the epitaxial process are available[22] and are constantly being refined. Computer simulations are considered in detail in Chapter 10. Such programs are a useful adjunct to the factorial design experimental method in setting up a CVD process.

Historically, the silicon CVD process has been performed at atmospheric pressure (760 Torr). However, operation in the range from 50 to 100 Torr has several advantages.[33, 34] First, vertical and lateral autodoping effects (Fig. 16) are significantly reduced. Second, pattern shift is also substantially reduced.[34]

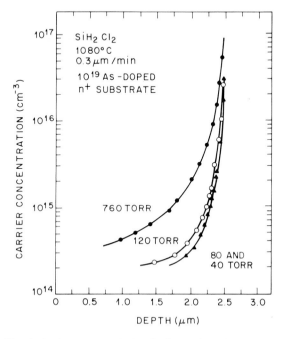

Fig. 16 Doping profiles obtained over an arsenic-doped substrate for various reactor pressures. *(After Herring, Ref. 34.)*

As in oxidation, diffusion, and LPCVD processes, wafers are cleaned before the expitaxy process begins. All organic and metallic residues on the wafers must be removed.[35] Particles are removed by using ultrasonic agitation in the cleaning baths, by brush scrubbing with water, or by high-pressure water jets. Clean wafers must be handled carefully to prevent recontamination, especially by particles. To prevent particle contamination, the entire reactor or load station is usually installed in a clean room. A second method is to use a clean-air hood at the loading station.

2.2.5 Buried Layers

To fabricate bipolar ICs, usually one or more diffusions are applied to the substrate to create the necessary isolation, collector, emitter, or base functions (Fig. 1). These diffusions are applied to the substrate prior to epitaxy using the lithographic, oxidation, diffusion, or ion-implantation processes discussed in other chapters. The diffusions are called buried layers or diffusions under film. The presence of a buried layer complicates the epitaxial process because of its effect on autodoping (vertical and lateral), defects, pattern shifting, and pattern distortion.

The pre-epitaxial process leaves a step of 500 to 1000 Å around the perimeter of the buried layer that marks its location (A of Fig. 1). Subsequent masking levels must be properly aligned with the buried layer pattern. Unfortunately, the deposition process shifts the pattern (B of Fig. 1b). Lithographic masks must compensate for the amount of the shift (L of Fig. 1b). A separate but related effect is pattern distortion or washout, which alters the shape of the feature in the layer. Figure 1a also illustrates the nature of pattern distortion. The pattern in the epitaxial layer is thus misplaced and

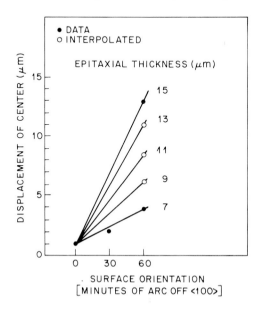

Fig. 17 Pattern shift for a $\langle 100 \rangle$ orientation with various amounts of misorientation. *(After Drum and Clark, Ref. 37.)*

misshaped relative to its original configuration in the substrate. These effects place limitations on the design of high-density circuits, and are a complicated function of substrate orientation, growth rate and temperature, and silicon source.[36]

The crystal orientation has a profound effect on pattern shift.[37] Since the layer does not grow normal to the substrate but rather by additions to microsteps (Fig. 9), the macrostep marking the diffusion is shifted. As a result the microscopic growth processes are altered by the orientation of the wafer. Current practice is to misorient $\langle 111 \rangle$ wafers by 2 to 5° towards the nearest $\langle 110 \rangle$ direction and to orient $\langle 100 \rangle$ wafers exactly on the orientation. Figure 17 illustrates the $\langle 100 \rangle$ case; note that the pattern shift changes with epitaxial thickness. As shown in Fig. 18, pattern shift is independent of reactor design,[38] but does show a pronounced dependence on growth rate and

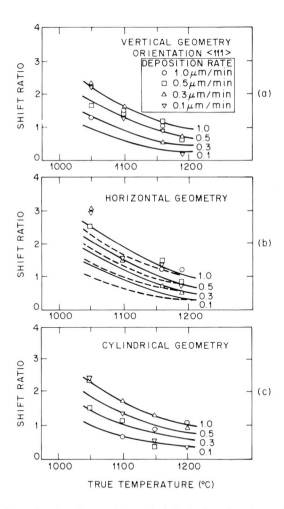

Fig. 18 Pattern shift as a function of reactor [(a) vertical, (b) horizontal, and (c) cylindrical (or barrel)], temperature, and growth rate. *(After Lee et al., Ref. 38.)*

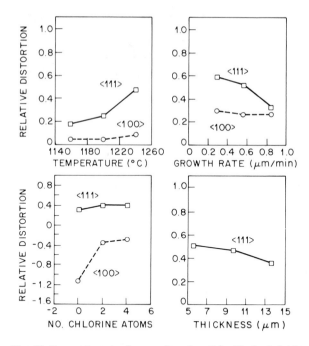

Fig. 19 Parametric study of pattern distortion. *(After Weeks, Ref. 36.)*

temperature. Pattern shift increases with growth rate and reduced deposition temperature. The magnitude of the shift is largely equal for both $\langle 111 \rangle$ and $\langle 100 \rangle$ orientations. These results are for an epitaxial process under atmospheric pressure. The pattern shift is substantially reduced as the reactor pressure is lowered.[34] Pattern distortion[36] exhibits an opposite relationship to the parameters previously mentioned, as shown in Fig. 19. For example, pattern shift is reduced at higher growth temperatures, but distortion increases and is more dependant on orientation.

A complete explanation of how all the variables affect pattern shift is not available, but includes the following elements. The step face (Fig. 9) exposes a number of crystal planes, which exhibit different growth rates. The anisotropy of growth increases at lower temperatures as does the pattern shift. The growth rate dependence of pattern shift is similar. The anisotropic nature of the layer growth rate increases with growth velocity. The effects of pressure and the silicon source are less clear, but apparently interrelated. Less chlorine as a by-product in the form of HCl correlates with less distortion and shift. Reduced pressure operation would aid the escape of HCl across the boundary layer.

The discussions of Section 2.2.2 relating to autodoping also apply to the vertical autodoping profile above a buried layer. However, an effect termed lateral autodoping can be observed in such structures. Figure 20a shows that lateral autodoping is a front-surface autodoping phenomenon involving the transport of dopant to regions adjacent to the diffusion. Figure 20b details the doping profiles on and off the buried layer. The off-profile is totally attributable to gas-phase autodoping. Dopant in these

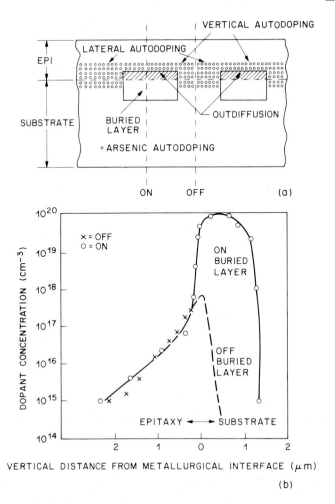

Fig. 20 Lateral autodoping effect. (a) Cross section of epitaxial wafer showing location of lateral autodoping as adjacent to the buried layer. (b) Doping profiles above and adjacent to the buried layer. *(After Srinivasan, Ref. 16.)*

regions is detrimental, because it produces an electrical short circuit between the adjacent devices,[16, 19] if it is not eliminated by a boron isolation diffusion. The peak concentration in the lateral autodoping profile is a function of the surface concentration in the buried layer and processing conditions such as HCl etch time, temperature, growth rate, and silicon source.

2.2.6 Epitaxial Defects

The crystal perfection of the layer never exceeds that of the substrate and is often inferior.[39, 40] The crystal perfection is a function of the properties of the substrate wafer and the epitaxial process itself. Defects arising from the substrate wafer can be

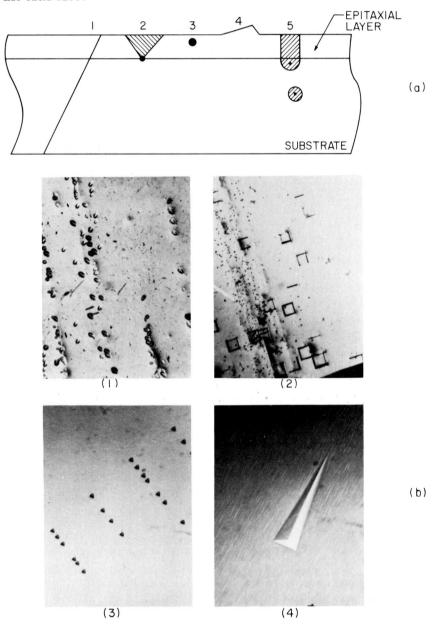

Fig. 21 Common defects occurring in epitaxial layers. (a) Schematic representation of line (or edge) dislocation initially present in the substrate and extending into the epitaxial layer (item 1), an epitaxial stacking fault nucleated by an impurity precipitate on the substrate surface (item 2), an impurity precipitate caused by epitaxial process continuation (item 3), growth hillock (item 4), and bulk stacking faults one of which intersects the substrate surface thereby being extended into the layer (item 5). (b) Photographs of defects in actual wafers. Dislocations revealed as circular etch pits by Secco etching in a region of slip on a $\langle 100 \rangle$ wafer (item 1), epitaxial stacking faults on a $\langle 100 \rangle$ wafer (item 2), dislocations revealed by Sirtl etching in a $\langle 111 \rangle$ wafer (item 3), and a growth hillock on a $\langle 111 \rangle$ wafer, visible without etching (item 4).

related to the bulk properties of the wafer or its surface finish. Item 1 in Fig. 21a is an example of an existing line dislocation continuing into the epitaxial layer. Impurity precipitates[41, 42] are one kind of surface defect that nucleate on an epitaxial stacking fault (item 2). Process-related defects include slip and impurity precipitates from contamination (item 3). Slip is a displacement of crystal planes past each other as the result of stress. Dislocations accompany the formation of slip. Contamination from the susceptor and the tweezers used in handling also contaminate the layer and substrate and form precipitates that act as defect nuclei in subsequent processing.[43] Tripyramids, hillocks, and other growth features (item 4) can be related to the process[44] or the surface finish of the wafer. Item 5 is an example of defect (bulk stacking fault) created in a pre-epitaxial process, such as buried layer fabrication. These defects in turn nucleate defects in the epitaxial layer. Figure 21b is a series of photographs of defects in actual wafers. In general, the quality of the deposit is strongly related to the quality of the substrate wafer, its cleaning, layer growth rate, and temperature.[45] For example, as the deposition temperature is lowered, minor flaws in the substrate surface act as points of preferential nucleation giving rise to stacking faults and pyramids. Higher growth rates aggravate the problem as discussed in the next section.

A temperature gradient exists normal to the substrate in an rf-heated reactor.[46] Slip due to this gradient (during epitaxy) is produced in the following manner. Heat flow from the susceptor through the wafer equals subsequent radiation from the front surface

$$EkT^4 = \frac{KdT}{dx} \tag{15}$$

where K is the thermal conductivity of silicon, dT/dx the normal temperature gradient, E the emissivity of silicon, k the radiation constant, and T the nominal wafer temperature. A front-to-rear temperature difference of only a few degrees causes a differential expansion of the wafer. In effect, the wafer curls up on the susceptor. When the wafer edge loses contact with the susceptor, the edge temperature drops, causing still further bowing. This radial temperature gradient results in sufficient stress to create dislocations the wafer (see the section on thermal stress in Chapter 1). The inverted heat flow of the radiant-heated reactor minimizes this problem.[25]

Another class of defects are misfit dislocations caused by lattice mismatch when the substrate is highly doped.[47] The resultant strain between the layer and substrate is relieved by the formation of dislocations.

2.2.7 Microscopic Growth Processes

A final point to consider in the CVD process is the conditions under which single-crystal films are obtained and the mechanism of their growth.[48] Figure 22 illustrates the maximum attainable growth rates for atmospheric pressure epitaxy. The activation energy obtained from that Arrhenius plot is 5 eV, which is equal to that of silicon self-diffusion. The physical explanation is that silicon atoms are absorbed on the surface of the substrate after a chemical reaction takes place. These atoms must migrate across the surface to find a crystallographically favorable site where they can be

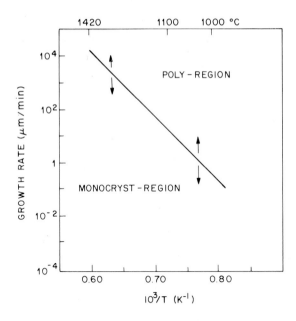

Fig. 22 Maximum growth rate for which monocrystalline silicon can be obtained as a function of temperature. *(After Bloem, Ref. 48.)*

incorporated into the lattice (Fig. 9). At high growth rates, insufficient time is allowed for surface migration, resulting in polycrystalline growth. The favorable sites are positioned at the leading edge of atomic height steps. Thus, the growth is not vertical, but lateral. This effect accounts for the variation in growth rate with surface orientation, availability and movement of steps being orientation dependent.[44] The adsorbed silicon atoms compete with dopant atoms, hydrogen, chlorine, and foreign atoms for these sites. Dopant atom concentration is usually low enough to be ignored, but impurities such as carbon (initially present on the surface) affect how silicon is incorporated and nucleate a stacking fault or tripyramid defect. This growth mechanism accounts for the effects that were discussed under pattern shift and distortion, and is an additional reason to misorient $\langle 111 \rangle$ wafers. Growth of epitaxial layers on $\langle 111 \rangle$ results in mounds being formed.[44]

2.3 MOLECULAR BEAM EPITAXY

Molecular beam expitaxy (MBE) is a non-CVD epitaxial process that uses an evaporation method. Although the method has been known since the early 1960s, it has only recently been seriously considered a suitable technology for silicon device fabrication.[49] The two major reasons why MBE was not used are that, historically, the quality was not commensurate with device needs, and no industrial equipment existed. MBE has a number of inherent advantages compared to CVD techniques.[49] Its main advantage for VLSI use is low-temperature processing. Low-temperature processing minimizes outdiffusion and autodoping, a limitation in thin layers prepared by CVD. Another advantage is the precise control of doping that MBE allows. Because doping in MBE is not affected by time-constant considerations unlike CVD epitaxy, complicated doping profiles can be generated. Presently, these advantages are not being exploited for IC fabrication, but they have found application in discrete microwave and photonic devices. For example, the C-V characteristic of a diode with homogeneous doping is nonlinear with aspect to reverse bias. Varactor diodes used as FM modulators could advantageously employ a linear dependence of capacitance on voltage. This linear voltage-capacitance relationship can be achieved with a linear doping profile, which is easily obtained with MBE.

2.3.1 Process Description

In contrast to CVD processes, MBE is not complicated by boundary-layer transport effects, nor are there chemical reactions to consider. The essence of the process is an evaporation of silicon and one or more dopants as depicted in Fig. 23. The evaporated species are transported at a relatively high velocity in a vacuum to the substrate. The relatively low vapor pressure of silicon and the dopants ensures condensation on a low-temperature substrate. Usually, silicon MBE is performed under ultra-high vacuum (UHV) conditions of 10^{-8} to 10^{-10} Torr, where the mean free path of the atoms[50, 51] is given by

$$L = 5 \times 10^{-3}/p \qquad (16)$$

where L is the mean free path in cm, and p is the system pressure in Torr. At a system pressure of 10^{-9} Torr L would be 5×10^6 cm.

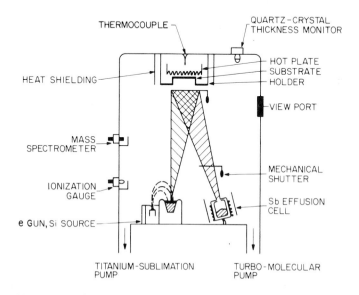

Fig. 23 Schematic of MBE growth system. *(After Konig, Kibbel, and Kasper, Ref. 54.)*

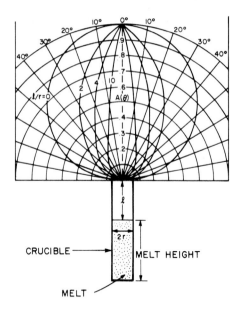

Fig. 24 Angular distribution of flux from a crucible of radius r and melt height l referenced from the top of the crucible. *(After Luscher and Collins, Ref. 52.)*

Because collisions between atoms are unimportant in a high vacuum, transport velocity is controlled more by thermal energy effects than by diffusion effects, and deposition and its uniformity can be controlled by the source characteristics.[52] Accordingly, evaporation from a crucible produces a flux of material varying with time and angle, as shown in Fig. 24. The lack of intermediate reactions and diffusion effects, along with relatively high thermal velocities, results in film properties changing rapidly with any change at the source.

A conventional temperature range for MBE is from 400 to 800°C. Higher-temperature processes are technically feasible, but the advantages of reduced outdiffusion and autodoping are lost. Growth rates in the range 0.01 to 0.3 μm/min have been reported.[49, 50] The higher value is comparable to those obtained in CVD epitaxy.

In-situ cleaning for MBE is done in two ways. High-temperature baking between 1000 and 1250°C for up to tens of minutes[53] decomposes the native oxide and removes other adsorbed species (notably carbon) by evaporation or diffusion into the wafer. A better approach is to use a low-energy beam of an inert gas to sputter clean the surface. A short anneal at 800 to 900°C is sufficient to reorder the surface.

MBE doping has several distinguishing features. A wider choice of dopants can be used, compared to CVD epitaxy, and more control of the doping profile is possible. Two doping processes are available. Under ideal conditions the doping process is like the growth process. In practice an absorbed layer of dopant will complicate the doping process.[85] A flux of evaporated dopant atoms arrives at the growing interface, finds a favorable lattice site, and is incorporated. The doping level is controlled by adjusting the dopant flux to the flux of silicon atoms. In practice a Knudsen effusion cell[52] is used to evaporate dopants. But desirable dopants such

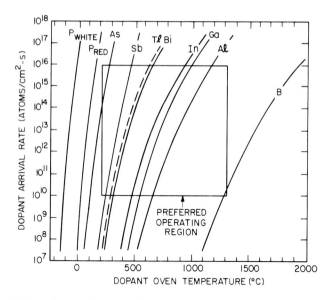

Fig. 25 Flux of various dopant species versus oven temperature. *(After Bean, Ref. 50.)*

as As, P, and B evaporate too rapidly or too slowly for controlled use. As a result most workers use Sb, Ga, or Al for dopants which compare favorably to other dopants as shown in Fig. 25. Another complication is the temperature-dependent sticking coefficient shown in Fig. 26. A low value means re-evaporation occurs readily and incorporation of the dopant is more difficult. This temperature dependence requires precise control of substrate heating. A wide latitude in doping by evaporation has been demonstrated.[54, 55] Values in the range 10^{13} to 10^{19} atoms/cm^3 have been reported with 1% radial uniformities.

Another doping technique uses ion implantation[56] (see Chapter 6). This technique uses a low-current (1 μA), low energy (0.1- to 3-keV) ion beam to implant dopant as the layer is growing. The low energy beam places the dopant just below the growing interface, ensuring incorporation. Doping profiles not obtainable with CVD processes can be produced with ion implantation as depicted in Fig. 27. This technique also allows the use of dopants such as B, P, and As. Since MBE is a vacuum process, it is very adaptable to ion-implant doping, and in-situ monitoring of the beam is feasible.[52]

2.3.2 Equipment

An elementary MBE system is depicted in Fig. 23. It is, in essence, a UHV chamber where furnaces holding electronic-grade silicon and dopant direct a flux of material to a heated substrate. In fact an early MBE system was made by modifying a bell-jar apparatus. Now, commercially designed and built equipment, although expensive and complicated, has become available. Figure 28 illustrates the many components of a comprehensive system. A distinguishing feature of MBE is the ability to use sophisticated analytical techniques in-situ to monitor the process.

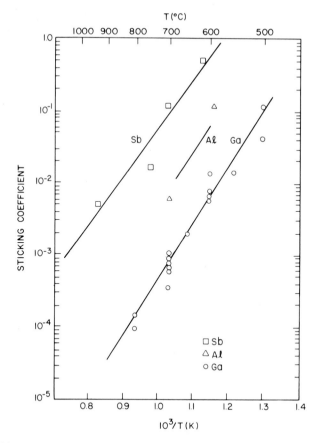

Fig. 26 Sticking coefficient for Sb, Al, and Ga versus temperature. *(After Bean, Ref. 50.)*

In contrast to the CVD process, MBE does not require the extensive safety precautions, although solid arsenic dopant must be handled carefully.

The vacuum system is the heart of the apparatus. To consistently attain a vacuum level in the 10^{-10}-Torr range, materials and construction must be carefully considered. Materials should have low vapor pressure and low sticking coefficients. Repeated exposure to air is detrimental to a UHV system because of the long bakes needed to desorb atmospheric species from the system walls. A load lock system minimizes this problem. Consistent low base pressure is needed to ensure overall film perfection and purity. These needs are best met with an oil-free pump design, such as a cryogenic pump.

Because of its high melting point, silicon is not volatilized by heating in the furnace, but rather by electron-beam heating. Dopants are heated in a furnace. A constant flux is assured by the use of closed-loop temperature control. Baffles and shutters shape and control the flux, so uniformity of doping and depostion can be attained without boundary layer effects being considered.

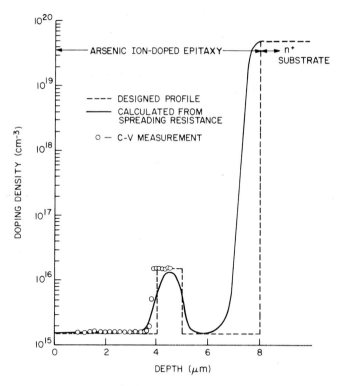

Fig. 27 Doping profile obtained by ion implantation during MBE growth. *(After Ota, Ref. 56.)*

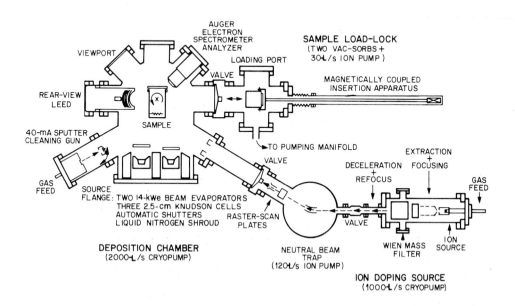

Fig. 28 Schematic of practical MBE system. *(After Bean, Ref. 49.)*

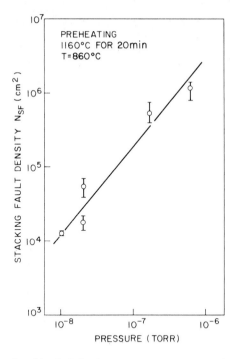

Fig. 29 The dependence of stacking fault density on system pressure at a substrate temperature of 860°C. *(After Sugiura and Yamaguchi, Ref. 51.)*

Substrates are best heated when they are placed in proximity to a resistance heater with closed-loop temperature control. Resistance heating generates temperatures over the range of 400 to 1100°C. A wide choice of temperature-sensing methods is available, including thermocouples, optical pyrometry, and infrared detection.

2.3.3 Film Characteristics

The preparation of high-quality films by MBE requires an in-situ cleaning process to remove absorbed contaminants and oxide films. Low base pressure is also a requirement to keep the surface clean. Figure 29 shows the effect of pressure on stacking fault density. Lowering the pressure lessens the concentration of contaminants absorbed on the substrate. These species would obstruct the single-crystal growth and nucleate a fault as discussed in Section 2.2.7 on nucleation. The effects on dislocation density can also be seen in the pre-heat time for substrate bakeout prior to growth and in the temperature of growth (see Figs. 30 and 31).

2.4 SILICON ON INSULATORS

An all-silicon device structure has inherent problems that are associated with parasitic circuit elements arising from junction capacitance. These effects are more of a prob-

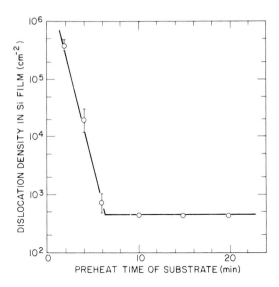

Fig. 30 The dependence of film quality on predeposition heating time. *(After Sugiura and Yamaguchi, Ref. 53.)*

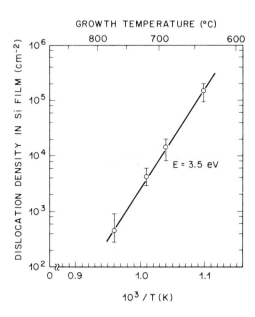

Fig. 31 The dependence of film perfection on growth temperature. *(After Sugiura and Yamaguchi, Ref. 53.)*

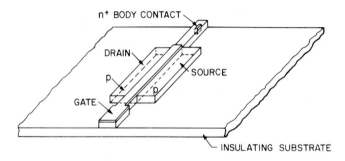

Fig. 32 MOSFET device fabricated in silicon island on sapphire substrate. *(After Schlotter, Ref. 59.)*

problem as devices are made smaller (see Chapter 11). A way to circumvent the problem is to fabricate devices in small islands of silicon on an insulating substrate as shown in Fig. 32. The initial approach to fabricating such a structure was to grow silicon epitaxially on a substrate of sapphire (Al_2O_3) or spinel ($MgAl_2O_4$). Since the substrate material differs from the layer, the process is termed heteroepitaxy. A more recent approach, yet to be perfected, is silicon on amorphous substrates.

2.4.1 Silicon on Sapphire

The processes and equipment used for silicon on sapphire (SOS) epitaxy are essentially identical to those employed for homoepitaxial growth. Silane is the favorite choice for the silicon source according to the pyrolysis reaction

$$SiH_4 => Si + 2H_2 \tag{17}$$

in a carrier gas of hydrogen. Silane is chosen mainly for its low-temperature deposition capability, which is used in SOS to control autodoping of aluminum from the substrate. Common process parameters are deposition temperatures between 1000 and 1050°C and growth rates of 0.5 μm/min. Film thicknesses are on the order of 1 μm or less with film doping in the range of 10^{14} to 10^{16} atoms/cm^3. Various substrate orientations such as $\langle 01\bar{1}2 \rangle$, $\langle 10\bar{1}2 \rangle$, and $\langle 1\bar{1}02 \rangle$, have been used to grow $\langle 100 \rangle$ oriented silicon layers.[57, 58, 59] Significant problems, however, exist with the technology. Aluminum autodoping from the substrate restricts the choice of doping level, and the films are usually characterized by a high defect density. The latter characteristic results in very low minority-carrier lifetimes (1 to 10 ns).[59] As a result only majority-carrier devices are practical. Both CMOS and NMOS circuits have been fabricated by using SOS epitaxy. The low minority-carrier lifetime also means that junction leakage currents could be higher than in comparable circuits in bulk wafers.

The defect structure of SOS devices has been studied by a number of workers.[57, 58] The films are generally characterized by high densities of various defects such as stacking faults, misfit dislocations, and dislocations. A key finding was that defect density varies inversely with distance from the substrate (Fig. 33). This effect is related to the lattice mismatch between the layer and substrate. The strain caused by lattice mismatch is somewhat relieved by the formation of misfit dislocations near the

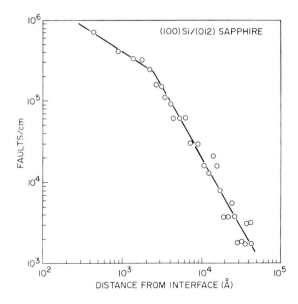

Fig. 33 Stacking fault density as a function of distance above the substrate for an SOS structure. *(After Abrahams and Buiocchi, Ref. 58.)*

original layer substrate interface. The transition layer between the epitaxial layer and the substrate is complicated, involving the formation of aluminum silicate from the outdiffusion of aluminum from the substrate. Such a layer is unavoidable in heteroepitaxy.[59] Another fundamental problem in SOS epitaxy is the thermal mismatch between the layer and the substrate. The thermal expansion of sapphire is approximately twice that of silicon. This difference in thermal expansion causes a strain-induced change in the band structure upon cooling that limits the carrier mobility to 80% of the bulk value. The carrier mobility is also reduced by the high defect densities.

These deficiencies have resulted in various attempts to improve SOS film quality. MBE is one solution, becuase its lower process temperature reduces autodoping and stress. Some workers[60, 61] have used laser annealing to improve the quality by melting and recrystallizing the layer. For example, a Q-switched ruby laser with energy densities greater than 1 J/cm^2 is used to reduce defect density and improve mobility.[60]

2.4.2 Silicon on Amorphous Substrates

Silicon on insulator (SOI) is a recent nonepitaxial approach to providing single-crystal silicon. With this technology, amorphous or polycrystalline silicon is recrystallized on an amorphous substrate. Figure 34 shows a setup for recrystallization using a strip heater. The process is considered nonepitaxial as this silicon film is not single crystal as-deposited. Energy for the process can also be supplied by electron beam[62] or laser.[63] The resultant structure is functionally similar to the heteroepitaxial SOS configuration, but without the attendant disadvantages just discussed. The recrystallized

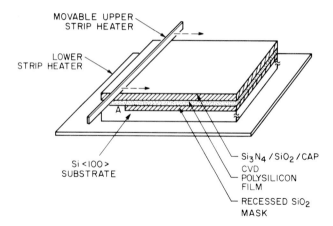

Fig. 34 Schematic of one technique used to recrystallize polysilicon on SiO$_2$. Region A acts as a seed for the lateral recrystallization when the heater moves to the right. *(After Tsuar et al., Ref. 65.)*

layers are potentially the equal of homoepitaxial silicon. SOI is not used commercially at present, but possible device applications include VLSI circuits, photovoltaic solar-energy conversion, and even three-dimensional ICs.

Several methods of preparing SOI have been investigated. Substrates can be conventional silicon wafers covered with silicon nitride or silicon dioxide or even fused quartz substrates.[64] The last method would be the most cost effective. If the conventional silicon wafer is used, it is processed in a manner which yields a pattern of exposed silicon areas, whose surface is coplanar with the surrounding oxide. This substrate is then coated with polysilicon in a low-pressure CVD process to a thickness of 0.5 μm. A movable strip heater (Fig. 34), positioned above one of the openings to the substrate,[65] melts the polysilicon through to the substrate. The heater is then moved laterally, and, with the substrate acting as a seed, single-crystal silicon is grown laterally over the oxide-covered regions. The thermal stability of the molten zone is improved if it is capped with oxide and nitride layers. Capping also prevents contamination of the film. This technique is suitable for recrystallizing large areas, such as an entire wafer. Similar procedures using a scanned CW argon laser have been reported.[66]

Another approach is to pattern a polysilicon layer on an amorphous substrate.[63] A laser is then rastered across the wafer to recrystallize the individual islands of silicon. This method does not need seeding from the substrate. Adjusting the energy parameters of the laser and its scan rate induces the islands to crystallize in a $\langle 100 \rangle$ orientation. High-quality n-channel depletion mode MOSFETs have been fabricated[67] in recrystallized silicon. The device structure (Fig. 35) is similar to that of SOS devices, but better. In particular, the surface electron mobility was reported at 600 to 700 cm^2/V-s, a value near that of devices fabricated in single-crystal silicon. These results are better than those obtained from SOS devices.

Additional work is needed before this technology is the equal of homoepitaxial silicon technology, but it has the potential to revolutionize device design and fabrication.

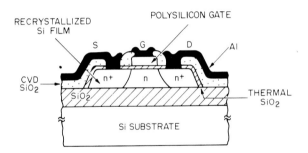

Fig. 35 Cross section of MOSFET formed in recrystallized polysilicon. *(After Tsuar et al., Ref. 67.)*

2.5 EPITAXIAL EVALUATION

To evaluate epitaxial slices layer doping and thickness, which are easily quantified, are measured. Additionally, a cosmetic inspection is usually performed even though this evaluation is somewhat subjective. The prime requisites for routine measurements are high speed and repeatability. In an industrial environment, information is needed at relatively short intervals ($<$1 h) to maintain process control. Absolute accuracy is of lesser concern, because the material requirements are usually adjusted on an empirical basis to satisfy device needs. Only a few evaluation methods are commonly used.[68]

2.5.1 Epitaxial Thickness

Lightly doped silicon is transparent in the near infrared region and heavily doped silicon ($>$1 $\times$ 10^{18} atoms/cm^3) is an absorber. However, increased doping reduces the index of refraction (Fig. 36) below that of lightly doped silicon ($n = 3.42$). As a result interference fringes in the 5- to 50-μm wavelength range can be observed in the reflection spectra on a conventional infrared spectrophotometer. The epitaxial layer thickness can be computed using the formula[69]

$$t = \frac{(P_n - \frac{1}{2} + P_i)\, W_n}{2(n^2 - \sin^2 \theta)} \tag{18}$$

where W_n is the position of the maxima or minima in the spectra in micrometers, n is the index of refraction, θ is the angle of the incident light, P_n is the order of the maxima or minima, and P_i is a correction factor that depends on the substrate used.

An automated approach to the measurement employs a Michaelson interferometer. This instrument samples all wavelengths simultaneously. Its output is called an interferogram, which is the Fourier transform of the reflectance spectra obtained on a spectrophotometer. A computer controls the interferometer and collects the data. The thickness can be computed from the interferogram, or the computer can calculate the Fourier transform and then proceed to calculate the thickness using Eq. 18. Equipment for this second method is commercially available. Such equipment can measure thickness from less than 1 μm to more than several hundred micrometers. Measurement time is about 5 s with a measurement repeatability of $\pm$0.05 μm.

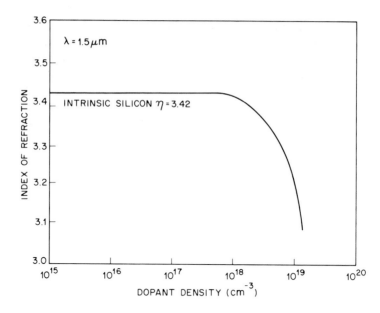

Fig. 36 Typical index of refraction versus doping level for silicon at one wavelength.

The equivalent point of reflection of infrared measurements is at a heavily doped point on the outdiffusion tail. For common processing conditions, this point is usually near the epi-substrate interface. Thus, an infrared measurement is a reasonable monitor of the thickness added to the substrate, but is relatively insensitive to the shape or extent of the outdiffusion autodoping tail.[70]

For structures that are not amenable to infrared measurements, there are several alternatives. The length of the side of an epitaxial stacking fault, nucleated at the substrate, is linearly related to the layer thickness[71]

$$t = C_1 L \tag{19}$$

where t is the layer thickness, L is the size of the fault, and C_1 is an orientation-dependent constant which is 0.707 for $\langle 100 \rangle$ and 0.816 for $\langle 111 \rangle$.

Wafers can also be sectioned and stained with a number of chemical solutions to delineate the layer.[72] Spreading resistance[73, 74] profiling (see Chapter 5) is particularly useful for structures that have multiple layers or structures where the total impurity profile is important.

2.5.2 Epitaxial Doping

The uncertainties of doping kinetics, background effects, and autodoping effects do not allow the doping in the layer to be established simply on the basis of the flows into the reactor. Three types of electrical measurements—sheet resistance, diode capacitance voltage, and spreading resistance—are used to measure doping levels.[75]

The control wafer technique is a widely used method that requires simple equipment. It involves placing in the reactor a lightly doped slice of a conductivity type opposite to the layer to be grown. After deposition a four-point probe measures the sheet resistance of the layer (see Chapter 5). The sheet resistance is converted to resistivity using the infrared thickness of an adjacent product slice.[76] This method is highly inaccurate in some cases.[75] Its suitability must be determined by correlating its measurements to measurements made on product slices by another method. In other cases, no correlation is possible due to a strong predeposition of substrate dopant onto the control wafer. The control wafer technique is expensive and often wasteful of reactor capacity.

The second method, the preferred approach, is the use of diode C-V measurements (see Chapter 5). Implicit in the capacitance versus voltage characteristic of a reverse-biased diode is the doping profile of the material according to the relationships

$$N(x) = C^3 \left[\frac{dC}{dV} \right] CF_1 CF_2 qA^2 \epsilon_s \tag{20}$$

$$x = \epsilon_s A / C \tag{21}$$

where C is capacitance, V is voltage, q is charge, A is the diode area, ϵ_s is the dielectric permittivity of silicon, N is the doping density, and x is depth. CF_1 and CF_2 are correction factors for diffused-junction and depletion-layer widening effects.[77, 78]

C-V measurements of a Schottky barrier diode, formed by using a mercury contact,[75, 79] are a rapid nondestructive way to determine slice doping. If the depletion layer can be spread to the substrate, some information on the autodoping tail can be obtained. The measurement can also be performed on mesa or planar junction diodes as a means of calibrating other measurements.[76] The principle drawbacks of a C-V measurement are its high sensitivity to small errors in area and capacitance.[80]

The third method, spreading resistance measurements, was previously mentioned as a profiling technique. This method can determine a wafer's resistivity by measuring on the surface. The major difficulties are in maintaining accurate calibration as the probes wear with repeated usage and in overcoming the influence of surface effects that affect the measured resistance.

2.5.3 Cosmetic Inspection and Perfection Evaluation

The wafer is usually examined with the unaided eye under high-intensity illumination to judge the quality of the deposit. Wafers may be rejected for any departure from a specular, smooth surface, including projections which are seen as bright spots of light, stains, haze, or scratches. The acceptance criteria is usually set empirically based on the type of device being fabricated. Attempts to automate this inspection using scanned laser or collimated light to detect light scattering centers have generally been unsuccessful. Additional inspection may be made at magnifications of from 50 to 200 to evaluate microdefect densities such as stacking faults and tripyramids. Nomarski phase contrast microscopy is preferred for this inspection. Another useful technique is to etch wafers[81] in solutions such as Secco's or Sirtl's etch to determine dislocation and saucer pit densities. The latter indicate that contamination is present in the process.

2.5.4 Lifetime

The lifetime of minority carriers is generally not a consideration in structures intended for IC fabrication, but could be of interest in some devices such as dynamic RAMs. Several measurements involving the transient response of diodes or MOS capacitors are applicable to epitaxial layers.[82, 83] However, the diffusion length of carriers is often many times that of the layer thickness. This complicates the interpretation of the measurement results.

2.6 SUMMARY AND FUTURE TRENDS

Epitaxy as a process will remain integral to circuit manufacture. It offers doping profiles and material properties not obtainable otherwise. Homoepitaxial silicon structures will remain popular design choices in the foreseeable future. The advantages of SOI technologies are compelling for high-density and high-speed circuits. In particular, if silicon-on-SiO_2 can be perfected, it will offer the advantages of SOS without the problems. Lateral-seeded SOI will undoubtedly receive considerable research attention. MBE would be advantageous in fully ion-implanted VLSI circuits in which the total thermal cycle is minimized so that the doping capabilities of MBE can be exploited.

Although presently available equipment is adequate for most needs, several aspects of the epitaxial process could be improved. In keeping with automation elsewhere in the fabrication process, an autoloading epitaxial reactor remains a desirable objective. This equipment could take the form of a cassette-fed machine processing a single wafer at a time. Conceptually, a uniwafer reaction chamber could be optimized for temperature and gas flows to produce wafers having exceptional uniformity. The throughput of epitaxial reactors is less than that of LPCVD processes (Chapter 3) by a factor of 5 to 10. However, monocrystalline silicon cannot be grown in LPCVD equipment. One difficulty is the low growth rates in the usual LPCVD temperature ranges (Fig. 8). An alternative reactor design,[84] termed the rotary disc, is similar in load configuration to LPCVD equipment, and offers high capacity and efficiency. Large-scale use of MBE will require equipment with throughputs comparable to present-day epitaxial reactors.

Although epitaxial processes are well characterized and understood, the trend to thinner layers for bipolar and unipolar ICs will result in incremental process improvements and the continued study of autodoping effects. Additionally, contamination, responsible for precipitates in epitaxial layers, needs to be reduced commensurate with the requirements of VLSI devices. Contamination-free epitaxy will be a worthwhile process improvement.

REFERENCES

[1] H. C. Theuerer et al., "Epitaxial Diffused Transistors," *Proc. IRE*, 48, 1642 (1960).

[2] F. E. Holmes and C. A. T. Salama, "VMOS—A New MOS Integrated Circuit Technology," *Solid State Electron.*, 17, 791 (1974).

[3] D. S. Yaney and C. W. Pearce, "The Use of Thin Epitaxial Layers for MOS VLSI," *Proceedings of the 1981 International Electron Device Meeting*, IEEE, 1981, p. 236.

[4] V. S. Ban, "Mass Spectrometric Studies of Chemical Reactions and Transport Phenomena in Silicon Epitaxy," *Proceedings of the Sixth International Conference on Chemical Vapor Deposition 1977*, Electrochem. Soc., 1977, p. 66.

[5] R. M. Olsen, *Essentials of Engineering Fluid Flow*, International Textbook, Scranton, Pennsylvania, 1966.

[6] C. W. Manke and L. F. Donaghey, "Numerical Simulation of Transport Processes in Vertical Cylinder Epitaxy Reactors," *Proceedings of the Sixth International Conference on Chemical Vapor Deposition 1977*, Electrochem. Soc., 1977 p. 151.

[7] J. Bloem, "Silicon Epitaxy from Mixtures of SiH_4 and HCl," *J. Electrochem. Soc.*, 117, 1397 (1970).

[8] E. Sirtl, L. P Hunt, and D. H. Sawyer, "High Temperature Reactions in the Silicon-Hydrogen-Chlorine System," *J. Electrochem. Soc.*, 121, 919 (1974).

[9] J. Nishizawa and M. Saito, "Growth Mechanism of Chemical Vapor Deposition of Silicon," *Proceedings of the Eighth International Conference on Chemical Vapor Deposition 1981*, Electrochem. Soc., p. 317.

[10] F. C. Eversteyn, "Chemical-Reaction Engineering in the Semiconductor Industry," *Philips Res. Rep.*, 29, 45 (1974).

[11] McD. Robinson, in F. F. Y. Wang, Ed., *Impurity Doping Processes in Silicon*, North-Holland, Amsterdam, 1981.

[12] J. Bloem, "The Effect of Trace Amounts of Water Vapor on Boron Doping in Epitaxially Grown Silicon," *J. Electrochem. Soc.*, 118, 1837 (1971).

[13] R. Reif, T. I. Kamins, and K. C. Saraswat, "A Model for Dopant Incorporation into Growing Silicon Epitaxial Films," *J. Electrochem. Soc.*, 126, 644 and 653 (1979).

[14] H. Basseches, R. C. Manz, C. O. Thomas, and S. K. Tung, *AIME Semiconductor Metallurgy Conference*, Interscience, New York, 1961, p. 69.

[15] A. S. Grove, A. Roder, and C. T. Sah, "Impurity Distribution in Epitaxial Growth," *J. Appl. Phys.*, 36, 802 (1965).

[16] G. R. Srinivasan, "Autodoping Effects in Silicon Epitaxy," *J. Electrochem. Soc.*, 127, 1334 (1980).

[17] P. H. Langer and J. I. Goldstein, "Boron Autodoping during Silane Epitaxy," *J. Electrochem. Soc.*, 124, 592 (1977).

[18] G. Skelly and A. C. Adams, "Impurity Atom Transfer during Epitaxial Deposition of Silicon," *J. Electrochem. Soc.*, 120, 116 (1973).

[19] G. R. Srinivasan, "Kinetics of Lateral Autodoping in Silicon Epitaxy," *J. Electrochem. Soc.*, 125, 146 (1978).

[20] B. A. Joyce, J. C. Weaver, and D. J. Maule, "Impurity Redistribution Processes in Epitaxial Layers," *J. Electrochem. Soc.*, 112, 1100 (1965).

[21] W. H. Shepard, "Autodoping of Epitaxial Silicon," *J. Electrochem. Soc.*, 115, 652 (1968).

[22] R. Reif and R. W. Dutton, "Computer Simulation in Silicon Epitaxy," *J. Electrochem. Soc.*, 128, 909 (1981).

[23] L. V. Gregor, P. Balk, and F. J. Campagna, "Vapor-Phase Polishing of Silicon with H_2—HBr Gas Mixtures," *IBM J. Res. Dev.*, 9, 327 (1965).

[24] M. L. Hammond, "Silicon Epitaxy," *Solid State Technol.*, 21, 68 (1978).

[25] R. C. Rossi and K. K. Scheregraf, "Glassy Carbon-Coated Susceptors for Semiconductor CVD Processes," *Semicond. Int.*, 4, 99 (1981).

[26] B. T. Murphy, V. J. Glinski, P. A. Gary, and R. A. Pedersen, "Collector-Diffusion Isolated Integrated Circuits," *Proc. IEEE*, 57, 1523 (1969).

[27] J. Bloem, L. J. Giling, and M. W. M. Graef, "The Incorporation of Phosphorous in Silicon Epitaxial Layer Growth," *J. Electrochem. Soc.*, 121, 1354 (1974).

[28] P. Rai-Choudhury and E. I. Salkovitz, "Doping of Epitaxial Silicon," *J. Cryst. Growth*, 7, 361 (1970).

[29] J. Borkowicz, J. Korec, and E. Nossarzewska-Orlowska, "Optimum Growth Conditions in Silicon Vapour Epitaxy," *Phys. Status Solidi A*, 48, 225 (1978).

[30] H. C. Theuerer, "Epitaxial Films by the Hydrogen Reduction of $SiCl_4$," *J. Electrochem. Soc.*, 108, 649 (1961).

[31] A. Lekholm, "Epitaxial Growth of Silicon from Dichorosilane," *J. Electrochem. Soc.*, 120, 1122 (1973).

[32] G. Kosza, F. A. Kuznetsov, T. Kormany, and L. Nagy, "Optimization of Si Epitaxial Growth," *J. Cryst. Growth*, 52, 207 (1981).

[33] M. Ogirima, H. Saida, M. Suzuki, and M. Maki, "Low Pressure Silicon Epitaxy," *J. Electrochem. Soc.*, 124, 903 (1977).

[34] R. B. Herring, "Advances in Reduced Pressure Silicon Epitaxy," *Solid State Technol.*, 22, 75 (1979).

[35] W. Kern and D. A. Pustinen, "Cleaning Solutions Based on Hydrogen Peroxide for Use in Silicon Semiconductor Technology," *RCA Rev.*, 34, 188 (1970)

[36] S. P. Weeks, "Pattern Shift and Pattern Distortion during CVD Epitaxy on $\langle 111 \rangle$ and $\langle 100 \rangle$ Silicon," *Solid State Technol.*, 24, 111 (1981)

[37] C. M. Drum and C. A. Clark, "Anisotropy of Macrostep Motion and Pattern Edge Displacements on Silicon near $\langle 100 \rangle$," *J. Electrochem. Soc.*, 115, 664 (1968) and 117, 1401 (1970).

[38] P. H. Lee, M. T. Wauk, R. S. Rosler, and W. C. Benzing, "Epitaxial Pattern Shift Comparison in Vertical, Horizontal, and Cylindrical Reactor Geometries," *J. Electrochem. Soc.*, 124, 1824 (1977).

[39] K. V. Ravi, *Imperfection and Impurities in Semiconductor Silicon*, Wiley, New York, 1981.

[40] C. M. Melliar-Smith, *Treatise of Materials Science and Technology*, Academic, New York, 1977, Vol. II.

[41] G. A. Rozgonyi, R. P. Deysher, and C. W. Pearce, "The Identification, Annihilation and Suppression of Nucleation Sites Responsible for Silicon Epitaxial Stacking Faults," *J. Electrochem. Soc.*, 123, 1910 (1976).

[42] L. E. Katz and D. W. Hill, "High Oxygen Czochralski Silicon Crystal Growth to Epitaxial Stacking Faults," *J. Electrochem. Soc.*, 125, 1151 (1978).

[43] C. W. Pearce and R. G. MacMahon, "Role of Metallic Contamination in the Formation of Saucer Pit Defects in Epitaxial Silicon," *J. Vac. Sci. Technol.*, 14, 40 (1977).

[44] S. K. Tung, "The Effects of Substrate Orientation on Epitaxial Growth," *J. Electrochem. Soc.*, 112, 436 (1965).

[45] B. J. Baliga, "Defect Control During Silicon Epitaxial Growth Using Dichlorosilane," *J. Electrochem. Soc.*, 129, 1078 (1982).

[46] J. Bloem and A. H. Goemans, "Slip in Silicon Epitaxy," *J. Appl. Phys.*, 43, 1281 (1972).

[47] Y. Sugita, M. Tamura, and K. Sugawara, "Misfit Dislocations in Bicrystals of Epitaxially Grown Silicon on Boron-Doped Silicon Substrates," *J. Appl. Phys.*, 40, 3089 (1969).

[48] J. Bloem, "Nucleation and Growth of Silicon by CVD," *J. Cryst. Growth*, 50, 581 (1980).

[49] J. C. Bean, "Silicon Molecular Beam Epitaxy as a VLSI Processing Technique," *IEEE Proc. Int. Electron Device Meet.*, IEEE, 1981, p.6.

[50] J. C. Bean, in F. F. Y. Wang, Ed., *Impurity Doping Processes in Silicon*, North-Holland, Amsterdam, 1981.

[51] H. Sugiura and M. Yamaguchi, "Growth of Dislocation-Free Silicon Films by Molecular Beam Epitaxy," *J. Vac. Sci. Technol.*, 19, 157 (1981).

[52] P. E. Luscher and D. M. Collins, in B. R. Pamplin, Ed., *Design Considerations for Molecular Beam Epitaxy Systems*, Pergamon, London, 1981.

[53] H. Sugiura and M. Yamaguchi, "Crystal Defects of Silicon Films Formed by Molecular Beam Epitaxy," *Jpn. J. Appl. Phys.*, 19, 583 (1980).

[54] U. Konig, H. Kibbel, and E. Kasper, "MBE: Growth and Sb Doping," *J. Vac. Sci. Technol.*, 16, 985 (1979).

[55] Y. Ota, "Si Molecular Beam Epitaxy (n on n^+) with Wide Range Doping Control," *J. Electrochem. Soc.*, 124, 1795 (1977).

[56] Y. Ota, "N-type Doping Techniques in Silicon Molecular Beam Epitaxy by Simultaneous Arsenic Ion Implantation and by Antimony Evaporation," *J. Electrochem. Soc.*, 126, 1761 (1979).

[57] M. S. Abrahams, C. J. Buiocchi, J. F. Corby, Jr., and G. W. Cullen, "Misfit Dislocation in Heteroepitaxial Si on Sapphire," *Appl. Phys. Lett.*, 28, 275 (1976).

[58] M. S. Abrahams and C. J. Buiocchi, "Cross-Sectional Electron Microscopy of Silicon on Sapphire," *Appl. Phys. Lett.*, 27, 325 (1975).

[59] H. Schlotter, "Interface Properties of Sapphire and Spinel," *J. Vac. Sci. Technol.*, 13, 29 (1976).

[60] Y. Kobayaski, T. Suzuki, and M. Tamura, "Improvement of Crystalline Quality of SOS with Laser Irradiation Techniques," *Jpn. J. Appl. Phys.*, 20, L249 (1981).

[61] G. A. Sai-Halary, F. F. Fang, T. O. Sedgwick, and A. Segmuller, "Stress-Relieved Regrowth of Silicon on Sapphire by Laser Annealing," *Appl. Phys. Lett.*, 36, 419 (1980).

[62] K. Shibata, T. Inoue, and T. Takigawa, "Grain Growth of Polycrystalline Silicon Films on SiO_2 by CW Scanning Electron Beam Annealing," *Appl. Phys. Lett.*, 39, 645 (1981).

[63] D. K. Biegelsen, N. M. Johnson, D. J. Bartelink, and M. D. Moyer, "Laser-Induced Crystallization of Silicon Islands on Amorphous Substrates: Multilayer Structures," *Appl. Phys. Lett.*, 38, 150 (1981).

[64] R. A. Lemons and M. A. Bosch, "Periodic Motion of the Crystallization Front during Beam Annealing of Si Films," *Appl. Phys. Lett.*, 39, 343 (1981).

[65] B-Y. Tsuar, J. C. C. Fan, M. W. Geis, D. J. Silversmith, and R. W. Mountain, "Improved Techniques for Growth of Large Area Single Crystal Si Sheets over SiO_2 Using Lateral Epitaxy by Seeded Solidification," *Appl. Phys. Lett.*, 39, 561 (1981).

[66] T. I. Kamins and P. A. Pianetta, "MOSFETs in Laser-Recrystallized Polysilicon on Quartz," *IEEE Electron. Device Lett.*, EDL-1, 214 (1980).

[67] B-Y. Tsuar, M. W. Geis, J. C. C. Fan, D. J. Silversmith, and R. W. Mountain, "N-Channel Deep-Depletion Metal-Oxide Semiconductor Transistors Fabricated in Zone-Melting-Recrystallized Polycrystalline Si Films in SiO_2 ," *Appl. Phys. Lett.*, 39, 909 (1981).

[68] P. H. Langer and C. W. Pearce, "Epitaxial Resistivity," *J. Test. Eval.*, 1, 305 (1973).

[69] Am. Soc. Test. Mater., ASTM Standard, F95, Part 43.

[70] K. Sato, Y. Ishikawa, and K. Sugawara, "Infrared Interference Spectra Observed in Silicon Epitaxial Wafers," *Solid State Electron.*, 9, 771 (1966).

[71] Am. Soc. Test. Mater., ASTM Standard, F143, Part 43.

[72] Am. Soc. Test. Mater., ASTM Standard, F110, Part 43.

[73] National Bureau of Standards, Special Publication 400-10, "Spreading Resistance Symposium," December 1974.

[74] Y. Isda, H. Abe, and M. Kondo, "Impurity Profile Measurements of Thin Epitaxial Wafers by Multilayer Spreading Resistance Analysis," *J. Electrochem. Soc.*, 124, 1118 (1977).

[75] D. L. Rehrig and C. W. Pearce, "Production Mercury Probe Capacitance-Voltage Testing," *Semicond. Int.*, 3, 151 (1980)

[76] Am. Soc. Test. Mater., ASTM Standard, F374, Part 43.

[77] M. G. Buehler, "Peripheral and Diffused Layer Effects on Doping Profiles," *IEEE Trans. Electron Devices*, ED-19, 1171 (1972).

[78] J. A. Copeland, "Diode Edge Effect on Doping-Profile Measurements," *IEEE Trans. Electron Devices*, ED-17, 404 (1970).

[79] P. J. Severin and G. J. Poodt, "Capacitance-Voltage Measurements with a Mercury-Silicon Diode," *J. Electrochem. Soc.*, 119, 1384 (1972).

[80] I. Amron, "Errors in Dopant Concentration Profiles Determined by Differential Capacitance Measurements," *Electrochem. Technol.*, 5, 94 (1967).

[81] D. G. Schimmel, "A Comparison of Chemical Etches for Revealing $\langle 100 \rangle$ Silicon Crystal Defects," *J. Electrochem. Soc.*, 123, 734 (1976).

[82] P. G. Wilson, "Recombination in P-I-N Diodes," *Solid State Electron.*, 10, 145 (1967).

[83] K. H. Zaininger and F. P. Herman, "The C-V Technique as an Analytical Tool," *Solid State Technol.*, 13, 46 (1970*).

[84] V. S. Ban and E. P. Miller, "A New Reactor for Silicon Epitaxy," *Proceedings of the 7th International Conference on Chemical Vapor Deposition 1979*, Electrochem. Soc., 1979, p. 102.

[85] S.S. Iyer, R.A. Metzger, and F.G. Allen, "Sharp Profiles with High and Low Doping Levels in Silicon Grown by Molecular Beam Epitaxy," *J. Appl. Phys.*, 52, 5608 (1982).

PROBLEMS

1 In a 1-h process at 1100°C using dichlorosilane, a 10-μm layer is grown on 20 substrates of 100-mm diameter in a horizontal reactor. Estimate the energy in kilowatthours for the process. Assume a growth rate of 1 μm/min.

2 Determine the amount of mask compensation needed for an epitaxial wafer of ⟨100⟩ orientation containing an antimony buried layer with an epitaxial thickness of 7 μm.

3 Using the figures in the chapter estimate the temperature of zero growth rate for each silicon source. Compare these temperatures to the nucleation temperature curve. What do you conclude?

4 Calculate activation energies from the Arrhenius plots for growth rate versus temperature and nucleation versus temperature. What do you conclude about the process?

5 A reverse-biased diode has a voltage capacitance characteristic defined by the relation $VC^2 = N$. What would you conclude about the shape of the doping profile? Suggest a graphical way to determine doping density from the C-V curve.

6 Using the diffusivity of boron at 1100°C (Chapter 5) calculate the minimum growth rate such that the condition of Eq. 12 is satisfied given a deposition time of 10 min.

7 Calculate the number of liters of hydrogen at STP that would be needed to be supplied *into* the reactor for the process of Problem 1. What do you conclude?

8 Does the thickness of the epitaxial wafer pose a problem in epitaxial processing from a stress viewpoint? Discuss your answer.

THREE

DIELECTRIC AND POLYSILICON FILM DEPOSITION

A. C. ADAMS

3.1 INTRODUCTION

Deposited films are widely used in the fabrication of modern VLSI circuits. These films provide conducting regions within the device, electrical insulation between metals, and protection from the environment. Deposited films must meet many requirements. The film thickness must be uniform over each device and over the large number of wafers processed at one time. The structure and composition of the film must be controlled and reproducible. Finally, the method for depositing the film must be safe, reproducible, easily automated, and inexpensive.

The most widely used materials are polycrystalline silicon, silicon dioxide, stoichiometric silicon nitride, and plasma-deposited silicon nitride. The most common deposition methods are atmospheric-pressure chemical vapor deposition (CVD), low-pressure chemical vapor deposition (LPCVD), and plasma-assisted chemical vapor deposition (PCVD or plasma deposition). Several reviews of these materials and their preparation are available.[1-3]

Polycrystalline silicon, usually referred to as polysilicon, is prepared by pyrolyzing silane at 600 to 650°C. Polysilicon is used as the gate electrode material in MOS devices, as a conducting material for multilevel metallization, and as a contact material for devices with shallow junctions. Polysilicon is usually deposited without dopants. The doping elements, arsenic, phosphorus, or boron, are added subsequently by diffusion or ion implantation. The dopants can also be added during deposition, which is advantageous for some device structures. Polysilicon containing several percent oxygen is a semi-insulating material that is used for circuit passivation.

Dielectric materials are used for insulation between conducting layers, for diffusion and ion implantation masks, for diffusion from doped oxides, for capping doped films to prevent the loss of dopants, for gettering impurities, and for passivation to protect devices from impurities, moisture, and scratches. Phosphorus-doped silicon dioxide (P-glass, phosphosilicate glass, or PSG) is especially useful, because it inhibits the diffusion of sodium impurities and because it softens and flows at 1000 to 1100°C, creating a smooth topography that is beneficial for subsequent metallization.

Silicon nitride is a barrier to sodium diffusion, is nearly impervious to moisture, and has a very low oxidation rate. Stoichiometric silicon nitride (Si_3N_4), deposited at 700 to 900°C, is used as an oxidation mask to create planar structures and as a gate dielectric in conjunction with thermally grown silicon dioxide in dual dielectric devices. Plasma-deposited silicon nitride (plasma nitride or SiN) is formed at much lower temperatures, 200 to 350°C, and is used as a passivation layer and for protection against scratches. The low deposition temperature allows this material to be used over aluminum or gold metallization.

Many methods are available for depositing thin films.[4] However, various CVD techniques are most frequently used for semiconductor processing. These chemical depositions occur under a large variety of conditions. Deposition temperatures vary from 100 to 1000°C and pressures range from atmospheric down to about 7 Pa (0.05 Torr). The energy for the reaction can be supplied thermally, by photons (photochemically), or by a glow discharge.

Historically, dielectric and polysilicon films have been deposited at atmospheric pressure by using a variety of reactor geometries.[1, 2, 5] These include horizontal reactors with the wafers lying on a hot suspector and the reactant gases flowing over the surfaces, usually at very high velocities. The suspector is heated by radiation using high-intensity lamps, by radio frequency induction, or by electrical resistance. Various vertical reactors also exist, usually consisting of a bell-jar reaction chamber with samples oriented in a vertical direction on a rotating assembly. As in the horizontal reactors, the suspector is heated by radiation, induction, or resistance. All of these atmospheric pressure reactors tend to have low wafer throughput, require extensive wafer handling during loading and unloading, and provide thickness uniformities that are usually no better than ±10%. As a consequence, they have been replaced by low-pressure, hot-wall reactors. Plasma-assisted depositions in hot-wall reactors or with parallel-plate geometries are also available for applications that require very low sample temperatures, 100 to 350°C.

The potential advantages of the low-pressure deposition processes are: (1) uniform step coverage, (2) precise control of composition and structure, (3) low-temperature processing, (4) fast deposition rates, (5) high throughput, and (6) low processing costs. Compromises and trade-offs are made among these properties. For instance, low deposition rates may be tolerated to achieve low deposition temperatures. A goal in developing a deposition process is to best use the advantages of CVD and to find the optimum compromise for specific device structures.

3.2 DEPOSITION PROCESSES

3.2.1 Reactions

Table 1 lists some typical reactions that may be used to deposit films on device wafers. The choice of a particular reaction is often determined by the deposition tem-

Table 1 Typical reactions for depositing dielectrics and polysilicon

Product	Reactants	Deposition temperature (°C)
Silicon dioxide	$SiH_4 + CO_2 + H_2$	850−950
	$SiCl_2H_2 + N_2O$	850−900
	$SiH_4 + N_2O$	750−850
	$SiH_4 + NO$	650−750
	$Si(OC_2H_5)_4$	650−750
	$SiH_4 + O_2$	400−450
Silicon nitride	$SiH_4 + NH_3$	700−900
	$SiCl_2H_2 + NH_3$	650−750
Plasma silicon nitride	$SiH_4 + NH_3$	200−350
	$SiH_4 + N_2$	200−350
Plasma silicon dioxide	$SiH_4 + N_2O$	200−350
Polysilicon	SiH_4	600−650

perature (which must be compatible with the device materials), the film properties, and certain engineering aspects of the deposition (wafer throughput, safety, and reactor maintenance).

The most common reactions for depositing silicon dioxide for VLSI circuits are oxidizing silane with oxygen at 400 to 450°C, decomposing tetraethoxysilane at 650 to 750°C, and reacting dichlorosilane with nitrous oxide at 850 to 900°C. Doped oxides are prepared by adding a dopant to the deposition reaction. The hydrides arsine, phosphine, or diborane are often used because they are readily available gases; however, halides and organic compounds can also be used. Silicon nitride is prepared by reacting silane and ammonia at atmospheric pressure at 700 to 900°C, or by reacting dichlorosilane and ammonia at reduced pressure at about 700°C. Plasma-deposited silicon nitride is deposited by reacting silane with ammonia or nitrogen in a glow discharge between 200 and 350°C. This reaction is useful for passivation where higher temperatures cause unwanted reactions between the silicon and the metal conductors. Similarly, plasma-deposited silicon dioxide is formed from silane and nitrous oxide in a glow discharge. Polysilicon is prepared by pyrolyzing silane at 600 to 650°C.

3.2.2 Equipment

Figures 1 and 2 give schematics of four reactors commonly used for depositions. Figure 1a shows a hot-wall, reduced-pressure reactor, used to deposit polysilicon, silicon dioxide, and silicon nitride. The reactor consists of a quartz tube heated by a three-zone furnace, with gas introduced in one end and pumped out the other. The mechanical pump is sometimes augmented with a Roots blower. Pressures in the reaction chamber are typically 30 to 250 Pa (0.25 to 2.0 Torr); temperatures range between 300 and 900°C; and gas flows are between 100 and 1000 std. cm^3/min. Wafers stand vertically, perpendicular to the gas flow, in a quartz holder. Each run processes 50 to 200 wafers. Special inserts that alter the gas flow dynamics are sometimes used.

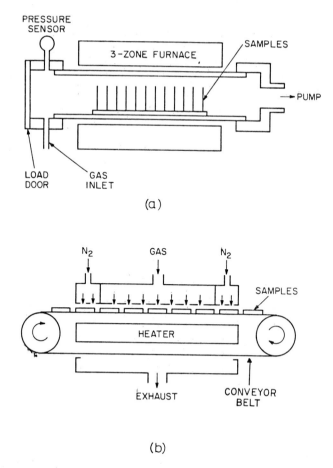

Fig. 1 Schematic diagrams of CVD reactors. (a) Hot-wall, reduced-pressure reactor. (b) Continuous, atmospheric-pressure reactor.

Thickness uniformities are within ±5%. Hot-wall, reduced-pressure reactors can be easily scaled to hold 150-mm-diameter wafers. The major advantages of these reactors are excellent uniformity, large load size, and ability to accommodate large diameter wafers. The disadvantages are low deposition rates and the frequent use of toxic, corrosive, or flammable gases.

Figure 1b shows a continuous throughput, atmospheric-pressure reactor used to deposit silicon dioxide. The samples are carried through the reactor on a conveyor belt. Reactant gases flowing through the center of the reactor are contained by gas curtains formed by a very fast flow of nitrogen. The samples are heated by convection. The advantages of these continuous throughput reactors are high throughput, good uniformity, and ability to handle large-diameter wafers. The major disadvantages are that very fast gas flows are required and these reactors must be cleaned frequently.

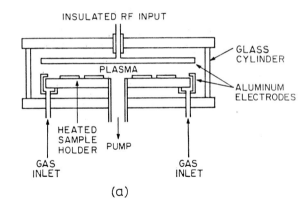

(a)

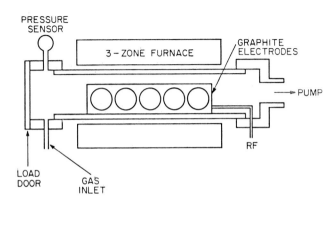

(b)

Fig. 2 Schematic diagrams of plasma deposition reactors. (a) Parallel-plate. (b) Hot-wall.

Figure 2a shows a radial-flow, parallel-plate, plasma-assisted CVD reactor. The reaction chamber is a cylinder, usually glass or aluminum, with aluminum plates on the top and bottom. Samples lie on the bottom electrode, which is grounded. A radio frequency voltage is applied to the top electrode to create a glow discharge between the two plates. Gases flow radially through the discharge. They are usually introduced at the outer edge and flow towards the center, although the opposite flow pattern can be used. The gases are pumped with a Roots blower backed by a mechanical pump. The grounded electrode is heated to a temperature between 100 and 400°C by resistance heaters or high-intensity lamps. This reactor is used for the plasma-assisted deposition of silicon dioxide and silicon nitride. Its main advantage is low deposition temperature, while there are three major disadvantages. Capacity is limited, especially for large-diameter wafers. Wafers must be loaded and unloaded individually, and wafers may be contaminated by loosely adhering deposits falling on them.

The hot-wall, plasma-deposition reactor shown in Fig. 2b solves many of the problems encountered in the radial-flow reactor. The reaction takes place in a quartz tube heated by a furnace. The samples are held vertically, parallel to the gas flow. The electrode assembly, which supports the samples, contains long graphite or aluminum slabs. Alternating slabs are connected to the power supply, which generates a discharge in the space between the electrodes. Advantages of this reactor are its high capacity and low deposition temperatures. Its drawbacks, however, are that particles can be formed while the electrode assembly is being inserted, and that wafers must be individually handled during loading and unloading.

3.2.3 Safety

Many of the gases used to deposit films are hazardous. The safety problems are more severe for low-pressure depositions because the processes often use concentrated gases. For instance, 100% silane is used for polysilicon depositions at reduced pressure, compared to only 3% silane in nitrogen for the same deposition at atmospheric pressure. Low-pressure depositions which use pumps have additional safety problems associated with them, because the gases can dissolve or react in the pump oil.

The hazardous gases fall into four general classes: poisonous; pyrophoric, flammable, or explosive; corrosive; and dangerous combinations of gases. Table 2 lists hazardous properties of common gases used in CVD. Examples of dangerous gas combinations that may be encountered are silane with halogens, silane with hydrogen, and oxygen with hydrogen.

Many of the flammable gases react with air to form solid products. Consequently, small leaks cause particles to form within the gas lines. These particles eventually plug the line or the gas metering equipment. The reactant gases and the reaction products also accumulate in the pumps and may present hazards during pump maintenance. Detailed safety precautions for CVD processes have been published.[6]

Table 2 Properties of common gases used in CVD

Gas	Properties
Silane	Toxic, flammable, pyrophoric
Dichlorosilane	Toxic, flammable, corrosive
Phosphine	Very toxic, flammable
Diborane	Very toxic, flammable
Arsine	Very toxic, flammable
Hydrogen chloride	Toxic, corrosive
Ammonia	Toxic, corrosive
Hydrogen	Nontoxic, flammable
Oxygen	Nontoxic, supports combustion
Nitrous oxide	Nontoxic, nonflammable
Nitrogen	Usually inert
Argon	Inert

3.3 POLYSILICON

Polysilicon is used as the gate electrode in MOS devices. It is also used for high-value resistors, diffusion sources to form shallow junctions, conductors, and to ensure ohmic contact to crystalline silicon. The polysilicon is deposited by pyrolyzing silane between 600 and 650°C in a low-pressure reactor (Fig. 1a). The chemical reaction is

$$SiH_4 \rightarrow Si + 2H_2 \tag{1}$$

Subsequent processing for polysilicon gates involves doping, etching, and oxidation. In some device structures a second polysilicon layer is deposited. This layer may be used as a contact material in small windows or as an interconnect between conducting features.

Two low-pressure processes are common for depositing polysilicon. One uses 100% silane at a pressure of 25 to 130 Pa (0.2 to 1.0 Torr). The other process is performed at the same total pressure but uses 20 to 30% silane diluted in nitrogen. Both processes deposit polysilicon on 100 to 200 wafers per run with thickness uniformities within 5%. The deposition rates are 100 to 200 Å/min.[7-9]

3.3.1 Deposition Variables

Temperature, pressure, silane concentration, and dopant concentration are important process variables in the deposition of polysilicon; wafer spacing and load size have only minor effects.[7, 10] Figure 3 shows that the deposition rate increases rapidly as the temperature increases. The activation energies, calculated from the slopes, are about 1.7 eV (40 kcal/mole), which is somewhat higher than the values observed for atmospheric-pressure depositions.[11] The difference is caused by changes in the desorption of the hydrogen produced in the reaction and by differences in the roles of mass transport and homogeneous reactions. Depositions at reduced pressure are limited to temperatures between 600 and 650°C. At higher temperatures, gas phase reactions, which result in a rough, loosely adhering deposit, and silane depletion, which causes poor uniformity, become significant.[12] At temperatures much lower than 600°C, the deposition rate is too slow to be practical.

Polysilicon depositions frequently use a temperature ramp with the rear furnace zone 5 to 15°C hotter than the front and center zones. The higher temperature increases the deposition rate, which compensates for the silane depletion. Under optimum conditions the increased deposition rate results in a uniform thickness throughout the deposition zone. However, the structure of polysilicon is strongly influenced by temperature, so a temperature ramp may cause a variation in structure and film properties.

Pressure can be varied in a low-pressure reactor by changing the gas flow into the reactor while keeping the pumping speed constant, or by changing the pumping speed at a constant inlet gas flow. If the inlet gas is a mixture of silane and nitrogen, the nitrogen flow can be changed while keeping the silane flow constant, or the silane and nitrogen can both be changed while keeping the ratio constant. All three methods,

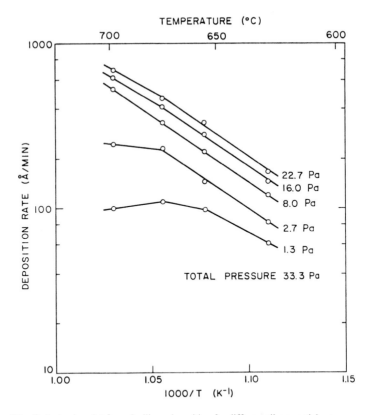

Fig. 3 Arrhenius plot for polysilicon deposition for different silane partial pressures.

changing pumping speed, changing nitrogen flow, or changing total gas flow with a constant ratio, are used to control the reactor pressure. If the total gas flow is varied (constant ratio and pump speed), the deposition rate is a linear function of pressure. But if the pumping speed or the nitrogen flow is changed, the rate only slightly depends on pressure (see Fig. 4). Deposition reproducibility is best when the inlet gas flows are kept constant and the pressure is controlled by the pumping speed.

The polysilicon deposition rate is usually not a linear function of the silane concentration.[10, 12] Figure 5 gives representative data for four deposition temperatures and for a total pressure of 33 Pa (0.25 Torr). The nonlinear behavior may be caused by mass transport effects, homogeneous reactions, or adsorbed hydrogen.[10–12] Gas phase nucleation occurs at high silane concentrations, thus imposing upper limits to the concentration and the deposition rate at a given temperature and pressure.

Polysilicon can be doped during deposition by adding phosphine, arsine, or diborane to the reactants. Figure 6 shows how the dopant affects the deposition rate. Adding diborane causes a large increase in the deposition rate. In contrast, adding phosphine or arsine causes a rapid decrease in the deposition rate. Similar effects have been observed for depositions at atmospheric pressure.[13] The thickness uniformity across a single wafer degrades when dopants are added. Uniformity can be maintained by using an insert to control the flow of reactant gases around the samples.

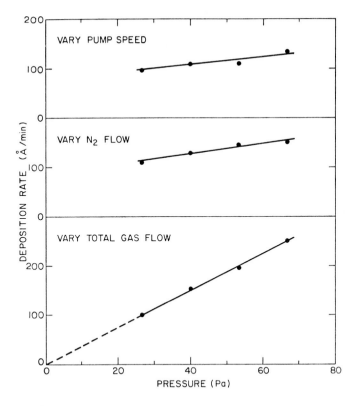

Fig. 4 The effect of total pressure on the polysilicon deposition rate.

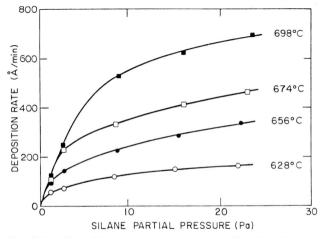

Fig. 5 The effect of silane concentration on the polysilicon deposition rate.

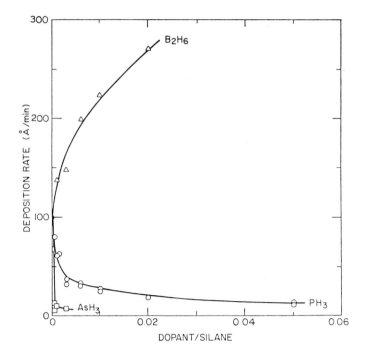

Fig. 6 The effect of dopants on the polysilicon deposition rate at 610°C.

3.3.2 Structure

The structure of polysilicon is strongly influenced by dopants or impurities, deposition temperature, and post-deposition heat cycles. Polysilicon deposited below 575°C is amorphous with no detectable structure.[14, 15] Polysilicon deposited above 625°C is polycrystalline and has a columnar structure. Crystallization and grain growth occur when either amorphous or columnar polysilicon is heated.[14, 15] Figure 7 illustrates all three structures, showing transmission electron microscope (TEM) cross sections of polysilicon deposited at 605°C (amorphous), 630°C (columnar), and annealed at 700°C (crystalline grains). After high-temperature heat cycles, there are no significant structural differences between polysilicon that is initially amorphous or columnar.

The deposition temperature at which the transition from amorphous to columnar structure occurs is well defined but depends on many variables, such as deposition rate, partial pressure of hydrogen, total pressure, presence of dopants, and presence of impurities (O, N, or C). The transition temperature is between 575 and 625°C for depositions in an LPCVD reactor.[14, 15] Polysilicon recrystallizes when heated; however, the crystallization temperature is also strongly influenced by dopants and impurities. Oxygen, nitrogen, and carbon impurities stabilize the amorphous structure to temperatures above 1000°C, and arsenic stabilizes the columnar structure to 900°C.

The average diameter of the column, that is, the columnar grains, can be measured by TEM surface replication. The diameter, which depends on film thickness, is typically between 0.03 and 0.3 μm and is often reported as grain size.[14-16] The grain size after crystallization depends on heating time, temperature, and dopant concentra-

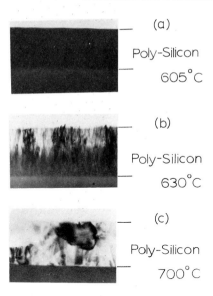

Fig. 7 TEM cross sections (60,000X) of polysilicon. (a) Amorphous structure deposited at 605°C. (b) Columnar structure deposited at 630°C. (c) Crystalline grains formed by annealing an amorphous sample at 700°C.

tion. Polysilicon doped with a high concentration of phosphorus and heated between 900 and 1000°C for 20 min has an average grain size of 1 μm.[16]

Polysilicon deposited at 600 to 650°C has a {110}-preferred orientation.[14-16] At higher deposition temperatures the {100} orientation predominates, but the structure contains significant contributions from other orientations, such as {110}, {111}, {311}, and {331}.[15] Dopants and impurities, as well as temperature, also influence the preferred orientation.

The structural changes in polysilicon during typical device processing can be summarized as follows. Polysilicon deposited between 600 and 650°C has a columnar structure with grain sizes between 0.03 and 0.3 μm and a {110}-preferred orientation. During phosphorus diffusion at 950°C the structure changes to crystallites with an average size of 0.5 to 1.0 μm. The grains grow during oxidation at 1050°C to a final size of 1 to 3 μm. Polysilicon deposited at temperatures below 600°C behaves similarly, except the initial film is amorphous.

3.3.3 Doping Polysilicon

Polysilicon can be doped by diffusion, implantation, or the addition of dopant gases during deposition (in-situ doping). All three methods are used for device fabrication. Figure 8 shows the resistivity of polysilicon doped with phosphorus by these three methods. The diffusion data, taken from Ref. 17, show the resistivity after a 1-h diffusion at the indicated temperature. The implantation data, taken from Ref. 18, show the resistivity after a 1-h, 1100°C activation. The resistivities for the in-situ doped samples are measured after deposition at 600°C and after a 30-min anneal at the indicated temperature. Diffusion is a high-temperature process that results in very low resistivities. The dopant concentration in diffused polysilicon often exceeds the solid

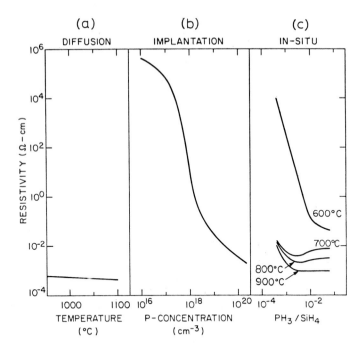

Fig. 8 Resistivity of P-doped polysilicon. (a) Diffusion. 1 h at the indicated temperature. *(After Kamins, Ref. 17.)* (b) Implantation. 1-h anneal at 1100°C. *(After Mandurah, Saraswat, and Kamins, Ref. 18.)* (c) In-situ. As-deposited at 600°C and after a 30-min anneal at the indicated temperature. *(After A. C. Adams, unpublished data.)*

solubility limit, with the excess dopant segregated at the grain boundaries.[19] A good correlation is found between the resistivity of diffused polysilicon and the dopant solubility.[17] Diffusion of dopants is faster in polysilicon than in single-crystal silicon; and lateral diffusion along a polysilicon film is faster than diffusion perpendicular to the surface.[20, 21] Hall mobilities for heavily diffused polysilicon are usually 30 to 40 cm^2/V-s.[17, 22, 23]

The resistivity of implanted polysilicon depends primarily on implant dose, annealing temperature, and annealing time.[18, 19, 24] The very high resistivity in lightly implanted polysilicon (Fig. 8) is caused by carrier traps at the grain boundaries.[18, 19, 24] Once these traps have been saturated with dopants, the resistivity decreases rapidly and approaches the resistivity for implanted single-crystal silicon.[18] The mobility for heavily implanted polysilicon is about 30 to 40 cm^2/V-s,[18] similar to the values for diffused polysilicon. Implanted polysilicon has about ten times higher resistivity than diffused polysilicon, because of the differences in dopant concentrations: approximately 10^{20} cm^{-3} for a heavy implant and greater than 10^{21} cm^{-3} for a heavy diffusion.

Polysilicon films that are doped during deposition by adding phosphine, arsine, or diborane have resistivities that are strong functions of deposition temperature,

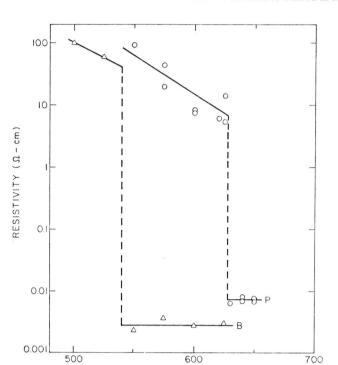

Fig. 9 Resistivity of in-situ doped polysilicon deposited at different temperatures. The triangles denote boron-doped polysilicon and the circles denote phosphorus-doped polysilicon.

dopant concentration, and annealing temperature. Figure 9 shows the resistivities for in-situ doped polysilicon deposited at different temperatures. The transition from high resistivity at low deposition temperature to low resistivity at high temperature corresponds to the change from an amorphous to columnar structure. The phosphorus-doped films in the figure (denoted by the circles) change structure at 625°C; the boron-doped polysilicon (denoted by the triangles) changes at temperatures between 525 and 550°C. The resistivity of doped amorphous polysilicon decreases during annealing, mainly because of crystallization (Fig. 8). After annealing the resistivity is not a strong function of the initial dopant concentration. Doped polysilicon that is crystalline when deposited shows almost no change in resistivity after annealing. The dopant concentration in in-situ doped polysilicon is high, 10^{20} to 10^{21} cm^{-3}, but the mobility is often low, 10 to 30 cm^2/V-s.[25] The low mobility gives a higher resistivity than expected for the high dopant concentration.

A comparison of the three doping processes shows that the major differences are lower resistivity for diffusion, lower dopant concentration for implantation, and lower mobility for in-situ doping. Implantation and in-situ doping, however, offer the advantage of lower processing temperatures, which is often the dominant consideration in VLSI processing.

3.3.4 Oxidation of Polysilicon

The details of polysilicon oxidation are discussed in Chapter 4. Polysilicon is usually oxidized in dry oxygen at temperatures between 900 and 1000°C to form an insulator between the doped-polysilicon gate and other conducting layers. Under these conditions, oxidation is controlled by surface reactions. Undoped or lightly doped silicon oxidizes at a rate between the rates for (111)- and (100)-crystalline silicon. Phosphorus-doped polysilicon oxidizes faster than undoped polysilicon, and the rate of oxidation is determined by the carrier concentration at the polysilicon surface. At very high phosphorus concentrations, the oxidation rate saturates, because the solubility limit of phosphorus in silicon has been reached.[26]

The silicon dioxide grown on polysilicon has lower breakdown fields, higher leakage currents, and higher stress than oxides grown on single-crystal silicon. The degraded oxide properties are related to the rough polysilicon-oxide interface, which is caused by different oxidation rates at the polysilicon grain boundaries.

3.3.5 Properties of Polysilicon

The chemical and physical properties of polysilicon often depend on the film structure (amorphous or crystalline) or on the dopant concentration. The etch rate of polysilicon in a plasma and its thermal oxidation rate depend on the dopant concentration. Polysilicon which is heavily phosphorus-doped etches and oxidizes at higher rates than undoped or lightly doped polysilicon. The reaction rates for oxidation and etching are determined by the free carrier concentration at the doped-polysilicon surface.

Polysilicon's optical properties depend on its structure. The imaginary part of the dielectric function is particularly structure-sensitive.[27] Crystalline polysilicon has sharp maxima in the dielectric function near 2950 and 3650 Å (4.2 and 3.4 eV). Amorphous polysilicon has a broad maximum without sharp structure. In addition, amorphous polysilicon has a higher refractive index throughout the visible region than crystalline polysilicon.[15, 28, 29]

Other reported properties of polysilicon are its density, 2.3 g/cm^3; coefficient of thermal expansion, $2 \times 10^{-6}/°C$; and temperature coefficient of resistance, $1 \times 10^{-3}/°C$. These are useful for modeling heat dissipation in devices.

3.4 SILICON DIOXIDE

Silicon dioxide films can be deposited with or without dopants. Undoped silicon dioxide is used as an insulating layer between multilevel metallizations, as an ion-implantation and diffusion mask, as a capping layer over doped regions to prevent outdiffusion during heat cycles, and to increase the thickness of field oxides. Phosphorus-doped silicon dioxide is used as an insulator between metal layers, as a final passivation over devices, and as a gettering source. Oxides doped with phosphorus, arsenic, or boron are occasionally used as diffusion sources. The deposition of oxide films has been reviewed.[30]

The processing sequence for silicon dioxide depends on its specific use in the device. Oxides used as insulators between conducting layers are deposited, densified by annealing, and plasma-etched to open windows. In the flowed glass process, phosphorus-doped silicon dioxide is heated to a temperature between 1000 and 1100°C so the oxide softens and flows, providing a smooth topography which improves the step coverage of the subsequent metallization. Phosphorus-doped oxides used for passivation are deposited at temperatures lower than 500°C, and areas for bonding are opened by etching.

3.4.1 Deposition Methods

Several deposition methods are used to produce silicon dioxide. They are characterized by different chemical reactions, reactors, and temperatures. Films deposited at low temperatures, lower than 500°C, are formed by reacting silane, dopant, and oxygen.[7, 30, 31] The chemical reactions for phosphorus-doped oxides are

$$SiH_4 + O_2 \rightarrow SiO_2 + 2H_2 \qquad (2)$$

$$4PH_3 + 5O_2 \rightarrow 2P_2O_5 + 6H_2 \qquad (3)$$

Under normal deposition conditions, hydrogen is formed rather than water. The deposition can be carried out at atmospheric pressure in a continuous reactor (Fig. 1b) or at reduced pressure in an LPCVD reactor (Fig. 1a). The main advantage of silane-oxygen reactions is the low deposition temperature, which allows films to be deposited over aluminum metallization. Consequently, these films can be used for passivation coatings over the final device and for insulation between aluminum levels. The main disadvantages of silane-oxygen reactions are poor step coverage and particles caused by loosely adhering deposits on the reactor walls.

Silicon dioxide is also deposited at 650 to 750°C in an LPCVD reactor by decomposing tetraethoxysilane, $Si(OC_2H_5)_4$.[32, 33] This compound, also called tetraethyl orthosilicate and abbreviated TEOS, is vaporized from a liquid source. The overall reaction is

$$Si(OC_2H_5)_4 \rightarrow SiO_2 + \text{by-products} \qquad (4)$$

where the by-products are a complex mixture of organic and organosilicon compounds. The decomposition of TEOS is useful for depositing insulators over polysilicon gates, but the high temperature required precludes its use over aluminum. The advantages of TEOS deposition are excellent uniformity, conformal step coverage, and good film properties. The disadvantages are the high-temperature and liquid source requirements.

Silicon dioxide is also deposited at temperatures near 900°C and at reduced pressure by reacting dichlorosilane with nitrous oxide[7, 8, 34]

$$SiCl_2H_2 + 2N_2O \rightarrow SiO_2 + 2N_2 + 2HCl \qquad (5)$$

This deposition, which gives excellent uniformity, is used to deposit insulating layers over polysilicon; however, this oxide frequently contains small amounts of chlorine which may react with the polysilicon or cause film cracking.[34]

Doping is achieved by adding small amounts of the dopant hydrides (phosphine, arsine, or diborane) during the deposition. Other dopant compounds, such as halides or organic compounds, can also be used, but they are not as convenient because they must usually be vaporized from solids or liquids.

Dopant concentrations are reported by weight percent (wt. %), atom percent (at. %), or mole percent (mol %). The relationships between these units for phosphorus-doped oxides are

$$\text{mole \% } P_2O_5 = \frac{6010 \ W}{6200 - 81.9 \ W} \tag{6}$$

$$\text{atom \% } P = \frac{12000 \ W}{18,600 - 81.9 \ W} \tag{7}$$

where W is weight percent of phosphorus. Occasionally weight percent, atom percent, and mole percent are written as w/o, a/o, and m/o.

The doped oxides used as diffusion sources contain 5 to 15 wt. % of the dopant. Doped oxides used for passivation or for interlevel insulation contain 2 to 8 wt. % phosphorus. Doped oxides used for the P-glass flow process (described in Section 3.4.4) contain 6 to 8 wt. % phosphorus. Glass with lower phosphorus concentrations will not soften and flow, and higher concentrations react slowly with atmospheric moisture to form acid products, which corrode the aluminum metallization.

3.4.2 Deposition Variables

The deposition of silicon dioxide depends on the same variables that are important for polysilicon, that is, temperature, pressure, reactant concentration, and presence of dopants. In addition, other variables, such as wafer spacing and total gas flow, are important for some silicon dioxide depositions. Deposition variables for the reaction of silane with oxygen at atmospheric pressure have been reviewed.[30] The deposition rate increases with temperature, but the apparent activation energy is very low, less than 0.4 eV (10 kcal/mole). This activation energy is much less than the values usually observed for chemical reactions and is similar to the values for adsorption on a surface or for gas phase diffusion. The deposition has a complicated dependence on oxygen concentration. Namely, if the oxygen concentration is varied at a constant temperature, the deposition rate increases rapidly, goes through a maximum, and then slowly decreases. Figure 10 gives representative data, compiled from Ref. 35. This relation has been explained by assuming surface-catalyzed reactions.[35] At high concentrations the oxygen adsorbs on the surface and blocks further silane reactions. When phosphine is added to the reaction, the rate rapidly decreases and then slowly increases.[36] This deposition behavior may also be attributable to surface adsorption effects.

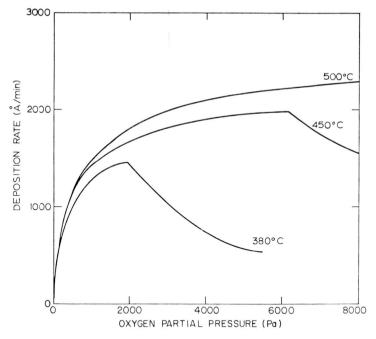

Fig. 10 The deposition rate of silicon dioxide at atmospheric pressure for different oxygen concentrations. *(After Maeda and Nakamura, Ref. 35.)*

The reaction between silane and oxygen at reduced pressure follows similar trends. The activation energy is very low, less than 0.4 eV (10 kcal/mole).[31] The deposition rate is a linear function of the silane partial pressure. At high partial pressures of silane the deposited silicon dioxide is hazy, probably because of gas-phase reactions.[7, 31] The deposition rate goes through a maximum as oxygen partial pressure is varied, which is similar to the result observed at atmospheric pressure.[37] The gas-phase transport of material to the wafer surface is very important. A special sample holder which directs the gas to the wafers is required for uniform depositions.[31, 37] The deposition rate also depends on wafer spacing.[7, 31]

The deposition of silicon dioxide by decomposing TEOS occurs at temperatures between 650 and 750°C. Figure 11 shows deposition rate as a function of temperature for the TEOS decomposition (from Ref. 33) and for the silane-oxygen reaction (from Ref. 31). The activation energy for the TEOS reaction is about 1.9 eV (45 kcal/mole), which decreases to 1.4 eV (32 kcal/mole) when phosphorus doping compounds are present.[33] Note the contrast between these activation energies and the very low activation energies required in silane-oxygen reactions. Figure 12 shows how the deposition rate depends on the TEOS partial pressure. The data points are taken from Ref. 33 and the solid line is from Ref. 32. The nonlinear behavior, which is similar to polysilicon deposition, has been explained by assuming surface catalyzed reactions.[32] At low TEOS partial pressures the deposition rate is determined by the rate of the sur-

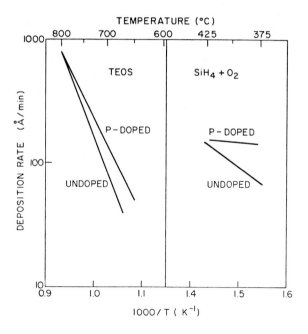

Fig. 11 Arrhenius plots for the low-pressure deposition of SiO_2. *(After Adams and Capio, Ref. 33, for the TEOS data and after Logar, Wauk, and Rosler, Ref. 31, for the silane-oxygen data.)*

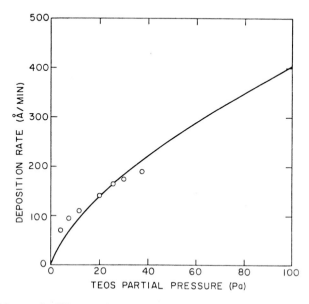

Fig. 12 Deposition rate for different TEOS concentrations. *(After Adams and Capio, Ref. 33, for the data points and after Huppertz and Engl, Ref. 32, for the solid line.)*

face reaction. At very high partial pressures, the surface becomes nearly saturated with adsorbed TEOS, and the deposition rate becomes independent of the TEOS pressure. The TEOS deposition also depends on the total pressure; however, this pressure dependence has not been adequately explained.[32]

The deposition of silicon dioxide at 900°C using dichlorosilane and nitrous oxide has a strong nonlinear pressure dependence, which is a function of wafer position in the reactor. Gas transport and depletion are significant in this deposition.[7, 8]

Phosphorus-doped oxides are deposited by adding phosphorus compounds, usually phosphine, to the silane-oxygen or TEOS reaction. Doping is difficult with the dichlorosilane-nitrous oxide reaction because of the high deposition temperature. Adding phosphorus to the low-pressure depositions causes the thickness uniformity to degrade. The deposition of phosphorus-doped silicon dioxide requires inserts, which ensure uniform gas flow over the wafer surfaces.

3.4.3 Step Coverage

Three general types of step coverage are observed for deposited silicon dioxide. They are schematically diagrammed in Fig. 13. Figure 13a shows a completely conformal step coverage; the film thickness along the walls is the same as the film thickness at

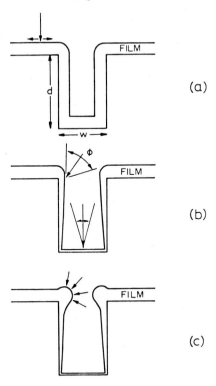

Fig. 13 Step coverage of deposited films. (a) Conformal coverage resulting from rapid surface migration. (b) Nonconformal step coverage for long mean-free path and no surface migration. (c) Nonconformal step coverage for short mean-free path and no surface migration.

the bottom of the step. Conformal step coverage results when reactants or reactive intermediates adsorb on the surface and then rapidly migrate along the surface before reacting. The rapid migration results in a uniform surface concentration, regardless of the topography, and gives a completely uniform thickness.

When the reactants adsorb and react without significant surface migration, the deposition rate is proportional to the arrival angle of the gas molecules. Figure 13b gives an example where the mean-free path of the gas is much larger than the dimensions of the step. The arrival angle in two dimensions at the top horizontal surface is 180°. At the top of the vertical surface, the arrival angle is only 90° so the film thickness is reduced by half. Along the vertical walls the arrival angle, ϕ, is determined by the width of the opening, and the film thickness, which is proportional to the arrival angle, can be calculated from

$$\phi = \arctan\frac{w}{d} \qquad (8)$$

where w is the width of the opening and d is the distance from the top surface. This type of step coverage is thin along the vertical walls and may have a crack at the bottom of the step caused by self-shadowing.

Figure 13c gives a diagram for no surface migration and for a short mean-free path. Here the arrival angle at the top of the step is 270°, giving a thicker deposit. The arrival angle at the bottom of the step is only 90° and the film is very thin. Gas depletion effects are also observed along the step walls. The thick cusp at the top of the step and the thin crevice at the bottom combine to give a concave shape which is particularly difficult to cover with metal.

Figure 14 gives actual examples of the different types of step coverage. The samples are prepared by etching (110) single-crystal silicon in hot potassium hydroxide to form vertical grooves 5 μm wide and 50 μm deep. Approximately 1 μm of oxide is deposited. The samples are cleaved and a cross section examined to determine the step coverage. A nearly conformal coverage is observed for the TEOS deposition at reduced pressure (Fig. 14a). The mean-free path at the deposition conditions (700°C and 30 Pa) is several hundred micrometers, much larger than the dimensions of the groove. Consequently, gas-phase diffusion into the groove is negligible. However, surface migration is very rapid, resulting in the conformal coverage.

Figure 14b shows silicon dioxide deposited from silane and oxygen at reduced pressure. The mean-free path is still large, several hundred microns, but no surface migration takes place, and the step coverage is determined by the arrival angle. Silicon dioxide deposited at atmospheric pressure by reacting silane and oxygen builds up at the top of the step because of the very short mean-free path at atmospheric pressure (less than 0.1 μm). Figure 14c shows this step coverage. The nonconformal step coverage shown in Figs. 14b and c causes metallization failures because of the concave shape. The region at the bottom of the step often etches rapidly, causing additional serious problems in subsequent processing.

Other materials besides deposited silicon dioxide have the types of step coverage shown in Figs. 13 and 14. Most evaporated or sputtered metals have step coverage similar to that shown in Fig. 14b. Chemically deposited polysilicon and silicon nitride have conformal coverage. Plasma-deposited silicon dioxide is similar to

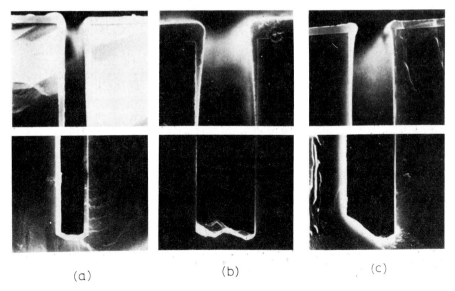

(a)   (b)   (c)

Fig. 14 SEM cross sections (5000X) showing step coverage of deposited oxides. (a) TEOS deposition at 700°C. (b) Silane-oxygen reaction at 450°C and reduced pressure. (c) Silane-oxygen reaction at 480°C and atmospheric pressure.

Fig. 14b, and plasma-deposited silicon nitride is intermediate between Figs. 14a and b. In this intermediate case the film is thin along the vertical walls, but somewhat thicker than expected for no surface migration.

3.4.4 P-Glass Flow

Phosphorus-doped silicon dioxide is frequently used as an insulator between polysilicon gates and the top level metallization. A concave shape in the oxide going over the polysilicon gate can cause an opening in the metal film, resulting in device failure. The poor step coverage of the phosphorus-doped silicon dioxide can be corrected by heating the samples until the oxide softens and flows. This process is called P-glass flow.

P-glass flow is illustrated in the scanning electron microscope (SEM) photographs in Figs. 15 and 16. Figure 15 shows a polysilicon line crossing an oxide step with the entire surface covered with 4.6 wt. % P-glass. The samples have been heated in steam at 1100°C for four different lengths of time between 0 (Fig. 15a) and 60 min (Fig. 15d). Flow is indicated by the progressive loss of detail. The SEM cross sections in Fig. 16 show P-glass covering polysilicon. The samples contain between 0 and 7.2 wt. % phosphorus and have been heated in steam at 1100°C for 20 min. Samples with no phosphorus do not flow (Fig. 16a). The concave shape, thick at the top and thin at the bottom, is easily seen. As the phosphorus concentration in the oxide increases, flow increases, decreasing the angle made by the P-glass going over the step. As these figures demonstrate, P-glass flow is a time-dependent phenomenon.

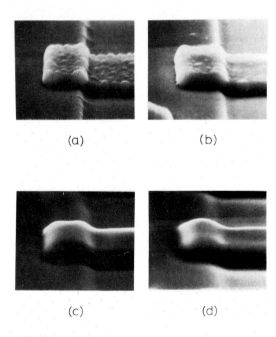

(a)   (b)

(c)   (d)

Fig. 15 SEM photographs (3200X) showing surfaces of 4.6 wt. % P-glass annealed in steam at 1100°C for the following times: (a) 0 min; (b) 20 min; (c) 40 min; (d) 60 min. *(After Adams and Capio, Ref. 38. Reprinted by permission of the publisher, The Electrochemical Society, Inc.)*

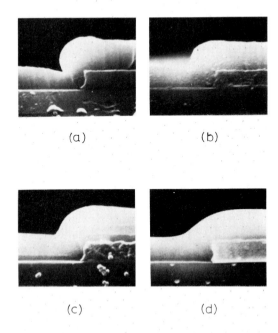

(a)   (b)

(c)   (d)

Fig. 16 SEM cross-sections (10,000X) of samples annealed in steam at 1100°C for 20 min for the following weight percent of phosphorus: (a) 0.0 wt. % P; (b) 2.2 wt. % P; (c) 4.6 wt. % P; (d) 7.2 wt. % P. *(After Adams and Capio, Ref. 38. Reprinted by permission of the publisher, The Electrochemical Society, Inc.)*

Samples usually do not reach an equilibrium state during flow. Flow depends on several variables: annealing time, temperature, rate of heating, phosphorus concentration, and annealing ambient.

Figure 17, which summarizes many of these effects, shows the angle made by the P-glass going over a step after different flow treatments and for different phosphorus concentrations. As deposited the steps are concave with 120° angles. Flow is measured by the decrease in the angle. P-glass flow is greatest for high phosphorus concentrations, steam ambient, and high temperatures.[38]

The P-glass flow process requires temperatures to be as high as 1000 to 1100°C. It also requires phosphorus concentrations of 6 to 8 wt. %. Less concentrated glasses do not flow readily. Phosphorus concentrations greater than 8 wt. % may cause corrosion of the aluminum metallization by the acid products formed from the reaction between the phosphorus in the oxide and atmospheric moisture.

3.4.5 Properties of Silicon Dioxide

Table 3 summarizes properties of silicon dioxide deposited by different techniques, including the plasma-assisted deposition of silicon dioxide. In general, oxides deposited at higher temperatures resemble thermally grown silicon dioxide. However, high-temperature oxides can not be deposited over aluminum and therefore can not be

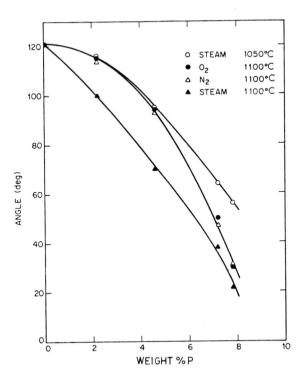

Fig. 17 Step angles made by P-glass after different flow treatments. *(After Adams and Capio, Ref. 38. Reprinted by permission of the publisher, The Electrochemical Society, Inc.)*

Table 3 Properties of deposited silicon dioxide

Deposition	Plasma	$SiH_4 + O_2$	TEOS	$SiCl_2H_2 + N_2O$
Temperature (°C)	200	450	700	900
Composition	$SiO_{1.9}(H)$	$SiO_2(H)$	SiO_2	$SiO_2(Cl)$
Step coverage	Nonconformal	Nonconformal	Conformal	Conformal
Thermal stability	Looses H	Densifies	Stable	Looses Cl
Density (g/cm³)	2.3	2.1	2.2	2.2
Refractive index	1.47	1.44	1.46	1.46
Stress (10^9 dyn/cm²)	3C−3T	3T	1C	3C
Dielectric strength 10^6 V/cm	3−6	8	10	10
Etch rate (Å/min) (100:1 H₂O:HF)	400	60	30	30

used for the final device passivation. Consequently, the low-temperature, phosphorus-doped oxides are used for final passivation in spite of their poor step coverage and somewhat inferior film properties.

Composition Silicon dioxide deposited at low temperatures, (400−500°C) contains hydrogen. This hydrogen is bonded within the silicon-oxygen network as silanol (Si—OH), hydride (Si—H), or water (H₂O). The bonded hydrogen can be observed by infrared spectroscopy.[39] Silicon dioxide deposited between 400 and 500°C typically contains 1 to 4 wt. % SiOH and less than 0.5 wt. % SiH. The amount of water in the film, which depends on the deposition temperature, increases with exposure to atmospheric moisture. Silicon dioxide films deposited at 700°C by TEOS decomposition, or at 900°C by the dichlorosilane-nitrous oxide reaction, do not contain hydrogen that is detectable by infrared absorption. The films formed from dichlorosilane, however, contain chlorine.[34] The chlorine can react with the silicon substrate or evolve from the film during high-temperature anneals.

Phosphorus concentrations in doped silicon dioxide can be measured by infrared absorption, neutron activation, x-ray emission spectroscopy, sheet resistance of diffused layers, etch-rate variation, the refractive index, or an electron microprobe. Several of these techniques have been compared.[40, 41] Figure 18 gives curves relating sheet resistance and infrared absorption to the phosphorus concentration. For the sheet-resistance method, the phosphorus-doped oxide is deposited on a lightly doped p-type substrate. After deposition the sample is heated at 1100°C for 20 min, the oxide film is removed by etching, and the sheet resistance of the n-type diffused layer is measured. This method is useful in a processing facility where furnace and etching operations are available; however, if the diffusion is performed at a different temperature or time, the calibration curve in the figure is displaced. The infrared technique is convenient in a laboratory environment. It requires measuring the ratio of the P—O absorption at 1325 cm^{-1} and the Si—O absorption at 805 cm^{-1}. Concentrations of

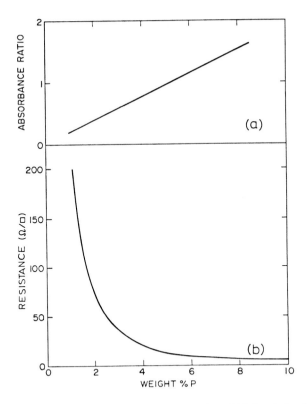

Fig. 18 Calibration curves for measuring phosphorus concentration. (a) Infrared absorbance ratio. *(After R. M. Levin and A. C. Adams, unpublished data.)* (b) Sheet resistance. *(After Adams and Murarka, Ref. 40.)*

other dopants, such as boron and arsenic, can also be measured by sheet resistance or infrared absorption.

Thickness Film thickness can be measured by a stylus instrument, reflectance spectroscopy, ellipsometry, or a prism coupler. Automated instruments, suitable for routine use, are available for all these techniques. Figure 19 compares them and shows that they all attain similar accuracy and precision.[42] While all four techniques are generally suitable for measuring silicon dioxide films, they each have specific limitations. Stylus measurement requires etching a step or masking part of the substrate during deposition. Prism coupling can not be used on oxide films that are less than 4000 Å thick. Ellipsometry requires the oxide thickness to be known to within 2500 Å, since the measured ellipsometric quantities are periodic functions of the thickness. Reflectance spectroscopy requires empirical calibration or accurate values for the film refractive index.

Structure Deposited silicon dioxide has an amorphous structure consisting of SiO_4 tetrahedra. Its structure is similar to that of fused silica. The film density ranges

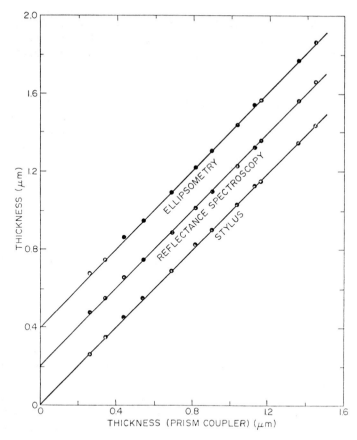

Fig. 19 Correlation plot showing oxide thickness measured by four techniques. The points have been separated by adding 0.2 μm to the reflectance spectroscopy thickness and 0.4 μm to the ellipsometry thickness. *(After Adams, Schinke, and Capio, Ref. 42. Reprinted by permission of the publisher, The Electrochemical Society, Inc.)*

between 2.0 and 2.2 g/cm^3. The lower densities occur in films deposited below 500°C. Heating deposited silicon dioxide at temperatures between 600 and 1000°C causes densification; the oxide thickness decreases and the density increases to 2.2 g/cm^3. During densification the amorphous structure is maintained; however, the arrangement of the SiO$_4$ tetrahedra becomes more regular.[43] Densification causes deposited silicon dioxide to take on many of the characteristic properties of thermally grown oxides.

Reactivity Silicon dioxide deposited at a low temperature reacts with atmospheric moisture, especially if the oxide contains phosphorus. The phosphorus-oxygen double bond undergoes a reversible hydrolysis. This effect can be minimized by densification at 800 to 900°C.

The etch rates of deposited oxides in a hydrofluoric acid solution depend on deposition temperature, annealing history, and dopant concentration.[39] These etch rates are important because solutions containing fluoride are frequently used for cleaning. An etchant containing nitric acid, hydrofluoric acid, and water is useful for evaluating and comparing deposited oxides. Etch rates in this solution (often called P-etch) are sensitive to film density, porosity, and composition.[39]

Refractive index and stress The refractive index of silicon dioxide is 1.458 at a wavelength of 0.6328 μm. Deposited oxides with refractive indices above 1.46 are usually silicon-rich. Oxides with lower indices are porous. An example is the oxide from the silane-oxygen deposition, which has a refractive index of about 1.44.

Stress in silicon dioxide depends on deposition temperature, deposition rate, annealing treatments, dopant concentration, water content, and film porosity. Undoped silicon dioxide deposited at a temperature between 400 and 500°C usually has a tensile stress of 1 to 4×10^9 dyn/cm^2. Undoped oxides deposited at a temperature between 650 and 750°C have a very low compressive stress, 0 to 1×10^9 dyn/cm^2, and oxides deposited at 900°C have a slightly higher compressive stress, 2 to 3×10^9 dyn/cm^2. The stress is usually more compressive when phosphorus is added.

3.5 SILICON NITRIDE

Stoichiometric silicon nitride (Si_3N_4) is used for passivating silicon devices, because it serves as an extremely good barrier to the diffusion of water and sodium. These impurities cause devices to corrode or become unstable. Silicon nitride is also used as a mask for the selective oxidation of silicon. The silicon nitride is patterned and the exposed silicon substrate is oxidized. The silicon nitride oxidizes very slowly and prevents the underlying silicon from oxidizing. This process of selective oxidation is used to produce nearly planar device structures.[44]

Silicon nitride is chemically deposited by reacting silane and ammonia at atmospheric pressure at temperatures between 700 and 900°C or by reacting dichlorosilane and ammonia at reduced pressure at temperatures between 700 and 800°C. The chemical reactions are

$$3SiH_4 + 4NH_3 \rightarrow Si_3N_4 + 12H_2 \tag{9}$$

$$3SiCl_2H_2 + 4NH_3 \rightarrow Si_3N_4 + 6HCl + 6H_2 \tag{10}$$

The reduced-pressure technique has the advantage of very good uniformity and high wafer throughput.[7, 8, 30] Thermal growth of silicon nitride by exposing silicon to ammonia at temperatures between 1000 and 1100°C has been investigated. The resulting films contain oxygen and are very thin.

3.5.1 Deposition Variables

Silicon nitride depositions at reduced pressure are controlled by temperature, total pressure, reactant concentrations, and temperature gradients in the furnace. The temperature dependence of the deposition rate is similar to that of polysilicon. The activation energy for the silicon nitride deposition is about 1.8 eV (41 kcal/mole). The deposition rate increases with increasing total pressure or dichlorosilane partial pressure, and decreases with an increasing ammonia to dichlorosilane ratio. A temperature ramp, with the furnace tube hotter at the exhaust end, is required for uniform depositions. (See Section 3.3.1 for a discussion of temperature ramps.)

3.5.2 Properties of Silicon Nitride

Silicon nitride, chemically deposited at temperatures between 700 and 900°C, is an amorphous dielectric containing up to 8 at. % hydrogen.[45] The hydrogen is bonded to the nitrogen and to the silicon. The amount of bonded hydrogen depends on the deposition temperature and on the ratio of reactants. More hydrogen is incorporated at low deposition temperatures or at high ammonia to dichlorosilane ratios. Silicon nitride deposited at low ammonia to dichlorosilane ratios contains excess silicon, which decreases the electrical resistivity.

Silicon nitride has a refractive index of 2.01 and an etch rate in buffered hydrofluoric acid of less than 10 Å/min. Both measurements are used to check the quality of deposited nitrides. High refractive indices indicate a silicon-rich film; low indices are caused by oxygen impurities. Oxygen impurities in the film also cause a higher etch rate. Silicon nitride has a very high tensile stress, about 1×10^{10} dyn/cm^2. Films thicker than 2000 Å sometimes crack because of the very high stress.

The resistivity of silicon nitride at room temperature is about 10^{16} Ω-cm. The electrical conduction depends on the deposition temperature, ratio of reactants, amount of the hydrogen in the film, and presence of oxygen impurities.

Silicon nitride is an excellent barrier to sodium diffusion. Its effectiveness is usually tested by evaporating radioactive sodium chloride ($Na^{22}Cl$) on the silicon nitride and then heating the samples at 600°C for 22 h. The sodium is counted as the silicon nitride is removed by step etching. Typically less than 10% of the original sodium diffuses more than 50 Å into the film.[46]

Table 4 summarizes the properties of silicon nitride and plasma-deposited nitride.

3.6 PLASMA-ASSISTED DEPOSITIONS

Plasma-assisted depositions provide films at very low sample temperatures. They do this by reacting the gases in a glow discharge, which supplies much of the energy for the reaction. Although the electron temperature in the discharge may be near 10^5 °C, the sample temperature is between 100 and 400°C. This technique, often referred to as plasma deposition, has been thoroughly reviewed.[4, 30, 47]

A large number of inorganic and organic materials have been deposited by plasma deposition but only two are useful in VLSI technology: plasma-deposited silicon nitride (SiN) and plasma-deposited silicon dioxide. Plasma-deposited silicon nitride is used as the encapsulating material for the final passivation of devices. The plasma-deposited nitride provides excellent scratch protection, serves as a moisture barrier, and prevents sodium diffusion. Because of the low deposition temperature,

Table 4 Properties of silicon nitride

Deposition	LPCVD	Plasma
Temperature (°C)	700−800	250−350
Composition	$Si_3N_4(H)$	SiN_xH_y
GSi/N ratio	0.75	0.8−1.2
At. % H	4−8	20−25
Refractive index	2.01	1.8−2.5
Density (g/cm^3)	2.9−3.1	2.4−2.8
Dielectric constant	6−7	6−9
Resistivity (Ω-cm)	10^{16}	$10^6−10^{15}$
Dielectric strength (10^6 V/cm)	10	5
Energy gap (eV)	5	4−5
Stress (10^9 dyn/cm^2)	10T	2C−5T

300 to 350°C, the nitride can be deposited over the final device. Plasma-deposited nitride and oxide are both used as insulators between metallization levels. They are particularly useful when the bottom metal level is aluminum or gold.

3.6.1 Deposition Variables

Silicon dioxide films are deposited by reacting silane and nitrous oxide in an argon plasma. Silicon nitride is formed by reacting silane and ammonia in an argon plasma or by reacting silane in a nitrogen discharge. The reactions are often assumed to be

$$SiH_4 + 4N_2O \rightarrow SiO_2 + 4N_2 + 2H_2O \tag{11}$$

$$SiH_4 + NH_3 \rightarrow SiNH + 3H_2 \tag{12}$$

$$2SiH_4 + N_2 \rightarrow 2SiNH + 3H_2 \tag{13}$$

However, the products depend strongly on the deposition conditions. The radial-flow, parallel-plate reactor (Fig. 2a) and the hot-wall plasma reactor (Fig. 2b) are commonly used for device processing.

Many variables must be controlled during a plasma deposition, such as frequency, electrode spacing, power, total pressure, reactant partial pressures, pumping speed, sample temperature, electrode materials, and reactor geometry.[48] Some variables have a predictable effect on the deposition. For instance, the deposition rate generally increases with increasing temperature, power, or reactant pressure. In many cases, however, variables interact so measuring and interpreting the effect of a specific variable becomes difficult. In other cases variables affect the deposition and film properties, but the effects are difficult to explain. For instance, silicon nitride that has been plasma-deposited at a frequency of 13.56 MHz has a tensile stress of about 4×10^9 dyn/cm^2, whereas a similar film deposited at a frequency of 50 kHz has a compressive stress of 2×10^9 dyn/cm^2. The strong dependence on deposition conditions makes it very difficult to compare films from different reactors. All deposition conditions must be carefully specified when discussing the properties of plasma-deposited films.

3.6.2 Properties of Plasma-Deposited Films

Plasma-deposited films contain large hydrogen concentrations, which depend on the deposition conditions.[49-51] Plasma silicon nitride may contain between 10 and 35 at. % hydrogen; however, most of the plasma nitride used in semiconductor processing contains 20 to 25 at. % hydrogen. The hydrogen is bonded to the silicon as Si—H and to the nitrogen as N—H.[45, 51] The plasma silicon nitride often contains 0.5 to 2.0 at. % oxygen as an impurity. Figure 20, compiled from Ref. 49, shows how the plasma nitride composition varies with different deposition conditions. The relative concentrations of Si—H and N—H change by large amounts, but the total hydrogen concentration remains nearly constant except at low temperatures. The silicon to nitrogen ratio, which varies between 0.7 and 1.7, also strongly depends on the deposition conditions.

Figure 21 gives similar data for the hydrogen concentration in plasma-deposited silicon dioxide. The hydrogen is bonded to silicon as Si—H and to oxygen as Si—OH

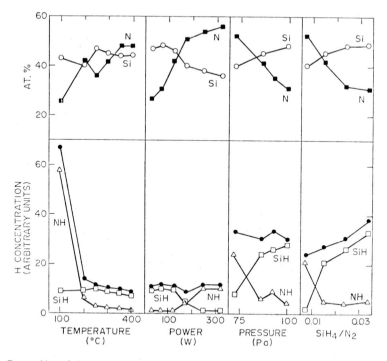

Fig. 20 Composition of plasma nitride for different deposition conditions: solid circles denote total H, open squares denote SiH, open triangles denote NH, open circles denote Si, and closed squares denote N. *(After Dun et al., Ref. 49.)*

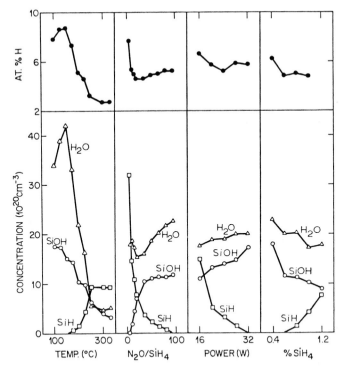

Fig. 21 The concentration of hydrogen groups and the total at. % H in plasma SiO_2 for different deposition conditions: triangles denote H_2O, circles denote SiOH, squares denote SiH, and closed circles denote total hydrogen. *(After Adams et al., Ref. 50.)*

and H_2O. The relative concentration of hydrogen in the three bonding sites strongly depends on deposition conditions; however, the total hydrogen only varies between 2 and 9 at. %. The data in Figs. 20 and 21 show that the composition of the plasma-deposited films depends on the specific deposition conditions. The subsequent variations in composition cause large changes in the film properties.

Stress is one of the most important properties of plasma silicon nitride, since high stress can cause cracking during bonding operations. Films with low tensile stress, about 2×10^9 dyn/cm^2, can be prepared, but the stress depends on nearly every deposition variable. Plasma nitride films deposited by reacting silane in a nitrogen plasma are more compressive than plasma nitride produced from silane and ammonia. In addition, films deposited at low frequencies are compressive, rather than tensile.

Plasma-deposited silicon nitride has a large range of resistivities (10^5 to 10^{21} Ω-cm) and of breakdown fields (1 to 6×10^6 V/cm). Figure 22 shows resistivity data for plasma nitride, taken from Refs. 49 and 52. The correlation between resistivity and film composition is excellent, even over resistivity changes of many orders of magnitude. Correlations of dielectric breakdown field with film composition and deposition conditions have also been made for plasma nitride and oxide.[49, 50, 52] Tables 3 and 4 list general properties of plasma-deposited silicon nitride and silicon dioxide.

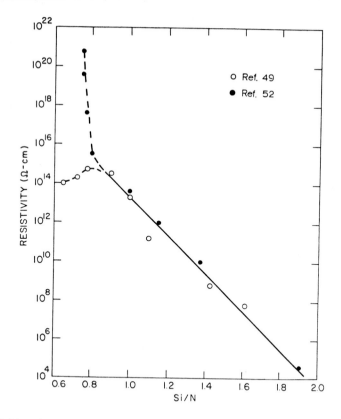

Fig. 22 Resistivity of plasma silicon nitride. *(After Dun et al., Ref. 49, and after Sinha and Smith, Ref. 52.)*

3.7 OTHER MATERIALS

Several insulating materials have been investigated for IC applications, primarily for passivation or for dual dielectric MOS devices. Silicon oxynitride is deposited by reacting silane, nitric oxide, and ammonia or by reacting silane, carbon dioxide, ammonia, and hydrogen.[53, 54] By adjusting the deposition conditions, any film composition between SiO_2 and Si_3N_4 can be obtained. Since silicon dioxide has a compressive stress and silicon nitride is in tension, they form an intermediate composition of silicon oxynitride with zero stress. This composition is useful for passivation in some applications.

Boro-phosphosilicate glass and lead silicate glass may also be useful for passivation. Both can be deposited at low temperatures (300 to 500°C), and both soften and flow at temperatures below 1000°C. Aluminum oxide, aluminum nitride, and titanium oxide have been evaluated as dielectrics for MOS applications. These films, which are chemically deposited between 800 and 1100°C, have high resistivities, high dielectric constants, and high breakdown fields.

Semi-insulating polysilicon (SIPOS) is deposited between 600 and 700°C by reacting silane and nitrous oxide. The deposited film, which contains 20 to 40 at. % oxygen, may be a multiphase mixture containing amorphous silicon, crystalline silicon, silicon dioxide, and silicon monoxide. This material is useful for passivation.

Various organic compounds, usually polyimides, have been used as insulators between metal levels. These compounds are applied by spinning and then are baked above their softening temperature. This process produces a planar surface that is ideal for metallization. The organic compounds have limited thermal stability and are very porous to moisture penetration.

3.8 SUMMARY AND FUTURE TRENDS

Table 5 summarizes current techniques for depositing dielectric and polysilicon films. The low-temperature processes for depositing P-glass and SiN are particularly attractive for passivation, since the films can be deposited over aluminum or gold metallization. The poor step coverage, however, is a severe disadvantage if these processes are used to deposit an insulating film between conducting layers. A higher temperature process (500−900°C), with conformal step coverage is generally much better. Other deposition methods for dielectric and polysilicon films are available, such as evaporation, sputtering, anodization, and molecular beam techniques, but they are not widely used for VLSI processing. Their major problems include defects caused by excessive wafer handling, low throughput, poor step coverage, and nonuniform depositions over many wafers.

Table 5 Comparison of different deposition methods

Deposition properties	Methods			
	Atmospheric-pressure CVD	Low-temperature LPCVD	Medium-temperature LPCVD	Plasma-assisted CVD
Temperature (°C)	300−500	300−500	500−900	100−350
Materials	SiO_2 P-glass	SiO_2 P-glass	Poly-Si SiO_2 P-glass Si_3N_4	SiN SiO_2
Uses	Passivation, insulation	Passivation, insulation	Gate metal, insulation, passivation	Passivation, insulation
Throughput	High	High	High	Low
Step coverage	Poor	Poor	Conformal	Poor
Particles	Many	Few	Few	Many
Film properties	Good	Good	Excellent	Poor
Low temperature	Yes	Yes	No	Yes

VLSI devices with very small dimensions require precise lithography, pattern transfer with anisotropic etching, and very shallow junctions. These conditions impose new requirements on the film deposition process. The major requirements are low processing temperatures to prevent movement of the shallow junctions, conformal step coverage over the anisotropically etched features, low process-induced defects (mainly particles generated during wafer handling and loading), and high wafer throughput to reduce cost. These requirements are met by hot-wall, low-pressure depositions (chemical or plasma). The reactors for this type of deposition are easily scaled to accommodate 125- or 150-mm wafers. In contrast, atmospheric-pressure depositions and physical-deposition techniques are much more difficult to scale and do not have the high throughput or the low defect densities. Consequently these techniques are being replaced by LPCVD and plasma-assisted depositions as critical device dimensions decrease and wafer size increases.

Low-temperature depositions will continue to increase in importance, because the maximum processing temperature for devices with shallow junctions is about 900 to 950°C. Depositions at very low temperatures, 30 to 200°C, have been investigated and they will probably find applications in new device technologies. These low-temperature techniques include plasma-assisted depositions of organosilicon compounds and photo-induced depositions of silicon dioxide and silicon nitride. The photo-induced reactions occur at about 100°C and introduce almost no radiation damage in devices.

REFERENCES

[1] W. Kern and V. S. Ban, "Chemical Vapor Deposition of Inorganic Thin Films," in J. L. Vossen and W. Kern, Eds., *Thin Film Processes*, Academic, New York, 1978, pp. 257–331.

[2] W. Kern and G. L. Schnable, "Low-Pressure Chemical Vapor Deposition for Very Large-Scale Integration Processing—A Review," *IEEE Trans. Electron Devices*, ED-26, 647 (1979).

[3] E. C. Douglas, "Advanced Process Technology for VLSI Circuits," *Sold State Technol.*, 24, 65 (May 1981).

[4] J. L. Vossen and W. Kern, "Thin-Film Formation," *Phys. Today*, 33, 26 (May 1980).

[5] M. L. Hammond, "Introduction to Chemical Vapor Deposition," *Solid State Technol.*, 22, 61 (December 1979).

[6] M. L. Hammond, "Safety in Chemical Vapor Deposition," *Solid State Technol.*, 23, 104 (December 1980).

[7] R. S. Rosler, "Low Pressure CVD Production Processes for Poly, Nitride, and Oxide," *Solid State Technol.*, 20, 63 (April 1977).

[8] W. A. Brown and T. I. Kamins, "An Analysis of LPCVD System Parameters for Polysilicon, Silicon Nitride and Silicon Dioxide Deposition," *Solid State Technol.*, 22, 51 (July 1979).

[9] R. J. Gieske, J. J. McMullen, and L. F. Donaghey, "Low Pressure Chemical Vapor Deposition of Polysilicon," in L. F. Donaghey, P. Rai-Choudhury, and R. N. Tauber, Eds., *Chemical Vapor Deposition—Sixth International Conference*, Electrochemical Society, Princeton, N.J., 1977, pp. 183–194.

[10] M. L. Hitchman, "Kinetics and Mechanism of Low Pressure CVD of Polysilicon," in T. O. Sedgwick and H. Lydtin, Eds., *Chemical Vapor Deposition—Seventh International Conference*, Electrochemical Society, Princeton, N.J., 1979, pp. 59–76.

[11] W. A. Bryant, "The Kinetics of the Deposition of Silicon by Silane Pyrolysis at Low Temperatures and Atmospheric Pressure," *Thin Solid Films*, 60, 19 (1979).

[12] C. H. J. Van Den Brekel and L. J. M. Bollen, "Low Pressure Deposition of Polycrystalline Silicon from Silane," *J. Cryst. Growth*, 54, 310 (1981).

[13] F. C. Eversteyn and B. H. Put, "Influence of AsH_3, PH_3, and B_2H_6 on the Growth Rate and Resistivity of Polycrystalline Silicon Films Deposited from a SiH_4 — H_2 Mixture," *J. Electrochem. Soc.*, 120, 106 (1973).

[14] T. I. Kamins, M. M. Mandurah, and K. C. Saraswat, "Structure and Stability of Low Pressure Chemically Vapor-Deposited Silicon Films," *J. Electrochem. Soc.*, 125, 927 (1978).

[15] T. I. Kamins, "Structure and Properties of LPCVD Silicon Films," *J. Electrochem. Soc.*, 127, 686 (1980).

[16] Y. Wada and S. Nishimatsu, "Grain Growth Mechanism of Heavily Phosphorus-Implanted Polycrystalline Silicon," *J. Electrochem. Soc.*, 125, 1499 (1978).

[17] T. I. Kamins, "Resistivity of LPCVD Polycrystalline-Silicon Films," *J. Electrochem. Soc.*, 126, 833 (1979).

[18] M. M. Mandurah, K. C. Saraswat, and T. I. Kamins, "Phosphorus Doping of Low Pressure Chemically Vapor-Deposited Silicon Films," *J. Electrochem. Soc.*, 126, 1019 (1979).

[19] M. M. Mandurah, K. C. Saraswat, C. R. Helms, and T. I. Kamins, "Dopant Segregation in Polycrystalline Silicon," *J. Appl. Phys.*, 51, 5755 (1980).

[20] T. I. Kamins, J. Manoliu, and R. N. Tucker, "Diffusion of Impurities in Polycrystalline Silicon," *J. Appl. Phys.*, 43, 83 (1972).

[21] D. J. Coe, "The Lateral Diffusion of Boron in Polycrystalline Silicon and its Influence on the Fabrication of Sub-Micron MOSTS," *Solid State Electron.*, 20, 985 (1977).

[22] T. I. Kamins, "Hall Mobility in Chemically Deposited Polycrystalline Silicon," *J. Appl. Phys.*, 42, 4357 (1971).

[23] S. Horiuchi, "Electrical Characteristics of Boron Diffused Polycrystalline Silicon Layers," *Solid State Electron.*, 18, 659 (1975).

[24] G. Yaron, "Characterization of Phosphorus Implanted Low Pressure Chemical Vapor Deposited Polycrystalline Silicon," *Solid State Electron.*, 22, 1017 (1979).

[25] M. Kuisl and W. Langheinrich, "Preparation and Properties of Phosphorus-Doped Polycrystalline Silicon Films," in J. M. Blocher, Jr., H. E. Hintermann, and L. H. Hall, Eds., *Chemical Vapor Deposition—Fifth International Conference*, Electrochemical Society, Princeton, N.J., 1975, pp. 380—389.

[26] T. I. Kamins, "Oxidation of Phosphorus-Doped Low Pressure and Atmospheric Pressure CVD Polycrystalline-Silicon Films," *J. Electrochem. Soc.*, 126, 838 (1979).

[27] B. G. Bagley, D. E. Aspnes, A. C. Adams, and C. J. Mogab, "Optical Properties of Low-Pressure Chemically Vapor Deposited Silicon Over the Energy Range 3.0—6.0 eV," *Appl. Phys. Lett.*, 38, 56 (1981).

[28] Ch. Kuhl, H. Schlotterer, and F. Schwidefsky, "Optical Investigation of Different Silicon Films," *J. Electrochem. Soc.*, 121, 1496 (1974).

[29] M. Hirose, M. Taniguchi, and Y. Osaka, "Electronic Properties of Chemically Deposited Polycrystalline Silicon," *J. Appl. Phys.*, 50, 377 (1979).

[30] W. Kern and R. S. Rosler, "Advances in Deposition Processes for Passivation Films," *J. Vac. Sci. Technol.*, 14, 1082 (1977).

[31] R. E. Logar, M. T. Wauk, and R. S. Rosler, "Low Pressure Deposition of Phosphorus-Doped Silicon Dioxide at 400°C in a Hot Wall Furnace," in L. F. Donaghey, P. Rai-Choudhury, and R. N. Tauber, Eds., *Chemical Vapor Deposition—Sixth International Conference*, Electrochemical Society, Princeton, N.J., 1977, pp. 195—202.

[32] H. Huppertz and W. L. Engl, "Modeling of Low-Pressure Deposition of SiO_2 by Decomposition of TEOS," *IEEE Trans. Electron Devices*, ED-26, 658 (1979).

[33] A. C. Adams and C. D. Capio, "The Deposition of Silicon Dioxide Films at Reduced Pressure," *J. Electrochem. Soc.*, 126, 1042 (1979).

[34] K. Watanabe, T. Tanigaki, and S. Wakayama, "The Properties of LPCVD SiO_2 Film Deposited by SiH_2Cl_2 and N_2O Mixtures," *J. Electrochem. Soc.*, 128, 2630 (1981).

[35] M. Maeda and H. Nakamura, "Deposition Kinetics of SiO_2 Film," *J. Appl. Phys.*, 52, 6651 (1981).

[36] M. Shibata, T. Yashimi, and K. Sugawara, "Deposition Rate and Phosphorus Concentration of Phosphosilicate Glass Films in Relation to PH_3/SiH_4+PH_3 Mole Fraction," *J. Electrochem. Soc.*, 122, 157 (1975).

[37] P. J. Tobin, J. B. Price, and L. M. Campbell, "Gas Phase Composition in the Low Pressure Chemical Vapor Deposition of Silicon Dioxide," *J. Electrochem. Soc.*, 127, 2222 (1980).

[38] A. C. Adams and C. D. Capio, "Planarization of Phosphorus-Doped Silicon Dioxide," *J. Electrochem. Soc.*, 128, 423 (1981).

[39] W. A. Pliskin, "Comparison of Properties of Dielectric Films Deposited by Various Methods," *J. Vac. Sci. Technol.*, 14, 1064 (1977).

[40] A. C. Adams and S. P. Murarka, "Measuring the Phosphorus Concentration in Deposited Phosphosilicate Films," *J. Electrochem. Soc.*, 126, 334 (1979).

[41] K. Chow and L. G. Garrison, "Phosphorus Concentration of Chemical Vapor Deposited Phosphosilicate Glass," *J. Electrochem. Soc.*, 124, 1133 (1977).

[42] A. C. Adams, D. P. Schinke, and C. D. Capio, "An Evaluation of the Prism Coupler for Measuring the Thickness and the Refractive Index of Dielectric Films on Silicon Substrates," *J. Electrochem. Soc.*, 126, 1539 (1979).

[43] N. Nagasima, "Structure Analysis of Silicon Dioxide Films Formed by Oxidation of Silane," *J. Appl. Phys.*, 43, 3378 (1972).

[44] J. A. Appels, E. Kooi, M. M. Paffen, J. J. H. Schatorje, and W. H. C. G. Verkuylen, "Local Oxidation of Silicon and its Application in Semiconductor Device Technology," *Philips Res. Rep.*, 25, 118 (1970).

[45] P. S. Peercy, H. J. Stein, B. L. Doyle, and S. T. Picraux, "Hydrogen Concentration Profiles and Chemical Bonding in Silicon Nitride," *J. Electron. Mat.*, 8, 11 (1979).

[46] J. V. Dalton and J. Drobek, "Structure and Sodium Migration in Silicon Nitride Films," *J. Electrochem. Soc.* 115, 865 (1968).

[47] J. R. Hollahan and R. S. Rosler, "Plasma Deposition of Inorganic Thin Films," in J. L. Vossen and W. Kern, Eds., *Thin Film Processes*, Academic, New York, 1978, pp. 335-360.

[48] M. J. Rand, "Plasma-Promoted Deposition of Thin Inorganic Films," *J. Vac. Sci. Technol.*, 16, 420 (1979).

[49] H. Dun, P. Pan, F. R. White, and R. W. Douse, "Mechanisms of Plasma-Enhanced Silicon Nitride Deposition Using SiH_4/N_2 Mixture," *J. Electrochem. Soc.*, 128, 1555 (1981).

[50] A. C. Adams, F. B. Alexander, C. D. Capio, and T. E. Smith, "Characterization of Plasma-Deposited Silicon Dioxide," *J. Electrochem. Soc.*, 128, 1545 (1981).

[51] W. A. Lanford and M. J. Rand, "The Hydrogen Content of Plasma-Deposited Silicon Nitride," *J. Appl. Phys.*, 49, 2474 (1978).

[52] A. K. Sinha and T. E. Smith, "Electrical Properties of Si-N Films Deposited on Silicon from Reactive Plasma," *J. Appl. Phys.*, 49, 2756 (1978).

[53] M. J. Rand and J. F. Roberts, "Silicon Oxynitride Films from the $NO—NH_3—SiH_4$ Reaction," *J. Electrochem. Soc.*, 120, 446 (1973).

[54] A. K. Gaind and E. W. Hearn, "Physiochemical Properties of Chemical Vapor-Deposited Silicon Oxynitride from a $SiH_4—CO_2—NH_3—H_2$ System," *J. Electrochem. Soc.*, 125, 139 (1978).

PROBLEMS

1 Find the empirical formula for plasma silicon nitride containing 25 at. % H and having a Si/N ratio of 1.1. Find the empirical formula for LPCVD silicon nitride containing 5 at. % H and having a Si/N ratio of 0.75.

2 If plasma-deposited SiO_2 contains 3×10^{21} H/cm^3, find the at. % H and the empirical formula.

3 If the average chlorine concentration within the first 1000 Å of a deposited SiO_2 is 1×10^{19} Cl/cm^3, what is the at. % Cl in this region?

4 Derive the relationship between wt. % B, at. % B, and mol % B_2O_3 for boron-doped SiO_2.

5 A polysilicon deposition uses 30 % silane in nitrogen at 625°C and 53.3 Pa (0.4 Torr). The total gas flow is 500 std. cm^3/min. The volume of the LPCVD reactor is 20 L, its length is 150 cm, and a cross-sectional area between the wafers and the walls is 45 cm². What is the partial pressure of the silane, and the linear velocity and residence time of the gas?

6 If the reactor in problem 5 has an effective area of 4000 cm² and 100 wafers have a total area of 15,000 cm², how much silane is required to deposit 0.5 μm of polysilicon if the reaction efficiency is 20%?

7 Consider SiO_2 deposited at 100 Å/min at 450°C with an activation energy of 10 kcal/mole. How much must the temperature be increased to double the rate? Repeat the calculation for a deposition at 700°C and an activation energy of 45 kcal/mole.

8 Sketch the step coverage expected for a conformal coating over a window 1 μm deep and 2 μm wide. Use film thicknesses of 0.5, 1.0, 1.5, and 2.0 μm. Repeat the calculation for a deposition with no surface migration, such as plasma oxide.

FOUR

OXIDATION

L. E. KATZ

4.1 INTRODUCTION

The oxidation of silicon is necessary during the entire process of fabricating modern integrated circuits. The production of high-quality ICs requires not only an understanding of the basic oxidation mechanism, but the ability to form, in a controlled and repeatable manner, a high-quality oxide. In addition, to ensure the reliability of the ICs, the electrical properties of the oxide must be understood.

Silicon dioxide has several uses: to serve as a mask against implant or diffusion of dopant into silicon, to provide surface passivation, to isolate one device from another (dielectric isolation as opposed to junction isolation), to act as a component in MOS structures, and to provide electrical isolation of multilevel metallization systems. Several techniques for forming the oxide layers have been developed such as thermal oxidation, wet anodization, vapor phase technique [chemical vapor deposition (CVD)], and plasma anodization or oxidation. When the interface between the oxide and the silicon is required to have a low charge density level, thermal oxidation has been the preferred technique. However, since the masking oxide is generally removed, this consideration is not as important in the case of masking against diffusion of dopant into silicon. Obviously when the oxide layer is required on top of a metal layer, as in the case of a multilevel metallization structure, the vapor phase technique is uniquely suited. This chapter concentrates on thermal silicon oxidation, because it is the principal technique used in IC processing.

In this chapter we describe the oxidation process to provide a foundation for understanding the kinetics of growth and interface properties. Section 4.2 examines the oxidation model and its fit to experimental data; the effect of orientation, dopant concentration, and HCl addition to the ambient; and surface damage on the kinetics of oxidation. Section 4.3 describes standard thermal oxidation techniques, such as dry, wet, and HCl dry as well as the less familiar high-pressure and plasma oxidation techniques. It also describes the cleaning processes needed to remove surface contamination prior to oxidation. Section 4.4 covers the characteristics and properties of oxides, with emphasis on oxide masking, oxide charges, and stresses in thermal oxides. Sections 4.5 and 4.6 examine the redistribution of dopants at the Si-SiO_2 interface during thermal oxidation and during oxidation of polysilicon, respectively. Section 4.7 considers oxidation-induced stacking faults and oxide isolation defects. A summary and a discussion of the future trends are presented in the last section.

4.2 GROWTH MECHANISM AND KINETICS

Since a silicon surface has a high affinity for oxygen, an oxide layer rapidly forms when it is exposed to an oxidizing ambient. The chemical reactions describing the thermal oxidation of silicon[1] in oxygen or water vapor are given in Eqs. 1 and 2, respectively.

$$Si(solid) + O_2 \quad \rightarrow \quad SiO_2(solid) \tag{1}$$

$$Si(solid) + 2H_2O \quad \rightarrow \quad SiO_2(solid) + 2H_2 \tag{2}$$

The basic process involves shared valence electrons between silicon and oxygen; the silicon-oxygen bond structure is covalent. During the course of the oxidation process the Si-SiO_2 interface moves into the silicon; however, the volume expands, resulting in the external SiO_2 surface not being coplanar with the original silicon surface. Based on the densities and molecular weights of Si and SiO_2, we can show that for growth of an oxide of thickness d, a layer of silicon $0.44d$ thick is consumed (Fig. 1).

The framework of a model to describe silicon oxidation has been created. Radioactive tracer,[1] marker,[2] and infared isotope shift[3] experiments have established that oxidation proceeds by the diffusion of the oxidizing species through the oxide to the Si-SiO_2 interface, where the oxidation reaction occurs. Uncertainties exist, however, as evidenced by controversies in the literature as to whether charged or neutral

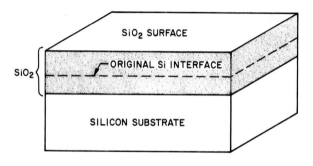

Fig. 1 Growth of SiO_2.

species are transported through the oxide, and on the details of the reaction at the Si-SiO$_2$ interface.

4.2.1 Silicon Oxidation Model

Deal and Grove's model describes the kinetics of silicon oxidation.[4] It is generally valid for temperatures between 700 and 1300°C, partial pressures between 0.2 and 1.0 atm (perhaps higher), and oxide thicknesses between 300 and 20,000 Å for oxygen and water ambients. Figure 2 shows the silicon substrate covered by an oxide layer that is in contact with the gas phase. The oxidizing species (1) are transported from the bulk of the gas phase to the gas-oxide interface with flux F_1 (the flux is the number of atoms or molecules crossing a unit area in a unit time), (2) are transported across the existing oxide toward the silicon with flux F_2, and (3) react at the Si-SiO$_2$ interface with the silicon with flux F_3.

For steady state, $F_1 = F_2 = F_3$. The gas-phase flux F_1 can be linearly approximated by assuming that the flux of oxidant from the bulk of the gas phase to the gas-oxide interface is proportional to the difference between the oxidant concentration in the bulk of the gas C_G and the oxidant concentration adjacent to the oxide surface C_S.

$$F_1 = h_G (C_G - C_S) \tag{3}$$

where h_G is the gas-phase mass-transfer coefficient.

To relate the equilibrium oxidizing species concentration in the oxide to that in the gas phase, we invoke Henry's law,

$$C_0 = Hp_s \tag{4}$$

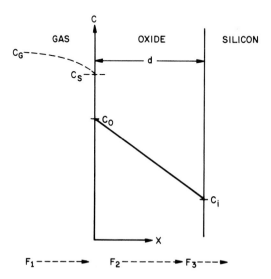

Fig. 2 Basic model for thermal oxidation of silicon. *(After Deal and Grove, Ref. 4.)*

and

$$C^* = Hp_G \tag{5}$$

where C_0 is the equilibrium concentration in the oxide at the outer surface, C^* is the equilibrium bulk concentration in the oxide, p_s is the partial pressure in the gas adjacent to the oxide surface, p_G is the partial pressure in the bulk of the gas, and H is Henry's law constant. Using Henry's law along with the ideal gas law [5] allows us to rewrite C_G and C_C

$$C_G = \frac{p_G}{kT} \tag{6a}$$

$$C_S = \frac{p_s}{kT} \tag{6b}$$

Combining Eqs. 3 to 6 gives

$$F_1 = h(C^* - C_0) \tag{7}$$

where h is the gas-phase mass-transfer coefficient in terms of concentration in the solid, given by $h = h_G / HkT$. When the concentration of the oxidant in the oxide at the oxide-gas interface C_0 is less than the equilibrium bulk oxide concentration, F_1 is positive. Oxidation is a nonequilibrium process with the driving force being the deviation of concentration from equilibrium.[6] Henry's law is valid only in the absence of dissociation effects at the gas-oxide interface. This implication is that the species moving through the oxide is molecular.

The flux of this oxidizing species across the oxide is taken to follow Fick's law

$$F_2 = -D \frac{dC}{dd} \tag{8}$$

at any point d in the oxide layer. D is the diffusion coefficient and dC / dd is the concentration gradient of the oxidizing species in the oxide. Following the steady-state assumption, F_2 must be the same at any point within the oxide (i.e., $dF_2 / dd = 0$) resulting in

$$F_2 = \frac{D(C_0 - C_i)}{d_0} \tag{9}$$

where C_i is the oxidizing species concentration in the oxide adjacent to the oxide-silicon interface and d_0 is the oxide thickness.

Assuming that the flux corresponding to the Si-SiO$_2$ interface reaction is proportional to C_i

$$F_3 = k_s C_i \tag{10}$$

where k_s is the rate constant of chemical surface reaction for silicon oxidation.

After setting $F_1 = F_2 = F_3$, as dictated by steady-state conditions, and solving simultaneous equations, we obtain the following expressions for C_i and C_0:

$$C_i = \frac{C^*}{1 + \dfrac{k_s}{h} + \dfrac{k_s\, d_0}{D}} \tag{11}$$

$$C_0 = \frac{\left[1 + \dfrac{k_s\, d_0}{D}\right] C^*}{1 + \dfrac{k_s}{h} + \dfrac{k_s\, d_0}{D}} \tag{12}$$

The limiting cases of Eqs. 11 and 12 arise when the diffusivity is either very small or very large. When the diffusivity is very small, $C_i \rightarrow 0$ and $C_0 \rightarrow C^*$. This case is called the diffusion-controlled case. It results from the flux of oxidant through the oxide being small (due to D being small) compared to the flux corresponding to the Si-SiO$_2$ interface reaction. Hence the oxidation rate depends on the supply of oxidant to the interface, as opposed to the reaction at the interface.

In the second limiting case, where D is large, $C_i = C_0 = C^*/(1 + k_s/h)$. This is called the reaction-controlled case, because an abundant supply of oxidant is provided at the Si-SiO$_2$ interface, and the oxidation rate is controlled by the reaction rate constant k_s and by C_i (which equals C_0).

To calculate the rate of oxide growth, we define N_1 as the number of oxidant molecules incorporated into a unit volume of the oxide layer. Since the oxide has 2.2×10^{22} SiO$_2$ molecules/cm^3 and one O$_2$ molecule is incorporated into each SiO$_2$ molecule, whereas two H$_2$O molecules are incorporated into each SiO$_2$ molecule, N_1 equals 2.2×10^{22} cm^{-3} for dry oxygen and twice this number for water-vapor oxidation. Combining Eqs. 10 and 11 along with the definition of flux, the flux of oxidant reaching the oxide-silicon interface is given by

$$N_1 \frac{dd_0}{dt} = F_3 = \frac{k_s\, C^*}{1 + \dfrac{k_s}{h} + \dfrac{k_s\, d_0}{D}} \tag{13}$$

We solve this differential equation assuming that an oxide may initially be present from a previous processing step or it may grow before the assumptions in the model are valid, that is, $d_0 = d_i$ at $t = 0$. The solution of Eq. 13 is

$$d_0^2 + A d_0 = B(t + \tau) \tag{14}$$

where

$$A = 2D \left[\frac{1}{k_s} + \frac{1}{h} \right] \tag{14a}$$

$$B = \frac{2DC^*}{N_1} \tag{14b}$$

$$\tau = \frac{d_i^2 + A d_i}{B} \tag{14c}$$

The quantity τ represents a shift in the time coordinate to account for the presence of the initial oxide layer d_i. Equation 14 is the well-known, mixed linear-parabolic relationship.[7]

Solving Eq. 14 for d_0 as a function of time gives

$$\frac{d_0}{A/2} = \left[1 + \frac{t + \tau}{A^2/4B} \right]^{\frac{1}{2}} - 1 \tag{15}$$

One limiting case occurs for long oxidation times when $t \gg \tau$.

$$d_0^2 = Bt \tag{16}$$

Equation 16 is the parabolic law, where B is the parabolic rate constant. The other limiting case occurs for short oxidation times when $(t + \tau) \ll A^2/4B$.

$$d_0 = \frac{B}{A}(t + \tau) \tag{17}$$

Equation 17 is the linear law, where B/A is the linear rate constant given by

$$\frac{B}{A} = \left[\frac{k_s h}{k_s + h} \right] \left[\frac{C^*}{N_1} \right] \tag{18}$$

Equations 16 and 17 are the diffusion-controlled and reaction-controlled cases, respectively.

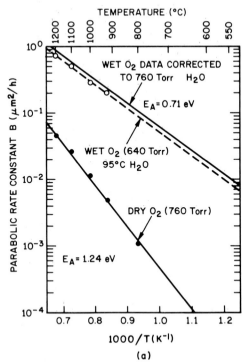

Fig. 3(a) The effect of temperature on the parabolic rate constant for dry and wet oxygen.

4.2.2 Experimental Fit

This section compares Deal and Grove's model to their experimental measurements. Deal and Grove used (111) oriented, lightly boron-doped silicon wafers, that were cleaned prior to oxidation in purified dry oxygen (less than 5-ppm water content) or in wet oxygen (the partial pressure of water was 640 Torr). For wet oxygen oxidation they found that $d_i = 0$ at $t = 0$ by plotting oxide thickness versus oxidation time. Algebraically manipulating Eq. 14 and using the plot of wet oxygen data, they graphically obtained the rate constants. Table 1 lists the values of these rate constants for wet oxidation of silicon.[4] The absolute value of A increases with decreasing temperature, while the parabolic rate constant B decreases with decreasing temperature (Figs. 3a and b).

For dry O_2 a plot of oxide thickness versus oxidation time does not extrapolate to zero initial thickness, but instead to a value which equals about 250 Å for data spanning a range of 700 to 1200°C. The faster initial oxidation rate during the initial phase of oxidation implies a different mechanism in this region. Thus use of Eq. 14 for dry oxidation requires a value for τ that can be generated graphically by extrapolating the linear region back to the time axis. Problems arise at higher temperatures where the linear-parabolic or parabolic ranges are encountered, in which case the

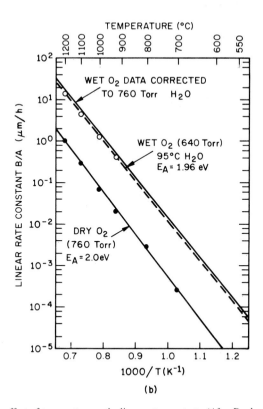

Fig. 3((b) The effect of temperature on the linear rate constant. *(After Deal and Grove, Ref. 4.)*

Table 1 Rate constants for wet oxidation of silicon

Oxidation temperature (°C)	A (μm)	Parabolic rate constant B (μm²/h)	Linear rate constant B/A (μm/h)	τ (h)
1200	0.05	0.720	14.40	0
1100	0.11	0.510	4.64	0
1000	0.226	0.287	1.27	0
920	0.50	0.203	0.406	0

value of τ as defined in Eq. 14 must be used. Table 2 lists the values of rate constants for dry oxidation of silicon.[4]

Examination of Eq. 14b reveals that B is expected to be proportional to C^*, which, according to Henry's law, is proportional to the partial pressure of the oxidizing species. However, A should be independent of the partial pressure. This has indeed been confirmed experimentally for both wet and dry oxidations[4, 8] in the temperature range between 1000 and 1200°C and between 0.1 and 1 atm. The pressure independence of A means that the linear rate constant B/A has the same linear pressure dependence as B.

Figure 3a shows the effect of temperature[4] on the parabolic rate constant B for both dry and wet oxygen at 640 Torr and for wet oxygen normalized to 760 Torr using the linear pressure dependence. As might be expected from Eq. 14, the temperature dependence of B is similar to that of D, that is, B increases exponentially with temperature. For dry oxygen the activation energy for B is 1.24 eV, which is comparable to the value of 1.17 eV for the diffusivity of oxygen through fused silica (similar in structure to thermal SiO_2). The wet oxygen activation energy (0.71 eV) also compares favorably with the activation energy for the diffusivity of water in fused silicon (0.80 eV).

Figure 3b shows the temperature dependence of the linear rate constant B/A for both dry and wet oxygen at 640 Torr and for wet oxygen normalized to 760 Torr. Once again an exponential dependence is observed with activation energies 1.96 and 2.0 eV for wet and dry oxidation, respectively. Deal and Grove[4] show that these

Table 2 Rate constants for dry oxidation of silicon

Oxidation temperature (°C)	A (μm)	Parabolic rate constant B (μm²/h)	Linear rate constant B/A (μ/h)	τ (h)
1200	0.040	0.045	1.12	0.027
1100	0.090	0.027	0.30	0.076
1000	0.165	0.0117	0.071	0.37
920	0.235	0.0049	0.0208	1.40
800	0.370	0.0011	0.0030	9.0
700	⋯	⋯	0.00026	81.0

Table 3 C^* **values in SiO$_2$ at 1000°C**

Species	C^* (cm^{-3})
O$_2$	5.2×10^{16}
H$_2$O	3.0×10^{19}

values reflect the temperature dependence of the interface reaction-rate constant k_s. As stated previously, in the linear range the reaction is reaction controlled. Similar values were obtained for the linear rate constants for both dry and wet oxidations, indicating a similar reaction or surface control mechanism. Interestingly, the above values are comparable to the 1.83 eV required to break a Si—Si bond.

The equilibrium concentration C^* of the oxidizing species in SiO$_2$ can be calculated from Eq. 14b by using appropriate values for B, D, and N_1. Table 3 gives an example.[4]

Even though the diffusivity of water in SiO$_2$ is lower than that of oxygen,[4] the parabolic rate constant B is substantially larger for wet oxidation than for dry. This is the major reason why the parabolic oxidation rate in steam is faster than in dry oxygen; the flux of oxidant, and hence B, is proportional to C^*, which is approximately three orders of magnitude greater for water than for oxygen (see Table 3). Furthermore, since the linear rate constant B/A also is related to B and hence C^*, we can also attribute the faster linear oxidation rate for wet oxidation to the above mechanism.

Deal and Grove's simple model (Eqs. 14 and 15) for thermal silicon oxidation provides excellent agreement with various normalized experimental data[4] for both wet and dry oxidations. The only exception is for SiO$_2$ films less than about 300 Å thick grown in dry oxygen. In this case an anomalously high oxidation rate is observed with respect to the model.

4.2.3 Diffusing Species and Interface Considerations

The excellent agreement between the model and experimental observations supports the use of Henry's law. This implies the lack of dissociative effects at the gas-SiO$_2$ interface indicating the species diffusing in the oxide is molecular for both oxygen (dry) and steam (water-vapor) oxidations. Additional results[9] indicate that the oxidant is molecular for both water and oxygen oxidations, since good agreement is obtained between the calculated (for fused silica) and measured oxidation rates (for oxidation of silicon) with respect to absolute rate and pressure, and with respect to temperature dependence.

A proposed modification[10] to the Deal-Grove model provides an excellent fit to the experimental data, including the thin, dry oxidation regime where the Deal-Grove model breaks down. The physical basis of the proposed model is that while diffusion through the oxide is still by molecular oxygen, the oxidation of silicon occurs by the reaction of a small concentration of atomic oxygen.

As stated earlier, the question of whether the oxidizing species is charged or neutral is still a subject of controversy. While the above discussion favors diffusion of a molecular species, supportive evidence[2] for a charged species arises from experiments showing that an applied electric field can influence the oxidation rate, either accelerating it or retarding it, depending on whether the silicon is positive or negative with respect to the oxide-gas interface. Another work,[11] based on studies of electrical conduction at elevated temperatures, concludes that the species responsible for ionic conduction is doubly negative interstitial oxygen ions (O_i^{2-}).

We now shift our discussion from the unresolved question of the nature of the diffusant to the Si-SiO$_2$ interface. The structure and oxidation mechanism at the interface is particularly important since what occurs here from an atomistic point of view can influence not only the oxidation kinetics but also allied areas of interest, such as diffusion. Both the interface structure and its oxidation mechanisms are complicated and a continuing source of discussion in the literature.

The controversies as to whether charged or neutral species are transported through the oxide have been reviewed[12, 14] followed by the proposal of a model that appears consistent with much of the earlier data. The model is based on a large molecular volume difference between Si and SiO$_2$. This difference must be accommodated to allow a newly formed SiO$_2$ molecule to fit into the normal SiO$_2$ structure. This leads to the proposal of an interface transition region, which consists of a network of extra half planes that terminate at the Si side of the Si-SiO$_2$ interface. Movement of this interface requires a supply of vacancies from the silicon to the interface, the movement of Si interstitials from the interface into the bulk Si, or free volume influx from the SiO$_2$ (i.e., viscous flow).[12−14] An additional proposal[14] relating to the interface suggests that silicon is transformed to α-cristobalite plus interstitial Si ions. Subsequent oxidation of the interstitials produces lattice distortion and transformation to vitrous silica. Hence the crystalline SiO$_2$ phase exists only as a buffer between the Si and vitrous SiO$_2$. The proposed interface mechanisms are consistent with qualitative explanations related to oxidation-enhanced diffusion, stacking fault formation, interface charge, and oxidation velocity.

Additional mechanisms have been proposed, which attempt to explain point-defect-related interface phenomena. The presence of doubly negative interstitial oxygen ions (O_i^{2-}) was discussed previously. Such ions at the Si-SiO$_2$ interface may react with silicon and displace it to an interstitial position in the lattice to form Si_I—O, which can combine to form SiO$_2$. A silicon interstitial flux can occur if the Si_I—O pair dissociates before forming SiO$_2$.[15] Such an incomplete oxidation occurs for one out of every thousand silicon atoms.[16] Although the interface reaction generally goes to completion, even a small flux of silicon interstitials into the silicon can have a large effect on defect formation or diffusion. The case of O_i^{2-} reacting with vacancies, as supplied from the silicon substrate, could lead to a vacancy flux. Such a process may be significant in the case of heavy doping.

4.2.4 Thin Oxide Growth

As noted earlier, the structure of the oxide very close to the silicon-oxide interface and the oxidation process itself both involve uncertainties. Our understanding is further complicated by the observation of an initial rapid oxidation for the case of dry oxide growth, which causes the linear portion of the oxide growth versus time curve to extrapolate to an initial thickness of about 200 Å.

With advanced MOS structures the ability to grow, with reproducible results, thin ≤300-Å, uniform, high-quality gate oxides has become increasingly important. In another application, thin pad oxides of thicknesses between 50 and 1000 Å have been used routinely under masking nitride layers to prevent stress-induced defects in the underlying silicon. The discussion in this section concentrates on the techniques and properties of thin oxides.

The technology for thin oxide growth is still emerging with a variety of techniques being used. Aside from the kinetics of oxide growth, other properties studied typically include refractive index, oxide composition, etch rate, density, susceptibility to pinholes, stress, and dielectric breakdown.

From a practical point of view, thin oxide growth must be slow enough to obtain uniformity and reproducibility. Various growth techniques include dry oxidation, dry oxidation with HCl, sequential oxidations using different temperatures and ambients, wet oxidation, reduced pressure techniques, and even high-pressure, low-temperature oxidations. The oxidation rate will, of course, be lower at lower temperatures and reduced pressures. Ultrathin oxides (<50 Å) have been produced using hot nitric acid, boiling water, and air at room temperature. For whatever technique is chosen, the desired properties must be obtained.

In discussing the techniques used and properties obtained it should be emphasized that thin oxide growth is influenced by the cleaning techniques used[17] and the purity of the gases used (especially moisture content). Figure 4 shows an example of thin oxide growth versus oxidation time in dry oxygen.[18] This data demonstrates that a set

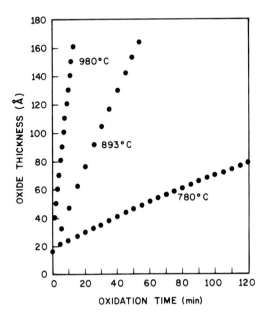

Fig. 4 Oxide thickness versus oxidation time at 780, 893, and 980°C in dry oxygen. *(After Irene, Ref. 18.)*

of time-temperature conditions can be chosen to grow thin oxides compatible with reasonable throughput.

Processing conditions have an important impact on oxide properties. For example, oxide density increases as the oxidation temperature is reduced.[19] Additionally, HCl ambients have typically been used to passivate ionic sodium, improve the breakdown voltage, and getter impurities and defects in the silicon. This passivation effect only begins to occur in the higher temperature range. A two-step process sequence has been devised[20] in which a uniform, reproducible oxide of small defect density is formed at a moderate temperature (1000°C or less) using a dry O_2, HCl ambient. The second step consists of a heat treatment in N_2, O_2, and HCl at 1150°C to provide passivation and to bring the oxide thickness to the desired level. Such a processing scheme takes advantage of beneficial effects occurring in both the lower and higher temperature ranges.

Reduced pressure oxidation offers an attractive way of growing thin oxides in a controlled manner. Oxides between 30 and 140 Å thick have been grown in a CVD reactor at 900 to 1000°C using oxygen at a pressure of 0.25 to 2.0 Torr.[21] The observed kinetics are parabolic, and the rate constants agree with values extrapolated from atmospheric pressure. Oxides obtained by this technique etch at the same rate as dry oxides obtained at 950°C and 1 atm. The equal etch rate indicates a similar oxide composition and structure between the two oxides. The intrinsic breakdown fields are high (10 to 13 MV/cm) and distributed over a narrow range. All indications are that the reduced pressure oxides are very uniform, homogeneous, and similar to thicker oxides prepared at atmospheric pressure.

As a final example of thin oxide growth, we consider the use of high-pressure, low-temperature steam oxidation of silicon. At 10-atm pressure and 750°C, a 300-Å-thick oxide can be grown in 30 min. Obviously the time, temperature, and pressure can be changed to vary the thickness. Such a technique has been applied to the growth of a thin gate oxide in the process to fabricate MOS dynamic RAMs.[22] At the same time the thin oxide layer was grown, a thick oxide layer was grown over a doped polysilicon layer as a result of concentration-enhanced oxidation. The properties of the oxides depended on the oxidation temperature rather than pressure. For example, oxide density and refractive index decreased whereas chemcial etch rate and residual stress increased with increasing temperature. The temperature and pressure ranges were 700 to 1000°C and approximately 5 to 10 atm, respectively.

4.2.5 Orientation Dependence of Oxidation Rates

Experiments have indicated that the oxidation kinetics are a function of the crystallographic orientation of the silicon surface.[23] This relationship is attributed to the orientation dependence of k_s (Eqs. 10 and 14a) and manifests itself in an orientation-dependent linear rate constant. The linear rate constant is related to the interface reaction kinetics and depends on the rate at which silicon atoms are incorporated into the oxide. This depends on the silicon surface atom concentration, which is orientation dependent. As might be expected, the parabolic rate constant B is independent of silicon surface orientation,[24] since B is diffusion limited. Figure 5 shows[25] oxide

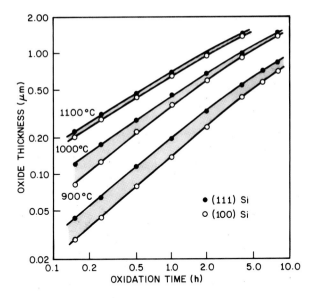

Fig. 5 Oxide thickness versus oxidation time for silicon in H_2O at 640 Torr. *(After Deal, Ref. 25.)*

thickness as a function of oxidation time in water at 640 Torr for both (111) and (100) oriented silicon. Table 4 gives rate constants obtained from this data.[25] This data along with that for oxygen yields linear rate constants for (111) silicon that are 1.68 times those for (100) silicon at corresponding temperatures.

A model has been presented[23] to explain how the linear oxidation rate of the silicon depends on the orientation of the silicon surface. According to this model, a direct reaction occurs between a water molecule in the silica and a silicon-silicon bond at the $Si-SiO_2$ interface. At this interface all the silicon atoms are partially

Table 4 Rate constants for silicon oxidation in H_2O (640 Torr)

Oxidation temperature (°C)	Orientation	A (μm)	Parabolic rate constant B (μm²/h)	Linear rate constant B/A (μm/h)	B/A ratio (111)/(100)
900	(100)	0.95	0.143	0.150	
	(111)	0.60	0.151	0.252	1.68
950	(100)	0.74	0.231	0.311	
	(111)	0.44	0.231	0.524	1.68
1000	(100)	0.48	0.314	0.664	
	(111)	0.27	0.314	1.163	1.75
1050	(100)	0.295	0.413	1.400	
	(111)	0.18	0.415	2.307	1.65
1100	(100)	0.175	0.521	2.977	
	(111)	0.105	0.517	4.926	1.65
					Average 1.68

Table 5 Calculated properties of silicon crystal planes

Orientation	Area of unit cell (cm²)	Si atoms in area	Si bonds in area	Bonds available	Bonds × 10^{14} cm^{-2}	Available bonds, N (× 10^{14} cm^{-2})	N relative to (110)
(110)	$\sqrt{2}\ a^2$	4	8	4	19.18	9.59	1.000
(311)	$1/8\sqrt{11}\ a^2$	1.5	3	2	24.54	16.36	1.707
(111)	$1/2\sqrt{3}\ a^2$	2	4	3	15.68	11.76	1.227
(100)	a^2	2	4	2	13.55	6.77	0.707

bonded to silicon atoms below and to oxygen atoms above. The orientation dependence of the oxidation rate comes from terms representing the activation energy for oxidation and the concentration of reaction sites. This concentration depends on the concentration of silicon-silicon bonds *available* for reaction at a given time. The bond is directional so its availability depends on its angle relative to the surface plane, its position with respect to adjacent atoms, and the water molecule size being such that when reacting with some angled silicon-silicon bonds, it can screen adjacent bonds from other water molecules.[6] These and other geometric effects are called steric hindrances and result in the linear oxidation rate being orientation dependent. Table 5 lists calculated properties of four silicon planes.[23] The orientation dependence is related to the available bond density N and the orientation dependence of the activation energy.

As might be expected, steric hindrance results in higher activation energy. Experimental data has been analyzed to determine the apparent activation energy, which is the sum of two components: a term related to the enthalpy of solution of water in the silica films and the orientation-dependent term related to the activation energy of oxidation. Table 6 lists the values of some apparent activation energies.[23] The interaction between the available bond density and the activation energy determines the orientation dependence of the linear oxidation rates. Experiments[23] show that the oxidation rate v in steam is ordered in the following manner

$$v_{110} > v_{311} > v_{111}$$

with a slower rate predicted for the (100) orientation. Additional measurements[24] in steam show the following oxidation rate sequence

$$v_{111} > v_{110} > v_{311} > v_{100}$$

However, this set of measurements was made at a higher temperature than the former set.[22]

For dry oxidation a similar argument for steric hindrance can be made. The following sequence is experimentally obtained[26]

$$v_{110} \geq v_{111} > v_{100}$$

for the linear oxidation rate.

Table 6 Apparent activation energies

Orientation	Activation energy (eV)
(110)	1.23 ± 0.02
(311)	1.30 ± 0.03
(111)	1.29 ± 0.03

4.2.6 Effect of Impurities and Damage on the Oxidation Rate

Because wet oxidation occurs at a substantially greater rate than for dry oxidation, any unintentional moisture accelerates the dry oxidation. In fact, both the linear and parabolic oxidation rates are sensitive to the presence of water and other impurities. The effects of some of these impurities are discussed in the following sections.

Water Experiments were done to study the effect of intentionally adding 15-ppm water vapor to a process that normally used less than 1-ppm water.[25] A significant acceleration in the oxidation rate was observed. For example, an 800°C oxidation of (100) silicon for 700 minutes grew an oxide approximately 300 Å thick with less than 1-ppm moisture and an oxide approximately 370 Å thick with 25-ppm moisture. In these experiments the oxygen was from a liquid source and the oxidation chamber was a double-wall, fused-silica tube with N_2 flowing between the walls. A precombustor and cold trap were used to achieve the less than 1-ppm moisture level.

Sodium High concentrations of sodium influence the oxidation rate by changing the bond structure in the oxide, thereby enhancing the diffusion and concentration of the oxygen molecules in the oxide.[6]

Group III and V elements The common dopant elements in this group, when present in silicon at high concentration levels, can enhance the oxidation behavior. The dopant impurities are redistributed at the growing Si-SiO₂ interface.[27] This effect is discussed in greater detail in Section 4.5, but we consider it from a mechanism standpoint here. The effect results in a discontinuous concentration profile at the interface, that is, the dopant either segregates into the silicon or into the oxide. The redistribution of the impurity at the interface influences the oxidation behavior. If the dopant segregates into the oxide and remains there (which is the case for boron in an oxidizing ambient), the bond structure in the silica weakens. This weakened structure permits an increased incorporation and diffusivity of the oxidizing species through the oxide, thus enhancing the oxidation rate. Impurities that segregate into the oxide but then diffuse rapidly through it (such as aluminum, gallium, and indium) exhibit the same oxidation kinetics as lightly doped silicon. Figure 6 shows oxidation rate curves for various concentrations of boron for wet oxygen.[28] From the above discussion it is not surprising that an enhancement in the oxidation kinetics is observed where diffusion control predominates. For oxidation of phosphorous-doped silicon in wet oxygen,[28] a concentration dependence is observed only at the lower temperatures where

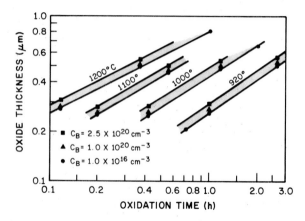

Fig. 6 Oxidation of boron-doped silicon in wet oxygen (95°C H_2O) as a function of temperature and concentration. *(After Deal and Sklar, Ref. 28.)*

the surface reaction becomes important (Fig. 7). This dependence may be the result of phosphorus being segregated into the silicon. Figure 8 shows the oxidation rate constants for dry oxygen as a function of phosphorus doping level.[29] Here B/A increases substantially at high concentrations, thus reflecting the reaction-rate control, whereas B is relatively independent of concentration, thus reflecting the diffusion-limited control.

The oxidizing interface is a complicated and not fully understood region. Its high concentration of dopant provides further complications. A theoretical model has been developed[30] to explain concentration enhancement. According to the model, the high doping levels shift the position of the Fermi level, which results in enhanced vacancy concentrations. These point defects may provide reaction sites for the chemical reaction converting Si to SiO_2, thereby increasing the reaction rate.

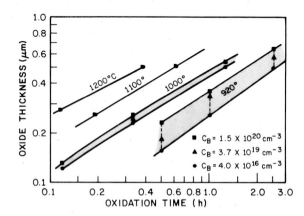

Fig. 7 Oxidation of phosphorus-doped silicon in wet oxygen (95°C H_2O) as a function of temperature and concentration. *(After Deal and Sklar, Ref. 28.)*

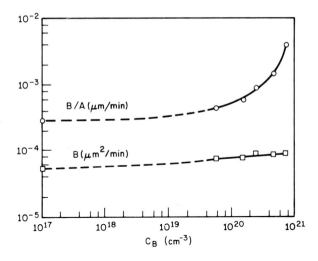

Fig. 8 Oxidation rate constants for dry oxygen as a function of phosphorus doping level at 900°C. *(After Ho et al., Ref. 29.)*

Figure 7 shows the large increase in oxidation thickness obtained for oxidation of heavily doped phosphorus in wet oxygen at lower temperatures. A dramatic example of this effect is seen in Fig. 9, which shows a bulk phosphorous-doped silicon wafer ($\sim 7 \times 10^{19}/cm^3$) after oxidation at 750°C in steam at 20-atm pressure to accelerate the kinetics. The wafer was not preferentially etched. Phosphorus dopant variations (striations), incorporated into the Czochralski crystal during solidification (see Section 1.2), appear as color variations representing oxide thickness variations. These striations clearly correspond to the concentration-enhanced oxidation of the more heavily phosphorous-doped regions in the crystal.

Halogen Certain halogen species are intentionally introduced into the oxidation ambient to improve both the oxide and the underlying silicon properties. Oxide improvements include a reduction in sodium ion contamination, an increased dielectric breakdown strength, and a reduced interface trap density. At or near the Si-SiO$_2$ interface, chlorine is instrumental in converting certain impurities in the silicon to volatile chlorides, resulting in a gettering effect. A reduction in oxidation-induced stacking faults is also observed. Chlorine is typically introduced into dry oxygen ambients in the form of chlorine gas, anhydrous HCl, or trichlorethylene.

Experimental results[31] for dry O_2−HCl mixtures show that HCl additions increase the oxidation rate. Typical HCl additions range from 1 to 5%. The parabolic rate constant B increases linearly with HCl additions above 1%. At 1000 and 1100°C large increases in B are initially observed. The linear rate constant B/A shows an initial increase when 1% HCl is added, but no further increase with subsequent HCl additions. The mechanisms associated with this enhanced growth rate are not fully understood. However, the generation of water upon adding HCl to dry oxygen does not account fully for the increased oxidation rate, since a similar increase occurs when chlorine[31] is added (even though no water is generated in that case).

Fig. 9 Concentration-enhanced oxidation, showing dopant variations in a heavily doped phosphorus substrate.

For thermal oxidation of silicon in H_2O, adding 5 volume % HCl decreases the silicon oxidation rate by about 5%, apparently because of the reduced H_2O vapor pressure.[25] Although it is not common practice to add HCl to H_2O ambients, such addition appears to reduce impurity contamination from the oxidation system.

Thermal oxidation of silicon at 1100°C with additions of up to 1% trichlorethylene (TCE) yield oxidation rates comparable to similar concentrations of chlorine. At lower temperatures the values for O_2/TCE are larger. The mechanisms involved are complicated and not fully understood.

Finally, a word of caution. Care must be taken in handling and using the halogens mentioned since the system's metallic parts and exhaust can corrode. Additionally, high concentrations of halogens at high temperatures can pit the silicon surface.

Effect of damage on oxidation rate Determining how damage to the silicon affects the oxidation rate is not easy. To study these effects, the silicon is usually intentionally damaged by ion implantation of a nonelectrically active species (Si or A), or of a group III or V dopant. Separating damage effects from dopant effects is difficult.

Enhanced thermal oxidation of implanted silicon as a function of ion species and concentration has been studied.[32] Implanted into (100) silicon were 80-keV arsenic, 60-keV boron, 106-keV antimony, and 48-keV argon with ion doses ranging from 4×10^{14} to 1×10^{16} cm^{-2}. For wet oxidation at 900°C the maximum enhancement of the oxidation rate was a factor of 1.1 for boron, 1.3 for argon, 3.5 for antimony, and 7.5 for arsenic. The higher enhancements occurred for the higher doses. The enhancement for argon is attributed to the damage effect; for the other cases the presence of the impurity atoms certainly contributes to the enhancement. Another study[33] found a retardation effect for oxidation, following implantation of Ge, Si, and Ga into silicon. It also found an enhancement for B, Al, P, As, and Sb.

4.3 OXIDATION TECHNIQUES AND SYSTEMS

The oxidation technique chosen depends upon the thickness and oxide properties required. Oxides that are relatively thin and those that require low charge at the interface are typically grown in dry oxygen. When sodium ion contamination is of concern, $HCl-O_2$ is the preferred technique. Where thick oxides (i.e., >0.5 μm) are desired steam is used ($\sim$1 atm or an elevated presure of up to 25 atm). Higher pressure allows thick oxide growth to be achieved at moderate temperatures in reasonable amounts of time.

One-atmosphere oxide growth, the most commonly used technique, is carried out in a quartz or silicon diffusion tube with the silicon wafers held vertically in a slotted paddle (boat) made of quartz or silicon. Typical oxidation temperatures range from 800 to 1200°C and should be held to within ±1°C to ensure uniformity. In a standard procedure the wafers are cleaned, dried, placed on the paddle, and automatically inserted into an 800 to 900°C furnace, which is then ramped up to temperature. Ramping is used to prevent wafer warpage. Following oxidation, the furnace is ramped down and the wafers are removed.

Eliminating particles during oxidation is necessary to grow high-quality, reproducible oxides. In earlier procedures the paddle rested directly on the tube during insertion and withdrawal or an integrated roller paddle design was used. In either case particulates were generated. Innovative designs now use a cantilevered arrangement; the paddle is inserted into the oxidation tube in a contactless manner and then lowered onto the tube. It is removed by reversing the steps.

4.3.1 Preoxidation Cleaning

Before placing wafers in a high-temperature furnace they must be cleaned to eliminate both organic and inorganic contamination arising from previous processing steps and handling. Such contamination, if not removed, can degrade the electrical characteristics of the devices as well as contribute to reliability problems.

Particulate matter is removed by either mechanical or ultrasonic scrubbing. Immersion processing techniques were the preferred chemical cleaning methods, until the development of centrifugal spray methods which eliminate the build up of contaminants as cleaning progresses. The chemical cleaning procedure usually involves removing the organic contamination, followed by inorganic ion and atom removal.

A common cleaning procedure[34] uses a $H_2O-H_2O_2-NH_4OH$ mixture to remove organic contamination by the solvating action of the ammonium hydroxide and the oxidizing effect of the peroxide. This process can also complex some group I and II metals. To remove heavy metals a $H_2O-H_2O_2-HCl$ solution is commonly used.

This solution prevents replating by forming soluble complexes with the removed ions and is performed between 75 and 85°C for 10 to 20 minutes, followed by a quench, rinse, and spin dry.[34]

Many "optimum" cleaning procedures have evolved over the years. Reference 35 reviews the necessary considerations for optimizing the cleaning procedure for silicon wafers prior to high-temperature operations.

Modern diffusion (oxidation) furnaces are microprocessor controlled to provide repeatable sequencing, temperature control, and gas flow (mass flow control). The entire procedure previously described, from boat loading to boat withdrawal, is programmed. The microprocessor control provides a feedback loop for comparing the various parameters to the desired ones and for making the appropriate changes. For example, the actual temperature of operation may change when the gas flow is changed. Direct digital control compares this temperature to the programmed temperature and automatically makes any necessary power changes.[36]

4.3.2 Dry, Wet, HCl Dry Oxidation

Dry oxidation or HC1 dry oxidation is straightforward using microprocessor-controlled equipment. The desired insertion and withdrawal rates, ramp rates, gas flows, and temperatures are all programmable. Care must be taken in handling HC1 especially with the exhaust because HC1 corrodes metal parts. Also remember that trace amounts of water vapor can drastically affect the oxidation rate.

Wet oxidation can be conveniently carried out by the pyrogenic technique, which reacts H_2 and O_2 to form water vapor. The microprocessor controls the H_2/O_2 mixture. The pyrogenic technique assures high-purity steam, provided high-purity gases are used. If wet oxidation by the bubbler technique is used, a carrier gas is typically flowed through a water bubbler maintained at 95°C. This temperature corresponds to a vapor pressure of approximately 640 Torr.

4.3.3 High Pressure

As we saw in Eq. 14b, the parabolic rate constant B is directly proportional to C^*, the equilibrium bulk concentration in the oxide, which in turn is proportional to the partial pressure of the oxidizing species in the gas phase. Oxidation in high-pressure steam produces a substantial acceleration in the growth rate.

High-pressure oxidation of silicon is particularly attractive, because thermal oxide layers can grow at relatively low temperatures in run times comparable to typical high-temperature, 1-atm conditions. The movement of previously diffused impurities can be minimized. Low-temperature operating conditions also minimize lateral diffusion, which is of great importance as device dimensions get smaller. Another advantage is that oxidation-induced defects are suppressed (see Section 4.7). For higher-temperature, high-pressure oxidations, the oxidation time is reduced significantly.

High-pressure oxidation has been under investigation since the early 1960s.[23, 37] Both experimental and production equipment are now available, along with device applications. For example, a high-speed, high-density, oxide-isolated bipolar pro-

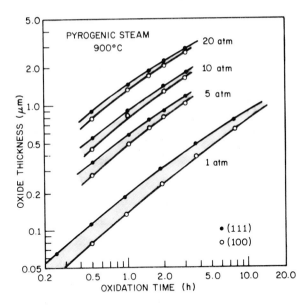

Fig. 10 Oxidation thickness versus oxidation time for pyrogenic steam at 900°C for (100) and (111) silicon and pressures up to 20 atm. *(After Razouk, Lie, and Deal, Ref. 40.)*

cess[38] has been described. In the MOS arena application has been successfully made to the growth of a thick-field oxide layer in a dynamic RAM.[39] The high-pressure technique is very promising and is beginning to be used more extensively.

Figure 10 shows oxide thickness versus time data[40] for steam oxidation at various pressures and 900°C. The substantial acceleration in the oxidation rate caused by the increased pressure is apparent. In analyzing the kinetics of oxidation at elevated pressure, several complications arise such as: continuous variations in pressure during pressurization, slightly variable pressurization times, small temperature variations that occur during pressurization and during the early part of the oxidation at full pressure, varying partial pressure of steam during depressurization, and thickness variations from run to run and across a wafer. A linear-parabolic model was used to analyze the data shown in Figure 10. A linear pressure dependence[40] was observed for both the linear and parabolic rate constants. Figure 11 shows the results for the parabolic rate constant,[40] where the dotted lines represent 5, 10, 15, and 20 times the parabolic rate constant at 1 atm. The figure shows that the rate constant is proportional to pressure, and also indicates the presence of a second activation energy below 900°C. This may be related to structural changes in the oxide.[40] A typical 10-atm oxidation cycle[41] is shown in Figure 12.

Both pyrogenic and water-pumped equipment can provide steam oxidation to 25 atm pressure and 1100°C.[41] The water-pumped system alleviates the concern associated with using hydrogen at high pressure and temperature, but requires extra attention to purity since the water quality and pumping apparatus determine the steam quality. Equipment for growing dry oxides at pressures up to 700 atm is in the developmental stages.

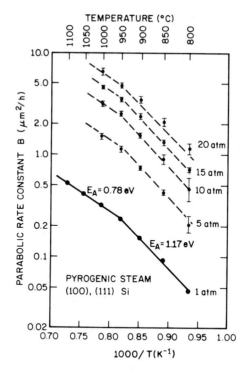

Fig. 11 Parabolic rate constant versus $1000/T$ for (100) and (111) silicon oxidized at pressures of 1, 5, 10, 15, and 20 atm in pyrogenic steam. *(After Razouk, Lie, and Deal, Ref. 40.)*

4.3.4 Plasma Oxidation

The anodic plasma-oxidation process offers the possibility of growing high-quality oxides at temperatures even lower than those achieved with the high-pressure technique. This process[42] has all the advantages associated with low-temperature processing, such as movement of previous diffusions and suppression of defect formation. Anodic plasma oxidation can grow reasonably thick oxides (on the order of 1 μm) at low temperatures (<600°C) at growth rates up to about 1 μm/h.

Plasma oxidation is a low-temperature vacuum process usually carried out in a pure oxygen discharge. The plasma is produced either by a high-frequency discharge or a DC electron source. Placing the wafer in a uniform density region of the plasma and biasing it positively below the plasma potential allows it to collect active charged oxygen species. The growth rate of the oxide typically increases with increasing substrate temperature, plasma density, and substrate dopant concentration.

The mechanisms involved with plasma oxidation are not fully understood. Uncertainty exists as to whether the oxide grows by the inward migration of oxygen species or by other, more complicated mechanisms. One model proposes that silicon and oxygen ions and/or their vacancies move across the oxide in opposite directions as a result of the applied electric field across the oxide.[43]

The beneficial effect of plasma oxidation will occur with selective oxidation

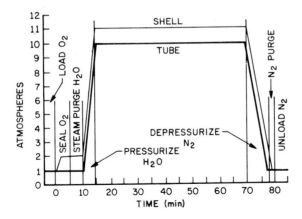

Fig. 12 Typical 10-atm steam oxidation cycle. *(After Katz et al., Ref. 41.)*

techniques (where portions of the wafer are masked against oxidation). Appropriate oxidation masks include aluminum oxide, magnesium oxide, and silicon nitride patterned by the photolithographic technique. Oxide properties, specifically the etch rate, refractive index, stress, fixed charge, interface states, and breakdown strength of plasma oxides grown at 500°C compare favorably to the properties of thermal oxides grown at 1100°C.[43-45]

4.4 OXIDE PROPERTIES

Although the literature quotes specific values for various oxide properties, it is becoming apparent that these values are affected by the experimental conditions of oxide growth. For example, the index of refraction of dry oxides[19] decreases with increasing temperature, saturating above 1190°C at an index of 1.4620. Additionally, the apparent density of oxides grown at 800°C is 3% greater than those grown[19] above 1190°C. The etch rate of thermal oxides at room temperature in buffered HF (49%) is generally quoted at about 1000 Å/min but varies with temperature and etch solution. The etch rate also varies with oxide density and thus with oxidation temperature. Measurements show that high-pressure oxides grown at 725°C and 20 atm exhibit a higher index of refraction, higher density, and slower etch rate in buffered HF than steam oxide grown at 900°C and 1 atm.[46] This difference is partially caused by the oxidation temperature effect.

For thin oxides the role of the interface in determining oxide properties is important. Unanswered questions involve the effect of lattice mismatch on oxide structure, optical properties, oxide kinetics, and oxide defects such as pinholes.

4.4.1 Masking Properties of SiO$_2$

A silicon dioxide layer can provide a selective mask against the diffusion of dopant atoms at elevated temperatures, a very useful property in IC processing. A predepo-

Table 7 Diffusion constants in SiO$_2$

Dopants	Diffusion constants at 1100°C (cm^2/s)
B	3.4×10^{-17} to 2.0×10^{-14}
Ga	5.3×10^{-11}
P	2.9×10^{-16} to 2.0×10^{-13}
As	1.2×10^{-16} to 3.5×10^{-15}
Sb	9.9×10^{-17}

sition of dopant, by ion implantation, chemical diffusion, or spin-on techniques, typically results in a dopant source at or near the surface of the oxide. During the high-temperature drive-in step, diffusion in the oxide must be slow enough with respect to diffusion in the silicon that the dopants do not diffuse through the oxide in the masked region and reach the silicon surface. The required thickness may be determined by experimentally measuring, at a particular temperature and time, the oxide thickness necessary to prevent the inversion of a lightly doped silicon substrate of opposite conductivity. A safety factor is added to this value. The impurity masking properties result when the oxide is converted into a silica impurity oxide "glass" phase.

The values of diffusion constants for various dopants in SiO$_2$ depend on the concentration, properties, and structure of the SiO$_2$. Not surprisingly quoted values may vary significantly. Table 7 lists diffusion constants for various common dopants.[47]

The commonly used n-type impurities P, Sb, and As, as well as the most frequently used p-type impurity B, all have very small oxide diffusion coefficients and are compatible with oxide masking. This is not true for gallium or aluminum (Al data not shown). Typically, oxides used for masking common impurities in conventional device processing are 0.5 to 0.7 μm thick.

4.4.2 Oxide Charges

The Si-SiO$_2$ interface contains a transition region, both in terms of atom position and stoichiometry, between the crystalline silicon and amorphous silica. Various charges and traps are associated with the thermally oxidized silicon, some of which are related to the transition region. A charge at the interface can induce a charge of the opposite polarity in the underlying silicon, thereby affecting the ideal characteristics of the MOS device. This results in both yield and reliability problems.

Figure 13 shows general types of charges.[48] These charges are described by $N = Q/q$ where Q is the net effective charge per unit area (coulombs/cm^2) at the Si-SiO$_2$ interface, N is the net number of charges per unit area (number/cm^2) at the Si-SiO$_2$ interface, and q is the electric charge. A brief description of the various charges follows.

Located at the Si-SiO$_2$ interface, interface-trapped charges Q_{it} have energy states in the silicon forbidden bandgap and can interact electrically with the underlying silicon. These charges are thought to result from several sources, including structural

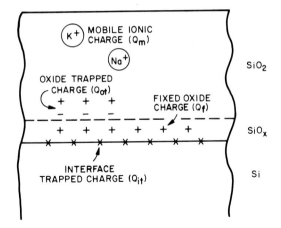

Fig. 13 Charges in thermally oxidized silicon. *(After Deal, Ref. 48.)*

defects related to the oxidation process, metallic impurities, or bond breaking processes. A low-temperature hydrogen anneal (450°C) effectively neutralizes interface-trapped charge.[48] The density of these charges is usually expressed in terms of unit area and energy in the silicon badgap (number/cm^2-eV). Capacitance-voltage (high frequency, low frequency, or quasistatic) and conductance-voltage techniques are typically used to determine the interface-trapped charges.[6] Values of 10^{10}/cm^2-eV and lower have been observed.

The fixed oxide charge Q_f (usually positive) is located in the oxide within approximately 30 Å of the Si-SiO$_2$ interface. Q_f cannot be charged or discharged. Its density ranges from 10^{10}/cm^2 to 10^{12}/cm^2, depending on oxidation and annealing conditions as well as orientation. Q_f is related to the oxidation process itself. For electrical measurements Q_f can be considered as a charge sheet at the Si-SiO$_2$ interface. The value of this charge can be determined using the capacitance-voltage (C-V) analysis technique and the following equation

$$\frac{Q_f}{q} = (-V_{FB} + \phi_{MS})\frac{C_0}{q} = (-V_{FB} + \phi_{MS})\frac{\epsilon_s}{qd_0} \tag{19}$$

where V_{FB} is the flatboard voltage, ϕ_{MS} is the metal silicon work-function difference, ϵ_s is the dielectric permittivity of the semiconductor, d_0 is the oxide thickness, and C_0 is the oxide capacitance per unit area. Q_f values for (100) oriented silicon are less than those for (111) silicon. This difference is apparently related to the number of available bonds per unit area of silicon surface.

From a processing standpoint both temperature and ambient determine Q_f.[49] In an oxygen ambient, the last high-temperature treatment determines Q_f; rapid cooling from high temperatures results in low values. Inert ambient annealing also results in low Q_f, however; at low temperatures enough time must be allowed for equilibrium to be reached.

The mobile ionic charge Q_m is attributed to alkali ions, such as sodium, potassium, and lithium, in the oxide as well as to negative ions and heavy metals. The alkali ions are mobile even at room temperature when electric fields are present. Densities range from $10^{10}/cm^2$ to $10^{12}/cm^2$ or higher and are related to processing materials, chemicals, ambient, or handling. Because of larger ionic radii and lower mobility, the heavier elements contributing to this charge drift at a slower rate than the lighter elements. Measurements can be made by using the C-V technique which involves a change in the silicon surface potential or current flow in the oxide as a result of ionic motion. Both the interface-trapped charge and oxide-trapped charge must be annealed to ensure that they do not contribute to the mobile ionic charge. Since alkali ions can be present at various places in the oxide, the MOS capacitor is subjected to a temperature-bias stress test which is compared to the standard C-V plot. The shift in flat-band voltage between the two curves allows the mobile ionic charge to be calculated. Common techniques to minimize this charge include cleaning the furnace tube in a chlorine ambient, gettering with phosphosilicate glass, and using masking layers such as silicon nitride. Although chlorine in the oxidation ambient and hence in the oxide can complex sodium, the temperatures at which this is effective are higher than the normal processing temperatures.

Oxide-trapped charge Q_{ot} may be positive or negative due to holes or electrons trapped in the bulk of the oxide. This charge is associated with defects in the SiO_2, may result from ionizing radiation, avalanche injection, or high currents in the oxide, and can be annealed out by low-temperature treatment (although neutral traps may remain).[48] Densities range from less than $10^9/cm^2$ to $10^{13}/cm^2$. Again the C-V technique can be used to measure the charge.

In addition to the earlier concerns, such as exposure of devices to ionizing radiation encountered in space flights, additional concerns arise from the newer device-processing techniques such as ion implantation, e-beam metallization, plasma or reactive-sputter etching, and e-beam or x-ray lithography.

4.4.3 Oxide Stress

Understanding the stress associated with a film is important, because high stress levels can contribute to wafer warpage, film cracking, and defect formation in the underlying Si. Room temperature measurements following thermal oxidation of silicon show SiO_2 to be in a state of compression on the surface. Stress values of 3×10^9 dynes/cm^2 are reported[50] with the stress attributed to the differences in thermal expansion for Si and SiO_2. Viscous (shear) flow of thermally grown SiO_2 occurs at temperatures as low as 960°C, evidenced by the inability of the oxide-silicon structure (oxide on one side only) to remain thermally warped above that temperature.[51] In one experiment the stress present in thermal (wet) SiO_2 during growth was measured as a function of growth temperature[52] in the range of 850 to 1030°C. Growth at 950°C and below resulted in a compressive stress of approximately 7×10^9 dynes/cm^2 in the SiO_2. This at-temperature stress value is somewhat higher than the room temperature value of 3×10^9 dynes/cm^2 given above, indicating the possibility of some stress relief during cool down. Stress-free growth at 975 and 1000°C was achieved.

During device processing, windows are cut into the oxide resulting in a complex stress distribution. At these discontinuities exceedingly high stress levels can occur.

Typically such stress would be relieved by plastic flow or other stress-relief mechanisms. The stress reduction is further accomplished by shear components which average the load over adjacent areas.[50]

The possibility of structural damage in the silicon is very real. Shear stresses at the interface are comparable to the values of compressive stress given above.[51] These shear stresses are substantially higher than the values of 3.2×10^7 dynes/cm^2 to 4.3×10^7 dynes/cm^2 given for the critical stress of shear flow for silicon at 800°C.[53] This leads to the possibility of plastic deformation in the silicon. The deleterious effect of structural damage in the silicon (particularly when decorated with impurities) on junction leakage and on other device properties is well documented. Additionally, viscous shear flow has been related to hole traps at the interface.

4.5 REDISTRIBUTION OF DOPANTS AT INTERFACE

When silicon is thermally oxidized, an interface is formed separating the silicon from the SiO$_2$. As oxidation proceeds this interface advances into the silicon. A doping impurity (initially present in the silicon) will redistribute at the interface until its chemical potential is the same on each side of the interface. This redistribution may result in an abrupt change in impurity concentration across the interface. The ratio of the equilibrium concentration of the impurity (dopant) in silicon to that in SiO$_2$ at the interface is called the equilibrium segregation coefficient. (Note: In some literature an inverse definition is used, so care must be taken in using published values.) The experimentally determined segregation coefficient may differ from the equilibrium segregation coefficient. This will primarily be determined by the chemical potential differences and the kinetics of redistribution at the interface.

Two additional factors that influence the redistribution process are the diffusivity of the impurity in the oxide (if large, the dopant can diffuse through the oxide rapidly, affecting the profile near the Si-SiO$_2$ interface) and the rate at which the interface moves with respect to the diffusion rate. Figure 14 shows four different possibilities of impurity segregation.[54]

The segregation coefficient determined experimentally is called the effective or interface segregation coefficient. It is particularly important to understand the concentration profile at the interface since electrical characteristics are affected. In extreme cases inversion can occur.

Typically, to determine the segregation coefficient experimentally, a model for diffusion has been formulated, diffusion profiles experimentally determined in the silicon, and a segregation coefficient chosen to force the data to fit the model. Direct determination of the segregation coefficient is possible using the secondary-ion mass spectrometry (SIMS) technique to obtain concentration values in the oxide and in the silicon.

Most of the effort in segregation coefficient determination has been related to boron. The segregation coefficient, as defined above, increases with increasing temperature, and is orientation dependent with values for (100) orientation being greater than for (111) orientation. Reported coefficients[55-57] are generally 0.1 to approxi-

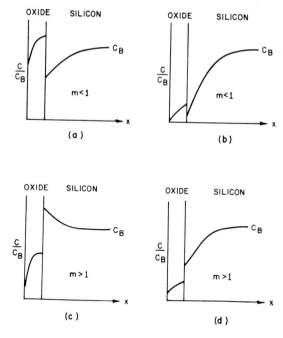

Fig. 14 Impurity segregation at the Si-SiO$_2$ interface resulting from thermal oxidation. (a) Diffusion in oxide slow (boron); (b) diffusion in oxide fast (boron-H$_2$ ambient); (c) diffusion in oxide slow (phosphorus); and (d) diffusion in oxide fast (gallium). *(After Grove, Leistiko, and Sah, Ref. 54.)*

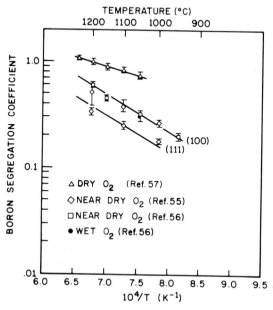

Fig. 15 Boron segregation coefficient as a function of temperature for dry, near dry, and wet oxidations. *(After Fair and Tsai, Ref. 56.)*

mately 1.0 over the temperature range 850 to 1200°C, although values greater than 1 have been obtained in special cases.[56] Figure 15 shows the results of some boron segregation determinations. Because very small amounts of moisture can greatly affect the segregation coefficient, a distinction must be made between dry, near dry, and wet oxidations. A "dry oxidation" containing even 20-ppm moisture exhibits a segregation coefficient similar to that of wet oxidation. The data in Fig. 15 shows that "near dry" oxidation (obtained in a furnace without special drying precautions) and wet oxidation give virtually identical segregation coefficients. Larger segregation coefficient values are obtained when special drying precautions are taken.[57] Additionally, boron implanted through oxide even when subsequent oxidations are performed in ambients with trace amounts of H_2O has segregation coefficients equal to those for dry O_2. These effects are particularly important at lower temperatures. For example, at 900°C the surface concentration following a "near dry" oxidation is approximately one-half that of pure dry oxidation.[56] Quoted effective segregation coefficients (m_{eff}) for boron in silicon are[56]

1. Pure dry O_2, orientation independent

$$m_{eff} = 13.4 \exp \frac{-0.33 \text{ eV}}{kT} \tag{20}$$

2. Near dry or wet O_2

$$m_{111(eff)} = 65.2 \exp \frac{-0.66 \text{ eV}}{kT} \tag{21}$$

$$m_{100(eff)} = 104.0 \exp \frac{-0.66 \text{ eV}}{kT} \tag{22}$$

For phosphorous, arsenic, and antimony, where the dopant segregates into the silicon (pile-up), segregation coefficient values of approximately 10 are usually quoted,[54] although higher values (up to 800 at 1050°C) have been determined for arsenic. With gallium, which diffuses rapidly in the oxide, a value of approximately 20 is given.[54]

4.6 OXIDATION OF POLYSILICON

Polycrystalline silicon has been used in IC technology to provide conducting lines between devices and gates. Thermal oxidation of polycrystalline silicon provides electrical isolation which can be employed as an interlevel dielectric for multilayer structures. An understanding of the oxidation mechanisms is necessary since device reliability depends on the quality of the oxide. Various parameters of polycrystalline silicon including growth method, growth temperature, doping level, grain size, and morphology have been studied with respect to oxidation rate and oxide properties, such as electrical conductivity and breakdown. Typically, comparisons are made with oxides grown on single-crystal silicon.

In one study,[58] using CVD doped and oxidized polycrystalline films, the atmospheric-pressure polysilicon (deposited at 960°C) oxidized at the same rate as low-presure polysilicon (deposited at 625°C). However, a substantial difference with respect to single-crystal silicon was observed. At moderate doping levels, the electrically active carrier concentration at the surface controlled the oxidation rate. While the total amount of dopant introduced into polysilicon and single-crystal samples was the same, the dopant diffused more deeply in the polysilicon reducing the oxidation rate with respect to single-crystal silicon. This result should not be too surprising in light of our previous discussion of concentration-enhanced oxidation. Following a phosphorus "predeposition," having 70-$\Omega/\square$ sheet resistance and 850°C steam oxidation, oxide thickness values of approximately 3000 to 3200 Å on polysilicon, approximately 3850 Å on (100) single crystal, and approximately 4250 Å on (111) single crystal were obtained.[58] The ratio of polysilicon-consumed oxidation to oxide grown was about the same as for single-crystal silicon (0.45).

In another study,[59] using CVD (at 625°C) undoped polysilicon and lightly doped single-crystal silicon, the oxidation rate increased in the following order: (100), (111), polysilicon, and (110). These observations are consistent with the transmission electron microscope determination that the polysilicon was oriented between (111) and (110). For heavily phosphorous-doped polysilicon, the parabolic rate constant is saturated at concentrations greater than 2×10^{20} cm^{-3} while the linear rate constant continues to increase.

If the oxidation rate of polysilicon depends on the random orientation of the grains, which is true in the surface or reaction-controlled region, then a surface roughening would be expected. Surface roughening, however, is not as pronounced for oxidations at higher temperatures where diffusion control is predominant. Transmission electron microscope results[60] show that the oxide exhibits thickness undulations coincident with previous grain boundaries. The oxide is thinner over grain boundaries by approximately 25% and forms intergranularly in addition to forming on the surface. For higher-temperature oxidations, the thickness undulations are less severe because the oxide and silicon can flow and the reaction can enter the diffusion-controlled region.

Device reliability may be affected when the oxide is removed to open the contacts to the polysilicon; the oxide in the intergranular regions may also be removed unintentionally. Subsequent metallization can form a conducting path along the exposed regions between the grains in the polysilicon, and electrical shorts.[60]

4.7 OXIDATION-INDUCED DEFECTS

4.7.1 Oxidation-Induced Stacking Faults

Thermal oxidation of silicon can produce stacking faults lying on (111) planes. These planar faults are structural defects in the silicon lattice that are extrinsic in nature and are bounded by partial dislocations. The growth mechanism generally invoked involves the coalescence of excess silicon atoms in the silicon lattice on nucleation sites such as defects grown in during crystal growth, surface mechanical damage present prior to oxidation, chemical contamination, or defects referred to as saucer pits or hillocks. As a result of the oxidation process, excess interstitial silicon is present near the Si-SiO$_2$ interface. A small fraction of these silicon atoms flow into the bulk silicon. The silicon interstitial supersaturation in the silicon determines the stacking fault growth rate.[61] An alternative explanation involves a decreased vacancy concentration in the silicon near the Si-SiO$_2$ interface.

The deleterious nature of oxidation-induced stacking faults is well known. Examples include degraded junction characteristics in the form of increased reverse leakage current, and storage time degradation in MOS structures. These problems occur when the stacking faults are electrically active as the result of being decorated with impurities, typically heavy metals. The decoration occurs both on the stacking fault itself and on the bounding dislocations. The dislocations, in particular, are favorable clustering sites because they represent a disarrayed high-energy region in the lattice. Diffusing impurity atoms prefer to reside in such a region because they distort the lattice less here than in the perfect lattice; that is, the high-energy region is energetically more favorable.

The growth of oxidation-induced stacking faults is a strong function of substrate orientation, conductivity type, and defect nuclei present. Observations show that the growth rate is greater for (100) than (111) substrates. Additionally, the density is greater for n-type conductivity than for p-type conductivity. Figure 16 shows that stacking fault length is a strong function of oxidation temperature.[62] The activation energy in the growth region is 2.3 eV independent of surface orientation and ambient (dry or wet). In the retrogrowth temperature range, stacking faults initially grow and then begin to shrink as oxidation proceeds. Typically the distribution of surface stacking fault lengths is very tight, except for an anomalous few percent which exhibit substantially greater lengths. Shorter-length stacking faults are usually bulk-nucleated stacking faults intersecting the surface. The length to depth ratio of the surface-oxidation stacking fault is approximately 3 to 10.

The curves in Fig. 16 clearly show two regions: a growth region and a retro-growth region. In the retrogrowth region, stacking fault formation is suppressed while preexisting stacking faults shrink. The addition of HCl to the ambient can also suppress stacking fault formation.[63] Additional observations show, for comparable oxide thickness, shorter stacking faults are grown (in the growth region) when the oxidation temperature is lower. Indeed even for oxides as thick as 1 μm, stacking fault formation is completely suppressed when the temperature is reduced below 950°C.[64] Shrinkage of preexisting stacking faults can also be accomplished by high-temperature inert ambient heat treatment, N$_2$ for example, with an activation energy of approximately 5 eV (which is almost equal to the activation energy of silicon self-diffusion). This indicates that the shrinkage is probably related to the diffusion of silicon atoms.

Experimental observations show that at comparable temperature and time, the oxidation stacking fault length is greater for steam ambients than for dry ambients. This suggests that the oxidation rate strongly influences the point-defect mechanism

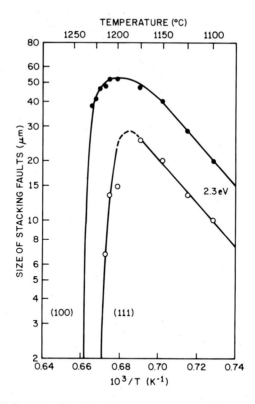

Fig. 16 Growth of oxidation-induced stacking faults versus temperature; for 3 h of dry oxidation. *(After Hu, Ref. 62.)*

responsible for stacking fault growth. Equation 23 is a proposed model[61] in which the oxidation rate is the controlling parameter in oxidation stacking fault length.

$$\frac{dl}{dt} = K_1 \left[\frac{dT_{ox}}{dt} \right]^n - K_2 \tag{23}$$

where l is the stacking fault length, T_{ox} is the oxide thickness, t is the time, n is the power dependence, K_1 is related to the growth mechanism and defect generation at the Si-SiO$_2$ interface, and K_2 is related to the retrogrowth mechanism. Applying this equation to experimental data gives values for n, K_1, and K_2. A 0.4th power dependence is observed.[61] This less-than-linear dependence of oxidation stacking fault growth rate on the oxidation rate means that smaller stacking faults will result for a higher oxidation rate at the same temperature for the same oxide thickness. This, of course, is the case with high-pressure oxidation where the oxidation rate is increased. Figure 17 shows an experimental result for a 950 to 1100°C temperature range at both 1- and 6.4-atm pressure.[64] The above results confirm the proposed model. Additional results[65] at 700°C and 20-atm pressure show complete stacking fault suppression for all thicknesses studies (up to 5 μm). Preexisting stacking faults tend to grow during

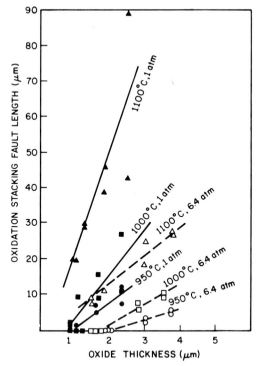

Fig. 17 Length of oxidation-induced stacking faults versus oxide thickness for 1-atm and 6.4-atm steam oxidations. *(After Tsubouchi, Miyoshi, and Abe, Ref. 64.)*

this high-pressure, low-temperature oxidation. However, the net length of the stacking fault is reduced by the consumption of silicon.

4.7.2 Oxide Isolation Defects

Selective oxidation of silicon represents an important process in IC processing. For VLSI, oxide isolation is preferred to junction isolation. Stress along the edge of an oxidized area especially in recessed oxides (that is, where the silicon has been trenched prior to oxidation to produce a reasonably planar surface) may produce severe damage in the silicon. Such defects result in increased leakage in nearby devices. The stress generated by the growing oxide, whose volume is over twice that of the consumed silicon, must be relieved without damaging the silicon. Various parameters have been examined for recessed isolation processes with the conclusion that the oxidation temperature must be sufficiently high to allow the stress in the oxide to be relieved by viscous flow. Temperatures (around 950°C) will prevent stress-induced defect formation in a recessed structure (recess approximately 1 μm and oxide growth approximately 2.2 μm). This critical temperature correlates well with that for stress-free growth in oxides at 1 atm.

4.8 SUMMARY AND FUTURE TRENDS

The ability to mathematically describe the oxidation process reasonably well in its simplest form has been demonstrated. Our understanding of the oxidizing species and the point-defect mechanisms in the vicinity of the oxidizing interface is still evolving. We can determine experimentally the effect of impurity species, dopant concentration, and orientation on the oxidation kinetics, but are somewhat less able to explain some of the mechanisms involved.

An understanding of oxide charge is necessary in order to fabricate highly reliable devices. This is particularly important with the new processing techniques used for VLSI fabrication. An understanding of how to form oxides without damaging the underlying silicon is necessary when fabricating advanced structures, such as dielectrically isolated devices that may require thick recessed oxides. Oxide viscosity is a first-order effect, and oxidation temperatures above 950°C minimize stress-related defect formation.

Polycrystalline silicon usage has become increasingly important and has attracted more study recently both in its formation and oxidation. The polysilicon deposition technique, polysilicon grain size, orientation, and doping level all affect oxidation. Formation of oxide in intergranular regions and its removal when contacts to the polysilicon are opened leads to the possibility of electrical shorts during metallization.

The impact of continually shrunken vertical and lateral dimensions, tighter design rules, and lower-temperature processing cannot be overlooked in future research. The recent availability of commercial high-pressure oxidation equipment allows thick oxides to be grown at low to moderate temperatures. As an added bonus, suppression of oxidation-induced stacking is obtained. This technique has not been exploited to any great extent, but more utilization is undoubtedly in the offing.

A low-temperature technique for growing reasonably thick oxides (~ 1 μm in 1 h), the anodic plasma-oxidation technique, offers vast potential. The low-temperature processing suppresses defect formation and minimizes movement of previous diffusions. Oxide properties are comparable to those of thermally grown oxides. Uses should proliferate when commercial equipment becomes available.

The prevention of the bird's beak during selective oxidation is another area which is receiving much attention and has a potentially big payoff. Bird's beak is associated with the thin "pad" oxide necessary to prevent defect formation. The oxide is between the silicon and masking nitride layer and results from the diffusion of oxygen and growth of SiO_2. Success has been demonstrated when the silicon is selectively trenched, processed so that nitride is present on the trench sidewall, and subsequently oxidized.[66] Encouraging results have also been obtained with the anodic plasma-oxidation technique for nonrecessed oxides.[43]

Oxide requirements on advanced structures are changing. As discussed earlier, these requirements range from highly reliable thin oxides to thick isolation oxides that can be grown at moderate temperatures. Renewed emphasis on oxidation techniques, such as high-pressure and plasma oxidation, has occurred. It is inevitable that further advances will be made in growth techniques, processing schemes, and understanding of oxidation mechanisms.

REFERENCES

[1] M. M. Atalla, "Semiconductor Surfaces and Films; the Si-SiO$_2$ System," *Properties of Elemental and Compound Semiconductors*, H. Gatos, Ed., Interscience, New York, 1960, Vol. 5, pp. 163−181.

[2] P. J. Jorgensen, "Effect of an Electric Field on Silicon Oxidation," *J. Chem Phys.*, 37, 874 (1962).

[3] J. R. Ligenza and W. G. Spitzer, "The Mechanisms for Silicon Oxidation in Steam and Oxygen," *J. Phys. Chem. Solids*, 14, 131 (1960).

[4] B. E. Deal and A. S. Grove, "General Relationship for the Thermal Oxidation of Silicon," *J. Appl. Phys.*, 36, 3770 (1965).

[5] A. S. Grove, *Physics and Technology of Semiconductor Devices*, Wiley, New York, 1967, Chapter 2.

[6] E. H. Nicollian and J. R. Breuws, *MOS Physics and Technology*, Wiley, New York, 1982.

[7] U. R. Evans, "The Relationship Between Tarnishing and Corrosion," *Trans. Electrochem. Soc.*, 46, 247 (1924).

[8] P. S. Flint, "The Rates of Oxidation of Silicon," Abstract 94, *The Electrochem. Soc. Extended Abs.*, *Spring Meeting*, Los Angeles, May 6−10, 1962.

[9] R. H. Doremus, "Oxidation of Silicon by Water and Oxygen and Diffusion in Fused Silica," *J. Phys. Chem.*, 80, 1773 (1976).

[10] J. Blanc, "A Revised Model for the Oxidation of Si by Oxygen," *Appl. Phys. Lett.*, 33, 424 (1978).

[11] T. G. Mills and F. A. Kroger, "Electrical Conduction at Elevated Temp. in Thermally Grown SiO$_2$ Films," *J. Electrochem. Soc.*, 120, 1582 (1973).

[12] W. A. Tiller, "On the Kinetics of the Thermal Oxidation of Silicon. I. A Theoretical Perspective," *J. Electrochem. Soc.*, 127, 619 (1980).

[13] W. A. Tiller, "On the Kinetics of the Thermal Oxidation of Silicon. II. Some Theoretical Evaluations," *J. Electrochem. Soc.*, 127, 625 (1980).

[14] W. A. Tiller, "On the Kinetics of the Thermal Oxidation of Silicon. III. Coupling With Other Key Phenomena," *J. Electrochem. Soc.*, 128, 689 (1981).

[15] R. B. Fair, "Oxidation, Impurity Diffusion, and Defect Growth in Silicon—An Overview," *J. Electrochem. Soc.*, 128, 1361 (1981).

[16] S. M. Hu, "Formation of Stacking Faults and Enhanced Diffusion in the Oxidation of Silicon," *J. Appl. Phys.*, 45, 1567 (1974).

[17] F. N. Schwettmann, K. L. Chiang, and W. A. Brown, "Variation of Silicon Dioxide Growth Rate with Pre-Oxidation Clean," Abstract 276, *The Electrochem. Soc. Extended Abs., Spring Meeting*, Seattle, Washington, May 1978.

[18] E. A. Irene, "Silicon Oxidation Studies: Some Aspects of the Initial Oxidation Regime," *J. Electrochem. Soc.*, 125, 1708 (1978).

[19] E. A. Taft, "The Optical Constants of Silicon and Dry Oxygen Oxides," *J. Electrochem. Soc.*, 125, 968 (1978).

[20] C. Hashimoto, S. Muramoto, N. Shiomo, and O. Nakajima, "A Method of Forming Thin and Highly Reliable Gate Oxides," *J. Electrochem. Soc.*, 127, 129 (1980).

[21] A. C. Adams, T. E. Smith, and C. C. Chang, "The Growth and Characterization of Very Thin Silicon Dioxide Films," *J. Electrochem. Soc.*, 127, 1787 (1980).

[22] M. Hirayama, H. Miyoshi, N. Tsubouchi, and H. Abe, "High Pressure Oxidation for Thin Gate Insulator Process," *IEEE Trans. Electron Devices*, ED-29, 503 (1982).

[23] J. R. Ligenza, "Effect of Crystal Orientation on Oxidation Rates in High Pressure Steam," *Phys. Chem.*, 65, 2011 (1961).

[24] W. A. Pliskin, "Separation of the Linear and Parabolic Terms in the Steam Oxidation of Si," *IBM J. Res. Dev.*, 10, 198 (1966).

[25] B. E. Deal, "Thermal Oxidation Kinetics of Silicon in Pyrogenic H$_2$O and 5% HCl/H$_2$O Mixtures," *J. Electrochem. Soc.*, 125, 576 (1978).

[26] E. A. Irene, "The Effects of Trace Amounts of Water of the Thermal Oxidation of Si in Oxygen," *J. Electrochem. Soc.*, 121, 1613 (1974).

[27] M. M. Atalla and E. Tannenbaum, "Impurity Redistribution and Junction Formation in Silicon by Thermal Oxidation," *Bell Syst. Tech. J.*, 39, 933 (1960).

[28] B. E. Deal and M. Sklar, "Thermal Oxidation of Heavily Doped Silicon," *J. Electrochem. Soc.*, 112, 430 (1965).

[29] C. P. Ho, J. D. Plummer, J. D. Meindl, and B. E. Deal, "Thermal Oxidation of Heavily Phosphorus Doped Silicon," *J. Electrochem. Soc.*, 125, 665 (1978).

[30] C. P. Ho and J. D. Plummer, "Si-SiO$_2$ Interface Oxidation Kinetics: A Physical Model for the Influence of High Substrate Doping Levels. I. Theory," *J. Electrochem. Soc.*, 126, 1516 (1979); "II. Comparison With Experiment and Discussion," *J. Electrochem. Soc.*, 126, 1523 (1979).

[31] D. W. Hess and B. E. Deal, "Kinetics of the Thermal Oxidation of Silicon in O$_2$/HCl Mixtures," *J. Electrochem. Soc.*, 124, 735 (1977).

[32] J. F. Gotzlich, K. Haberger, H. Ryssel, H. Kranz, and E. Traumuller, "Dopant Dependence of the Oxidation Rate of Ion Implanted Silicon," *Radiat. Eff.*, 47, 203 (1980).

[33] G. Mezey, T. Nagy, J. Gyulai, E. Kotai, A. Manuaba, T. Lohner, and J. W. Mayer, "Enhanced and Inhibited Oxidation of Implanted Silicon," in *Ion Implantation in Semiconductors*, F. Chernow, J. A. Borders, and D. K. Brice, Eds., Plenum Press, New York, 1977, p. 49.

[34] W. Kern and D. A. Puotinen, "Cleaning Solutions Based on Hydrogen Peroxide for Use in Silicon Semiconductor Technology," *RCA Rev.* 31, 187 (1970).

[35] D. Burkman, "Optimizing the Cleaning Procedure for Silicon Wafers Prior to High Temperature Operations," *Semicond. Int.*, 4, 103 (1981).

[36] P. S. Burggraff, "The Case for Computerized Diffusion Control," *Semicond. Int.*, 4, 37 (1981).

[37] J. R. Ligenza, "Oxidation of Silicon by High Pressure Steam," *J. Electrochem. Soc.*, 109, 73 (1962).

[38] J. Agraz-Guerena, P. T. Panousis, and B. L. Morris, "OXIL, A Versatile Bipolar VLSI Technology," *IEEE Trans. Electron Devices*, ED-27, 1397 (1980).

[39] N. Tsubouchi, H. Miyoshi, H. Abe, and T. Enomoto, "The Applications of High Pressure Oxidation Process to the Fabrication of MOS LSI," *IEEE Trans. Electron Devices*, ED-26, 618 (1979).

[40] R. R. Razouk, L. N. Lie, and B. E. Deal, "Kinetics of High Pressure Oxidation of Silicon in Pyrogenic Steam," *J. Electrochem. Soc.*, 128, 2214 (1981).

[41] L. E. Katz, B. F. Howells, L. P. Adda, T. Thompson, and D. Carlson, "High Pressure Oxidation of Silicon by the Pyrogenic or Pumped Water Technique," *Solid State Technol.*, 24, 87 (1981).

[42] J. R. Ligenza, "Silicon Oxidation in an Oxidation Plasma Excited by Microwaves," *J. Appl. Phys.*, 36, 2703 (1965).

[43] V. Q. Ho and T. Sugano, "Selective Anodic Oxidation of Silicon in Oxygen Plasma," *IEEE Trans. Electron Devices*, ED-27, 1436 (1980).

[44] J. R. Ligenza and M. Kuhn, "DC Arc Anodic Plasma Oxidation," *Solid State Technol.*, 13, 33 (1970).

[45] A. K. Ray and A. Reisman, "The Formation of SiO$_2$ in an RF Generated Oxygen Plasma," *J. Electrochem. Soc.*, 128, 2466 (1981).

[46] L. E. Katz and B. F. Howells, "Low Temperature, High Pressure Steam Oxidation of Silicon," *J. Electrochem. Soc.*, 126, 1822 (1979).

[47] M. Ghezzo and D. M. Brown, "Diffusivity Summary of B, Ga, P, As, and Sb in SiO$_2$," *J. Electrochem. Soc.*, 120, 146 (1973).

[48] B. E. Deal, "Standardized Terminology for Oxide Charges Associated with Thermally Oxidized Silicon," *IEEE Trans. Electron Devices*, ED-27, 606 (1980).

[49] B. E. Deal, "The Current Understanding of Charges in the Thermally Oxidized Silicon Structure," *J. Electrochem. Soc.*, 121, 198C (1974).

[50] R. J. Jaccodine and W. A. Schlegel, "Measurement of Strains at Si-SiO$_2$ Interface," *J. Appl. Phys.*, 37, 2429 (1966).

[51] E. P. EerNisse, "Viscous Flow of Thermal SiO$_2$," *Appl. Phys. Lett.*, 30, 290 (1977).

[52] E. P. EerNisse, "Stress in Thermal SiO$_2$ During Growth," *Appl. Phys. Lett.*, 35, 8 (1979).

[53] S. M. Hu, "Temperature Dependence of Critical Stress in Oxygen Free Silicon," *J. Appl. Phys.*, 49, 5678 (1978).

[54] A. S. Grove, O. Leistiko, and C. T. Sah, "Redistribution of Acceptor and Donor Impurities During Thermal Oxidation of Silicon," *J. Appl. Phys.*, 35, 2695 (1964).

[55] J. W. Colby and L. E. Katz, "Boron Segregation at Si-SiO$_2$ Interface as a Function of Temperature and Orientation," *J. Electrochem. Soc.*, 123, 409 (1976).

[56] R. B. Fair and J. C. C. Tsai, "Theory and Direct Measurement of Boron Segregation in SiO_2 during Dry, Near Dry and Wet O_2 Oxidation," *J. Electrochem. Soc.*, 125, 2050 (1978).

[57] S. P. Murarka, "Diffusion and Segregation of Ion-Implanted Boron in Silicon in Dry Oxygen Ambients," *Phys. Rev. B*, 12, 2502 (1975).

[58] T. I. Kamins, "Oxidation of Phosphorous-Doped Low Pressure and Atmospheric Pressure CVD Polycrystalline-Silicon Films," *J. Electrochem. Soc.*, 126, 838 (1979).

[59] H. Sunami, "Thermal Oxidation of Phosphorus-Doped Polycrystalline Silicon in Wet Oxygen," *J. Electrochem. Soc.*, 125, 892 (1978).

[60] E. A. Irene, E. Tierney, and D. W. Dong, "Silicon Oxidation Studies: Morphological Aspects of the Oxidation of Polycrystalline Silicon," *J. Electrochem. Soc.*, 127, 705 (1980).

[61] A. Lin, R. W. Dutton, D. A. Antoniades, and W. A. Tiller, "The Growth of Oxidation Stacking Faults and the Point Defect Generation at Si-SiO$_2$ Interface during Thermal Oxidation of Silicon," *J. Electrochem. Soc.*, 128, 1121 (1981).

[62] S. M. Hu, "Anomalous Temperature Effect of Oxidation Stacking Faults in Silicon," *Appl. Phys. Lett.*, 17, 165 (1975).

[63] H. Shiraki, "Stacking Fault Generation, Suppression and Grown-In Defect Elimination in Dislocation Free Silicon Wafers by HCl Oxidation," *Jpn. J. Appl. Phys.*, 15, 1 (1976).

[64] N. Tsubouchi, H. Miyoshi, and H. Abe, "Suppression of Oxidation-Induced Stacking Fault Formation in Silicon by High Pressure Steam Oxidation," *Japan J. Appl. Phys.*, 17 Supp. *17-1*, 223 (1978).

[65] L. E. Katz and L. C. Kimerling, "Defect Formation During High Pressure, Low Temperature Steam Oxidation of Silicon," *J. Electrochem. Soc.*, 125, 1680 (1978).

[66] D. Kahng, T. A. Shankoff, T. T. Sheng, and S. E. Haszko, "A Method for Area Saving Planar Isolation Oxides Using Oxidation Protected Sidewalls," *J. Electrochem. Soc.*, 127, 2468 (1980).

4.9 PROBLEMS

1 Show from the densities and molecular weights of Si and SiO_2 that a layer of silicon of thickness $0.45\,d_0$ is consumed when a SiO_2 layer of thickness d_0 is formed. Use density values of 2.27 gm/cm^3 for SiO_2 and 2.33 gm/cm^3 for Si.

2 Show that in Eq. 14, $d_0^2 + Ad_0 = B(t + \tau)$ reduces to $d_0^2 = Bt$ for long times and to $d_0 = B/A\,(t + \tau)$ for short times.

3 (a) Show that in Eq. 14, $d_0^2 + Ad_0 = B(t + \tau)$ can be used graphically to obtain an equation describing the oxidation rate.

(b) Generate such a plot for the 1100°C oxidation data of Fig. 5. Use $\tau = 0$ and (100) orientation to obtain rate constants. Compare your results to those of Fig. 3.

4 Using Eq. 14 and Table 1, how long will it take to grow 2.0 μm of SiO_2 at 920°C and 25-atm steam pressure?

5 Define a set of conditions to minimize the chance of inverting the surface of an n-type substrate (containing a boron diffusion) when oxidizing the wafer.

6 List possible ways of growing an initial oxide on a substrate without forming oxidation-induced stacking faults.

7 Solve Eq. 14 for oxide thickness as $f(t, \tau, A, B)$.

8 Make use of the equation derived in Problem 7, and the data in Tables 1 and 2, to generate oxide thickness versus time curves for wet and dry oxidations at 1100°C. Assume $\tau = 0$.

9 Generate a model showing possible interface reaction and point-defect fluxes at the interface.

10 Devise a processing scheme to generate selectively a planar recessed oxide in silicon. Show how you might prevent lateral oxidation during the oxide growth.

FIVE

DIFFUSION

J. C. C. TSAI

5.1 INTRODUCTION

Diffusion of impurity atoms in silicon is important in VLSI processing. The idea of using diffusion techniques to alter the type of conductivity in silicon or germanium was disclosed in a patent in 1952 by Pfann.[1] Since then, various ideas on how to introduce dopants into silicon by diffusion have been studied with the goals of controlling the dopant concentration, uniformity, and reproducibility, and of processing a large number of device wafers in a batch to reduce the manufacturing costs. Diffusion is used to form bases, emitters, and resistors in bipolar device technology, to form source and drain regions, and to dope polysilicon in MOS device technology. Dopant atoms which span a wide range of concentrations can be introduced into silicon wafers in the following ways: (1) diffusion from a chemical source in a vapor form at high temperatures, (2) diffusion from a doped-oxide source, or (3) diffusion and annealing from an ion implanted layer. Annealing of ion implanted layers is for activating the implanted atoms and reducing the crystal damages from ion implantation. When the annealing is at a high temperature, diffusion also occurs. Since ion implantation provides more precise control of total dopants from 10^{11} cm^{-2} to greater than 10^{16} cm^{-2}, it is used to replace the chemical or doped-oxide source wherever possible. (Ion implantation and annealing properties are discussed in Chapter 6.) Ion implantation is extensively applied in VLSI device fabrication.

Another aspect of the study of diffusion attempts to develop improved models from experimental data for predicting diffusion results from theoretical analysis. The ultimate goal of diffusion studies is to calculate the electrical characteristics of a semiconductor device from the processing parameters. Diffusion theories have been developed from two major approaches, namely, the continuum theory of Fick's diffusion equation and the atomistic theory, which involves interactions between point defects, vacancies and interstitial atoms, and impurity atoms. The continuum theory describes the diffusion phenomenon from the solution of Fick's diffusion equation with appropriate diffusivities. (This chapter uses the terms "diffusivity" and "diffusion coefficient" interchangeably.) The diffusivities of a dopant element can be determined from experimental measurements, such as the surface concentration, junction depth, or the concentration profiles, and the solutions of Fick's diffusion equation. When impurity concentrations are low, the measured diffusion profiles are well behaved and agree with Fick's diffusion equation with a constant diffusivity which can be calculated readily. In these cases the detailed atomic movements do not have to be known. When impurity concentrations are high, the diffusion profiles deviate from the predictions of the simple diffusion theory, and the impurity diffusion is affected by other factors, which are not considered in Fick's simple diffusion laws. Since the diffusion profile measurements reveal concentration-dependent diffusion effects, we apply Fick's diffusion equation with concentration-dependent diffusivities to the high-concentration diffusions. The concentration-dependent diffusivities are determined by a Boltzmann-Matano analysis[4] or other formulations of profile analysis.

Various atomistic diffusion models based on defect-impurity interactions have been proposed to explain the experimental results from concentration-dependent diffusivities and other anomalous diffusions. The atomistic diffusion theory is still undergoing active development. Theoretical and experimental results on the diffusion of Group III and V elements in silicon have been incorporated into various process models. Chapter 11 discusses the process models in detail. Because process modeling is still developing we have to be aware of the model's limitations.

5.2 MODELS OF DIFFUSION IN SOLIDS

At high temperatures point defects, such as vacancies and interstitial atoms, are generated in a single-crystal solid. When a concentration gradient of the host or impurity atoms exists, the point defects affect atom movement (diffusion). Diffusion in a solid can be visualized as atomic movement of the diffusant in the crystal lattice by vacancies or self-interstitials. Figure 1 shows some common atomic diffusion models[2] in a solid, using a simplified two-dimensional crystal structure with lattice constant a. The circles represent the host atoms occupying the low-temperature lattice positions. The solid circles represent either host or impurity atoms. At elevated temperatures the lattice atoms vibrate around the equilibrium lattice sites. Occasionally a host atom acquires sufficient energy to leave the lattice site, becoming an interstitial atom and creating a vacancy. When a neighboring atom (either the host or the impurity atom) migrates to the vacancy site, the mechanism is called diffusion by a vacancy (Fig. 1a). If the migrating atom is a host atom the diffusion is referred to as self-diffusion; if it is an impurity atom the diffusion is impurity diffusion. [Self-diffusion experiments are usually conducted by introducing radioactive isotopes of the host atom (Fig. 1a).]

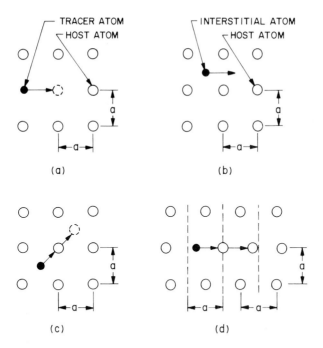

Fig. 1 Models of atomic diffusion mechanisms for a two-dimensional lattice. a is the lattice constant. (a) Vacancy mechanism. (b) Interstitial mechanism. (c) Interstitialcy mechanism. (d) Crowdion mechanism. *(After Tuck, Ref. 2.)*

If an interstitial atom moves from one place to another without occupying a lattice site (Fig. 1b), the mechanism is interstitial diffusion. An atom smaller than the host atom often moves interstitially. The activation energies required for diffusion of interstitial atoms are lower than those for diffusion of lattice atoms by a vacancy mechanism.

Figure 1c shows that the atomic movement of an interstitial atom (a host or an impurity atom) displaces a lattice atom, which in turn becomes an interstitial atom. This is an example of the extended interstitial mechanism, sometimes called the "interstitialcy" mechanism. A related interstitialcy mechanism is the crowdion mechanism, in which an interstitial atom located half-way between two lattice sites migrates into one of the lattice sites and displaces the lattice atom, which becomes an interstitial atom at the half-way position (Fig. 1d).

By applying statistical thermodynamics, the activation energies and concentrations of the point defects for a given crystal can be estimated and diffusion theory can be developed.[21] The theoretical results may then be compared with experimental findings. For example, in the case of silicon, Group III and V elements are generally considered to diffuse predominately by the vacancy mechanism. Group I and VIII elements have small ionic radii, and they are fast diffusers in silicon. They are usually considered to diffuse by an interstitial mechanism. These simple atomic mechanisms are not adequate for describing the diffusion when the impurity concentrations are high, dislocations are present, or other impurities are present in high concentrations. When the impurity concentration is low and the dislocation density is low, the impurity diffusion can be described by a phenomenological law of diffusion, that is, by using Fick's diffusion law with a constant diffusivity. Mathematical expressions are obtained by solving Fick's diffusion equation and the diffusivities of the diffusant are determined for different temperatures. For high-impurity concentrations, concentration-dependent diffusivities are related to an assumed atomistic-diffusion mechanism or mechanisms.

5.3 FICK'S ONE-DIMENSIONAL DIFFUSION EQUATIONS

In 1855 Fick published his theory on diffusion. He based his theory on the analogy between material transfer in a solution and heat transfer by conduction.[3] Fick assumed that in a dilute liquid or gaseous solution in the absence of convection, the transfer of solute atoms per unit area in an one-dimensional flow can be described by the following equation:

$$J = -D\frac{\partial C(x,t)}{\partial x} \tag{1}$$

where J is the rate of transfer of solute per unit area or the diffusion flux, C is the concentration of solute, which is assumed to be a function of x and t only, x is the coordinate axis in the direction of the solute flow, t is the diffusion time, and D is the diffusion coefficient.

Equation 1 states that the local rate of transfer (local diffusion rate) of solute per unit area per unit time is proportional to the concentration gradient of the solute and defines the proportionality constant as the diffusion coefficient of the solute. The negative sign on the right-hand side of Eq. 1 states that the matter flows in the direction of decreasing solute concentration (i.e., the gradient is negative). Equation 1 is called Fick's first law of diffusion.

From the law of conservation of matter, the change of solute concentration with time must be the same as the local decrease of the diffusion flux, that is,

$$\frac{\partial C(x,t)}{\partial t} = -\frac{\partial J(x,t)}{\partial x} \tag{2}$$

Substituting Eq. 1 into Eq. 2, yields Fick's second law of diffusion in one-dimensional form:

$$\frac{\partial C(x,t)}{\partial t} = \frac{\partial}{\partial x}\left[D\frac{\partial C(x,t)}{\partial x}\right] \tag{3}$$

When the concentration of the solute is low, the diffusion coefficient can be considered as a constant, and Eq. 3 becomes

$$\frac{\partial C(x,t)}{\partial t} = D\frac{\partial^2 C(x,t)}{\partial x^2} \tag{4}$$

Equation 4 is often referred to as Fick's simple diffusion equation. In Eq. 4, D is given in units of cm^2/s and $C(x,t)$ is in units of $atoms/cm^3$. Sometimes D is also expressed in $\mu m^2/h$. Solutions for Eq. 4 with various simple initial and boundary conditions have been obtained.[4, 5] The most commonly used solutions are given in the following section.

5.3.1 Constant Diffusivities

Impurity diffusion for junction formation can be achieved easily under two conditions, namely, a constant surface concentration condition and a constant total dopant condition. In the first case impurity atoms are transported from a source vapor onto the silicon surface and diffused into silicon wafers. The source vapor maintains a constant level of surface concentration during the entire diffusion period. In the case of a constant total dopant, a small amount of dopant is deposited onto the silicon surface. Mathematically, this instantaneous deposition of dopant is like a delta function. This condition can be achieved by diffusion at low temperatures, as in predeposition diffusion. Diffusion from an ion implanted layer is similar to the second case. This section gives solutions of Fick's diffusion equation, Eq. 4, for these two cases.

Constant surface concentration The initial condition at $t = 0$ is

$$C(x, 0) = 0 \tag{5}$$

The boundary conditions are

$$C(0,t) = C_s \tag{6}$$

and

$$C(\infty,t) = 0 \tag{7}$$

The solution of Eq. 4 that satisfies the initial and boundary conditions is given by

$$C(x,t) = C_s \text{ erfc} \left[\frac{x}{2\sqrt{Dt}} \right] \tag{8}$$

where C_s is the constant surface concentration (in $atoms/cm^3$), D is the constant diffusion coefficient (in cm^2/s), x is the distance coordinate (in cm), with $x = 0$ at the silicon surface, t is the diffusion time (in s), and erfc is the complementary error function.

Figure 2 shows the normalized concentration profile for a complementary error function distribution of Eq. 8. The position where the diffusant concentration equals the substrate concentration is defined as the metallurgical junction x_j, that is, $C(x_j) = C_{sub}$. Assuming that the substrate conductivity is opposite that of the diffusant, and since the ordinate is a logarithmic scale, $| C_{sub}/C_s |$ can be plotted to show the concentration of the net dopants $| N_D - N_A |$ near a p-n junction.

Constant total dopant Suppose that a thin layer of dopant is deposited onto the silicon surface with a fixed (or constant) total amount of dopant S per unit area, and that

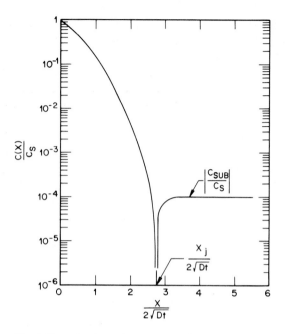

Fig. 2 Normalized complementary error function distribution.

the dopant diffuses into the silicon. The silicon substrate has an impurity concentration C_{sub} (in atoms/cm^3) of the opposite conductivity. The initial and boundary conditions and the solution of the diffusion equation (Eq. 4) that satisfies these conditions are given in Eqs. 9 through 13.

Initial condition:

$$C(x, 0) = 0 \qquad (9)$$

Boundary conditions:

$$\int_0^\infty C(x,t) \, dx = S \qquad (10)$$

$$C(x, \infty) = 0 \qquad (11)$$

The solution of the diffusion equation Eq. 4 that satisfies Eqs. 9 through 11 is

$$C(x,t) = \frac{S}{\sqrt{\pi Dt}} \exp\left[-\frac{x^2}{4Dt} \right] \qquad (12)$$

By setting $x = 0$ we obtain the surface concentration,

$$C_s = C(0,t) = \frac{S}{\sqrt{\pi Dt}} \qquad (13)$$

Equation 12 is often called the Gaussian distribution and the diffusion condition is referred to as the predeposition diffusion.

Redistribution diffusion In bipolar linear ICs, redistribution diffusion from a predeposition diffused layer is an important step. The redistribution diffusion in a nonoxidizing ambient has been studied extensively. In VLSI technology, no intentional rediffusion is applied in order to keep the diffusion depth shallow. From an ion implanted source, however, some redistribution diffusion can occur while thermally annealing the ion implanted region for electrical activation at temperatures greater than 1000°C. The solution to Fick's equation, Eq. 4, with an initial ion implanted Gaussian distribution has been obtained.[6]

The equation for redistribution diffusion in an oxidizing ambient involves a moving boundary problem and is more difficult to solve. No analytical solutions have been found. A mathematical formulation of diffusion in an oxidizing ambient from a given initial profile has been obtained;[7] however, the solution involves expressions that require numerical integration. Segregation of impurity atoms during oxidation between the growing oxide and silicon was discussed in Chapter 4. Since redistribution diffusion is not important in VLSI technology, we will not discuss it in this chapter.

5.3.2 Concentration-Dependent Diffusivities

At high concentrations, when the diffusion conditions are close to the constant surface concentration case or to the constant total dopant case, the measured impurity profiles deviate from Eqs. 8 and 12, respectively. In these high-concentration regions, the impurity profile can often be represented by concentration-dependent diffusivities. Equation 3 is used to determine the concentration-dependent diffusivities from the experimentally measured concentration profiles. This section considers diffusion under two conditions: constant surface concentration and a constant total dopant.

Constant surface concentration The one-dimensional diffusion equation with a concentration-dependent diffusion coefficient was given in Eq. 3. If D is only a function of the concentration C and the surface concentration is maintained at a constant value, Eq. 3 can be transformed into an ordinary differential equation[4] with a new variable η, where

$$\eta = \frac{x}{\sqrt{t}} \tag{14}$$

Thus, both D and C depend on x implicitly. After a change of variable to η, Eq. 15 can be obtained from Eq. 3:

$$D(C) = \frac{-\frac{1}{2}\int_{C_0}^{C} \eta \, dC}{\dfrac{dC}{d\eta}} \tag{15}$$

Equation 15 refers to an infinite system. To determine the concentration-dependent diffusivity from Eq. 15, we first plot the measured diffusion profile as concentration (or normalized concentration) versus η (see Fig. 3). We choose the origin of the abscissa so that the area under the profile on the left-hand side equals the area under

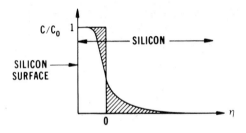

Fig. 3 The diffusion coordinate for the Boltzmann-Matano analysis for concentration-dependent diffusivity $D(C)$. Constant surface concentration.

the profile on the right-hand side. The concentration-dependent diffusivity can then be determined by performing the numerical integration

$$\int_{C_0}^{C} \eta \, dC \qquad \text{or} \qquad C_0 \int_{1}^{C/C_0} \eta \, d(C/C_0)$$

and calculating the slope $dC/d\eta$ for each value of η over the region where the diffusivity is not constant. To the left of the origin, η has negative values. Experimentally, the condition that C is only a function of η can be checked by plotting x versus $(t)^{1/2}$ for a given value of concentration. (We should observe a straight-line relationship.) The above derivation is called the Boltzmann transformation. Matano used this method to study the interdiffusion of alloys across the interface of two metals. Thus this method is also called the Boltzmann-Matano analysis.

Constant total dopants Equation 15 requires the concentration at a distance far to the left of $\eta=0$ (Fig. 3) to remain invariant with diffusion time. In most device fabrications, the diffusion is done after the introduction of the impurity into silicon; thus, Eq. 15 can not be used to determine the concentration-dependent diffusivity from the measured concentration profiles. For example, Eq. 15 is not applicable to redistribution diffusion from a high-concentration predeposition-diffused layer or an ion implanted layer at high ion doses. An alternative expression is used to remove the constant surface-concentration condition and it is replaced by a requirement that the total dopant remain invariant with diffusion time (i.e., constant total dopant).[8, 9] This requirement is expressed as

$$S = \int_0^{\infty} C(x,t) \, dx = \text{constant} \tag{16}$$

where S is the total dopant per unit area in the diffused layer and is independent of the diffusion time. Equation 16 has been applied to the redistribution diffusion of arsenic from an ion implanted layer.[10] The expression for determining the diffusion coefficient from the concentration profile is given by

$$D\left[\frac{C(x_0,t)}{C_s}\right] = \frac{-C(x_0,t)x_0}{2t\left[\dfrac{dC}{dx}\right]\Bigg|_{x=x_0}} \tag{17}$$

where C_s is the surface concentration, x_0 is the location at which D is determined, and $(dC/dx)_{x=x_0}$ is the concentration gradient at $x = x_0$.

For diffusion in an oxidizing ambient, assuming that the oxidation rate is a linear function of diffusion time, the equation corresponding to Eq. 17 is

$$D \left[\frac{C(x_0,t)}{C_s} \right] = \frac{-C(x_0,t)\,(x_0 + d)}{2t \left[\dfrac{dC}{dx} \right] \Bigg|_{x=x_0}} \tag{17a}$$

where d is the oxide thickness, which equals $2vt$, and v is the inward velocity of the oxidizing silicon surface. Since Eq. 17a is derived under the assumption of a constant total dopant, when impurity atoms are incorporated into the oxide layer, this assumption is violated. For example, for boron redistribution in an oxidizing ambient, the total amount of boron in silicon is not a constant (refer to Chapter 4) and thus Eq. 17a can not be used.

5.3.3 Temperature Dependence of the Diffusivities

The diffusion coefficients determined experimentally over a range of diffusion temperatures can often be expressed as

$$D = D_0 \exp \left[- \frac{E}{kT} \right] \tag{18}$$

where D_0 is the frequency factor (in cm^2/s), E is the activation energy (in eV), T is temperature (in K), and k is the Boltzmann constant (in eV/K). Thus when D is plotted versus $1/T$ on semilogarithmic coordinates, D is a straight line with slope E/kT. From the atomic diffusion theories involving the defect-impurity interactions, D_0 is related to the atomic jumping frequency or the lattice vibration frequency (typically 10^{13} Hz) and a jumping distance of an impurity, a defect, or defect-impurity pairs. At the diffusion temperatures D_0 can often be considered temperature independent. The activation energy E is related to the energies of motion and the energies of formation of defect-impurity complexes.

In metals and for some elements in silicon for a simple vacancy diffusion model, E is between 3 and 4 eV, while for the interstitial diffusion model E is between 0.6 and 1.2 eV. Thus by measuring the diffusivity as a function of temperature, we can determine whether the diffusion is dominated by an interstitial or vacancy mechanism. For fast diffusants, the measured activation energies are generally less than 2 eV and the diffusion mechanism is considered to be related to interstitial atom movements.

5.4 ATOMISTIC DIFFUSION MECHANISMS

The concept of point-defect impurity interaction and their effects on impurity diffusion are further developed in this section. Experimental results for impurity diffusion at low concentrations follow the phenomenological description of the diffusion process defined by Fick's diffusion law with a constant diffusion coefficient. The upper limit of the dopant concentration for which the diffusion coefficient is a constant can be estimated from the intrinsic carrier concentrations n_i at the diffusion temperature. When the impurity concentration $C(x)$ is less than n_i, the diffusion results can be described by a concentration-independent diffusion coefficient, and Eqs. 8 and 12 in Section 5.3, with the appropriate boundary conditions, can be used to determine the diffusion coefficients from the measured diffusion profiles. The diffusion coefficient at low concentrations is often referred to as the intrinsic diffusion coefficient D_i. When the impurity concentration, including both the substrate doping and the diffusant, is greater than $n_i(T)$, the silicon is considered as extrinsic silicon and the diffusivity is considered as the extrinsic diffusivity D_e. Experimentally measured values of D_i and D_e for boron, phosphorus, arsenic, and antimony are summarized in Section 5.5.

To understand the diffusion process at high-concentration levels and the physical mechanisms for the impurity diffusion at various concentration levels, atomic models of solid-state diffusion have been proposed and compared with experimental measurements. The atomic mechanism of solid-state diffusion was established from the diffusion study in metals. The vacancy mechanism is most probable in a cubic face-centered crystal.[11] Diffusion in silicon can be described by mechanisms involving impurity and point-defect interactions with the point defects at different charge states.

Point defects can become electrically active when they accept or lose electrons. A vacancy can be charged to act as an acceptor with a negative charge, V^-.

$$V + e \leftrightarrows V^- \tag{19}$$

Similarly an interstitial atom can be charged to act as an accepter I^-;

$$I + e \leftrightarrows I^- \tag{20}$$

where V represents a vacancy and I represents an interstitial. These concepts of ionized point defects have been applied to impurity diffusion in silicon with varied success. It has been found that both vacancy and interstitial atoms can be neutral, singly charged, or doubly charged. The probability of a charge state higher than 2 is very small. The exact mechanisms that dominate a diffusion process depend on the species under consideration; in many cases a consensus can not be reached.[12]

Equations 19 and 20 express equilibrium reactions, so the law of mass action can be applied to determine the equilibrium constants. The law of mass action states that the equilibrium constant of a chemical reaction in the gas phase can be expressed in terms of the chemical activity of the reactants and products. Consider a simple reversible chemical reaction

$$aA + bB \leftrightarrows cC \tag{21}$$

The equilibrium constant of the reaction towards the right-hand side is

$$K_c = \frac{a_A^a a_B^b}{a_C^c} \tag{22}$$

where K_c is the equilibrium constant, a_A is the chemical activity of element A, a_B is the chemical activity of element B, a_C is the chemical activity of the product C, and a, b, and c represent the mole concentration of elements A, B, and C of the reaction shown in Eq. 21. For a dilute solution (a near ideal solution), the activities can be replaced by the concentrations of the reactants and products according to Rault's law, and Eq. 22 becomes

$$K_c = \frac{[A]^a [B]^b}{[C]^c} \qquad (23)$$

where $[A]$ is the concentration of element A, $[B]$ is the concentration of element B, and $[C]$ is the concentration of element C.

The law of mass action has been applied to dilute solid solutions where point defects in a solid are considered as dilute solid solutions of defects in the crystal lattice. The law of mass action is applicable to a dilute solid solution when the reactions are in thermal equilibrium and sometimes applicable when the reactions are in quasi-thermal equilibrium.

Vacancy and interstitial concentrations can be determined from statistical thermodynamics. They are expressed in terms of entropies of formation ΔS and formation energies ΔH. For a neutral monovacancy in silicon, the concentration C_V^x can be expressed as

$$C_V^x = 5.5 \times 10^{20} \exp \left[\frac{\Delta S_V^x}{k} \right] \exp \left[\frac{-\Delta H_V^x}{kT} \right] \qquad (24)$$

where ΔS_V^x is the entropy of formation of a neutral monovacancy, ΔH_V^x is the formation energy of a neutral monovacancy (expressed in eV). The superscript x represents a neutral charge state of the defect. The subscript V denotes a vacancy defect.

For silicon, ΔH_V^x is estimated to be greater than or equal to 2.5 eV and ΔS_V^x is estimated to equal $1.1k$. Thus the intrinsic concentration of monovacancy at the diffusion temperatures of interest is rather low for silicon.

For an extrinsic silicon, the acceptor-type vacancy concentration can be expressed as[13]

$$C_V^- = \frac{1 + \frac{1}{2} \exp \left[\dfrac{E_V - E_i}{kT} \right]}{1 + \frac{1}{2} \exp \left[\dfrac{E_V - E_F}{kT} \right]} C_i(V^-)$$

$$\cong \frac{\exp \left[\dfrac{E_V - E_i}{kT} \right]}{\exp \left[\dfrac{E_V - E_F}{kT} \right]} C_i(V^-) \qquad (25)$$

for $(E_V - E_F) \gg kT$ and $(E_V - E_i) \gg kT$. C_V^- is the acceptor vacancy concentration in the extrinsic silicon, C_i (V^-) is the acceptor vacancy concentration in the intrinsic silicon, E_V is the acceptor vacancy energy level (in eV), E_i is the intrinsic Fermi level (in eV), and E_F is the Fermi level of the extrinsic silicon (in eV). Thus,

$$\frac{C_V^-}{C_i \ (V^-)} = \exp \left[\frac{E_F - E_i}{kT} \right] \tag{26}$$

But for the nondegenerate case, we obtain

$$n = n_i \ \exp \left[\frac{E_F - E_i}{kT} \right]$$

for n-type silicon, and Eq. 26 becomes

$$\frac{C_V^-}{C_i \ (V^-)} = \frac{n}{n_i} \tag{27}$$

If the impurity diffusion is dominated by the acceptor monovacancy mechanism, the diffusion coefficient is approximately proportional to the acceptor monovacancy concentration. Thus, we have

$$\frac{D}{D_i} = \frac{n}{n_i} \tag{28}$$

where D is the diffusion coefficient in extrinsic silicon, and D_i is the diffusion coefficient in intrinsic silicon.

The intrinsic carrier concentration n_i can be calculated using the following empirical formula:[14]

$$n_i^2 = 1.5 \times 10^{33} \ T^3 \ \exp[\ (-1.21 + \Delta E_g)/kT \ ] \tag{29}$$

where

$$\Delta E_g = -7.1 \times 10^{-10} \left[\frac{n_i}{T} \right]^{\frac{1}{2}} \tag{30}$$

and an assumed $E_g = 1.21$ eV.

Equation 28 states that the interaction of the impurity atoms with charged acceptor vacancies leads to a dependence of the diffusion coefficient on the Fermi level at the diffusion temperature. Since vacancies and interstitials can have various charge states, Eq. 28 can be generalized to include all possible combinations of impurity-point defect interactions.[15]

$$D = D^x + \sum_{r=1}^{m} (D^{-r}) \left| \frac{n}{n_i} \right|^r + \sum_{r=1}^{m} (D^{+r}) \left| \frac{n_i}{n} \right|^r \tag{31}$$

where D^x, x for neutral defects, (D^{-r}), and (D^{+r}) refer to the intrinsic impurity diffusivities associated with the particular charge states, r, of the point defects that affect the impurity diffusion and r is an integer 1, 2, 3, ..., m. For example, D^x represents the intrinsic diffusivity of impurity interaction with a neutral-point defect; (D^-), $(r =$

1) represents the intrinsic diffusivity of impurity interaction with a singly charged acceptor-point defect; and (D^+) represents the intrinsic diffusivity of impurity interaction with a singly charged donor-point defect. The exponent r in (D^{-r}) and (D^{+r}) corresponds to the charge state of the point defect. For example, (D^{-2}) (or D^{2-}) represents the intrinsic diffusivity of impurity interacting with doubly charged acceptor defects, and the corresponding contribution to the diffusivity is $D^{2-}(n/n_i)^2$. The superscript r in Eq. 31 does not represent an exponent for (D^{-r}) and (D^{+r}) but it is the exponent for $(n/n_i)^r$ and $(n_i/n)^r$ terms. Thus when Eq. 31 is used to fit experimental profiles with defects of different charge states, it does not specify the dominating diffusion mechanism or mechanisms. The exact mechanisms, either vacancy or self-interstitial type, involved in the impurity-defect interaction during the diffusion process have to be determined from other experimental evidence and/or theoretical considerations. We can therefore consider Eq. 31 as a phenomenological expression of the concentration dependence of the diffusion coefficients, which provides a description of diffusion phenomena by extending Fick's diffusion equation (Eq. 3).

The concentration-dependent diffusion coefficient can be determined from the experimental diffusion profiles without knowing the details of the atomic diffusion mechanisms. However, the measured diffusion coefficients as a function of diffusion temperature can sometimes be fitted to appropriate impurity-point defect interaction models.

Isolated point defects in silicon are generated at or below room temperatures by high-energy (≥ 1 MeV) electron, x-ray, or neutron irradiations. When these defects are in various charge states, their electronic states and annealing properties can be studied by electron paramagnetic resonance (EPR) measurements,[16] by infrared absorption spectra analysis[17] for neutral defects, and by other techniques. The deep level transient spectroscopy (DLTS) method has also been used to study electrically active defects in proton-bombarded silicon crystals.[18] Theoretical calculations of these defects have also been made, using various models of the charge states of these point defects and their annealing properties to explain the experimental observations. For vacancies in silicon, the EPR and optical absorption studies have identified four charge states (V^+, V^x, V^-, and V^{2-}), where V^+ is a donor vacancy, V^x a neutral vacancy, V^- an acceptor vacancy, and V^{2-} a doubly charged acceptor vacancy.[19]

Figure 4 shows a few examples of the geometrical configurations of vacancy and interstitial point defects which have been established from theoretical and experimental studies. These three-dimensional models can be used to calculate the activation energies and entropies of defects with different charge states. Figure 4a shows the atomic arrangements of two tetrahedra in a silicon crystal lattice, and Fig. 4b shows a simple vacancy. Figure 4c shows one of the possible configurations of a divacancy which has atoms missing from two neighboring bonds[19] (the dotted circles in Fig. 4c). Three kinds of interstitials have been used to calculate theoretically the characteristics of the observed defect configurations, namely, the simple tetrahedron (Fig. 4d), the bond centered (Fig. 4e), and the $\langle 110 \rangle$ split interstitial[20] (Fig. 4f). In a unit cell, the positions for the five interstitial sites are (1/2, 1/2, 1/2), (1/4, 1/4, 1/4), (1/4, 3/4, 3/4), (3/4, 1/4, 3/4), and (3/4, 3/4, 1/4).

The investigations of point-defect formation by studying the radiation effects have provided fundamental information on the defect configurations, energies, and

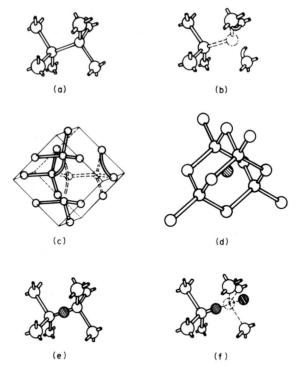

Fig. 4 Geometrical configurations of vacancy and interstitial point defects. (a) 8 Si atoms form two adjacent tetrahedral bonds. (b) A simple vacancy. (c) Divacancies. (d) A simple tetrahedral interstitial. (e) A bond centered interstitial. (f) An $\langle 100 \rangle$ split interstitial.

entropies of formation and migration. This information is used to analyze atomic diffusion mechanisms in silicon from the measured diffusion coefficients as a function of temperature.

Researchers have reasoned that silicon and Group III and V elements in silicon should have a similar diffusion mechanism. Thus extensive efforts have been made to study silicon self-diffusion. The measured silicon self-diffusion coefficients D_{Si} can be explained by a vacancy model[21] involving neutral vacancies V^x, singly charged acceptor vacancies V^-, doubly charged acceptor vacancies V^{2-}, and singly charged donor vacancies V^+:

$$D_{\text{Si}} = D_{\text{Si}}^x + D_{\text{Si}}^- \left[\frac{n}{n_i} \right] + D_{\text{Si}}^{2-} \left[\frac{n}{n_i} \right]^2 + D_{\text{Si}}^+ \left[\frac{n_i}{n} \right] \tag{32}$$

with

$$D_{\text{Si}}^x = 0.015 \exp \left[\frac{-3.89 \text{ eV}}{kT} \right] \tag{33}$$

$$D_{\text{Si}}^- = 16 \exp \left[\frac{-4.54 \text{ eV}}{kT} \right] \tag{34}$$

$$D_{Si}^+ = 1180 \exp \left[\frac{-5.09 \text{ eV}}{kT} \right] \tag{35}$$

$$D_{Si}^{2-} \cong 10 \exp \left[\frac{-5.1 \text{ eV}}{kT} \right] \tag{36}$$

The values for D_{Si}^{2-} in Eq. 36 are estimated values with activation energy close to that of D_{Si}^+ and a $D_0 = 10$. Figure 5 shows the silicon self-diffusivity versus temperature. The units for all the diffusivity expressions are cm^2/s. Thus the effect of D_{Si}^{2-} can be neglected in Eq. 32. Although the study of irradiation of silicon established the existence of V^{2-}, it contributes little to the silicon self-diffusivity. The V^{2-} becomes significant for high-concentration phosphorus diffusion, which we shall discuss later.

Similarly, over a narrow range of high temperatures, the silicon self-diffusion data can be expressed in terms of D_{Si}^x, D_{Si}^-, and D_{Si}^+ with activation energies 5.23, 4.84, and 3.91 ev, respectively.[15] Note that these values for the activation energies for D_{Si}^x and D_{Si}^+ are almost opposite to those of Eqs. 33 and 35. Calculations of diffusion coefficients using experimental data and Eq. 31 are very sensitive to the accuracy of these data, and thus, these results represent approximations. As the accuracy of the

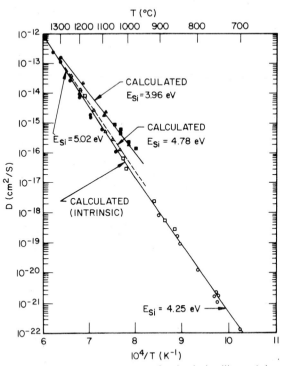

Fig. 5 Silicon self-diffusivity vs. temperature. ● and □ intrinsic silicon; △ boron doped to 2.5 $\times 10^{19}$ cm^{-3}; ▲ arsenic doped to 8×10^{19} cm^{-3}; ■ phosphorus doped; ○ nickel doped in intrinsic silicon.

(After Fair, Ref. 21.)

measurements improves, theoretical models also improve. A neutral self-interstitial mechanism for silicon self-diffusion over a temperature range of 1050° to 1380°C has also been proposed.[22] However, the mechanism for silicon self-diffusion has not been clearly resolved.

5.5 MEASUREMENT TECHNIQUES

Diffusivity, an important parameter in diffusion study, must be determined experimentally. This section discusses measurement techniques for determining diffusivities in diffusion study.

5.5.1 Junction Depth and Sheet Resistance

Diffusion results can be checked by two simple measurements, the junction depth and the sheet resistance of the diffused layer. The junction depth is commonly measured with a chemical staining of a beveled (1 to 5°) sample in a mixture of 100 cm^3 HF (49%) and a few drops of HNO$_3$. Sometimes HF alone is sufficient. If the sample is put under a strong illumination for a minute or two, the p-type region will be stained darker than the n-type region. With the aid of the interference-fringe techniques of Tolansky,[23] the junction depths can be measured accurately from 0.5 to over 100 μm.

The sheet resistance of a diffused layer can be measured by a four-point probe technique (Section 1.3). A geometric correction factor is required to convert the measured resistance V/I into sheet resistance (also called the sheet resistivity). This factor is a function of the sample size, shape, and the probe spacings. The sheet resistance R_s is given by Eq. 37, and the correction factors for simple circular, rectangular, and square samples are given[24] in Table 1.

$$R_s = \frac{V}{I} \text{ C.F.} \tag{37}$$

where R_s is the sheet resistance of a diffused layer (in $\Omega/\square$); V is the measured dc voltage across the voltage probes (in volts); I is the constant dc current passing through the current probes (in amperes); and C.F. is the correction factor that is a function of the sample geometry and the probe spacings.

The correction factors for a circular sample (with a diameter d) and a rectangular sample (with the side parallel to the probe line as a and that perpendicular to the probe line as d) are given in Table 1 (where s is the probe spacing). Note that for a large d/s, the correction factor approaches that of a two-dimensional sheet extending to infinity in both directions, that is, C.F. = 4.53. For the correction factors to be insensitive to the sample size and the positions of the probe points with respect to the sample edge, a large d/s is desirable. Equation 37 and the correction factors in Table 1 are valid only for shallow junctions which are diffused only on the front side of the sample. Diffusion from a chemical source will have the diffused region wrapped around the sample. The back side of the diffused layer has either to be removed or isolated from the front side; otherwise a different correction factor should be used.

When measuring a shallow diffused layer at low concentrations, reliable measurements free of noise are difficult to make. This problem is sometimes overcome by

Table 1 Correction factor C.F. for the measurement of sheet resistances with the four-point probe[24]

d/s	Circle diam d/s	Square $a/d=1$	Rectangle $a/d=2$	Rectangle $a/d=3$	Rectangle $a/d \geqslant 4$
1.0				0.9988	0.9994
1.25				1.2467	1.2248
1.5			1.4788	1.4893	1.4893
1.75			1.7196	1.7238	1.7238
2.0			1.9475	1.9475	1.9475
2.5			2.3532	2.3541	2.3541
3.0	2.2662	2.4575	2.7000	2.7005	2.7005
4.0	2.9289	3.1137	3.2246	3.2248	3.2248
5.0	3.3625	3.5098	3.5749	3.5750	3.5750
7.5	3.9273	4.0095	4.0361	4.0362	4.0362
10.0	4.1716	4.2209	4.2357	4.2357	4.2357
15.0	4.3646	4.3882	4.3947	4.3947	4.3947
20.0	4.4364	4.4516	4.4553	4.4553	4.4553
40.0	4.5076	4.5120	4.5129	4.5129	4.5129
∞	4.5324	4.5324	4.5325	4.5325	4.5324

measuring the voltages for current flowing in two directions, and then averaging the two readings. This average reading removes some of the effect of contact resistance. If the voltage differences are large, however, probe points and the cleanliness of the sample surface should be checked. To ensure that the readings are correct, the sheet resistances at two or three current levels can be measured. These measurements show whether the measured sheet resistances are constant over the range of measured currents. For high-resistivity silicon, annealing the sample in N_2 at 150°C for a few minutes improves the accuracy of readings. Always try to use as low current as possible to avoid ohmic heating or to avoid reaching the punchthrough voltage.

For a diffused layer, an average sheet resistance R_s is related to the junction depth x_j, the carrier mobility μ, and the impurity distribution $C(x)$ by the following expression:

$$R_s = \frac{1}{q \int_0^{x_j} \mu C(x)\, dx} \tag{38}$$

The depletion of charge carriers near x_j can be neglected in the above calculation. In general, the mobility is a function of the total impurity concentration, and often an effective mobility is defined as

$$\mu_{\text{eff}} = \frac{\int_0^{x_j} \mu[C(x)]C(x)\, dx}{\int_0^{x_j} C(x)\, dx} \tag{39}$$

Equation 38 can be expressed as

$$R_s = \frac{1}{q \, \mu_{\mathrm{eff}} \int_0^{x_j} C(x) \, dx} \tag{40}$$

For a given diffusion profile, the average resistivity, $\rho = R_s \, x_j$, is uniquely related to the surface concentration of the diffused layer and the substrate dopant concentration for an assumed diffusion profile. Design curves relating to the surface concentration and the average resistivity (or the average conductivity) have been calculated for simple diffusion profiles, such as exponential, Gaussian, or erfc distributions. They are often called the Irvin curves.[25] To use these curves, be sure that the diffusion profiles agree with the assumed profiles. For high concentration and shallow diffusions, the diffusion profiles cannot be represented by these simple functions. The measured sheet resistance and junction depth can not be used to find the impurity surface concentration or calculate the diffusivities of the diffused layer with the Irvin curves.

Since both the junction-depth measurement and the sheet-resistance measurement are simple and give important information about a diffused layer without using elaborate profile measurements, they are used routinely for monitoring diffusion processes. For ion implanted samples, sheet-resistance measurement is a simple method to check the electrical activity (the combined effects of mobilities and carrier concentrations) after the sample is annealed or diffused.

5.5.2 Profile Measurements

The diffusivities and the diffusion models that describe the diffusion results are self-consistent for the diffusion conditions for which the diffusion profiles are determined. The accuracy of the diffusion model and its associated diffusivities depends on the correctness of the diffusion profile measurements which are indispensable in diffusion studies. The simple measurements of the junction depth and the sheet resistance of a diffused layer, although useful for process monitoring, are grossly inadequate for diffusion study. A few commonly used techniques for diffusion profile measurements and their limitations are discussed in the following sections.

C-V technique From the p-n junction theory, the space-charge capacitance is a function of the reverse-bias voltage. For the depletion approximation, this capacitance can be treated as a parallel-plate capacitor. For an abrupt junction where the impurity concentration is very high on one side of the junction and decreases to a low value abruptly on the other side (i.e. an n^+p or p^+n junction), the following expression[26] can be derived:

$$C(x) = \frac{2}{q \, \epsilon_s \, \dfrac{d}{dV}\left[\dfrac{1}{C(V)}\right]^2} = \frac{C^3(V)}{q \, \epsilon_s} \, \frac{1}{\dfrac{dC}{dV}} \tag{41}$$

and

$$x = \frac{\epsilon_s}{C(V)} \tag{42}$$

where $C(x)$ is the impurity concentration at the space-charge layer edge, $C(V)$ is the junction reverse-bias capacitance per unit area at a reverse voltage V, and ϵ_s is the dielectric permittivity of silicon. To avoid confusion in the symbols, $C(x)$ means concentration and $C(V)$ means junction capacitance. Now,

$$V = V_R + V_{bi} \tag{43}$$

where V_R is the applied reverse bias, and V_{bi} is the built-in potential of the p-n junction.

$$V_{bi} \cong \frac{kT}{q} \ln \frac{C_A C_D}{n_i^2} \tag{44}$$

where C_A is the acceptor concentration, and C_D is the donor concentration. Thus,

$$C(V) = \left[\frac{q \epsilon_s}{2} C_B \right]^{1/2} \left[V_{bi} \pm V_R - \frac{2kT}{q} \right]^{-1/2}$$

$$= \frac{\epsilon_s}{\sqrt{2} L_D} (\beta V_{bi} \pm \beta V - 2)^{-1/2} \tag{45}$$

where C_B is the substrate doping concentration, $\beta \equiv q/kT$, and

$$L_D = \text{the Debye length} = \left[\frac{\epsilon_s}{qC_B} \frac{kT}{q} \right]^{1/2} \tag{46}$$

Thus, V_{bi} can be determined from the junction capacitance at zero reverse bias from Eq. 45. The C-V method is limited to a few L_D's away from the depletion layer edge at zero bias and it can not resolve the concentration distribution within a few L_D's. The impurity profile can be determined by measuring the reverse-bias capacitance as a function of the applied voltage from Eqs. 41, 42, and 45.

Figure 6 gives an example of the measured C-V profiles for phosphorus implanted then diffused samples by using a Schottky diode. The zero-bias space-charge width is close to 0.1 μm. The phosphorus concentration in this surface region can not be easily measured and has to be estimated. The diffusion was at 1100°C for 15, 30, and 60 min in O_2. Note that all three profiles (data points shown in Fig. 6) can be represented by a Gaussian distribution (Eq. 12) with a constant diffusion coefficient $D = 2.34 \times 10^{-13}$ cm^2/s and a total phosphorus concentration of

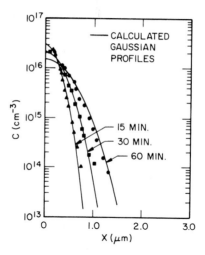

Fig. 6 Phosphorus profiles from C-V measurement. Phosphorus implantation: ion energy = 30 keV, ion dose = 10^{12} cm^{-2}. Diffusion at 1100°C in oxygen. Calculation $2Dt = 1.7 \times 10^{-9}$ cm^2, $\Delta R_p^2 = 1.7 \times 10^{-12}$ cm^2, $S = 8 \times 10^{11}$ cm^{-2}, $D = 2.3 \times 10^{-13}$ cm^2/s.

$S = 8 \times 10^{11}$ cm^{-2} which is within 20% of the implant dose. The ΔR_p^2 is one-thousandth of $2Dt$; thus the implanted profile can be assumed to be a delta function (i.e., all the implanted atoms are confined to a very thin sheet with a total phosphorus concentration S and less than 20% of the implant dose is incorporated into the oxide film which was grown during the diffusion).

Differential conductivity technique Differential conductivity is one of the oldest techniques for measuring the diffusion profiles in silicon by the electrical method.[27] This technique involves repeatedly measuring the sheet resistance of a diffused layer by the four-point probe measurement after removing a thin layer of silicon by anodic oxidation and etching the oxide off in HF solution. Because the anodic oxidation is at room temperature, the impurity atoms do not move in the diffused layer during oxidation and there is no segregation effect; hence, a true distribution profile can be determined. To use this technique, either the carrier mobility is measured by the Hall effect measurement or the resistivity versus impurity concentration curves are used.[28] Figure 23 of Chapter 1 gives the composite curves of the resistivities for boron- and phosphorus-doped silicon over a wide range of concentrations. The polynomial fittings for calculating the impurity concentration from resistivity measurements are given in Ref. 28. The differential conductivity technique is not suitable for diffusion study in VLSI process development.

Spreading resistance technique The C-V technique has a limited range of junction depths and dopant concentration that can be used for profile measurement, and the differential-conductivity technique is a time-consuming method for profiling diffused layers. Various techniques have been investigated to improve the spatial resolution and to reduce the measurement time, and as a result, the two-point probe spreading resistance technique has been developed[29] for diffusion profile measurement. Since a refined and improved instrument is commercially available, the spreading resistance technique for diffusion profile measurement is becoming a routine evaluation technique.

For a two-point probe arrangement, the total spreading resistance is given by

$$R_{sr} = \frac{\rho}{2a} \qquad (47)$$

where R_{sr} is the spreading resistance, ρ is the average resistivity near the probe points, and a is the probe radius. The spreading resistance technique is very sensitive to local impurity concentration variations, that is, it has high spatial resolution. However, measurements are also sensitive to the sample surface and the conditions of the probe points. Unless very elaborate measuring and checking procedures are conducted, this technique is best used to compare an unknown sample with a sample of known profile. For profile comparisons, this technique is often sufficient. Concentration profiles, however, should be checked with another method such as the differential conductivity method or the SIMS method to be discussed next. To convert spreading resistance into concentration, various correction factors have been derived for different boundary conditions. Because we have imprecise knowledge of these correction factors and varying probe conditions, empirical calibration curves have to be used. Often only the spreading resistance profiles are used for comparing different treatment results. Figure 7 shows an example of the spreading resistance profile of a transistor structure; the collector-base junction x_{cb} and the emitter-base junction x_{eb}

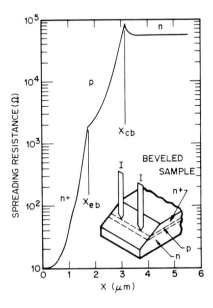

Fig. 7 The spreading resistance profile of an n-p-n transistor structure. x_{eb} = the emitter-base junction depth = 1.7 μm; x_{cb} = the collector-base junction depth = 3.2 μm.

are clearly shown. The emitter region n^+ is phosphorus diffused and shows a kink in the profile about 1.2 μm from the surface. This kink in the phosphorus profile has been extensively studied and it will be discussed in Section 5.6.3.

SIMS technique Chapter 12 discusses the principle and instrument design of the secondary ion mass spectroscope (SIMS), an important tool for diffusion profile measurement.[30] Since the SIMS technique is not a primary measurement, converting the secondary ion signal into concentration requires the use of either a standard sample or certain established procedures that are described in the following paragraphs.

Two methods are often used to convert the secondary ion signal into concentration: (1) using the ratio of the ion yield of the element of interest to that of the host element (^{30}Si in the present case), and (2) using samples with known concentrations as calibration standards. When the impurity concentration is high (10^{20} atoms/cm^3), the ion-ratio technique is accurate and convenient. This technique provides an internal standard; ion signals of the elements are collected under the same measurement conditions. Any change in the measurement or the equipment conditions will be seen as a change of the ion yield of the host element. Because of the limitation of the counting system and the presence of background ion counts, the range of the SIMS measurement is between 10^3 to 10^4. For example if the boron surface concentration is in the range of 10^{20} atoms/cm^3, the measurement limit will be between 10^{16} and 10^{17} atoms/cm^3, although the detection limit for boron in silicon is below 10^{16} atoms/cm^3.

In the second method, samples of known impurity concentrations are measured under the same conditions as the sample for which the diffusion profile is measured. The ratio of the ion counts are assumed to be proportional to the concentration ratios, and the ion counts are assumed proportional to the atomic concentrations. With the measurement conditions optimized, both assumptions have been verified for common impurity elements in silicon. Ion implanted samples provide a convenient set of standards over a wide range of ion doses. Experiments have shown that the ion counts at the peak concentration are a linear function of the ion doses, and that the integrated ion counts are also a linear function of the ion doses for samples implanted at the same energy. Both results establish the relationship that the secondary-ion counts are linear functions of the atomic concentration of the element. Figure 8 shows an example, for boron-implanted samples, of the peak ion counts versus ion doses and the ratio of the peak ion counts to ^{30}Si ion counts versus ion doses. For measuring the diffusion profiles of ion implanted samples, integrated ion counts are preferred over peak ion counts, because integrated ion counts are not sensitive to slight variations in the measurement conditions.

The SIMS technique measures the total impurity profile. Thus, other electrical methods should be used for determining the electrically active portions. Since the sputtering rates generally range from less than one angstrom per second to several tens of angstroms per second, this technique is suited for measuring diffusion profiles for depths less than 1 or 2 μm.

Figure 9 gives a few examples of measured profiles. Figure 9a shows the SIMS profile of a phosphorus-diffused layer and, for comparison, profile measured by the differential conductivity technique. Figure 9b shows the boron profile in SiO$_2$ and Si

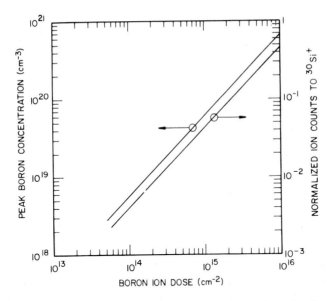

Fig. 8 SIMS analysis calibration curves. Peak boron concentration versus boron ion dose and normalized peak ion counts to ^{30}Si ion counts versus boron ion dose.

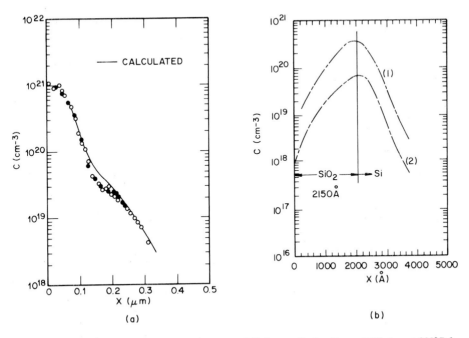

Fig. 9 Examples of SIMS profiles. (a) Phosphorus diffusion profile in silicon. Diffusion at 900°C for 30 min. POCl$_3$ source. ○ Differential conductivity data. ● SIMS data. *(After Fair, Ref. 21.)* (b) Boron implanted profile in SiO$_2$ and Si. (1) Ion dose = 5×10^{15} cm^{-2}, (2) Ion dose = 1×10^{15} cm^{-2}. Implant energy = 50 keV.

from a sample which has a 2150 Å thermal oxide and in which the boron was implanted at 50 keV. In this case, the boron concentration at the interface is nearly continuous. Nearly half of the implanted boron atoms are in the oxide layer. These examples show that the SIMS technique is a powerful tool for profile determination and will, therefore, be extensively applied to diffusion studies in VLSI technology.

Summary of profiling techniques Various other techniques have also been used for impurity profile measurements. These techniques often require special laboratory set-ups or special equipment, but are useful for independently determining the total impurity concentration profile and to verify the results from the electrical or SIMS measurement. Table 2 summarizes the measurement techniques discussed in the previous sections and others that were not discussed in detail but are mentioned here.

The Rutherford backscattering (RBS) technique has been used for measuring distributions of heavy elements (such as arsenic, platinum, gold, etc.)[32] in silicon but cannot be used for measuring boron or phosphorus profiles. In this technique high-energy helium ions (1 to 3 MeV) are used as the incident ion beam, and the backscattered He ion energy-loss spectra are analyzed. A few nuclear reaction processes have been used for measuring the boron atom distribution nondestructively. For example, thermal neutrons interact with ^{10}B, causing the emission of monoenergetic ^4He ions at 1471 keV.[33] By analyzing the energy losses of the helium ions, the depth of boron atoms can be determined from the specific-energy loss spectra of ^4He ions in silicon, which are measured experimentally. The boron concentration can be related to the ^7Li particle signals at 839 keV (94%) and 1014 keV (6%) that are generated in the

Table 2 Commonly used diffusion profile measurement techniques

Profile techniques	Characteristics	Ref.
Capacitance-Voltage	Carrier concentration at the edge of the depletion layer of a p-n junction. Maximum total dopants 2×10^{12} atoms/cm^2.	26
Differential conductance and Hall effect	Resistivity and mobility of net electrically active species. Requires thin layer removal. 10^{20} to 10^{18} atoms/cm^3.	27
Spreading resistance	Resistance on angle beveled sample. Good for comparison with known profiles and quick semi-quantitative evaluation. Depth >1 μm.	29
SIMS	High sensitivity on many elements, for B and As detection limit 5×10^{15} cm^{-3}. Capable of measuring profiles in 1000 Å range. Needs standards.	30
Radioactive tracer analysis	Total concentration. Lower limit 10^{15} cm^{-3}. Limited to radioactive elements with suitable half-life times: P, As, Sb, Na, Cu, Au, etc.	31
Rutherford backscattering	Only applicable for elements heavier than Si.	32
Nuclear reaction	Measures total boron through $^{10}B(n, {}^4He)^7Li$, or $^{11}B(p, \alpha)$. Needs Van de Graaff generator.	33 34

nuclear reaction of $^{10}\text{B}(n,\,^4\text{He})^7\text{Li}$. Another nuclear reaction for measuring boron profiles involves the use of a proton beam at 400 keV which reacts with ^{11}B in silicon.[34] The energy spectra of the α-particles from their reaction have been analyzed. This reaction is expressed as $^{11}\text{B}(p,\,\alpha)$. For boron implanted profiles, the results of this method and of the SIMS method agree.

5.6 DIFFUSIVITIES OF B, P, As, AND Sb

In VLSI technology, boron, phosphorus, arsenic and sometimes antimony are used as dopant elements for junction formations. Hence, the diffusivities of these elements are of interest and they are summarized in this section. We give both the intrinsic and extrinsic diffusivities. By applying the vacancy-impurity diffusion model for multiple charge states, we can tentatively identify the species contributing to the diffusivities. Since this diffusion theory is still being developed, the identification of these species has not been confirmed. Various effects on the diffusion results at high-concentration levels and impurity interactions are also discussed.

5.6.1 Low-Impurity Concentration Diffusion into Intrinsic Silicon

Table 3 shows the intrinsic diffusivities[21] of boron, phosphorus, arsenic, and antimony in terms of a frequency factor D_0 and an activation energy E. The expression of the diffusivity as a function of temperature was given in Eq. 18.

According to the multiple-charge-state vacancy model, the boron intrinsic diffusivity is dominated by the interaction of boron with the donor-type vacancy V^+ and is designated as $(D_i{}^+)_\text{B}$. For phosphorus, the intrinsic diffusivity is dominated by interaction of impurity atoms with the neutral vacancy V^x and is designated as $(D_i^x)_\text{P}$. For arsenic, three sets of D_0 and E are given in Table 3. Since each set of data represents the measured values for the experimental conditions studied and all of them

Table 3 Intrinsic diffusivity of B, P, As and Sb

	Unit	Boron	Phosphorus	Arsenic[*]			Antimony
				CS	PD	IS	
		$(D_i{}^+)_\text{B}$	$(D_i^x)_\text{P}$		$(D_i{}^-)_\text{As}$		$(D_i^x)_\text{Sb}$
D_0	cm^2/s	0.76	3.85	24	22.9	60	0.214
E	eV	3.46	3.66	4.08	4.1	4.2	3.65

[*]CS are results from chemical source and PD are results from predeposition diffusion of ion implanted ^{75}As and low-concentration predeposited layers (Ref. 10). IS are the results from iso-concentration diffusion experiments (Ref. 35).

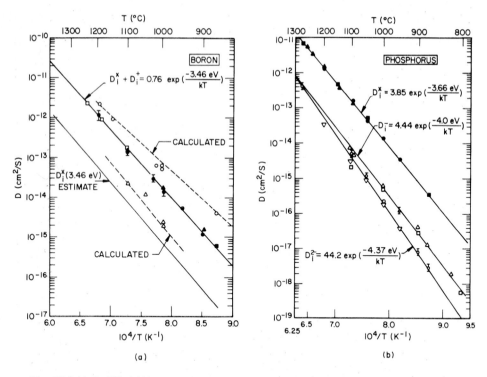

Fig. 10 Intrinsic diffusivities vs. temperature. (a) Boron: ●, ▲, □, ■ data from diffusion in intrinsic silicon; △ data from diffusion in n-type silicon doped to 1.5×10^{20} cm^{-3}; ○ data from diffusion in p-type silicon doped to 5×10^{19} cm^{-3}. (b) Phosphorus: ●, ▲, ■ data from diffusion in intrinsic silicon; ○, △, □ data from high-concentration P diffusion; ∇ data from diffusion in extrinsic silicon.

are within the scattering of the measurement, no attempt is made to express preferences for any of them. In Table 3, D_i represents the impurity intrinsic diffusivity, $(D_i)_B$ for boron, $(D_i)_P$ for phosphorus, and so on.

Figure 10a through d summarizes the diffusivities of boron, phosphorus, arsenic, and antimony as functions of diffusion temperatures. Detailed descriptions of the experimental data on which the parts of this figure are based are given in reference 21.

5.6.2 The Electric-Field Effect

When impurity atoms are ionized at the diffusion temperature, a local electric field is set up between the ionized impurity atoms and the electrons or holes. The concentration gradient of these ionized impurity atoms (donors or acceptors) produces an internal electric field that enhances the diffusivity of the ionized impurity atoms. This internal electric field is related to the electrical potential $\phi(x)$ as

$$E_x = -\frac{\partial}{\partial x} \phi(x,t) \tag{48}$$

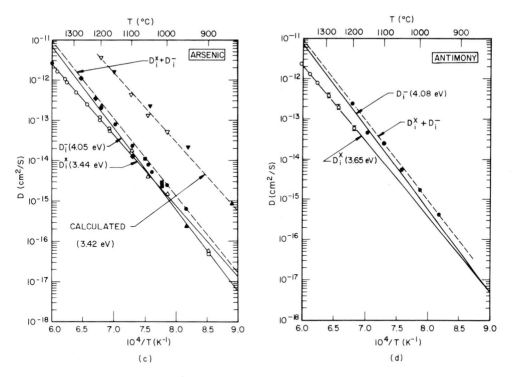

Fig. 10 (continued) (c) Arsenic: ●, ♦ , □, ○, △ data from diffusion in intrinsic silicon; ▼, ▽ data from diffusion in extrinsic silicon. (d) Antimony: ○ data from diffusion in intrinsic silicon; ● data from diffusion in extrinsic silicon. *(After Fair, Ref. 21.)*

For a donor impurity, $\phi(x,t)$ can be expressed as

$$\phi(x,t) = (E_C - E_F)/q \tag{49}$$

where E_C is the conduction band energy and E_F is the Fermi level. Assuming that a charge neutrality exists between the ionized donor and the electron and that all donor atoms N_D are ionized, we have $np = n_i^2$ and $N_D \cong n$. It can be shown that

$$E_x = \frac{kT}{q} \frac{\partial}{\partial x} \ln \left[\frac{N_D}{n_i} \right] \tag{50}$$

The diffusion flux in an electric field can be expressed as

$$J = -qD\frac{\partial N_D}{\partial x} - qZD\frac{q}{kT} N_D E_x \tag{51}$$

where Z is the charge state of the donor atoms. For a singly charged donor atom, $Z = 1$. By substituting Eq. 50 into Eq. 51 and by changing variables from $\partial/\partial x$ to $(\partial/\partial N_D)(\partial N_D/\partial x)$, Eq. 51 becomes

$$J = -qDh \frac{\partial N_D}{\partial x}$$

and

$$h \equiv 1 + Z N_D \frac{\partial}{\partial N_D} \ln \frac{N_D}{n_i} \tag{52}$$

where h is the electric-field enhancement factor. It can be shown that

$$h = 1 + \frac{N_D}{2n_i} \frac{1}{\left[\left[\frac{N_D}{2n_i}\right]^2 + 1\right]^{1/2}} \tag{53}$$

When $N_D / 2n_i \gg 1$, h equals 2 which means that the maximum enhancement of the diffusivity from the electric-field effect is 2. For an acceptor diffusion with the electric-field enhancement, N_A should be substituted for N_D in Eq. 53.

For phosphorus-diffused samples at temperatures below 900°C, an electric-field enhancement of the diffusivity has been observed in which neutral vacancies V^x dominate the diffusion and the measured D_e / D_i^x resembles h, as shown in Eq. 53. Figure 11 shows this electric-field enhancement for phosphorus.

5.6.3 High-Concentration Effects

This section briefly summarizes diffusion results of arsenic, boron, and phosphorus at high concentrations, when the surface concentrations are greater than n_i. Expressions for diffusivities which are derived from the impurity-defect interaction diffusion

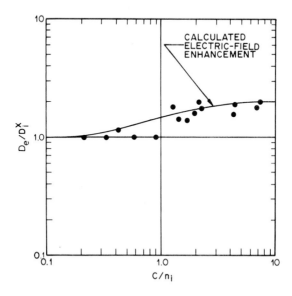

Fig. 11 Electric-field-enhanced diffusion of phosphorus in silicon at 900°C. ● data from diffusion of phosphorus in silicon in the temperature range 875−900°C. *(After Fair, Ref. 21.)*

model are given. For high-concentration arsenic, we discuss a model for the cluster formation of impurity atoms. This model explains the observation that only a portion of the diffused arsenic atoms is electrically active at room temperature. Similar results are also observed for high-concentration boron-diffused layers. Results from a phosphorus diffusion model are also given.

Arsenic According to the multi-charge-state, impurity-defect interaction model, the arsenic diffusivity can be expressed as[21]

$$D_{As} = (2n / n_i) (D_i)_{As} \tag{54}$$

Equation 54 is similar to Eq. 28; the factor 2 represents the electric-field effect. A similar expression based on interactions of charge vacancy with arsenic is[36]

$$D_{As} = \frac{1 + \gamma n / n_i}{1 + \gamma} (D_i)_{As} \tag{55}$$

with $\gamma \cong 100$ for donor-impurity diffusions. Thus, D_{As} calculated from Eq. 54 is almost twice that of D_{As} calculated from Eq. 55.

The electric activity of arsenic from ion implanted samples depends on the ion dose and the annealing or diffusion temperatures. For arsenic-ion doses below 1×10^{16} cm^{-2} and diffusion temperatures greater than 1000°C, nearly all of the arsenic atoms are ionized and contributing to the electrical activities.[10] However, for diffusion temperatures below 1000°C and an arsenic-ion dose greater than 10^{16} cm^{-2}, the concentration of ionized arsenic is a fraction of the total arsenic, and the differences become greater[37] as the diffusion temperature decreases below 900°C.

The difference between the ionized and the total arsenic can be explained by an arsenic clustering model. In this model, arsenic atoms form clusters that are partially active when their concentration is above 10^{20} cm^{-3}. The most recent clustering model consists of three arsenic atoms and one electron that are electrically active at the diffusion (or annealing) temperature and electrically neutral at room temperature.[37] The model is expressed as

$$3As^+ + e^- \underset{\xrightarrow{\hspace{1cm}}}{\overset{\text{high temp.}}{\xleftarrow{\hspace{1cm}}}} As_3^{+2} \overset{25°C}{\longrightarrow} As_3 \tag{56}$$

Applying the law of mass action to the high-temperature region, the equilibrium constant is

$$K_{eq} = \frac{[As_3^{+2}]}{[As^+]^3 n} \tag{57}$$

and the carrier concentration at the annealing/diffusion temperature is

$$n = [As^+] + 2[As_3^{+2}] \tag{58}$$

where $[As^+]$ is the carrier concentration from isolated arsenic atoms and $2[As_3^{+2}]$ is the carrier concentration as the arsenic clusters $[As_3^{+2}]$ at high temperatures. At room temperature, $[As_3^{+2}]$ is electrically neutral and the carrier concentration is

$[As^+] = C$. Thus, the total arsenic can be expressed as the sum of the unclustered arsenic $[As^+]$ and the clustered arsenic, which has three arsenic atoms per cluster:

$$C = [As^+] + 3[As_3^{+2}]$$

$$= C + \frac{K_{eq} C^4}{1 - 2K_{eq} C^3} \tag{59}$$

The second term on the right-hand side of Eq. 59 can be obtained from Eqs. 57 and 58. Limiting values for the electrically active arsenic are determined by letting

$$1 - 2K_{eq} C^3 = 0$$

or

$$C_{max} = (2K_{eq})^{-\frac{1}{3}} = 1.584 \times 10^{23} \exp\left[-\frac{0.687}{kT}\right] \tag{60}$$

A generalized model for cluster formation of arsenic atoms has been derived. The model considers m arsenic atoms interacting with k electrons[38] and anlyzes all the possibilities. The conclusions support the model shown in Eq. 56 where three arsenic atoms and one electron form a cluster at high arsenic concentrations. The expression for C_{max} is[38]

$$C_{max} = 1.896 \times 10^{22} \exp\left[-\frac{0.453}{kT}\right] \tag{61}$$

Equations 60 and 61 give comparable values at temperatures above 900°C, but Eq. 61 gives a better fit to experimental data at temperatures below 900°C.

Figure 12 shows the maximum carrier concentration C_{max} as a function of annealing/diffusion temperature for arsenic at high concentrations. Experimental results agree with this model rather well.

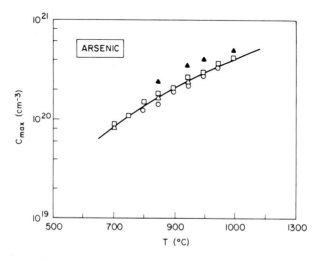

Fig. 12 Maximum carrier concentration of arsenic in silicon versus temperature. $\Delta, \square, \circ, \blacktriangle$, experimental data, curve fits Eq. 61. (*After Guerrero, et al., Ref. 38.*)

The diffusivity of arsenic clusters is negligible below 1000°C. At higher temperatures, these clusters separate first (decluster) and diffuse as separate arsenic species. At low diffusion temperatures (<1000°C), Eq. 54 or 55 is the diffusivity of the portion of arsenic atoms that did not form clusters.

Boron When the multi-charge-state impurity-defect interaction mechanism is applied to the experimental profiles, the diffusivity of boron at high concentrations can be expressed as[21]

$$D_B = (D_i^{+})_B \frac{p}{n_i} \tag{62}$$

by using Eq. 31 with $D^{+r}(r = 1)$. In ion implanted samples, when the boron concentration is above 10^{20} cm^{-3} the concentration of the electrically active boron is also less than that of the total boron in the high-concentration region.[39] The diffusivity of boron in the high-concentration region is reduced considerably, to nearly zero. The limiting values for the electrically active boron have been obtained experimentally; however, a physical model has not been developed. Figure 13 shows the experimental activity limits for boron at different temperatures.[39]

Phosphorus Phosphorus is not only useful as an emitter and base dopant, it also possesses the property of gettering fast-diffusing metallic contaminants such as Cu and Au. These contaminants, when precipitated out in crystal defects, cause junction leakage current problems. Thus, phosphorus is indispensable in VLSI technology. However, n-p-n transistors made with arsenic-diffused emitters have better low-current gain characteristics and better control of narrow base widths than those made with phosphorus-diffused emitters. Therefore, in VLSI, phosphorus as an active

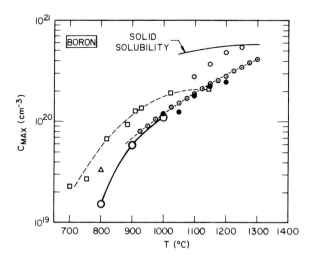

Fig. 13 Maximum carrier concentration of boron in silicon versus temperature. ⊙TEM data; ○, ○ nuclear reaction data; Δ, □, ● electrical data; -----, ———, ———— curves connecting data points. *(After Ryssel et al., Ref. 39.)*

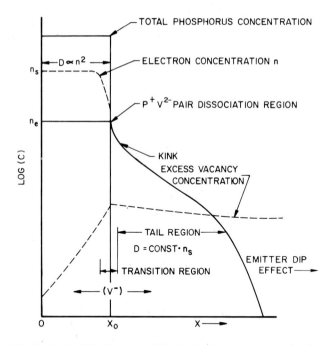

Fig. 14 A model for phosphorus diffusion in silicon. *(After Fair, Ref. 21.)*

dopant in small, shallow junctions and low-temperature processing will be limited to the base dopant of p-n-p transistors and as a gettering agent. Arsenic is the most used dopant for the source and drain regions in n-channel MOSFETs.

For completeness the diffusion model for phosphorus is discussed briefly. The characteristic profile of phosphorus can be described as consisting of three regions (Fig. 14): the high-concentration region, the transition region (often called the "kink" of the profile), and the low-concentration region (the tail region).

In the high-concentration region, a fraction of the phosphorus ion (P^+) pairs with V^{2-} vacancies as $(PV)^-$. The concentration of $(PV)^-$ is proportional to n_s^3, the surface electron concentration or the peak concentration for a Gaussian implanted profile. The n_s has to be determined experimentally. The diffusivity of phosphorus D_P, in this region, is proportional to n^2, the electron concentration, and is

$$D_P = [D_i^x + D_i^{2-} (n / n_i)^2]$$ (63)

where

$$D_i^x = 3.85 \exp(-3.66 / kT)$$ (64)

and

$$D_i^{2-} = 44.2 \exp(-4.37 / kT)$$ (65)

Near the transition region, the electron concentration decreases, and when the Fermi level is close to 0.11 eV below the conduction band edge, the $(PV)^-$ pairs show significant dissociations. The electron concentration in this transition region is

$$n_e = 4.65 \times 10^{21} \exp(0.39/kT) \tag{66}$$

The dissociation of $(PV)^-$ increases the vacancy concentration in the tail region which can be expressed as

$$(PV)^- \rightarrow (PV)^x + e^- \tag{67}$$

and

$$(PV)^x \rightleftarrows P^+ + V^- \tag{68}$$

The arrows shown in Fig. 14 next to (V^-) signify that at $n = n_e$ the excess vacancies diffuse into both directions from $X = X_0$. The diffusivity in the tail region increases as V^- is increased and is

$$D_{\text{tail}} = D_i^x + D_i^- \frac{n_s^3}{n_e^2 n_i} [1 + \exp(0.3 \text{ eV}/kT)] \tag{69}$$

where

$$D_i^- = 4.44 \exp(-4 \text{ eV}/kT) \tag{70}$$

The expression for the total phosphorus concentration and the electrically active phosphorus is

$$C_T = n + 2.4 \times 10^{-41} n^3 \tag{71}$$

for temperatures between 900 and 1050°C.

Emitter push effect In n-p-n narrow-base transistors using phosphorus-diffused emitter and boron-diffused base, the base region under the emitter (phosphorus) region is deeper than that outside the emitter region by 0.2 to 0.6 μm. This phenomenon is called the emitter push effect. Since the discovery of this phenomenon, researchers have proposed various physical mechanisms to explain it. However, a bandgap narrowing effect together with the phosphorus diffusion model shown in Fig. 14 adequately explains the emitter push effect. The results are summarized in the following paragraph. However, the derivations of the equations are omitted.

The dissociation of P^+V^{2-} pairs at the kink region of the phosphorus profile (Fig. 14) provides a mechanism for the enhanced diffusion of phosphorus in the tail region. The diffusivity of boron under the emitter region (the inner base) is enhanced by the dissociation of P^+V^{2-} pairs also. However, at phosphorus concentrations greater than 5×10^{20} atoms/cm³, misfit between silicon and phosphorus atoms induce a lattice strain (called the misfit-induced strain) and reduces the concentration of P^+V^{2-} pairs.

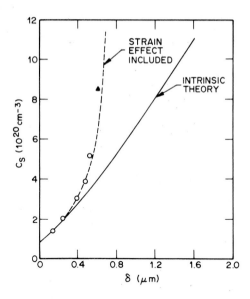

Fig. 15 Inner base push-out depth versus total phosphorus surface concentration. ○ junction measurement, ▲ SIMS measurement phosphorus diffusion at 1000°C for 60 min using POCl₃ diffusion source. The integrated initial base dopant $Q_0 = 1.6 \times 10^{14}$ cm^{-2} $C_{p0} = 1.7 \times 10^{18}$ cm^{-3}. (After Fair, Ref. 21.)

This reduced concentration is related to a bandgap narrowing effect. The combination of the bandgap narrowing effect[40] and the P^+V^{2-} dissociation explains the emitter push effect and agrees with experimental observations. The emitter push depth as a function of phosphorus surface concentration is shown in Fig. 15. Based on the mechanism of P^+V^{2-} of dissociation, the inner base (the base region under the emitter diffusion area) depth enhancement (inner base push out) will be a monotonic function of phosphorus surface concentration as given in Eq. 72.

$$\delta(t) = W_0 \left[\left[1 + \frac{2 D_{\text{in}}^B t}{W_0^2} \right]^{\frac{1}{2}} - \left[1 + \frac{2 D_i^B t}{W_0^2} \right]^{\frac{1}{2}} \right] \tag{72}$$

where $\delta(t)$ is the difference between the inner and outer base, and W_0 is a quantity relating to the integrated doping of a Gaussian profile of the base region prior to the emitter diffusion,

$$W_0 = 0.4 \frac{Q_0}{C_{p0}} \tag{73}$$

D_{in}^B is the diffusivity of the inner base which is assumed to increase from the intrinsic value by the same ratio as the diffusivities of the phosphorus tail.

$$D_{\text{in}}^B = (D_i)_B \frac{(D_i^x)_P + (D^-)_P}{(D_i^x)_P} \tag{74}$$

$D_i^B \equiv (D_i)_B$ which is the intrinsic diffusivity of boron, Q_0 is the integrated doping in the base, and C_{p0} is the peak concentration of the base dopant prior to the phosphorus diffusion. However, the reduction of the concentration of P^+V^{2-} pairs caused by the bandgap narrowing effect, which is induced by the lattice strain, limits the maximum depth of $\delta(t)$. For the data shown in Fig. 15, the maximum depth is close to 0.6 μm for emitter diffusion at 1000°C.

For phosphorus diffusion in the tail region, the lattice strain from bandgap narrowing effect on boron diffusivity D_{in}^B can be estimated because $(D^-)_P$ in Eq. 74 is proportional to $(n_s/n_e)^3$. The bandgap narrowing effect on the diffusivity is given in Eq. 74,[40] with

$$(D^-)_P \equiv (D^-)_s = D_i^- \frac{n_s^3}{n_e^2} \frac{1}{n_i} \left[1 + \exp \frac{0.3\text{eV}}{kT} \right] \exp \left[\frac{3\Delta Eg}{kT} \right] \quad (75)$$

Where $(D^-)_s$ represents the lattice strain effect and D_i^- is given in Eq. 70.

5.6.4 Analytical Expressions for Arsenic

Although the concentration-dependent diffusivities in As can be determined from the experimental diffusion profiles by numerical analysis, for some cases, approximate analytical expressions represent the experimental data rather well. These expressions are useful simplifications to estimate slight processing variations without the complications in using computer numerical analysis. Chebyshev orthogonal polynomials can be used to represent ion-implanted-diffused As profiles.[10] The expressions for As profiles are given in Eqs. 76 and 77.

$$\frac{C}{C_s} = 1 - 0.87Y - 0.45Y^2 \quad (76)$$

$$Y = x \left[8 \frac{C_s}{n_i} (D_i)_{As} t \right]^{-1/2} \quad (77)$$

Expressions for x_j, R_s, C_s, and Q_T can be derived from Eqs. 76 and 77:

$$x_j = 2 \left[\frac{Q_T}{n_i} (D_i)_{As} t \right]^{1/3} \quad (78)$$

$$R_s = \frac{1.76 \times 10^{10}}{Q_T^{7/9}} \left[\frac{n_i}{(D_i)_{As} t} \right]^{1/9} \quad (79)$$

$$C_s = 0.91 \left[\frac{Q_T^2 \, n_i}{(D_i)_{As} t} \right]^{1/3} \quad (80)$$

$$Q_T = 0.55 C_s \, x_j \quad (81)$$

where x_j is the junction location at a concentration equal to $0.01C_s$, Q_T is the ion dose in cm^{-2}, $(D_i)_{As}$ is the intrinsic diffusivity of arsenic in cm^2/s, R_s is the sheet resistance in $\Omega/\square$, and C_s is the surface concentration in cm^{-3}. In order to calculate R_s and Q_T, we arbitrarily select $0.01C_s$ for the location of x_j. Since the arsenic profiles have a steep concentration gradient for concentrations below $0.1C_s$, assuming the junction depth to be $0.01C_s$ introduces small errors in the estimation of sheet resistance R_s and total Q_T.

Equation 78 gives an estimate of the junction depth only, whereas the angle lap and staining technique gives a more accurate result; here the depth depends on the dopant concentration level where the junction is formed.

Design curves are available[41] for arsenic implantation at 100 keV into random equivalent direction on (100) oriented wafers. The curves are for ion doses from 1.2×10^{15} to 2.4×10^{16} cm^{-2} and diffusion temperatures between 925°C and 1000°C. The expression for x_j is the same as Eq. 78. The expression for the surface concentration C_s is

$$ C_s = 0.86Q_T^{2/3} \left[\frac{n_i}{D_i t} \right]^{1/3} \tag{82} $$

By assuming that mobility is proportional to $C^{-1/3}$, the expression for R_s is given in Eq. 83,

$$ R_s = \frac{1}{K_\mu q Q_T^{2/3} x_j^{1/3}} \tag{83} $$

where K_μ is the mobility proportional constant 2.82×10^8 $cm/V\text{-}s$, and Q_T is the total arsenic ion dose. Equations 82 and 83 extend the temperature range of arsenic implantation/annealing to 925°C. When the calculated C_s from either Eq. 80 or 82 is greater than that calculated from Eq. 61, the arsenic clustering effect should be taken into consideration and the approximations in this section will be subject to errors.

5.7 DIFFUSION IN SiO₂

VLSI and silicon planar device fabrication relies on the thermal oxide of silicon as a mask to prevent diffusion of impurity atoms into silicon. Therefore, understanding diffusion in SiO_2 films is important. The diffusivities in SiO_2 were deduced by measuring the dopants in silicon that diffused through the oxide, and by using the solutions of the diffusion equations from Fick's law with an assumed set of initial and boundary conditions. The impurity distribution at the Si-SiO_2 interface is assumed to be in equilibrium, and the concentration ratio is described by a segregation coefficient which was discussed in Chapter 4. Both the diffusivity and the segregation coefficient are unknown. The diffusivities are calculated, and the segregation coefficient is either assumed or deduced.

Since Group III and Group V elements are glass formers in SiO_2, they lower the melting temperature of the oxide film. The diffusivities of these elements depend strongly on their concentrations. For example, phosphorus at 3 to 6 at. % forms a thin viscous film on SiO_2 that flows at 800 to 900°C. (Phosphorus is used for planarization in VLSI circuits as discussed in Chapter 3.) However, outside the liquid-solid boundary, the phosphorus concentration becomes too low to show any diffusion. With P_2O_5 used as the diffusion source, a very-high-concentration layer is present on the thermal oxide which is used to mask phosphorus diffusion. This phosphorus layer can be considered as a liquid; the diffusion is from the liquid-solid interface into SiO_2 (the thermal oxide layer).

For phosphorus diffusion in SiO_2 from a doped-oxide source, the out-diffusion from the doped oxide at the ambient-oxide (phosphorus doped) interface and the in-diffusion from the doped oxide and a nondoped oxide interface under the doped oxide are represented by two diffusivities.[42] The diffusivity of the in-diffusion near the oxide-silicon interface depends on the mole percent (the phosphorus concentration) of phosphorus in the doped oxide. Although the in-diffusion profile can be fitted by an erfc function, the diffusivity depends on the phosphorus concentration in the oxide. The diffusivity also depends on the oxide structures, that is, the diffusivity is close to twice as large in a wet oxide as in a dry oxide.[42] This fitting of a measured profile to an erfc function (Eq. 8) with diffusivities varying with concentration is believed to be the result of imprecise profile measurements.

The diffusivity for the out-diffusion portion near the ambient-oxide (doped) interface has a larger value than the in-diffusion portion. At phosphorus concentrations below 0.5 mol % of P_2O_5 in SiO_2, the diffusivity is independent of the phosphorus concentration. The out-diffusion, which can be represented by

$$D_I = 7.23 \exp\left[\frac{-4.44}{kT}\right] \tag{84}$$

does not contribute to masking failure. The diffusion responsible for masking failure has a smaller diffusivity and depends on the phosphorus concentration and the properties of the oxide. Moisture has a significant effect on the masking properties of SiO_2 against phosphorus diffusion.

Similar concentration-dependent properties of boron diffusion in silicon oxide have been observed.[43] As a general rule, over the temperature range used in VLSI, diffusivities of these elements (B, As, P, and Sb) are very low when their concentrations are below 1 at. %. Hydrogen, He, OH, Na, O_2, and Ga are fast diffusants in SiO_2. At 900°C the diffusivities of these elements are greater than 10^{-13} cm^2/s. Table 4 shows the diffusivities of some elements used in VLSI technology.[44] These values represent the magnitudes of the deduced diffusivities for the diffusion conditions listed. The calculated diffusivities at 900°C are from the D_0 and E given in the table using Eq. 18 and are subject to errors. The values of arsenic diffusivities in SiO_2 are calculated from measured profiles of arsenic in SiO_2 films rather than deduced values from measurements in silicon.[45] Most of the deduced values of diffusivities in SiO_2 are rather good estimates.

Table 4 Diffusivities in SiO_2

Element	Ref.	D_0 (cm^2/s)	E (eV)	$D(900°C)$ (cm^2/s)	C_s (cm^{-3})	Source and ambient
Boron	44	7.23×10^{-6}	2.38	4.4×10^{-16}	$10^{19} -$ 2×10^{20}	$B_2 O_3$ vapor, $O_2 + N_2$
	44	1.23×10^{-4}	3.39	3.4×10^{-19}	6×10^{18}	$B_2 O_3$ vapor, Ar
	44	3.16×10^{-4}	3.53	2.2×10^{-19}	Below 3×10^{20}	Borosilicate
Gallium	44	1.04×10^5	4.17	1.3×10^{-13}	$\cdots$	$Ga_2 O_3$ vapor, $H_2 + N_2 + H_2O$
Phos-phorus	44	5.73×10^{-5}	2.30	7.7×10^{-15}	8×10^{20} to 10^{21}	$P_2 O_5$ vapor, N_2
	42	1.86×10^{-1}	4.03	9.3×10^{-19}	$8 \times 10^{17} -$ 8×10^{19}	Phosphosilicate, N_2
Arsenic	45	67.25	4.7	4.5×10^{-19}	$<5 \times 10^{20}$	Ion implant, N_2
	45	3.7×10^{-2}	3.7	4.8×10^{-18}	$<5 \times 10^{20}$	Ion implant, O_2
Antimony	44	1.31×10^{16}	8.75	3.6×10^{-22}	5×10^{19}	$Sb_2 O_5$ vapor, $O_2 + N_2$
Hydrogen (H_2)		5.65×10^{-4}	0.446	7×10^{-6}		
Helium		3×10^{-4}	0.24	2.8×10^{-5}		
Water		10^{-6}	0.79	4×10^{-10}		
Oxygen		2.7×10^{-4}	1.16	2.8×10^{-9}		
Gold		8.2×10^{-10}	0.8	3×10^{-13}		
Gold		1.52×10^{-7}	2.14	10^{-16}		
Platinum		1.2×10^{-13}	0.75	7.2×10^{-17}		
Sodium		6.9	1.3	1.8×10^{-5}		

Note: C_s = Surface concentration on silicon after diffusion from the specified source and ambient in the absence of an oxide barrier.

5.8 FAST DIFFUSANTS IN SILICON

Group I and VIII elements are fast diffusants in silicon. They form deep level traps and affect the minority-carrier life time and the junction-leakage currents. For example, gold and platinum are used to reduce the storage time of switching transistors. These elements diffuse mainly through an interstitial mechanism that is modified to account for the experimental results. Many factors affect the distribution and diffusion rate of these elements. These factors include the dislocation concentration, the precipitation and clustering of these elements near dislocations and point defects, the cooling rates, the presence of high concentrations of dopant elements such as phosphorus and boron, and the heat treatment history of the substrate silicon crystal. It is almost impossible to measure the diffusivities of these elements with any consistency. For instance, the distribution of gold throughout a silicon wafer resembles a U-shape with high concentrations near the front and the back surfaces of the silicon wafer, and a nearly uniform low-concentration distribution in the center of the wafer.

Table 5 shows the diffusivities, solubilities, and the distribution coefficients at melting temperature of the fast diffusants in silicon.[46] Diffusivities of hydrogen, oxygen, and recent values for Pt, Cr, and Co are also given.

5.9 DIFFUSION IN POLYCRYSTALLINE SILICON

Polysilicon films are used in VLSI for two major purposes: (1) as a polysilicon gate in a self-aligned structure; and (2) as an intermediate conductor in two-level structures. To reduce the resistivity of polysilicon it is often doped with boron, phosphorus, or arsenic. Since the gate electrode is over a thin oxide, typically 250 to 500 Å, it is very important that the dopant atoms in the polysilicon film not diffuse through the gate oxide or cause degradation of the gate oxide. To minimize this problem, the polysilicon film is deposited at a low temperature without doping elements. After the gate region is defined, the polysilicon film is doped. Dopant atoms are introduced by diffusion from a doped-oxide source, from a chemical source, or by ion implantation.

Impurity diffusion in polysilicon film can be explained qualitatively by a grain boundary diffusion model.[51] The diffusivity of impurity atoms that diffuse along grain boundaries can be about 100 times larger than the diffusivities in a single-crystal lattice. The polycrystal film is considered to be composed of single crystallites of varying sizes (from less than 1000 Å to a few tens of micrometers) that are separated by grain boundaries. Experimental results indicate that the impurity atoms inside each crystallite have diffusivities comparable to that found in the single crystal. Impurity atoms also diffuse along grain boundaries, so the diffusivity in a polysilicon film depends strongly on the textures of the film. The textures of the films are functions of the film deposition temperature, rate of deposition, thickness of the film, and composition of the substrate film which is an oxide layer, a silicon nitride film, or a single-crystal silicon surface.

Although diffusion results that are universally useful are difficult to present, some general observations can be made. Experimental profiles in polysilicon films resemble simple diffusion results such as a complementary error function or a Gaussian function which depends on the applicable diffusion conditions. Because of this resemblance, the diffusivities can be estimated from the measured junction depth and the surface concentration using Eq. 8 or 12.

The junction depths are measured by chemical staining of a beveled sample using the same staining solution as for the single-crystal Si (a few drops of HNO_3 in 100 cm^3 HF) or a chloroplatinic acid solution which consists of 0.5 to 1 g of $H_2 PtCl_6$ in 100 cm^3 HF (49%). The surface concentrations can be assumed to equal the measured concentrations of companion single-crystal samples that were diffused at the

Table 5 The diffusivity, solubility, and distribution coefficient at melting temperature of the fast diffusant in silicon

Element	Ref.	Diffusivity D_0 (cm²/s)	E (eV)	Solubility (cm^{-3})	Distribution coefficient
Li (25−1350°C)	46	2.3×10^{-3} $- 9.4 \times 10^{-4}$	0.63 −0.78	Max. 7×10^{19} (1200°C)	10^{-2}
Na (800−1100°C)	46	1.6×10^{-3}	0.76	$10^{18} - 9 \times 10^{18}$ (600−1200°C)	
K (800−1100°C)	46	1.1×10^{-3}	0.76	$9 \times 10^{17} - 7 \times 10^{18}$ (600−1200°C)	
Cu (800−1100°C)	46	4×10^{-2}	1.0	$5 \times 10^{15} - 3 \times 10^{18}$ (600−1300°C)	4×10^{-4}
(Cu)$_i$ (300−700°C)	46	4.7×10^{-3}	0.43		
Ag (1100−1350°C)	46	2×10^{-3}	1.6	$6.5 \times 10^{15} - 2 \times 10^{17}$ (1200−1350°C)	1.1×10^{-4}
Au (800−1200°C) (Au)$_i$ (Au)$_s$ (700−1300°C)	46	1.1×10^{-3} 2.4×10^{-4} 2.8×10^{-3}	1.12 0.39 2.04	$5 \times 10^{14} - 5 \times 10^{16}$ (900−1300°C)	2.5×10^{-5}
Pt (800−1000°C)	47	1.5×10^{2} -1.7×10^{2}	2.22 −2.15	$4 \times 10^{16} - 5 \times 10^{17}$ (800−1000°C)	
Fe (1100−1250°C)	46	6.2×10^{-3}	0.87	$10^{13} - 5 \times 10^{16}$ (900−1300°C)	8×10^{-6}
Ni (450−800°C)	46	0.1	1.9	6×10^{18} (1200−1300°C)	$\sim 10^{-4}$
Cr (1100−1250°C)	48	0.01	1.0	$2 \times 10^{13} - 2.5 \times 10^{15}$ (900−1280°C)	
Co (900−1200°C)	49	9.2×10^{4}	2.8	Max. 2.5×10^{16} (1300°C)	8×10^{-6}
O$_2$ (700−1240°C)	50	7×10^{-2}	2.44	$1.5 \times 10^{17} - 2 \times 10^{18}$ (1000−1400°C)	5×10^{-1}
H$_2$	46	9.4×10^{-3}	0.48	$2.4 \times 10^{21} \exp \dfrac{-1.86}{kT}$ at 1 atm	

same time. For boron an empirical resistivity versus concentration curve for a polysilicon film has been determined.[52] For arsenic-diffused samples, the Rutherford backscattering (RBS) method has been employed to measure the diffusion profiles and to determine the diffusivities.[53]

The polysilicon film, deposited by a CVD or evaporation technique, grows with a preferred grain orientation at substrate temperatures greater than 800°C. Below 800°C the grain growth demonstrates less orientation preference. Thick polyfilms show columnar grain structures which are oriented in the $\langle 110 \rangle$ direction. Thin films deposited at low temperatures have small grains and are more randomly oriented. After heat treatment at high temperatures, thin films also show grain growth in the $\langle 110 \rangle$ preferred orientations. Examination of a cleaved cross section of these films by a defect etch shows grain boundaries that are almost parallel to each other at a slanted angle with respect to the substrate surface, that is, the grains do not grow in a direction perpendicular to the substrate surface.

The electrical property of the As- and P-doped polyfilms indicates that these elements segregate at the grain boundaries. Heat treatment of these films between 800 and 900°C shows reversible change of the resistivities.[54] The resistivity of As- and P-doped polyfilms is influenced by both carrier trapping (electrons) and atom trapping (P or As) at the grain boundaries. The resistivity increases when the dopant atoms are trapped at the grain boundaries. However, boron atoms do not appear to segregate at the grain boundaries.

Table 6 gives a few examples of the diffusivities of As, B, and P in polysilicon films used in VLSI. Two values are given to stress that the diffusivities depend on polyfilm textures and other factors.

5.10 DIFFUSION ENHANCEMENTS AND RETARDATIONS

Diffusion study is complicated not only by the presence of defects or high-concentration effects but also by other processing factors. Diffusion in an oxidizing ambient and the lateral enhancement of diffusivity can significantly affect VLSI structures.

Table 6 Examples of diffusivities in polysilicon films

Elements	D_0 (cm²/s)	E (eV)	D (cm²/s)	T (°C)	Ref.
As	8.6×10^4	3.9	2.4×10^{-14}	800	53
As	0.63	3.2	3.2×10^{-14}	950	55
B	$(1.5-6) \times 10^{-3}$	2.4–2.5	9×10^{-14}	900	56
B			4×10^{-14}	925	52
P			6.9×10^{-13}	1000	51
P			7×10^{-13}	1000	51

5.10.1 Effect of Diffusion in Oxidizing Ambient

In addition to the high-concentration effect, such as the interactions between Group III and V elements and the bandgap narrowing, a few processing conditions have also been shown to enhance or retard diffusion. Among these, diffusion in an oxidizing ambient of boron, phosphorus, and arsenic have been investigated extensively. Most of the experimental data were obtained from samples that had been processed under conditions similar to those under which self-aligned gate MOS devices and circuits are fabricated.

The oxidation-enhanced diffusion (OED) of boron was first observed in high-concentration diffusions into both (100) and (111) oriented silicon wafers.[57] Some experiments attempted to separate the oxidation effect from the high-concentration effect by diffusing dopants at concentration levels below n_i at the diffusion temperature. This method introduces dopants at low concentrations to form a prediffused layer from a chemical source or an ion implanted source at low dopant levels. A thin oxide layer (100 to 500 Å) is grown, at low temperatures, to protect the silicon surface and is then covered by the deposition of a silicon nitride film 0.1 to 0.2 μm thick. The thin oxide layer between the silicon nitride and the silicon surface also serves to adjust the interface properties. The interface between a Si_3N_4 film and a silicon surface exhibits a charge storage effect, which causes surface leakage current and instabilities. Strips of silicon nitride and oxide films are removed by a selective photolithography and etching technique. These samples having alternating regions of free silicon surface and nitride-oxide protected surface are oxidized at different temperatures, in different ambients, for different time periods, and sometimes with both (100) and (111) oriented wafers. Most of the data are from (100) silicon. The enhancement or retardation is evaluated by measuring the junction depths, spreading resistance profiles, or concentration profiles by the differential conductivity method.

Figure 16a shows the cross section of the diffusion structures with adjacent oxidized and masked regions.[58] The junction depth on the right-hand side under the silicon nitride mask is shallower than the one on the left-hand side. All the junction depths are measured from the original sample surface prior to the oxidation but after the silicon nitride deposition. The enhancement or retardation depth Δx_j can be expressed as

$$\Delta x_j = (x_j)_{fo} - (x_j)_f \qquad (85)$$

where $(x_j)_i$ is the initial junction depth (Fig. 16a); $(x_j)_{fo}$ is the final junction depth under the oxide region; and $(x_j)_f$ is the final junction depth under the silicon nitride mask. Figure 16b shows an example of the measured Δx_j as a function of the oxidation time for boron at 1100°C. Since the concentration levels are below n_i, the diffusion under the masking nitride film is due to the intrinsic diffusivity and the diffusivity under the oxide can be expressed as

$$D_{OED} = D_i + \Delta D \ (T, \ t, \ P_{O_2}, \ \text{orientation}) \qquad (86)$$

where D_{OED} is the diffusivity for oxidation-enhanced diffusion; D_i is the intrinsic diffusivity or the diffusivity in a nonoxidizing ambient; and ΔD is the enhancement diffusivity that can depend on diffusion temperature, time, partial pressure of oxygen, P_{O_2}, and crystal orientations.

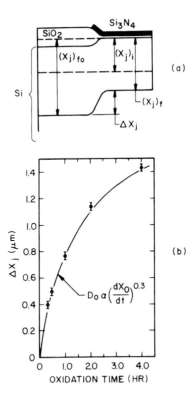

Fig. 16 Oxidation-enhanced diffusion. (a) Cross section of the experimental structure. (b) Δx_j versus oxidation time. Boron diffusion at 1100°C in wet oxygen. ♦ measured value and range. *(After Taniguchi, Kurosawa, and Kashiwagi, Ref. 58.)*

Since the observed enhancement showed a strong dependence on the diffusion time, but the measured results were from a given diffusion-oxidation period, an effective diffusivity, sometimes called the diffusion time average diffusivity, is used.[59] In this manner, the time dependence of the enhancement is approximated by

$$\langle D_A \rangle_{\text{eff}} = \frac{1}{t} \int_0^t D_A \ dt \tag{87}$$

where D_A, a function of the oxidation rate, is the diffusivity of the dopant at diffusion time t.

The diffusivity enhancement is proportional to a fractional power of the oxidation rate.[60]

$$\Delta D = \alpha \left[\frac{dX_{\text{ox}}}{dt} \right]^n \tag{88}$$

where α is a proportional constant that can be estimated from an assumed diffusion model, and n is between 0.4 and 0.6. Results on arsenic and phosphorus diffusion in dry oxygen showed that ΔD decreases as the oxidation temperature is increased.[61]

Oxidation enhancement of As and P have been investigated using prediffused samples.[62] In this case, the diffusion equation with a moving boundary during the oxidation was solved by the numerical method with a measured initial profile after the prediffusion and an assumed parabolic oxidation-rate relationship.[7] For this assumed relationship, the oxidation rate can be determined as follows:

$$\frac{dX_{ox}}{dt} = \frac{B}{2} (t)^{-\frac{1}{2}}$$ (89)

Equation 89 assumes that the initial thin oxide (20 to 30 Å) on silicon surfaces can be neglected and the linear growth portion is also negligible (see Chapter 4). The results can be summarized as follows:

1. In dry N_2, the diffusivities are the same in (100) and (111) oriented wafers for both arsenic and phosphorus. This result agreed with observations by others.[58, 60]
2. In dry O_2, the diffusivities are enhanced for As and P in (100) oriented wafers and for P in (111) wafers, but little enhancement was observed for As in (111) silicon.
3. The enhancement in (100) Si is greater than that in (111) Si.
4. Since the oxidation rates are higher at shorter oxidation times, the enhancement is larger for short oxidation time and decreases with increasing oxidation time.
5. The diffusivity enhancement $\Delta D = (D)_{O_2} - (D)_{N_2}$ can be expressed in terms of an effective oxidation rate $(X_{ox}/t)^n$.

A retardation of antimony diffusion in silicon during oxidation was observed.[63] In addition, the stress at the silicon-silicon nitride edge caused junction retardation under the silicon nitride film laterally to 20 to 30 μm inside the nitride film edge. The junction retardation is depicted in Fig. 17 which is traced from a photograph of angle lapped and stained sample. The retardation is a fraction of a micrometer for junction depths of 3 to 6 μm and diffusion temperatures between 1000° and 1200°C.

The observation of oxidation-induced stacking fault (OSF) and oxidation-enhanced diffusion (OED) has lead to the proposal of a dual diffusion mechanism. The diffusion of impurity under the nitride layer is considered to be dominated by the vacancy mechanism, and the oxidation enhancement of diffusion is attributed to the presence of silicon self-interstitials that also cause the extrinsic stacking faults to grow.[60] Interstitials are generated at the $Si-SiO_2$ interface during oxidation. By assuming that the vacancy concentration is constant during oxidation, the enhancement of boron and phosphorus diffusivities during oxidation is due to the excess of interstitials diffusing away from the oxide-silicon interface and that these elements are governed by a dual mechanism, vacancy and interstitialcy.[58, 64]

The observation of oxidation-retarded diffusion of antimony suggests that, during diffusion and oxidation, thermal equilibrium between vacancies and interstitial exists. The generation of interstitials at the oxide-silicon interface causes the depression of vacancy concentrations.[63] The diffusion retardation of antimony could be due to the reduction of vacancy concentrations; thus antimony diffusion is governed by a vacancy mechanism. By a similar reasoning since silicon interstitials are enhanced at the oxide-silicon interface, it has been suggested that both boron and phosphorus diffuse via the interstitialcy mechanism in either an oxidizing or neutral ambient.

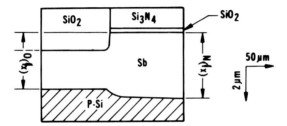

Fig. 17 Junction retardation of antimony during oxidation with the lateral effect shown under the Si_3N_4 layer. *(After Mizuo and Higuchi, Ref. 63.)*

5.10.2 Lateral Enhancement of Diffusivity

Another enhancement effect that is important in VLSI devices is the lateral enhanced diffusion at an oxide or silicon-nitride edge. Diffusion into narrow windows of silicon oxide can result in anomalous junction depths.[65] Various enhancements and retardations of the junction depth near the oxide window edges have been observed. These are the results of elastic strain fields near the window edges.

For boron diffusion in a structure similar to that shown in Fig. 18, lateral-enhanced diffusion extends under the nitride layer up to 30 μm.[66] Strips of silicon nitride layers with widths varying from 2.5 to 100 μm were separated with 100-μm windows without oxide. The samples were oxidized after boron implantation and annealing at 900°C. The junction depth at the center of the nitride-oxide strip was measured as a function of the widths of the strips. Figure 18 shows the results. For

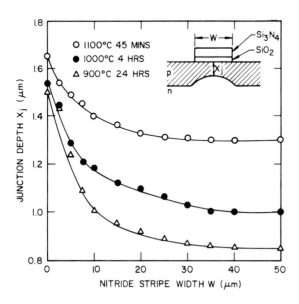

Fig. 18 Lateral enhancement of junction depth under a Si_3N_4 film during oxidation. *(After Lin, Dutton, and Antoniadis, Ref. 66.)*

narrow strips the lateral enhancement of the diffusivity is significant for VLSI device designs. In a narrow structure the junction depths under the silicon nitride film are enhanced and nonuniform.

5.11 SUMMARY AND FUTURE TRENDS

This chapter discusses diffusion results in silicon with emphasis on VLSI applications. Various factors affecting diffusion control are presented. Fick's classical diffusion laws with constant diffusivities are obeyed for Group III and V elements when the concentrations are below the intrinsic carrier concentration. When the concentrations are high, concentration-dependent diffusivities are required, and Fick's generalized diffusion equation with concentration-dependent diffusivities can be solved by numerical methods. The concentration-dependent diffusivities can be determined from the measured profiles using mathematical formulations of a Boltzmann transformation or modifications of it.

Atomic diffusion mechanisms are being developed to relate the impurity diffusion with lattice defects. Attempts have been made to construct diffusion models based on defect-impurity interactions. The diffusivities are functions of the concentration of the ionized point defects, vacancies, or interstitials. This approach is successful in explaining the high-concentration diffusions of the Group III and V elements, especially for phosphorus in silicon. Various other models have also been proposed and tested.

Diffusion in an oxidizing ambient also exhibits a time dependency because of the parabolic oxidation rate of silicon. Observations of the coexistence of the oxidation-enhanced diffusion and the formation of oxidation-induced stacking faults suggest that an extrinsic mechanism for generating silicon self-interstials near the silicon-oxide interface may also influence the impurity diffusivities. The oxidation-induced stacking faults are extrinsic in nature, that is, they grow by absorbing silicon self-interstitials. These observations have led to the proposal of a dual vacancy-interstitialcy diffusion mechanism. Hence, the atomic diffusion mechanism is still an area of active research.

The advancement of device technology and the development of complex circuits require more precise diffusion measurements and good theoretical models, so that circuit performances can be modeled from process parameters. In theoretical modeling, the dominating diffusion mechanism for Group III and V elements needs to be resolved. Further development of the dual vacancy-interstitialcy diffusion mechanism is also needed.

Theoretical studies rely on good experimental data. As the device size becomes smaller and smaller, the need for better measurement techniques becomes more urgent. At present, the spreading resistance technique is widely used for profile measurements. Unfortunately, it relies on a beveled sample technique that limits the junction depth to 1 μm and it is a comparative and semiquantitative method. The differential conductivity method has a comparable limitation in depth of less than 1 μm. Thus, both methods will be less attractive in VLSI development.

The SIMS analysis is a powerful tool for diffusion profile measurements. It can measure boron and arsenic concentrations as low as 5×10^{15} atoms/cm^3 and has a high depth resolution of a few tens of angstroms. Therefore, it is an ideal tool for measuring shallow diffusion profiles. This technique will provide the needed precision for profile measurements in VLSI structures.

Conceivably, some pulse annealing procedures for annealing ion implanted wafers in 5 to 20 s will be developed shortly through the use of an arc-lamp furnace or a graphite heater in a vacuum environment.[67] Experimental and theoretical investigations on the effect of temperature transients on the diffusion-annealing properties of these samples will definitely be addressed soon. Junction depths in the range of 1000 Å or less can be realized and the impurity profiles from short period annealing have to be evaluated. This need points to SIMS analysis technique again and further development of this technique is required.

REFERENCES

[1] W. G. Pfann, Semiconductor Signal Translating Device, U.S. Patent No. 2,597,028 (1952).

[2] B. Tuck, "Introduction to Diffusion in Semiconductors," *IEE Mono. Ser., London,* **16** 119 (1974).

[3] A. Fick, *Ann. Phys. Leipzig,* **170**, 59 (1855).

[4] J. Crank, *The Mathematics of Diffusion,* Oxford University Press, London, 1957.

[5] B. I. Boltaks, *Diffusion in Semiconductors,* Academic, New York, 1963.

[6] E. C. Douglas and A. G. F. Dingwall, "Ion Implantation For Threshold Control in COSMOS Circuits," *IEEE Trans. Electron Devices,* **ED-21**, 324 (1974).

[7] G. Masetti, P. Negrini, S. Solmi, and G. Soncini, "Boron Drive-in in Silicon in Oxidizing Atmosphere," *Alta Freq.,* **42**, 346 (1973).

[8] M. Ghezzo, "Diffusion from a Thin Layer into a Semi-Infinite Medium with Concentration Dependent Diffusion Coefficient," *J. Electrochem. Soc.,* **119**, 977 (1972).

[9] M. Ghezzo, "Diffusion from a Thin Layer into a Semi-Infinite Medium with Concentration Dependent Diffusion Coefficient, Part II," *J. Electrochem. Soc.,* **120**, 1123 (1973).

[10] R. B. Fair and J. C. C. Tsai, "The Diffusion of Ion Implanted Arsenic in Silicon," *J. Electrochem. Soc.,* **122**, 1689 (1975).

[11] P. G. Shewmon, *Diffusion in Solids,* McGraw-Hill, New York, 1963.

[12] U. Gosele and H. Strunk, "High Temperature Diffusion of Phosphorus and Boron in Silicon Via Vacancies or Via Self-Interstitials?" *Appl. Phys.,* **20**, 265 (1979).

[13] R. L. Longini and R. F. Green, "Ionization Interaction Between Impurities in Semiconductors and Insulators," *Phys. Rev.,* **102**, 992 (1956).

[14] F. J. Morin and J. P. Maita, "Electrical Properties of Silicon Containing Arsenic and Boron," *Phys. Rev.,* **96**, 28 (1954).

[15] D. Shaw, "Self and Impurity Diffusion in Ge and Si," *Phys. Status Solidi B,* **72**, 11 (1975).

[16] G. D. Watkins, J. R. Troxell, and H. P. Chatterjee, "Vacancies and Interstitials in Silicon," *Inst. Phys. Conf. Ser.,* **46**, 16 (1979).

[17] R. E. Whan and F. L. Vook, "Infrared Studies of Defect Production in N-Type Silicon: Irradiation-Temperature Dependence," *Phys. Rev.,* **153**, 814 (1967).

[18] L. C. Kimerling, P. Blood, and W. M. Gibson, "Defect States in Proton-Bombarded Silicon at T < 300 K," *Inst. Phys. Conf. Ser.,* **46**, 273 (1979).

[19] J. G. De Wit, C. A. J. Ammerlaan, and E. G. Sieverts, "An ENDOR Study of the Divacancy in Silicon," *Inst. Phys. Conf. Ser.,* **23**, 178 (1975).

[20] J. C. Bourgoin, "Ionization Effects on Impurity and Defect Migration in Semiconductors," *Inst. Phys. Conf. Ser.,* **23**, 149 (1975).

[21] R. B. Fair, "Concentration Profiles of Diffused Dopants in Silicon," in F. F. Y. Wang, Ed., *Impurity Doping Processes in Silicon*, North-Holland, New York, 1981, Chapter 7.

[22] A. Seeger, H. Foll, and W. Frank, "Self-Interstitials, Vacancies and Their Clusters in Silicon and Germanium," *Inst. Phys. Conf. Ser.*, **31**, 12 (1977).

[23] A. M. Smith, "Diffusion," in R. M. Burger and R. P. Donovan, Eds., *Fundamentals of Silicon Integrated Device Technology*, Prentice-Hall, Englewood Cliffs, N.J., 1967, pp. 309–324, Vol. 1.

[24] F. M. Smits, "Measurement of Sheet Resistivities with the Four Point Probe," *Bell Syst. Tech. J.*, **37**, 711 (1958).

[25] J. C. Irvin, "Resistivity of Bulk Silicon and Diffused Layers in Silicon," *Bell Syst. Tech. J.*, **41**, 387 (1962).

[26] C. P. Wu, E. C. Douglas, and C. W. Mueller, "Limitations of the CV Technique for Ion Implanted Profiles," *IEEE Trans. Electron Devices*, **ED-22**, 319 (1975).

[27] E. Tannenbaum, "Detailed Analysis of Thin Phosphorus Diffused Layers in P-Type Silicon," *Solid State Electron.*, **3**, 123 (1961).

[28] *Standard Practice for Conversion Between Resistivity and Dopant Density for Boron Doped and Phosphorus Doped Silicon*, ASTM Book of Standards, Part 43, F723, ASTM, Philadelphia (1981).

[29] R. G. Mazur and D. H. Dickey, "A Spreading Resistance Technique for Resistivity Measurements on Silicon," *J. Electrochem. Soc.*, **113**, 255 (1966).

[30] W. K. Hofker, "Implantation of Boron in Silicon," *Philips Res. Rep. Suppl.* **8**, 1-121 (1975).

[31] P. F. Kane and G. B. Larrabee, *Characterization of Semiconductor Materials*, McGraw-Hill, New York, 1970, Chapter 9, p. 278.

[32] W. K. Chu, J. W. Mayer, M-A Nicolet, T. M. Buck, G. Amsel, and F. Eisen, "Microanalysis of Surface, Thin Films and Layered Structures by Nuclear Backscattering and Reactions," in H. R. Huft and R. R. Burgess, Eds., *Semiconductor Silicon 1973*, Electrochem. Soc., New York, 1973, p. 416.

[33] J. F. Ziegler, G. W. Cole, and J. E. E. Baglin, "Technique for Determining Concentration Profiles of Boron Impurities in Substrates," *J. Appl. Phys.*, **43**, 3809 (1972).

[34] J. L. Combasson, J. Bernard, G. Guernet, N. Hilleret, and M. Bruel, "Physical Profile Measurements in Insulating Layers Using the Ion Analyzer," in B. L. Crowder, Ed., *Ion Implantation in Semiconductors and Other Materials*, Plenum, New York, 1973, p. 285.

[35] B. J. Masters and J. M. Fairfield, "Arsenic Isoconcentration Diffusion Studies in Silicon," *J. Appl. Phys.*, **40**, 2390 (1969).

[36] S. M. Hu and S. Schmidt, "Interactions in Sequential Diffusion Processes in Semiconductors," *J. Appl. Phys.*, **39**, 4272 (1968).

[37] M. Y. Tsai, F. F. Morehead, and J. E. E. Baglin, "Shallow Junctions by High Dose As Implants in Si: Experiments and Modeling," *J. Appl. Phys.*, **51**, 3230 (1980).

[38] E. Guerrero, H. Pötzl, R. Tielert, M. Grasserbauer, and G. Stingeder, "Generalized Model for the Clustering of As Dopants in Si," *J. Electrochem Soc.*, **129**, 1826 (1982).

[39] H. Ryssel, K. Muller, K. Haberger, R. Henkelmann, and F. Jahael, "High Concentration Effects of Ion Implanted Boron in Silicon," *Appl. Phys.*, **22**, 35 (1980).

[40] R. B. Fair, "The Effect of Strain-Induced Bandgap Narrowing on High Concentration Phosphorus Diffusion in Silicon," *J. Appl. Phys.*, **50**, 860 (1979).

[41] T. M. Liu and W. G. Oldham, "Sheet Resistance-Junction Depth Relationships in Implanted Arsenic Diffusion," *IEEE Electron Device Lett.*, **EDL-2**, 275 (1981).

[42] R. N. Ghoshtagore, "Silicon Dioxide Masking of Phosphorus Diffusion in Silicon," *Solid State Electron.*, **18**, 399 (1975).

[43] D. M. Brown and P. R. Kennicott, "Glass Source B Diffusion in Si and SiO_2," *J. Electrochem. Soc.*, **118**, 293 (1971).

[44] M. Ghezzo and D. M. Brown, "Diffusivity Summary of B, Ga, P, As, and Sb in SiO_2," *J. Electrochem. Soc.*, **120**, 146 (1973).

[45] Y. Wada and D. A. Antoniadis, "Anomalous Arsenic Diffusion in Silicon Dioxide," *J. Electrochem. Soc.*, **128**, 1317 (1981).

[46] B. L. Sharma, "Diffusion in Semiconductors," *Trans. Tech. Pub. Germany*, 87 (1970).

[47] R. F. Bailey and T. G. Mills, "Diffusion Parameters of Platinum in Silicon," in R. R. Habarecht and E. L. Kern, Eds., *Semiconductor Silicon 1969*, Electrochem. Soc., New York, 1969, p. 481.

[48] W. Wurker, K. Roy, and J. Hesse, "Diffusion and Solid Solubility of Chromium in Silicon," *Mater. Res. Bull., (U.S.A.)*, **9**, 971 (1974).

[49] H. Kitagano and K. Hashimoto, "Diffusion Coefficient of Cobalt in Silicon," *J. Appl. Phys. Jpn.*, **16**, 173 (1977).

[50] J. C. Mikkelsen, Jr., "Diffusivity of Oxygen in Silicon During Steam Oxidation," *Appl. Phys. Lett.*, **40**, 336 (1982).

[51] T. I. Kamins, J. Manolin, and R. N. Tucker, "Diffusion of Impurities in Polycrystalline Silicon," *J. Appl. Phys.*, **43**, 83 (1972).

[52] C. J. Coe, "The Lateral Diffusion of Boron in Polycrystalline Silicon and Its Influence on the Fabrication of Sub-Micron Mosts," *Solid State Electron.*, **20**, 985 (1977).

[53] B. Swaminathan, K. C. Saraswat, R. W. Dutton, and T. I. Kamins, "Diffusion of Arsenic in Polycrystalline Silicon," *Appl. Phys. Lett.*, **40**, 795 (1982).

[54] M. M. Mandurah, K. C. Saraswat, C. R. Helms, and T. I. Kamins, "Dopant Segregation in Polycrystalline Silicon," *J. Appl. Phys.*, **51**, 5755 (1980).

[55] K. Tsukamoto, Y. Akasaka, and K. Horie, "Arsenic Implantation into Polycrystalline Silicon and Diffusion to Silicon Substrate," *J. Appl. Phys.*, **48**, 1815 (1977).

[56] S. Horiuchi and R. Blanchard, "Boron Diffusion in Polycrystalline Silicon Layers," *Solid State Electron.*, **18**, 529 (1975).

[57] W. G. Allen and K. V. Anand, "Orientation Dependence of the Diffusion of Boron in Silicon," *Solid State Electron.*, **14**, 397 (1971).

[58] K. Taniguchi, K. Kurosawa, and M. Kashiwagi, "Oxidation Enhanced Diffusion of Boron and Phosphorus in (100) Silicon," *J. Electrochem. Soc.*, **127**, 2243 (1980).

[59] S. M. Hu, "Formation of Stacking Faults and Enhanced Diffusion in the Oxidation of Silicon," *J. Appl. Phys.*, **45**, 1567 (1974).

[60] A. M. R. Lin, D. A. Antoniadis, and R. W. Dutton, "The Oxidation Rate Dependence of Oxidation-Enhanced Diffusion of Boron and Phosphorus in Silicon," *J. Electrochem. Soc.*, **128**, 1131 (1981).

[61] D. A. Antoniadis, A. M. Lin, and R. W. Dutton, "Oxidation-Enhanced Diffusion of Arsenic and Phosphorus in Near Intrinsic $\langle 100 \rangle$ Silicon," *Appl. Phys. Lett.*, **33**, 1030 (1978).

[62] Y. Ishikawa, Y. Sakina, H. Tanaka, S. Matsumoto, and T. Niimi, "The Enhanced Diffusion of Arsenic and Phosphorus in Silicon by Thermal Oxidation," *J. Electrochem. Soc.*, **129**, 644 (1982).

[63] S. Mizuo and H. Higuchi, "Retardation of Sb Diffusion in Si During Thermal Oxidation," *J. Appl. Phys. Jpn.*, **20**, 739 (1981).

[64] T. Y. Tan and U. Gosele, "Oxidation-Enhanced or Retarded Diffusion and the Growth or Shrinkage of Oxidation-Induced Stacking Faults in Silicon," *Appl. Phys. Lett.*, **40**, 616 (1982).

[65] C. F. Gibbon, E. I. Povilonis, and D. R. Ketchow, "The Effect of Mask Edges on Dopant Diffusion into Semiconductors," *J. Electrochem. Soc.*, **119**, 767 (1972).

[66] A. M. Lin, R. W. Dutton, and D. A. Antoniadis, "The Lateral Effect of Oxidation on Boron Diffusion in $\langle 100 \rangle$ Silicon," *Appl. Phys. Lett.*, **35**, 799 (1979).

[67] D. F. Downey, C. J. Russo, and J. T. White, "Activation and Process Characteristics of Infrared Rapid Isothermal and Furnace Annealing Techniques," *Solid State Technol.*, **25**, No. 9, 87 (1982).

PROBLEMS

1 *(a)* Derive expressions of concentration gradients for the erfc and Gaussian distributions. If the substrate doping density is C_{sub}, derive the expressions of the junction depths.

(b) Assuming $C_s = 10^{19}$ cm^{-3} for an erfc distribution and $S = 1 \times 10^{13}$ atoms/cm^2 for a Gaussian distribution, $C_{sub} = 10^{15}$ atoms/cm^3, and $D = 1 \times 10^{-15}$ cm^2/s (which is close to the boron diffusivity at 900°C), calculate the junction depths and the concentration gradients for both distributions. Calculate the integrated dopants for the erfc distribution and the surface concentration for the Gaussian distribution for diffusion times of 10, 30, and 60 min.

(c) Compare and discuss the results of *(a)* and *(b)*.

2 Derive Eq. 15 from Eq. 3.

3 Derive Eq. 25 assuming that the ionized acceptor vacancy concentration can be expressed as a function of the Fermi level and E_V, the activation energy of the acceptor vacancy.

4 In order to determine if the intrinsic diffusivity of an impurity is applicable at a given diffusion temperature, one has to know the intrinsic carrier concentration, n_i. Thus the plot n_i versus temperature is a very useful curve. Using Eqs. 29 and 30, construct n_i versus T.

5 Using Eq. 48, derive the electric-field enhancement factor of Eq. 53.

6 A p-type (100)-oriented silicon wafer with a substrate doping at 10^{16} atoms/cm^3 has been implanted and diffused with arsenic to an ion dose of 1×10^{15} cm^{-2} at 30 keV and diffusion at 850°C for 30 min in nitrogen.

    (a) Calculate the sheet resistance from Eqs. 79 and 83.
    (b) Calculate the surface concentrations from Eqs. 80 and 82.
    (c) Find the surface concentration of the electrically active arsenic.
    (d) Discuss the results.

7 *(a)* Using Eqs. 76 and 77 for an arsenic implanted-diffused profile, derive the approximate expressions for Eqs. 78, 79, and 80.

    (b) For an arsenic dose less than 10^{16} cm^{-2} and a diffusion temperature greater than 1000°C, assuming the electrically active arsenic equals the total arsenic (neglect As clusters), derive Eq. 79 using Eq. 40 and an effective mobility of

$$\mu_{eff} \cong \frac{28.2 \times 10^7}{C_A^{1/3}} \ \text{cm}^2/\text{V-s}$$

for 10^{19} cm$^{-3} < C_A < 6 \times 10^{20}$ cm^{-3} where C_A is the concentration of the electrically active As.

8 For an acceptor type impurity, the diffusion current including the electric-field term is

$$J = -qD_A \frac{\partial C_A}{\partial x} + q \, \mu_A \, C_A \, E$$

where D_A is the diffusivity, C_A is the acceptor concentration, μ_A is the mobility, and E is the electric field. If

$$D_A = D_i \frac{p}{n_i}$$

show that

$$J = -q \frac{d}{dx} (D_A \, C_A)$$

The above expression reduces the computation time when it is used to numerically analyze diffusion profiles.

ION IMPLANTATION

T. E. SEIDEL

6.1 INTRODUCTION

Ion implantation is the introduction of ionized-projectile atoms into targets with enough energy to penetrate beyond surface regions. The most common application is the doping of silicon during device fabrication. The use of 3- to 500-keV energy for boron, phosphorus, or arsenic dopant ions is sufficient to implant the ions from about 100 to 10,000 Å below the silicon surface. These depths place the atoms beyond any surface layers of 30-Å native SiO_2 and therefore any barrier effect of the surface oxides during impurity introduction is avoided. The depth of implantation, which is nearly proportional to the ion energy, can be selected to meet a particular application.

The major advantage of ion implantation technology is the capability of precisely controlling the number of implanted dopant atoms. Upon annealing the target (heating to elevated temperatures of approximately 600 to 1000°C), precise dopant concentrations between 10^{14} to 10^{21} atoms/cm^3 in silicon are obtained. Furthermore, the dopant's depth distribution profile can be well controlled.

During the 1960s important research in the calculation and measurement of ion ranges,[1] of radiation damage effects, and of ion channeling[2] was carried out. Many radiation-induced point defects already were identified,[3] aiding in a rapid understanding of ion implantation phenomena. Device applications were also being reported in the later 1960s. Variable-capacitance p-n junction diodes (varactors) with rapidly varying doping concentrations and the first implanted self-aligned MOS transistor using aluminum metal gates[4] were reported by 1968.

It took about six years (from 1969 to 1975) for ion implantation phenomena to be well enough understood and documented so that it was routinely used in VLSI fabrication. A summary of research during this time may be found in the extensive collection of articles from conference proceedings for the First through the Fifth International Conferences on Ion Implantation,[5a-5e] and in review articles.[6-8]

Later work addresses topics important for the implementation of VLSI.[9-13] Shallow junctions with high concentration profiles are formed by using rapid annealing techniques (e.g., lasers),[14] often making use of solid phase epitaxy. As high-beam-current equipment became commercially available in the late 1970s, beam heating, sputtering, oxide charge-up during implantation, and gas-beam interactions received attention.

For dopant control in the 10^{14} to 10^{18} atoms/cm^3 range, implantation offers a clear advantage over chemical deposition techniques. Masks can be made of any convenient material used in VLSI fabrication such as photoresist, oxides, nitrides, polysilicon, etc. The implant process, which is done in a vacuum, is both clean and dry.

Special damage configurations can be generated by implanting with ions such as argon at high doses. Annealing then gives fine-grain polycrystalline layers and/or dislocation-rich regions, to which unwanted impurities diffuse. These implanted damage-induced defects are useful for capturing unwanted impurities, such as copper, from junction regions. This process is called gettering.

This chapter covers the implantation system and dose control techniques, ion and disorder distributions, annealing, gettering, other implantation effects, and future trends.

6.2 ION IMPLANT SYSTEM AND DOSE CONTROL

This section discusses ion accelerators and the features needed for good dose control.[15, 16] Figure 1 shows a schematic view of a commercial ion implantation system. Starting at the source end, (bottom center), we have:

1. A gaseous source of appropriate material, such as BF_3 or AsH_3, at high (accelerating) potential V. An adjustable valve controls the flow of gas to the ion source.
2. A power supply to energize the ion source, also at high potential.
3. An ion source containing an ion plasma with the species of interest: $^+As^{75}$, $^+B^{11}$, or $^+BF_2^{49}$, at pressures of approximately $\sim 10^{-3}$ torr. A source diffusion pump establishes lower pressures for beam transport with reduced ion-gas scattering.
4. An analyzer magnet that selects only the ion species of interest and rejects other species. The desired ion species passes through a resolving slit (aperture) and is then injected into the accelerator column.
5. An acceleration tube through which the beam passes. The beam is then ready for transport to the target.

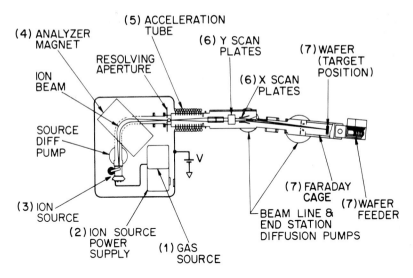

Fig. 1 Schematic diagram of a typical commercial ion implant system. *(After Varian-Extrion, DF-3000 brochure.)*

6. Sawtooth voltages applied to x and y (electrically rastered) deflection plates to scan the beam and give a uniform implantation. Beam-line and end-station diffusion pumps keep pressure low enough to avoid charge-exchange effects (see below).
7. A target chamber consisting of an area-defining aperture, Faraday cage, and wafer feed mechanism.

We now develop the idea of ion dose. Consider an ion beam with mass M, charge mq (m is the charge state of the ion and q is the electron charge), and energy E as it moves through a vacuum drift space toward a target (Fig. 2). The ion beam is swept by a charged-particle deflection scheme to obtain a uniform implantation over the target area. The swept beam is limited by an aperture of area A. Behind the aperture, the silicon wafer sample is placed inside the area of the aperture that is projected onto the metal target holder. The sample is in good electrical contact with the target holder which is connected in turn to a charge integrator. Electrons pass through the integrator and neutralize the implanted charges as they come to rest in the silicon. An integrated charge Q (coulombs) results in a dose ϕ defined by

$$\phi \equiv \frac{Q}{mqA} \text{ atoms}/\text{cm}^2 \tag{1}$$

The integrated charge is $Q = \int I \, dt$, where the beam current I (amps) is applied for time t (seconds). For example, a beam current of 10^{-9} A swept over a 1-cm^2 area for 1 second gives a dose of 0.6×10^{10} atoms/cm^2 for $m = 1$. If the width of the implanted layers is 600 Å, control of the doping concentration is possible at a level of

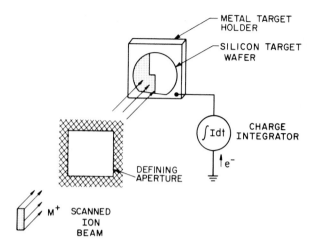

Fig. 2 Schematic of a rastered ion beam, showing the defining aperture and the target. The charge integrator measures the time-averaged swept beam current.

10^{15} atoms/cm^3. The use of milliamp beam currents for 100 seconds give doping levels up to the solid solubility values of approximately 10^{20} to 10^{21} atoms/cm^3 on 100-mm-diameter wafers.

The control of the dose and its uniformity may be compromised because of neutral beam species, charge exchange, secondary electron emission, and sputtering effects. These effects, which are discussed below, can be made negligible by use of good experimental techniques.

Neutral beam species are undeflected by charged-particle deflection schemes. If the aperture and target are "off-set" from the neutral beam, the neutral beam is stopped and trapped by a beam stop or chamber wall, and only charged species reach the target.

Charge exchange can neutralize the beam if the vacuum in the drift space is poor. In this process an ion collides with neutral gas and picks up an electron, leaving the ion neutral. Significant neutralization can occur if pressures are greater than about 10^{-5} torr.

In secondary electron emission, ions hitting the target eject low-energy electrons from the target. If these electrons (secondaries) are lost to chamber walls, there will be an error in the dose measurement. Errors due to secondary electron emission are minimized by the use of a Faraday cage (metal electrode configuration) that nearly surrounds the target and has an opening facing the aperture (Fig. 1). A bias of a few hundred volts can be applied to the Faraday cage relative to the target so most secondary electrons are returned to the target and the integrator circuit.

Sputtering of aperture material onto the sample will always occur. This effect is minimized if the aperture is made of low-sputter-coefficient or benign material such as carbon or silicon. If the aperture is constructed of Fe or Ta, then a few percent of heavy metal relative to the ion dose can be sputtered onto the target.

Sputtering of already implanted atoms from the target can be important at high doses ($\sim 10^{16}$ atoms/cm^2). One may expect a "saturated-sputter limited" dose where each new implanted atom removes one previously implanted ion by sputtering. In practice, however, there may be less saturation than expected because of channeling effects and thermal diffusion caused by beam heating. These effects place previously implanted atoms further away from the surface than simple range theory suggests, and tend to reduce the sputtering of already implanted atoms.

Other surface contamination during high-dose implantations can occur. The physisorption of hydrocarbons from diffusion pump oil followed by radiation-induced polymerization can occur.[17] This effect can result in high metal-to-semiconductor electrical contact resistance. Implantation through thin protective (screen) oxides, followed by contact lithography down to the bare silicon, can avoid the adverse effect of polymerized hydrocarbons on silicon surfaces.

Silicon wafers are often implanted with both thin protective coverings and thick SiO$_2$ layers (masks). Ion-induced charging of SiO$_2$ layers can have a deleterious effect on the quality of the SiO$_2$ if dielectric breakdown ($\mathscr{E} \sim 10^7$ V/cm) occurs. Techniques used to avoid this are the use of an added electron source to neutralize positive surface charge of the implanted beams, the use of thin conductive layers on the oxide, intentionally causing excess conductivity in the SiO$_2$, and modification of the oxide pattern by cutting bare silicon regions in the oxide (for example, in the grid regions between chips).

In summary, the control of many parasitic effects are essential for accurate and high-quality implantations. We now take up the question of ion mass selection.

A typical ion source produces many different elements, isotopes, and charge state species. Separation of these species is obtained with a mass spectrometric analyzer magnet. In a magnetic field B, an ion path takes on a radius of curvature R such that $RB = \sqrt{2VM/mq}$, where V is the acceleration voltage. By adjusting the magnetic field the species of interest is selected to pass through slits that define the radius of curvature. The selection of the desired species (element, isotope, and charge state) is certified by the signature of the entire mass spectrum. See Fig. 3, where the target beam current is plotted against magnetic-field strength. A straight line relates the mass and charge state to the magnetic-field strength. See Problem 3 for further discussion.

Use of a doubly ionized species extends the machine's energy capability by a factor of 2 ($E = 2qV$). To determine an atom dose for a doubly ionized species, we must count 2 electrons for every implanted atom, that is, in Eq. (1), $m = 2$. Doubly ionized species are usually less abundant and applications must recognize this limitation.

Many applications use ion deposition followed by a thermal drive-in to obtain a desired dopant distribution. Such applications use low, fixed energy. Relatively inexpensive machines with an ion extractor at the terminal voltage, and magnetic-mass spec analysis at low magnetic fields, form a class of "pre-deposition," dedicated accelerators (10 to 30 keV). Accelerators with higher energies usually are made with a variable energy range capability. A relatively large and expensive magnet is needed to analyze the accelerated species.

Machines with high beam current (~ 10 mA) are now commercially available.

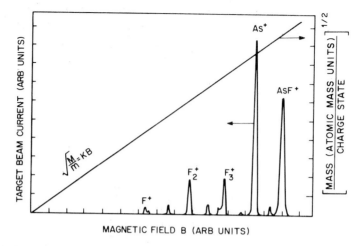

Fig. 3 A typical ion beam mass spectrum of a AsF_5 gas source. Target current and magnetic field are in arbitrary units. The major ionized species are labeled. The ratio of mass to charge state is determined from the straight line.

To avoid target heating, the beams are defocused, and samples are placed on rotating target plates and/or sometimes cooled. For rotating targets, the dose is defined by the integrated charge corrected by the fraction of time the ion beam hits the sample.

Ion sources have been developed using ovens, rf plasmas, hot cathodes, and arc discharges. Laser-pulsed and microwave-energized plasmas[18] provide larger fluxes. Implantation, when interfaced with molecular beam epitaxy equipment (MBE),[19] can yield impurity distributions at greater depths than is possible by implantation alone.

6.3 ION RANGES

6.3.1 Distribution Description

An individual implanted ion undergoes scattering events with electrons and atoms in the target, reducing the ion's energy until it comes to rest. Point defects and even small amorphous disorder zones may result (Fig. 4a). The total path length of the ion is called the range, R (Fig. 4b). A typical ion stops at a distance normal to the surface, called the projected range, R_p. Some ions are statistically "lucky"; they encounter fewer scattering events in a given distance in the target, and come to rest beyond the projected range. Other ions are "unlucky"; they have more than the average number of scattering events, and come to rest between the surface and the projected range. The fluctuation or straggle in the projected range is ΔR_p. There is also a fluctuation in the final ion's position perpendicular to the incident ion's direction, called the lateral straggle, $\Delta R_\perp$.

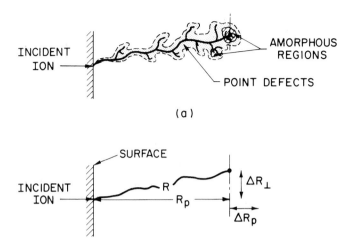

Fig. 4 (a) A "tree" of disorder for a typical implanted ion. (b) A schematic of the ion range R, projected range R_p, uncertainty in R_p or projected straggle ΔR_p, and the lateral straggle $\Delta R_\perp$.

The depth distribution or profile of stopped ions can be approximated by a symmetric Gaussian distribution function. The concentration of implanted atoms as a function of position is

$$n(x) = n(R_p) \exp \left[\frac{-(x - R_p)^2}{2\Delta R_p^2} \right] \qquad (2)$$

where the maximum concentration occurs at $x = R_p$, and ΔR_p is the standard deviation or "straggle" of the distribution. The integral $\int_0^\infty n(x)\, dx$ gives the dose ϕ, and the maximum concentration $n(R_p)$ can be written as

$$n(R_p) = \frac{\phi}{\sqrt{2\pi}\, \Delta R_p} \cong \frac{0.4\phi}{\Delta R_p} \qquad (3)$$

The projected range and straggle of the Gaussian distribution give a good first-order description of the implanted ions in amorphous or fine-grain polycrystalline substrates. The data of some implanted distributions can be fit rather well by the Gaussian distribution function; certain values are given in Table 1. Although the fit is almost always good near the peak, there is a pronounced skewness in the actual distributions. To account for the skewness and also any tailing character higher moment descriptions are needed.

A three-moment approach[20] uses two Gaussians, each with their own straggle ΔR_{p1} and ΔR_{p2}. The Gaussians are joined at their "modal range," R_M. From the three fundamental calculated parameters R_p, ΔR_p, and the third moment ratio CM_{3p},

Table 1 Gaussian and erfc values

$\dfrac{x-R_p}{\Delta R_p}$	$\exp\dfrac{-(x-R_p)^2}{2\Delta R_p^2}$	$\dfrac{y-a}{\Delta R_\perp}$	$0.5\,\text{erfc}\ \dfrac{y-a}{\sqrt{2}\,\Delta R_\perp}$
0	1.00	0	0.50
1.0	0.61	0.28	0.39
1.18	0.50	0.56	0.28
1.5	0.325	0.70	0.24
2.0	0.14	1.00	0.16
2.14	0.10	1.26	0.10
2.5	0.044	1.4	0.078
3.04	0.01	2.0	0.022
3.5	0.0022	2.33	0.01
3.72	0.001	2.4	0.008
4.0	0.00034	3.07	10^{-3}
4.3	10^{-4}	3.7	10^{-4}
4.8	10^{-5}	4.3	10^{-5}
5.25	10^{-6}	4.8	10^{-6}
5.67	10^{-7}	5.2	10^{-7}

it is possible to calculate ΔR_{p1}, ΔR_{p2}, and R_M. A distribution is then obtained by the joining of two Gaussians

$$n(x) = \frac{2\phi}{\sqrt{2\pi}\,(\Delta R_{p1} + \Delta R_{p2})} \exp\left[-\frac{(x - R_M)^2}{2\Delta R_{p1}^2}\right] \qquad x \geqslant R_M \quad (4a)$$

$$n(x) = \frac{2\phi}{\sqrt{2\pi}\,(\Delta R_{p1} + \Delta R_{p2})} \exp\left[-\frac{(x - R_M)^2}{2\Delta R_{p2}^2}\right] \qquad x \leqslant R_M \quad (4b)$$

A more exact description[21] uses the "four-moment" approach: first (R_p), second (ΔR_p), third—skewness—(γ_1), and fourth—kurtosis—(β). Kurtosis describes the tail character of the distribution. Several equations lead to the Pearson-IV-type distribution. Pearson distributions are based on the differential equation

$$\frac{dh(x)}{dx} = \frac{(x'-a)\,h(x')}{b_2 x'^2 + b_1 x' + b_0} \qquad (5)$$

where h is the normalized distribution function, $h(x)$ satisfies $\int_{-\infty}^{+\infty} h(x)\,dx = 1$, and $x' \equiv x - R_p$. Four constants, a, b_0, b_1, and b_2 are defined in terms of the four moments μ_1, μ_2, γ_1, and β where

$$\mu_1 \text{ (mean range)} = R_p = \int_{-\infty}^{+\infty} xh(x)\,dx \qquad (6a)$$

$$\mu_2 \text{ (straggle)} = \Delta R_p = \int_{-\infty}^{+\infty} (x - R_p)^2\,h(x)\,dx \qquad (6b)$$

$$\gamma_1 \ (\text{normalized skewness}) = \frac{\int_{-\infty}^{+\infty} (x - R_p)^3 \, h(x) \, dx}{\Delta R_p^3} \tag{6c}$$

$$\beta \ (\text{normalized kurtosis}) = \frac{\int_{-\infty}^{+\infty} (x - R_p)^4 \, h(x) \, dx}{\Delta R_p^4} \tag{6d}$$

Using these definitions, the four constants are related to the moments,

$$a = -\gamma_1 \mu_2 (\beta + 3)/A \tag{7a}$$

$$b_0 = -\mu_2^2 (4\beta - 3\gamma_1^2)/A \tag{7b}$$

$$b_1 = a \tag{7c}$$

$$b_2 = -(2\beta - 3\gamma_1^2 - 6)/A \tag{7d}$$

where $A = 10\beta - 12\gamma_1^2 - 18$. Only Pearson-IV solutions are applicable to implanted profiles. The solution is

$$\ln \frac{n(x)}{n_0} = \frac{1}{2b_2} \ln \left[b_2 x'^2 + b_1 x' + b_0 \right]$$

$$- \frac{\dfrac{b_1}{b_2} + 2a}{(4b_2 b_0 - b_1^2)^{1/2}} \ \arctan \frac{2b_2 x' + b_1}{(4b_2 b_0 - b_1^2)^{1/2}} \tag{8}$$

where $n_0 = \phi / \int_0^\infty h \, dx$. Using these four moments, excellent fits to the implanted distributions can be obtained.

Figure 5 shows measured, skewed boron atom profiles and their fitted four-moment distributions for 30 to 800 keV.[21] The implants were done into fine-grain polycrystalline silicon to avoid channeling effects (described in Section 6.3.3). Gaussians only fit the data well at low energy and over part of the profile. Pronounced skewness is evident toward the surface (γ_1 is negative so a maximum occurs at x greater than the mean distance R_p). Arsenic profiles, however, show skewness on the deep side of the implant profile (γ_1 is positive). The different skewness can be visualized by thinking of forward momentum. If light ions impact on the target atoms, they will have a relatively large amount of backward scattering. The result is a filling-in of the distribution on the surface side, as with boron. Conversely, if heavy ions impact on a target atoms, they will have a disproportionate amount of forward scattering. The result is a filling-up of the ion distribution on the deep side of the distribution, as with arsenic.

Lateral ion straggle effects are an extremely important, practical aspect of ion stopping. In applications of self-aligned implanted sources and drains (Chapter 11), the lateral ion straggle is a limiting fundamental factor which determines the doping between source and drain and therefore the electrical channel length.

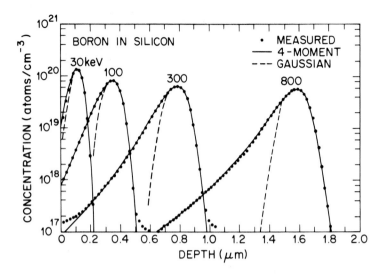

Fig. 5 Boron implanted atom distributions, with measured data points, and four-moment (Pearson-IV) and symmetric Gaussian curves. The boron was implanted into amorphous silicon without annealing. *(After Hofker, Ref. 21.)*

Figure 6a shows a "thick" ion mask (thickness $>> R_p + \Delta R_p$) with a vertical-slotted window. In this figure the slot's long direction (z) is into the paper, the short direction (y) has a slot width of $2a$, and x is the depth coordinate. The profile is given by

$$n(x,y) = \frac{n(x)}{2} \left[\text{erfc} \left[\frac{y-a}{\sqrt{2}\,\Delta R_\perp} \right] - \text{erfc} \left[\frac{y+a}{\sqrt{2}\,\Delta R_\perp} \right] \right] \quad (9)$$

where $n(x)$ is the depth distribution density far away from the mask edge. For $a >> \Delta R_\perp$, $n(x,a) \cong n(x)/2$, which is the result[22] expected for a "half-source." Figure 6a is for $a >> \Delta R_\perp$, the edges are far removed from each other. Figure 6b shows the contours of equal-ion concentration for 70-keV boron into a 1-μm slit. The lateral doping extends well under the mask edge and will effect the channel length of a short gate.

6.3.2 Theory of Ion Stopping

The range of an ion is determined by Lindhard, Scharff, and Schiott (LSS) theory,[1] where energy loss mechanisms are considered to be independent of each other and additive. For electronic and nuclear stopping, the energy loss per unit length is defined as

$$\left. \frac{dE}{dx} \right|_{total} = \left. \frac{dE}{dx} \right|_{nuclear} + \left. \frac{dE}{dx} \right|_{electronic} \quad (10)$$

The nuclear and electronic energy loss, are both functions of energy.

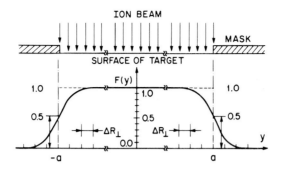

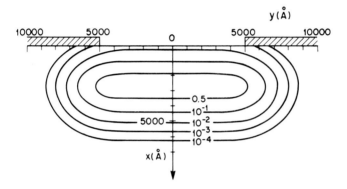

Fig. 6 Illustration of lateral profiles. (a) Ion concentration along the lateral direction (y) for a gate mask with a $\gg \Delta R_\perp$ and infinite extension in the z-direction. (b) Contours of equal-ion concentrations for 70-keV B^+ ($R_p = 2710$ Å, $\Delta R_p = 824$ Å, and $\Delta R_\perp = 1006$ Å) incident into silicon through a 1-μm slit. *(After Furukawa, Matsumura, and Ishiwara, Ref. 22.)*

The range R of ions is given by

$$R(E) = \int_0^E \frac{dE}{|dE/dx|_{\text{total}}} \equiv \frac{1}{N} \int_0^E \frac{dE}{S(E)} \tag{11}$$

where N is the number of target atoms/cm^3, $S(E)$ is the stopping power, and E is the initial incident ion energy.

A physical description of the scattering process is provided by classical mechanics.[23] The transferred energy T between an incoming ion (energy E_1, mass M_1, and atomic number Z_1) and the target atom (mass M_2 and atomic number Z_2), having a scattering angle θ in the center-of-mass system, is

$$T = \frac{E_1 4 M_1 M_2}{(M_1 + M_2)^2} \sin^2 \frac{\theta}{2} \equiv \gamma E_1 \sin^2 \frac{\theta}{2} \tag{12}$$

When $\theta = 180°$ (head-on), T is a maximum value. The scattering angle is obtained by integrating the equation of motion for the scattering trajectory, using an atomic scattering potential

$$V(r) = \frac{q^2 Z_1 Z_2}{r} f_s(r) \tag{13}$$

$V(r)$ is a coulombic- (Rutherford-)type potential with screening function f_s. (The Thomas-Fermi function is one example of f_s.) The nuclear energy loss is given by

$$\frac{dE}{dx} = N \int_0^{T_m} T \, d\sigma \equiv NS(E) \tag{14}$$

where $d\sigma$ is the differential cross section. LSS has introduced a number of reduced variables that make the integration tractable and also lead to a "universal" curve for the nuclear energy loss.[1] The reduced quantities are ϵ, ρ, and t:

$$\epsilon \text{ (energy)} = \left[\frac{M_2}{M_1 + M_2} \middle/ \frac{Z_1 Z_2 q^2}{a} \right] E \tag{15}$$

$$\equiv \epsilon_1 E$$

where a is the screening length equal to $0.88 a_0 / (Z_1^{2/3} + Z_2^{2/3})^{1/2}$ and a_0 is the Bohr radius.

$$\rho \text{ (distance)} = N \pi a^2 \gamma x$$

$$\equiv \rho_1 x \tag{16}$$

$$t \text{ (scattering parameter)} = \frac{\epsilon^2 T}{\gamma E} \tag{17}$$

The value $t^{1/2}$ is also used as an integration parameter for the energy loss. In reduced units the nuclear energy loss is then

$$\frac{d\epsilon}{d\rho} = \frac{\epsilon_1}{\rho_1} \frac{dE}{dx} = \frac{\epsilon_1}{\rho_1} N \int_0^{T_m} T d\sigma = \frac{1}{\epsilon} \int_0^{\epsilon} f(t^{1/2}) \, dt^{1/2} \tag{18}$$

where the following relations have been used:

$$d\sigma = \pi a^2 f(t^{1/2}) \, dt^{1/2} \tag{19a}$$

$$N \, d\sigma = \frac{\rho_1}{\gamma} \frac{dt}{2t^{3/2}} f(t^{1/2}) \tag{19b}$$

$$t^{1/2} = \epsilon \sin \frac{\theta}{2} \tag{19c}$$

The scattering function $f(t^{1/2})$ depends upon the form of $V(r)$. The use of Thomas-Fermi screening over the full range of t values results in a universal nuclear $d\epsilon/d\rho$

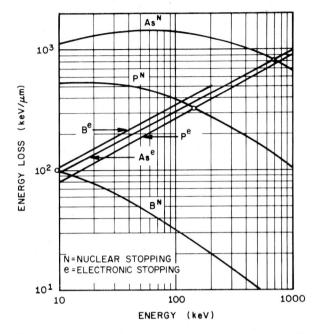

Fig. 7 Calculated values of dE/dx for As, P, and B at various energies. The nuclear (N) and electronic (e) components are shown. Note the points (o) at which nuclear and electronic stopping are equal. *(After Smith, Ref. 24.)*

energy loss curve.[1] Other functions may give better physical estimates of ranges, straggles, etc. The universal nuclear loss $d\epsilon/d\rho$ is independent of Z_1, Z_2, M_1, M_2, or N. In general, nuclear energy loss is relatively low at high energies. Fast particles have a smaller interaction time with the scatterer—that is, the cross section is reduced. At intermediate energies $dE/dx\,|_{\text{nuclear}}$ rises, and at the lowest energies, where screening effects reduce the effective value of the target's coulomb charge Z_2, the value of $dE/dx\,|_{\text{nuclear}}$ again is reduced.

The LSS electronic stopping, which is similar to stopping in a viscous medium, is proportional to the ion's velocity.

$$dE/dx\,|_{\text{electronic}} = k_e \sqrt{E} \tag{20}$$

The coefficient k_e is a relatively weak function of Z_1, Z_2, M_1, and M_2, the atomic charges and masses. Values[24] of actual nuclear and electronic dE/dx (keV/μm) for B, P, and As are plotted in Fig. 7. The values at electronic dE/dx are not monotonic with incident mass, but the values are based on experimental data. When the nuclear and electronic stopping curves are added, it is noted that the total value of dE/dx is nearly a constant over a very large range of energies. See Problem 5. As a result the range from Eq. (11) is nearly proportional to the initial incident ion energy.

Curves of projected range, R_p for B, P, and As in silicon and thermal SiO_2 are shown in Fig. 8. Figure 9 shows the projected straggle ΔR_p and the transverse straggle $\Delta R_\perp$ for the same elements.[24]

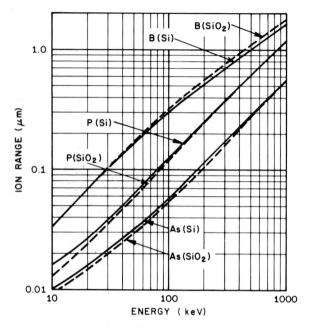

Fig. 8 Projected range, R_p, calculated for B, P, and As at various energies. The results pertain to amorphous silicon targets and thermal SiO_2 (2.27 g/cm^3). *(After Smith, Ref. 24.)*

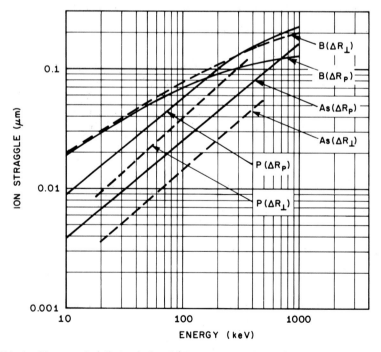

Fig. 9 Calculated ion straggle ΔR_p (vertical) and $\Delta R_\perp$ (transverse) for As, P, and B ions in silicon. *(After Smith, Ref. 24.)*

The two-moment description tells where the ions are located and roughly how broad the distribution is. Higher moments (skewness and kurtosis) determine the detailed shape. In practice we can make use of the Gaussian two-moment description to quickly estimate doping distributions and then "fine-tune" the dose or energy to obtain better results. However, we can anticipate the effects of a non-Gaussian behavior on device behavior. For example, the skewed boron implants result in "high doping" at the surface when deep p-wells are implanted (in CMOS technology), and skewed arsenic implants result in "deep junctions" when n^+ sources and drains are fabricated. The non-Gaussian effects are both an intrinsic property of the distribution and important for VLSI consideration.

Monte Carlo techniques have been used to calculate histograms for representing the profiles that are fitted to the Pearson-IV solution, Eq. 8. Table 2 lists a set of values[25] for R_p, ΔR_p, γ_1, and β.

As previously mentioned, the distributions of implanted atoms in amorphous and small ($\sim$100-Å) grain-size polycrystalline silicon are the same. All results discussed so far are for this case. However, implanted distributions into single-crystal and large-grain polycrystalline silicon show the effects of channeling.

Table 2 Values of four moments and $\Delta R_\perp$

Ion	Parameter	E (keV)			
		10	30	100	300
B	R_p	382	1065	3070	6620
	ΔR_p	190	390	690	1050
	γ_1	−0.32	−0.85	−1.12	−1.59
	β_2	3.2	4.49	5.49	8.35
	$\Delta R_\perp$	190	465	871	1523
P	R_p	150	420	1350	4060
	ΔR_p	78	195	535	1150
	γ_1	0.45	0.20	−0.37	−0.91
	β_2	3.4	3.1	3.26	4.89
	$\Delta R_\perp$	61	168	471	1097
As	R_p	110	233	678	1946
	ΔR_p	40	90	261	667
	γ_1	0.57	0.46	0.45	0.30
	β_2	3.6	3.4	3.4	3.16
	$\Delta R_\perp$	33	64	187	481
Sb	ΔR_p	100	208	507	1303
	ΔR_p	30	62	158	390
	γ_1	0.54	0.51	0.40	0.18
	β_2	3.5	3.5	3.3	3.1
	$\Delta R_\perp$	23	46	108	266

Note: $\Delta R_\perp$ Furukawa[22] scaled to ΔR_p of Fichtner.[25]

6.3.3 Ion Channeling

Channeling starts to occur when an incident ion finds entry into an open space between rows of atoms.[26] Once the ion is inserted, steering forces of the atomic row potentials become operative and steer the ion toward the center of the open space (channel). The ion is stably guided along the channel over considerable distances. The ion gradually loses energy through its gentle, glancing collisions on the edges of the channel, and eventually scatters out of the channel. Channeled penetration distances can be several times the penetration in amorphous targets, because the energy loss for channeled ions is low compared with non-channeled ions. If attempts are made to avoid channeling in single crystal silicon targets by the orientation to dense atom directions (e.g., the $\langle 763 \rangle$) then channeling effects are minimized but not eliminated (Fig. 10).

Profiles obtained by implantation into single crystals in such a way as to avoid channeling are characterized by tails of atoms [as determined by careful secondary ion mass spectroscopy, or SIMS (see Chapter 12) and tails of free charge doping (as determined by electrical data). The channeling tails often can be fit to an exponential function of position, $\exp(-x/\lambda)$, where λ is typically found to be ~ 0.1 μm. The tails are more prominent for phosphorus than for boron because the acceptance or critical angle for channeling is larger for heavier atoms. The critical angle is proportional to $Z_1^{1/2}$. For 50-keV phosphorus, $\psi_{crit} = 5.9°$, while for 50-keV boron, $\psi_{crit} = 4.8°$ along the $\langle 110 \rangle$ axis. The critical angle (for relatively high energies)[26] is given by

$$ \psi_{crit} = \left[\frac{2Z_1 Z_2 q^2}{Ed} \right]^{1/2} \tag{21} $$

where d is the atomic spacing along the aligned row.

The primary mechanism for tail formation in crystalline targets that are oriented off major axes is believed to be channeling into major axes. (An alternate proposal invokes an interstitial diffusion mechanism at implantation temperatures.) Experiments with phosphorus prove that the tail of the distribution is due to channeling.[27] Transmission of phosphorus ions through thinned silicon shows that ions emerge with measurable energies and therefore are unambiguously channeled. The crystals are about 0.5 μm thick and implanted with radioactive P^{32}. The ions corresponding to the

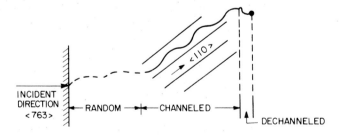

Fig. 10 Schematic diagram of an ion path in a single crystal for an ion incident in a "dense" $\langle 763 \rangle$ direction. The path shown has non-channeled and channeled behavior.

deep tail part of the distribution are collected at a second target in back of the thinned silicon. These results, obtained for targets aligned off any major axes, can only happen if the ions in the tail of the distribution are channeled (and the interstitial diffusion mechanism is not operative).

The practical use of the deeper-penetrating channeling ions has been considered. However, profile control and reproducibility are difficult because of the very critical control of orientation which is required. Figures 11a and 11b show the critical control that is needed for phosphorus and boron. The ions were implanted into silicon for various orientations away from major-index axes. After a relatively low temperature anneal (850°C), free-carrier profiles were measured using C-V techniques. Here, an angular variation of one degree is shown to be significant.[5b]

The self-aligned gate of an MOS transistor often uses polysilicon (or other polycrystalline materials) as a mask against implantation. If the range of channeled ions is larger than the thickness of the gate material, then channeled ions can arrive at the gate-oxide interface with enough energy to penetrate the gate oxide. A patchy-doping effect, observed under large-grain, polycrystalline silicon gates,[28] results in a small population of depletion-mode MOSFETs when the grain size is comparable to the electrical channel length. This effect can be avoided simply by selecting a gate thickness that is greater than the channeled range plus several straggle distances.

6.3.4 Knock-On Ranges

Device fabrication often uses processing in which surface coatings are present on the targets. When implantations are done through SiO_2 layers, oxygen and silicon atoms are knocked into the underlying silicon. The range and numbers of oxygen atoms recoiled from surface coatings are comparable to implanted arsenic concentrations. Experimental results show that the free-carrier mobility is not degraded for arsenic implanted into $\langle 100 \rangle$ surfaces through SiO_2 layers, although it is degraded for $\langle 111 \rangle$ surfaces.[29] When ions are implanted, a major part of the disorder production is due to the recoil or knock-on effects of target atoms. The multiple scattering and the final resting place of the recoils determine the radiation damage distribution.

6.4 DISORDER PRODUCTION

When ions enter a silicon crystal, they undergo electronic and nuclear scattering events, but only the nuclear interactions result in displaced silicon atoms. The sequence is as follows: In $\sim 10^{-13}$ second a given ion comes to rest (this is roughly the ion range divided by the average ion velocity $R_p / \sqrt{E / 2M}$), in $\sim 10^{-12}$ second thermal vibrations settle down to equilibrium values, and in $\sim 10^{-9}$ second the non-stable crystal disorder relaxes and some ordering occurs by a local diffusion process.

Light and heavy implanted ions have a qualitatively different "tree of disorder" along the stopping track. Light ions (e.g., B^{11}) which enter the surface initially suffer mostly electronic stopping. They gradually lose energy until nuclear stopping becomes dominant. While undergoing nuclear stopping they displace silicon atoms

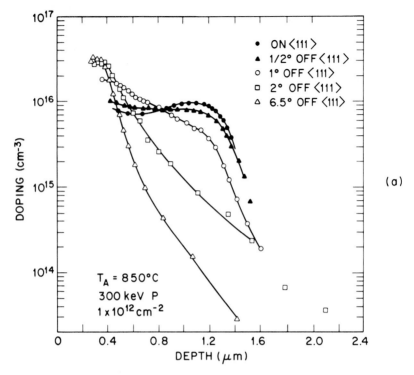

(a)

Fig. 11 (a) Donor free-carrier profiles for various orientations away from the ⟨111⟩ axis for 300-keV P³¹. *(After Moline and Reutlinger, in Ref. 5b.)*

and also change direction. The displaced atom profile has a buried peak concentration. In contrast to this heavy ions (P³¹ or As⁷⁵) enter the surface and immediately encounter a relatively higher fraction of nuclear stopping. They displace large numbers of silicon atoms close to the surface. As they slow down, the nuclear stopping power of the primary incident ion is nearly constant over most of the energy values but recoiled atoms transfer deposited energy to greater depths. The final damage density profile exhibits a broad buried peak which is a replica of the recoiled range distribution.

An individual energy transfer process can result in different displacement configurations. If the energy transferred to a given silicon atom, ΔE, is less than the displacement energy E_{do}, no displacement occurs. If the value of ΔE is greater than E_{do}, one displacement and simple isolated defects occur. If $\Delta E \gtrsim 2E_{do}$, we obtain stable defects and secondary displacements. If $\Delta E \gg E_{do}$, there are multiple secondary (recoil or knock-on) displacements accompanied by defect clusters. These highest-density disorder regions may be locally amorphous, especially for heavy mass ion implantations.

A complicated array of different kinds of defects along the ion track results because of the displacement profile. This inventory of defects consists of vacancies

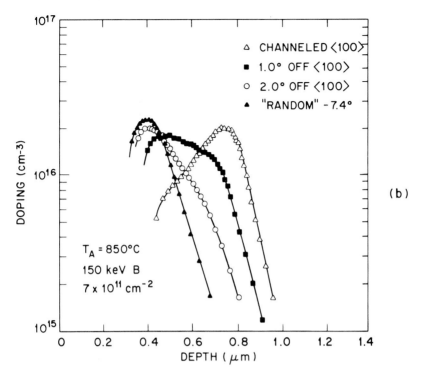

Fig. 11 (continued) (b) Acceptor profiles for various orientations away from the $\langle 100 \rangle$ axis for 150-keV B[11]. *(After Seidel, in Ref. 5b.)*

(V) and—at least before a reordering thermal relaxation can take place at a typical implantation temperature of 300 K—divacancies (V^2), higher-order vacancies (V^3, V^4), or vacancy-impurity complexes (V-Donor, V-Acceptor, V-Oxygen). In addition, if beam heating during implantation is severe and temperatures exceed about 500°C, dislocations will form. The actual inventory of defects is complicated and depends on position, thermal history, and impurity species. A fair number of the incident ions end up on substitutional sites when they come to rest. However, the damage in the absence of annealing produces a larger number of deep-level states than the implanted ion concentration.

The nature of the actual disorder is a complex topic and depends on many factors such as the crystal orientation and temperature of the target. The total energy deposited into atomic displacements (Q_D eV/Å per ion) has been calculated assuming that ion channeling, thermal diffusion, and saturation effects are negligible during the stopping process.[30] Figure 12 shows the calculated damage density ($Q_D R_p / E_d$) plotted against distance (x/R_p) for boron and arsenic implantations into silicon. The damage density and distance values are normalized. The density of displaced silicon N_{dis} atoms/cm³ is approximated by $\phi Q_D/E_{do}$, where ϕ is the ion dose (ions/cm²) and E_{do} is the target atom's displacement energy. For silicon, E_{do} is taken to be about 15 eV.

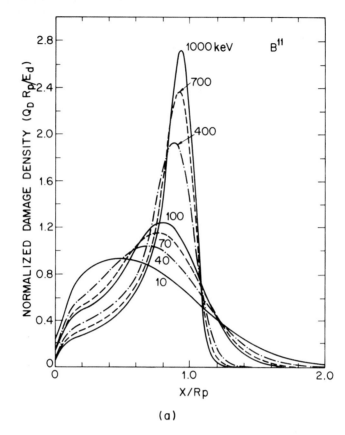

Fig. 12 Calculated damage density profiles of (a) boron and (b) arsenic. *(After Brice, Ref. 30.)* The values of deposited energy Q_D are obtained by multiplying the normalized damage density by E_d/R_p (eV/Å). Some values of E_d/R_p for B are 12.9 (10 keV) and 6.1 (100 keV); for As they are 91.4 (10 keV) and 99.3 (100 keV).

We can also attempt to simply estimate the number of displaced silicon atoms, both at the surface and for the peak value of the nuclear stopping. In this case,

$$N_{\text{dis}} \ (\text{cm}^{-3}) = \frac{\phi}{E_{do}} \ \frac{dE}{dx}\bigg|_{\text{nuclear}} \tag{22}$$

The value of the nuclear stopping power, $dE/dx\,|_{\text{nuclear}}$, can be taken from Fig. 7. For 300 keV, 10^{13} boron ions/cm^2 and $E_{do} = 15$ eV, we obtain surface disorder concentrations of about 10^{20}/cm^3, and a peak disorder concentration of about 7×10^{20}/cm^3. For arsenic the surface concentration is about 8×10^{21}/cm^3 and the peak concentration is about 10^{22}/cm^3. From this it is clear that the heavier ions will give displacement disorder concentrations approximately equal to the silicon density at doses of $\sim 10^{14}$/cm^2. These are order of magnitude estimates only and differ from the detailed calculations[30] by a factor of 2 to 5.

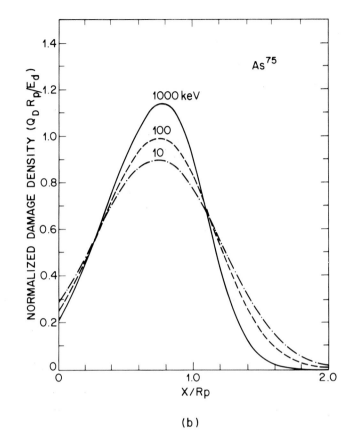

(b)

Fig. 12 (continued) (b) arsenic.

The total number of displaced atoms per incident ion is given roughly by[31]

$$N(E)_{\text{total}} \cong E_n / 2E_{do} \qquad (23)$$

where E_n is the incident energy available for nuclear stopping processes (i.e., the area under the $dE/dx \mid_{\text{nuclear}}$ curve). The buildup of disorder is linear with dose until saturation occurs, where a previously displaced atom again absorbs energy from another implanted ion. When the number of stably displaced silicon atoms reaches $N_{\text{Si}} = 5 \times 10^{22}/\text{cm}^3$, that is certainly by the time every target atom is stably displaced the material changes phase and becomes "amorphous."

Other views hold that there is a critical energy density[32, 33] that must be placed into the crystal to make it amorphous. This critical energy is

$$E_c = (f) N_{\text{Si}} E_{do}$$
$$\cong f \, 10^{24} \text{ eV}/\text{cm}^3 \qquad (24)$$

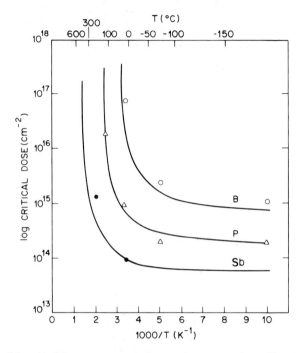

Fig. 13 A plot of the critical dose necessary to make a continuous amorphous silicon layer, against recipro-cal target temperature for various ions. Arsenic falls between P and Sb. The temperature at which silicon cannot be made amorphous is higher for higher-mass ions. *(After Morehead and Crowder, Ref. 34.)*

This result is applicable for low-temperature implantations and when the prefactor f for silicon is approximately 0.1–0.5. It is likely that an amorphous state will form before every atom is displaced. If the thermodynamic free energy of the damaged state equals that of the amorphous state, a transformation will occur. The critical dose to form an amorphous layer is given by $D_c = E_c / \langle dE/dx \big|_{\text{nuclear}} \rangle$.

The effects of substrate temperature on the accumulation of disorder is substan-tial. For example, consider an individual ion track with a locally disordered amor-phous region. Such a region was modeled to be amorphous in a cylinder with an orig-inal radius R_0, and the radius can shrink by the thermal motion of defect vacancies out of the core of the cylinder.[34] With the use of the vacancy out-diffusion idea, the tem-perature dependence of the critical dose for an amorphous layer formation can be estimated. For light ions such as B^{11}, a 50°C rise in temperature above room tempera-ture prevents the formation of amorphous material at any dose (Fig. 13). This is because boron implanted at room temperature produces only a few stably displaced silicon atoms during implantation, and a slight rise in temperature allows the recombi-nation of vacancy-silicon interstitial pairs.

The effects of non-uniform ion beam heating across wafers are interesting. If a wafer is uniformly heated during implantation the accumulation of disorder will be uniform and eventually a uniform buried amorphous layer will be formed. Since the

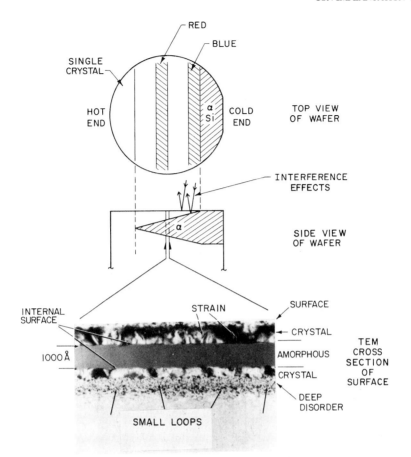

Fig. 14 Schematic showing color band effects. The transmission electron microscopy (TEM) cross-section shows a buried amorphous layer. The implant was 10^{16} argon / cm^2 at 200 keV and 0.5 mA. *(After Sheng and Seidel, unpublished.)*

index of refraction of amorphous layers is higher than that of crystalline layers we can obtain interference effects (a color) from the buried amorphous layers. However, if a wafer is non-uniformly heated (laterally across the wafer) the colder region will have a thicker buried amorphous layer. These non-uniformly heated wafers exhibit a rainbow of colors.[35] Figure 14 shows a schematic of a wafer with cold and hot ends labeled, a side view schematic of the buried disorder, and a transmission electron microscopy (TEM) cross section of a thin slice. The TEM structure is rich in detail, showing light-dark contrast in the strained surface crystalline region, the buried amorphous layer, a buried dislocation layer deeper than the amorphous layer, and evidence of very small dislocation loops in the tail of the disorder. Amorphous surface layers also give interference effects. High optical absorption in the visible range limits color effects to very thin amorphous surface layers. Disordered layers with mixed amorphous and crystalline phases also give visible interference effects.[36]

6.5 ANNEALING OF IMPLANTED DOPANT IMPURITIES

This section covers the increase in the free-carrier content and the decrease of disorder as a function of anneal temperature for boron and phosphorus. A study of the detailed annealing behavior leads to a unified view of annealing: The annealing of amorphous layers is contrasted with annealing of point and extended defects.

One of the fundamental questions in the field of ion implantation is: What minimal temperatures and times are required to achieve full donor or acceptor activity without leaving degrading residual defects? A related question is: Can complete electrical activity be obtained without significant atomic diffusion? The second question is prompted by the need for very shallow junctions for the micrometer-size designs of VLSI.

The systematics of annealing are both ion-dose- and ion-species-dependent. The annealing will be discussed first in terms of the spatially integrated electrically active charge (donors or acceptors) $N(\text{cm}^{-2}) \cong \int n(\text{cm}^{-3}) \, dx$. This data—the "areal" density—is approximately obtained using the Hall effect technique. The Hall effect measures an average, effective doping, which is an integral over local doping densities and local mobilities.

$$N_{\text{Hall}} = \frac{\left[\int_0^{x_j} \mu n \, dx \right]^2}{\int_0^{x_j} n\mu^2 \, dx} \tag{25}$$

where μ is the mobility and x_j is the junction depth.[37] For uniform doping layers, or where the mobility is not strongly dependent on position, the relation gives $N_{\text{Hall}} \approx \int_0^{x_j} n \, dx$. In the data presented below, the measured Hall density is normalized to the implanted dose ϕ. When all the atoms of the distribution become electrically active, $N_{\text{Hall}} \approx \phi$, which is taken to be the condition for "full electrical" activity. The Hall effect measures equilibrium majority-carrier concentrations and gives no direct information about minority-carrier effects.

The number of displaced silicon atoms is almost always greater than the number of implanted atoms. Thus the usual situation for a non-annealed sample is an electrical layer dominated by deep-level traps. If an implantation is done into either n- or p-type substrates of moderate doping (10^{16} atoms/cm^3), the result is a high-resistivity layer. Both electron and hole traps are produced.

6.5.1 Isochronal Boron Annealing

Figure 15 shows the isochronal (same time, different temperatures) annealing behavior for boron, implanted at 150 keV and at three different doses. Three annealing temperature regions are noted as I, II, and III. The low-dose case shows a monotonic increase in electrical activity, the two higher doses show a reverse anneal in region II between 500 and 600°C.

Region I is characterized by point-defect disorders that dominates the electrical free-carrier concentration. TEM shows no extended defects (dislocations) in this

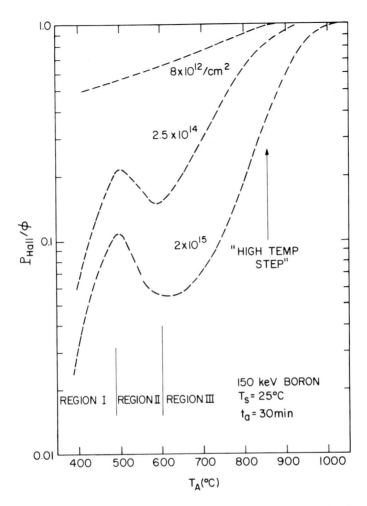

Fig. 15 Isochronal annealing behavior of boron. The ratio of free-carrier content (P_{Hall}) to dose (ϕ) is plotted against anneal temperature (T_A) for three doses of boron. At $T_A \geq 900°C$, the free carriers approach the dose. *(After Seidel and MacRae, in Ref. 5a.)*

region. Increasing the annealing temperature from room temperature to approximately 500°C results in the removal of point defects such as divacancies. The boron substitutional concentration also decreases up to approximately 500°C,[38] but by a factor of ~2, while the free-carrier concentration increases by orders of magnitude, which reflects the removal of trapping defects.

TEM studies show a dislocation structure in region II coincident with the removal of substitutional atoms.[39] Dislocations form above 500°C. Compared to the situation at 500°C, the final state of region II at 600°C is a smaller boron substitutional concentration and a larger nonsubstitutional boron concentration with an undefined[38] lattice location. Therefore, the boron may be precipitated on or near dislocations.

In region III, the substitutional concentration increases with approximately 5.0-eV activation energy.[40] This energy corresponds to the generation and migration of a silicon self-vacancy species at elevated temperatures. Vacancies are generated and then move to the nonsubstitutional boron (precipitate), allowing the boron to dissociate from the nonsubstitutional precipitate site. For lower doses of boron where no reverse annealing occurs, substitutional behavior may occur without the need for thermally generated vacancy species. At doses of approximately $10^{12}/cm^2$, complete annealing takes place at 800°C in minutes. A small but measurable diffusion is especially observable for lower-energy implantations where the straggle is only 250 Å. For higher doses of boron implanted at room temperature, complete electrical activity requires a higher temperature. For higher doses of boron implanted at room temperature amorphous layers are not formed unless doses are above $5 \times 10^{16}/cm^2$. However, an amorphous condition for $\sim 10^{15}$ boron/cm^2 can be obtained by reducing the target temperatures.

6.5.2 Isochronal Phosphorus Annealing

Phosphorus layers implanted in room-temperature substrates anneal in a qualitatively different way.[41] At doses up to about $10^{14}/cm^2$, the implanted layers are not amorphous. Increasing the dose from 3×10^{12} to $3 \times 10^{14}/cm^2$ requires increasingly higher annealing temperatures to anneal out the progressively more complex disorder, similar to the case with boron annealing. In Fig. 16 the dashed curves are for implantations where the damage is not amorphous and the solid curves represent amorphous layers. After the phosphorus-implanted layer becomes amorphous at doses greater than $3 \times 10^{14}/cm^2$, a different annealing mechanism comes into play. For all higher doses the annealing temperatures are essentially fixed at about 600°C. This temperature is lower than that for annealing the non-amorphous ($\sim 10^{14}/cm^2$) case! The effect is associated with the solid phase epitaxial regrowth process that goes on for an amorphous layer regrowing on a single-crystal substrate. Group V donor atoms are essentially indistinguishable from the silicon atoms in the regrowth process, so the implanted atoms are incorporated as substitutional during the recrystallization process.

When the amorphous layer is not continuous in depth but is buried, a more complex behavior occurs. Epitaxial annealing takes place at both interfaces and a mismatch can occur when the annealing interfaces meet. Another interesting feature occurs when a continuous amorphous layer is epitaxially annealed, for example at 600°C. We can now consider the different annealing behavior for different parts of the profile. Low concentration ($\sim 10^{16}/cm^3$) doping (locally equivalent to $\sim 10^{12}/cm^2$) in the tail of the phosphorus distribution is well annealed, but the doping in the subamorphous—intermediate concentration $\sim 5 \times 10^{17}/cm^3$—part of the profile has a low electrical activity. The low electrical activity is due to a high defect concentration between the as-implanted, amorphous crystal boundary and the low-concentration tail of the distribution. We will return to this feature when we compare implanted phosphorus and BF_2 annealed layers below.

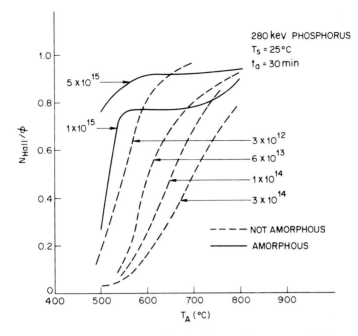

Fig. 16 The ratio of free-carrier content to dose plotted against anneal temperature (T_A) for various phosphorus doses. The solid curves represent amorphous layers that anneal by solid phase epitaxy. The dashed curves represent implantation where the damage is not amorphous. *(After Crowder and Morehead, Jr., Ref. 41.)*

The annealing behavior for room-temperature-implanted arsenic and antimony is similar to that of phosphorus, except that lower doses are required to make the layer amorphous.

6.5.3 Synthesis of Annealing Behavior

One way to test the basic explanations offered for the annealing of implanted boron and phosphorus is to implant boron into cold substrates to produce an amorphous layer and also to implant phosphorus into hot substrates to prevent the formation of amorphous layers. Figure 17 shows the result of such a study for boron and phosphorus.[42] The annealing behavior of both boron and phosphorus in the presence of amorphous layers is similar. There is no reverse anneal for boron at a dose of $10^{15}/cm^2$ when the layer is amorphous, and rather complete annealing occurs for 600°C epitaxial annealing. The annealing behavior of boron and phosphorus in non-amorphous layers is also similar. Figure 17b shows a reverse anneal for the electrical activity of the 200°C implant for high-dose ($5 \times 10^{15}/cm^2$) phosphorus. This occurs without the accompaniment of an amorphous to crystalline, solid phase epitaxial mechanism. Instead, as seen in TEM studies, a dislocation structure is associated with the reverse anneal of phosphorus implanted into hot targets.

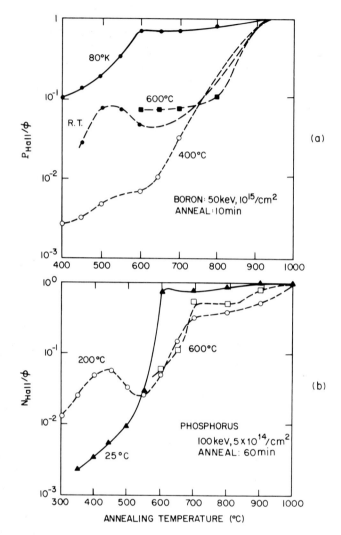

Fig. 17 (a) Isochronal annealing curves for boron implanted at various substrate temperatures. *(After Yoshihiro et al., Ref. 42.)* (b) Isochronal annealing curves for phosphorus implanted at various temperatures. *(After Tamura, Ikeda, and Tokuyama, in Ref. 5b.)*

In summary, the annealing behavior for boron and phosphorus is similar if amorphous layers are formed and solid phase epitaxy occurs. If no amorphous layer is formed, a reverse anneal occurs at about 500°C and this is accompanied by the formation of extended defects. These dislocations require 900 to 1000°C temperatures to be removed.

One other aspect for annealing of implanted impurities is the existence of small dislocations in the deep tail side of the distribution.[43] If arsenic is implanted (amorphous layer) for sources and drains, and annealed in the 600 to 850°C temperature range to make use of solid phase epitaxy, there will be small (~ 50-Å) dislocations in the tail of the distribution.

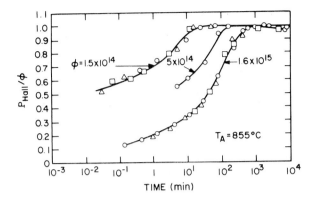

Fig. 18 Isothermal annealing of boron. *(After Seidel and MacRae, Ref. 40.)*

6.5.4 Isothermal Annealing (Kinetics)

Additional information can be obtained by annealing at fixed temperatures for various times. We will first discuss the example of a non-amorphous case: implanted boron annealed at temperatures above 600°C and for doses between 10^{14} and 10^{15} ions/cm². As time is increased, the doping increases rather "slowly," requiring several orders of magnitude of time to go from the initial fractions to >90% (Fig. 18). The shape results because the lower concentration part of the profile anneals first, and the central region anneals last. After approximately 35 minutes the 10^{15}/cm² profile has electrically active boron in the wings of the distribution and inactive boron in the central region[21] (Fig. 19). If the time constants for the annealing are plotted on a log τ versus T^{-1} (assuming $\tau \sim e^{-Ea/kT}$), straight lines give about a 5.0-eV activation energy.[40]

This high activation energy corresponds to the generation 3.4-eV and migration 1.6-eV energy of thermally generated vacancy species. The intrinsic silicon self-diffusion, interpreted as vacancy generation and migration, has been independently measured using radio tracer techniques.[44] Non-substitutional boron in the central part of the distribution is considered to be associated with both boron impurity disorder complexes and boron impurity pinning at dislocations. Thermally generated vacancies migrate to these nonsubstitutional locations, and substitutional behavior occurs.

Various mathematical models using specific point defect species can be developed.[45] One such model uses coupled diffusion equations with three species: boron substitutional concentrations (B^-), positively charged vacancy concentrations (V^+), and neutral boron-vacancy complexes ($B^- V^+$). The boron vacancy complex is viewed as electrically inactive but rapidly diffusing. By comparing channeling measurements with electrical activity, it is clear that most, if not all, of the substitutional boron is electrically active in the 700 to 1000°C annealing temperature range. Precipitation associated with boron on dislocations can be added to the model to account for the non-substitutional boron.

Amorphous implanted layers anneal by the solid phase epitaxy process. The rate of regrowth has been studied in detail for various crystal orientations and doping conditions.[46] When silicon is made amorphous by implanting silicon into silicon, the rates of regrowth are: approximately 100 Å/ min for $\langle 100 \rangle$ orientation and 3Å/ min for

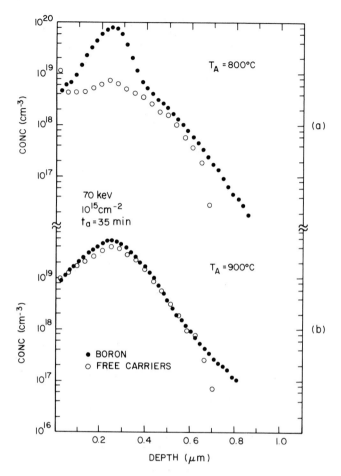

Fig. 19 Concentration profiles of boron atoms (SIMS—solid dots) and corresponding free-carrier concentrations (Hall data—open circles). (a) At 800°C and (b) at 900°C. *(After Hofker, Ref. 21.)*

$\langle 111 \rangle$ orientation at 550°C (Fig. 20). For the $\langle 111 \rangle$ surface, the slower rate is accompanied by defective (twinned) silicon. This is proposed to be due to growth along inclined (111) planes.

Plotting the regrowth rates against reciprocal temperature gives an activation energy of 2.3 eV. This low temperature process is associated with bond breaking to allow reordering at the interface. Adding impurities such as O, C, N, or Ar slows or disorganizes the regrowth, presumably because the impurities tie up (remove) broken bonds. Impurities such as B, P, or As increase the growth rate (by a factor of about 2 for 10^{20} impurity atoms/cm^3) because the substitutional impurities weaken and increase the likelihood of having broken bonds.

For certain cases (such as high-dose arsenic), annealing at temperatures near 550°C followed by a high temperature results in a more orderly recrystallization process.[47] If the formation of polycrystals or high-concentration dislocations can be

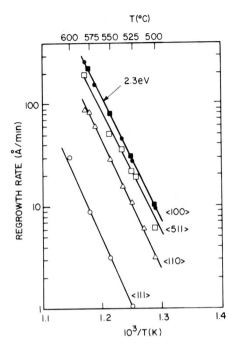

Fig. 20 The solid phase epitaxial regrowth rate of amorphous silicon as a function of temperature for various crystal orientations. *(After Csepregi, Mayer, and Sigmon, Ref. 46.)*

avoided (which could happen with a fast high-temperature anneal), then isolated dislocations can be annealed out by a second anneal[48] at ~1000°C.

6.5.5 Diffusion of Implanted Impurities

The diffusion of implanted impurities in silicon is complex even when there is no ion damage. The role of thermal silicon vacancies (their associated charge states) and silicon interstitials are important as are the effects of sinks or sources of these species (Chapter 5). Diffusion of implanted impurities requires consideration of damage-induced vacancies, interstitials, their vacancy-impurity species, and extended defects.

Consider the case of 10^{15} boron atoms/cm^2 implanted at room temperature, giving a non-amorphous layer.[21] Figure 21 shows that the profile broadens in the tail region (at 700 to 800°C in 35 minutes) while the peak concentration remains undiffused. This tail diffusion is anomalously high compared to published boron diffusion coefficients. This diffusivity may be enhanced because of the break-up of silicon-vacancy and interstitial-cluster species; vacancies should enhance the substitutional diffusivity, and silicon interstitials can replace substitutional boron resulting in a rapidly diffusing interstitial-boron species. The undiffused peak concentration has disorder that does not break up at 700 to 800°C.

At 900°C the peak concentration broadens, while the boron on the sides stays rather fixed. One probable explanation is that the dislocation disorder in the peak

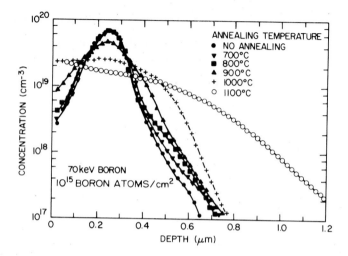

Fig. 21 Boron atom concentrations as a function of annealing at various temperatures. The anneal time is 35 minutes. *(After Hofker, Ref. 21.)*

concentration begins to anneal, giving silicon vacancies and interstitials that can promote diffusion. The profile of boron at 900°C and 35 minutes can be fit by an effective diffusion constant which is about three times that for the "normal" 900°C value (3×10^{-14} cm^2/s versus 1×10^{-14} cm^2/s).

At 1000°C additional thermal broadening occurs, but the effects can be described by ordinary diffusion theory. The diffusion constant is independent of position on the profile. Thermal vacancies and interstitials can participate in the diffusive motion.

If one assumes that the diffusion coefficient is constant, and therefore independent of position, time, defect concentration, etc., then a simple solution can be written for a Gaussian distribution.[49] The initial implanted distribution is taken to be a Gaussian, and the solution to a limited-source diffusion is also a Gaussian. Thus a solution to Fick's equation $dn/dt = D\, d^2n/dx^2$ is obtained if ΔR_p^2 is replaced by $\Delta R_p^2 + 2Dt$. The solution is:

$$ n(x,t) = \frac{\phi}{\sqrt{2\pi}\sqrt{\Delta R_p^2 + 2Dt}} \exp\left[\frac{-(x - R_p)^2}{2(\Delta R_p^2 + 2Dt)} \right] \qquad (26) $$

Boundary conditions also may be imposed, for example, which require no particle current to leave the surface during the diffusion. Solutions with oxides present have also been developed.[50]

6.5.6 Rapid Annealing

Implanted layers can be annealed using laser beams with energy densities of ~ 1–100 J/cm^2. Many potential advantages of this method have been proposed.[14, 51] Because of the short duration of the heat, profiles of implanted impurities may be annealed without appreciable diffusion. An implanted amorphous layer 1 kÅ thick is

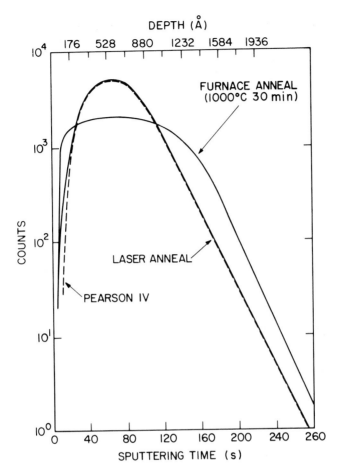

DEPTH (Å)

Fig. 22 Profile of arsenic implanted into silicon and annealed both with a CW laser and with a standard thermal anneal. The as-implanted Pearson-IV distribution and the laser annealed profiles are virtually identical. *(After Gat et al., Ref. 52.)*

annealed in a few seconds at 800°C using solid phase epitaxy. The diffusion length $\sqrt{Dt}$ of dopant impurities is only a few angstroms. Figure 22 shows a plot of the concentration (counts) from a SIMS measurement against depth for arsenic using a CW laser and solid phase epitaxy (SPE). Electrical measurements on the annealed layer shows that the electrical activity from sheet resistance is comparable to a 1000°C, 30-min standard thermal furnace anneal. The impurity profile of the laser annealed sample is identical to that of an "as-implanted" distribution.

The rapid annealing process is inherently clean—furnace contamination in the usual sense is not a problem. Laser energy may be localized over part of an IC chip so some junctions of the circuit can be diffused more, while others are not altered. One possible use would be the fabrication of a locally adjustable junction depth, or the production of different breakdown voltages on this same chip.

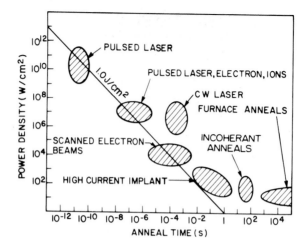

Fig. 23 Power density plotted against anneal time (pulse duration) for various rapid thermal annealing techniques. *(After Current and Pickar, Ref. 11.)*

An exciting discovery of the pulsed laser annealing technique is that after implanted amorphous layers are melted and undergo liquid-phase epitaxy, no extended defects are observed by TEM. This process is believed to involve a melting of the amorphous material and resolidification on the underlying single-crystal template. There are, however, substantial point defect concentrations that exist as a result of the rapid resolidification process. Low-temperature (400°C) annealing and use of hydrogen plasma ambients reduce the effect of point defect concentrations. Devices have been made with varying success, such as bipolar and MOS transistors and silicon solar cells. The performance of these laser annealed devices are generally comparable, but not substantially superior to their thermally annealed counterparts.

Rapid thermal annealing techniques now include pulsed lasers (with times down to a few picoseconds), pulsed electron and ion beams, scanned electron beams, CW (scanned) lasers, high-beam-current implants, and broad-band spectral sources (high-intensity lamps[52]) with "fast" (10-second) programmable anneals. These techniques are illustrated in Fig. 23, where the power density (W/cm^2) is plotted against the anneal time. Most of the techniques fall along a locus of 1.0-J/cm^2 energy density.

optical interference effects and still keep the advantages of rapid thermal annealing. The practical use of rapid thermal annealing appears to be close at hand.

6.5.7 Annealing in Oxygen Ambients

Annealing processes that result in the complete return of implanted ions to electrically active substitutional positions usually leave microdefects. These microdefects, which are observable by TEM, are referred to as secondary defects.

Studies show that if implantations (of B, Ne, or P) at room temperature are followed by thermal oxidation, any extrinsic microdefects are expanded into large dislo-

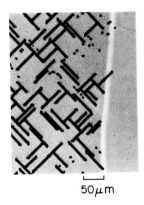

50 μm

Fig. 24 Photomicrograph of a silicon surface that was implanted with boron ($10^{14}/cm^2$) on the left-hand part, oxidized at 1150°C for 6 hours in oxygen and then Secco-etched. *(After Prussin, Ref. 53, and Robinson et al., Ref. 54.)*

cations and stacking faults. These defects, referred to as ternary defects,[53] are large enough to be seen with optical microscopes after chemical etching (Fig. 24).[54] Oxidation creates an excess concentration of silicon interstitials at and near the Si-SiO$_2$ interface. These interstitials "plate out" on any microdefect (nuclei), forming a stacking fault. Thus implantation provides defect nuclei which will grow when fed by a high concentration of silicon interstitials. These defects can degrade device performance. To avoid these defects, the recommended procedure is to anneal in neutral ambients (e.g., N, Ar) and then follow with any necessary oxidation.

6.6 SHALLOW JUNCTIONS (As, BF$_2$)

The requirements for VLSI shallow junctions for n$^+$ layers are rather easily met by the implantation of As. Arsenic has a very shallow range R_p (~300 Å) while using a convenient implantation energy, 50 keV. This moderate energy allows the use of relatively high beam currents in most accelerators. The heavy ion species results in an amorphous layer, so low-temperature solid phase epitaxy can be used to produce doped layers without appreciable atomic diffusion. If necessary, the arsenic layer can be annealed at 900°C with very little diffusion.

Future implementation of VLSI includes CMOS designs; therefore, shallow p$^+$ junctions are also of great importance. These junctions are not easily obtained using B$^+$ implantation, since high-dose B$^+$ implantation in room temperature targets does not give amorphous layers. Anneal temperatures > 900°C are required to get full electrical activity and considerable diffusion occurs. The implantation range at 30 keV, which is the lowest practical energy for obtaining high beam currents, is 100 Å and undesirably large.

The problems associated with boron are practically alleviated[55] by using the molecular species BF$_2$. The dissociation of BF$_2^+$ upon its first atomic scattering event gives a lower-energy boron atom. The energy of the boron atom is $(M_B / M_{BF_2})E_0 = (11/49)E_0$, where M_B and M_{BF_2} are the masses of the boron and BF_2 molecule, respectively, and E_0 is the incident energy of the BF_2 molecule. Thus, 50-keV BF_2 gives a boron range of ~300 Å.

Also, BF_2 provides an annealing advantage. Flourine ions are relatively heavy, giving an amorphous zone that contains most, but not all, of the boron (Fig. 25a).

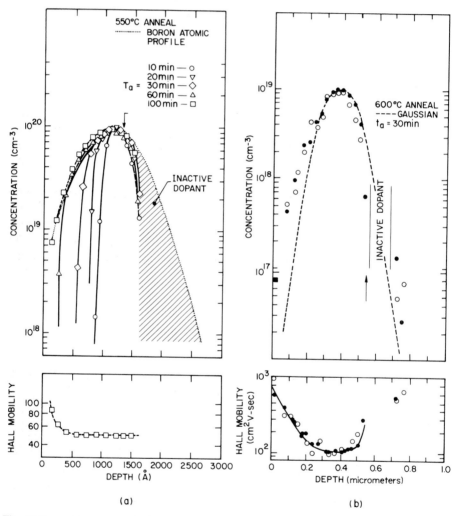

Fig. 25 Free-carrier concentration and mobility profiles for implanted layers which illustrate dopant incorporation by solid phase epitaxy (SPE). (a) Profiles for BF_2^+ implanted into $\langle 100 \rangle$ silicon at 150 keV and $10^{15}/cm^2$ after different isothermal anneals. The dotted curve is the as-implanted atomic profile from SIMS analysis. The original amorphous-crystalline interface is denoted by the arrow, SPE is complete after ~100 minutes. The hatched region is electrically inactive. *(After Tsai and Streetman, Ref. 55.)* (b) Profile for silicon implanted at 280 keV and 3E14 P^+/cm^2. The original amorphous-crystalline interface is denoted by the arrow. The inactive region is also noted. *(After Crowder, Ref. 56.)*

Solid phase epitaxy can be used to anneal the amorphous layer in the order of minutes at temperatures of 550 to 700°C. The portion of the B profile that is initially contained within the amorphous layer shows full electrical activity. The region between the tail (not measured in Fig. 25a) and the amorphous layer has free carriers compensated by defect traps, which need higher temperatures to anneal. Similar effects were found for the phosphorus P^+ implantations[56] previously mentioned (see Fig. 25b). The similarity and comparison of the two implanted layers should be noted. Finally BF_2^+ implants exhibit less ion channeling than B implants due to the formation of an amorphous region, and boron redistributes in the damaged and flourine-rich regions during annealing.

6.7 MINORITY-CARRIER EFFECTS

Various measurements characterize the effects of residual disorder on minority carriers. These measurements include junction leakage, bipolar transistor gain, forward-to-reverse bias recovery time, MOS pulse recovery technique, thermally stimulated currents, deep-level transient spectroscopy (DLTS),[57] and electron-beam-induced current (EBIC).

Junction leakage typically recovers to within about one order of magnitude of an unimplanted control, when implant-induced damage is annealed above 800°C.[58] This incomplete recovery is perfectly adequate for digital MOS, bipolar, and most memory applications. Present VLSI manufacture uses annealing conditions that diffuse the ions slightly beyond the original ion and damage distributions. Future annealing studies will need to consider the applicability of minimum annealing and thermal diffusion for devices with high lifetime requirements.

6.8 GETTERING

Physical phenomena which use the concepts of gettering (the removal of impurities and defects from junction regions) can help control leakage currents for very shallow junctions of VLSI application. The use of ion implantation damage for gettering of heavy metal impurities has been known for some time. Gettering action requires three physical effects, regardless of the specific method used: (1) the *release* of impurities or the decomposition of the constituents of extended defects (here we are distinguishing impurity removal from defect removal), (2) the *diffusion* of the impurities or constituents of a dislocation (that is silicon self-interstitials) to a capture zone, and (3) the *capture* of the impurities or self-interstitials at some sink.

Thus to get good gettering, "impurities" must be released, diffused, and captured. If any one of the three mechanisms is inoperative, then gettering will not be effective. For example, the capture mechanism can be perfect, so that every impurity atom which enters the capture environment is captured and no particles come back out. However, if no impurities are released or diffused then gettering will be poor and can be thought to be rate-limited by the release or diffusion effects. For implantation-induced disorder, the "sink" is either a dislocation array or polycrystalline grain boundaries.

We will now classify and discuss four major techniques for the gettering (capture) of impurities.

6.8.1 Ion Pairing

Phosphorus diffusion is an effective gettering technique. Impurities such as copper, which are known to be mainly interstitial in undoped silicon and to diffuse by an interstitial mechanism, take the shape of the diffused phosphorus profile.[59] Thus the diffused phosphorus in the silicon and Cu are correlated. It is further known that all the Cu is on substitutional lattice locations within the phosphorus profile. This has been demonstrated using analysis by He ion backscattering from copper combined with channeling. In this experiment the sample under study was first contaminated with Cu and then getter-diffused with phosphorus, so both gettered Cu and the P are near the surface. The probing ion (He) is then channeled into a silicon surface. The He ion backscattered yield versus backscattered energy is shown in Fig. 26a. The channeled yield is reduced from the random yield, the random yield is obtained with a non-channeled condition. Since the Cu is substitutional, the channeled probing He ions do not backscatter from the copper. The non-channeled or random spectra show $2 \times 10^{15}/cm^2$ gettered copper to be located in a layer about 2500 Å thick, which is the

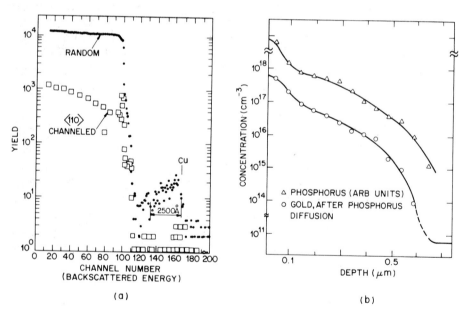

(a)   (b)

Fig. 26 Data for Cu and Au gettered in phosphorus diffused silicon. (a) Spectrum from Rutherford backscattering for Cu contaminated sample gettered by phosphorus diffusion at 1100°C. The channeled spectrum (□) shows the Cu is substitutional. *(After Seidel and Meek, Ref. 59.)* (b) Profiles of gold and phosphorus obtained by neutron activation analysis on a 900°C phosphorus-diffused gettered sample. *(After Lecrosnier et al., Ref. 60.)*

high-concentration part of the phosphorus profile. Phosphorus donates a large number of electrons to the substitutional acceptor charge state of copper, making it Cu^{-3}. This gives a large coulomb binding energy between substitutional Cu and P.

In summary, copper diffuses, as an interstitial atom with a very large diffusion coefficient, to reach the phosphorus in its diffusing profile, and then copper finds a vacancy next to the phosphorus atom and "ion pairs" as $P^+ Cu^{-3}$. The binding energy and diffusion are both species-dependent.

Results for iron or gold are only quantitatively different because of their singly charged state $(-)$ and their lower thermal diffusivity. Gold, which is gettered under the phosphorus-diffused profile, not only follows the phosphorus profile near the surface but also follows the shape of the phosphorus diffusion tail (Fig. 26b).[60]

Phosphorus diffusion gettering requires both a high concentration and a thick layer of phosphorus. Lowering diffusion temperatures to conform to VLSI applications reduces the concentration and depth of phosphorus. Below about 900°C, damage capture mechanisms can become better than phosphorus diffusion.[61]

6.8.2 Damage Gettering

Damage gettering has been demonstrated using: sandblasting, mechanical shot abrasion ("sound stressing"), laser-induced damage, and ion implantation.

Certain ions when implanted at high doses ($10^{16}/cm^2$) do not allow good solid phase epitaxy. Under annealing, the strain, precipitation, or defect character result in dislocations and polycrystalline material with grain boundaries. In particular, when inert gas ions are implanted (such as Ne, Ar, and Kr) and are annealed the gas coalesces to form internal bubbles with faceted crystallographic surfaces.[62] These surfaces form multiple platlet substrates upon which multiply seeded solid phase epitaxy takes place, resulting in a polycrystalline structure. The detailed damage is concentration- and species-dependent.

Although grain boundary gettering is quite efficient, it has been shown that isolated dislocations with large $\frac{1}{2}\langle 110 \rangle$ Burgers vectors efficiently getter at a lower dose[63] (Fig. 27). When dislocations overlap and relieve strain, their ability to getter is reduced. In Fig. 27 this effect occurs at a dose of approximately $3 \times 10^{15}/cm^2$.

An optimum gettering temperature has been reported, again for specific cases; this may be the manifestation of the idea that too low a temperature gives a diffusion-limited gettering and too high a temperature results in too much thermal energy to hold the gettered impurities in the "sink" provided[64] (Fig. 28). Figure 28 also shows substantial improvement in minority-carrier lifetime for argon implantation annealed at 850°C.

6.8.3 Intrinsic Gettering

We can use the bulk silicon substrate to getter if there are SiO_x precipitates and associated dislocations in the sample. In this case one starts with oxygen concentrations close to the solid solubility[65] ($\sim 10^{18}/cm^3$). Upon heat treatment in neutral or oxidizing ambients at ~ 1100°C, the surface regions become denuded of oxygen by the out-

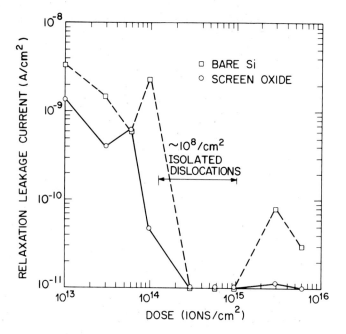

Fig. 27 Relaxation leakage current as a function of dose for Xe^{132} implanted into bare silicon and SiO_2 on the back of the wafer. "Annealing" was performed at 1000°C in dry oxygen for 80 minutes. Optimum gettering is at doses in the mid-$10^{14}/cm^2$ dose range. *(After Geipel and Tice, Ref. 63.)*

diffusion of oxygen. The sample is then annealed at ~800°C, where the oxygen in the interior is super-saturated and the SiO_x precipitates form. These precipitates "punch out" dislocations to act as sinks for heavy metal impurities, while the surface regions are denuded of defects. Junction regions near the surface are free of defects, while the interior of the silicon is filled with gettering sites.

6.8.4 Ambient Gettering

It is possible to clean the furnaces and wafer surfaces of heavy metal contaminants when oxidation is done in the presence of HCl. The heavy metal chlorides (e.g., CuCl) are volatile and are swept away from the wafers and out of the furnace tube. A shrinkage of stacking faults is also seen[65a] when oxidations are done in the presence of Cl.

6.9 EFFECTS IN VLSI PROCESSING

There are many known effects which may play a role in an emerging VLSI technology. After implantation, the Si, SiO_2, photoresist, or metal target is modified and can change the behavior of subsequent process steps. We briefly discuss some of these effects.

High doses of nitrogen considerably decrease the oxidation rate because of the

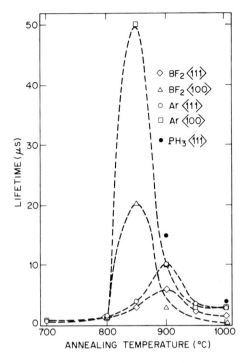

Fig. 28 Lifetime versus gettering temperature for argon, BF_2, and PH_3 diffusion in $\langle 100 \rangle$ and $\langle 111 \rangle$ silicon. The implants were 150 keV, $10^{16}/cm^2$; and anneals or diffusions were done for 30 minutes. *(After Ryssel and Ruge, Ref. 64.)*

formation of "nitride," while damage introduced by B, Ar, As, and Sb can increase the oxidation rate by various amounts.[66] These effects may be used to modify oxide thicknesses on different parts of a VLSI device. In another application, oxides with surface damage have been used to taper the edge of etched windows, and the surface region etches more rapidly than the undamaged region.

Implantation into oxides results in broken bonds, with displaced oxygen and silicon atoms. After annealing, an implant-species-dependent electron trapping effect is observed. Understanding[67] of electron trapping effects continues to be important since scaling to small dimensions brings high electric fields in the drain region of MOS devices. Some VLSI devices will be operated at the onset of avalanche multiplication which might supply electrons to the traps.

When photoresist is used as an implantation mask it is damaged during implantation. The result is bond-breaking and evaporation of the volatile components such as hydrogen and nitrogen. A carbon-rich layer is obtained which can be removed with oxygen plasma, ozone, or an oxidation processes.

Implantation into metal-silicon interfaces leads to interface reactions. For example Mo films on silicon were converted to MoSi by a sufficiently high dose of phosphorus.[68] The contact resistance was reduced, presumably due to the knock-on effects at the interface, which have been referred to as ion-beam mixing.

Some silicide formations result in the segregation of previously implanted dopants. For example, arsenic is driven ahead of platinum and palladium during PtSi and Pd_2 Si formation.[69] Segregation behavior can result in lower contact resistance for VLSI contacts.

6.10 SUMMARY AND FUTURE TRENDS

Ion implantation is now being used in every doping step of a typical VLSI process. We can distinguish between low-dose and high-dose applications. Low-dose applications include: the threshold voltage control for MOS devices, resistors, n-well and p-well doping for CMOS devices, control of the vertical dimension of the space charge width, and base doping for bipolar transistors. High-dose applications include self-aligned source and drain for MOS devices, high-conductance resistors, buried layers, and emitters for bipolar transistors.

We now look forward to the possible widespread use of shallow junction depths ($\sim$1000 Å) in MOS devices where most of the doping is contained in a 200 Å-thick layer. MOS devices with electrical channel lengths as small as 0.5 μm seem feasible using the principles we have discussed. The channeling effects and atomic diffusion must be reproducibly minimized to give shallow and controlled junctions. Preamorphizing the silicon eliminates channeling effects on the implanted profile. Impurity diffusion may be altered and possibly reduced by the interactions of impurities with point and extended defects.

Devices with ultra-narrow depletion widths of approximately 200 Å will show hot-electron effects. If the electron mean free path ($\sim$100 Å) is of the order of thickness of the depletion widths, ballistic effects set in. Electrons are accelerated through the thin layers without scattering and can reach high velocities. Various hot-electron transistor devices have been proposed. These are an extension of the concept that very shallow implanted layers can lower and narrow the barrier of Schottky devices.[70] Bipolar VLSI will continue to make use of the Schottky barrier modifications because logic level differences are directly generated from the Schottky barrier differences.

In addition, high concentrations of impurities confined to very narrow widths can give low-temperature-coefficient, high-sheet-resistance monolithic resistors. The ionized impurity and lattice scattering temperature dependencies offset each other and $\sim$4 kΩ/$\square$ sheet resistances with < 300-ppm/°C temperature coefficients are obtained. Finally, if the depletion layer thickness of a depletion MOS device can be made thin most of the carriers can be easily controlled by the gate to obtain relatively high transconductance.[70]

Presently the ultimate device dimensions are not known. However, for gate dimensions of 0.1 $\times$ 0.1 μm and for a 10^{18}/cm^3 channel doping in an enhancement MOSFET, we will have statistical doping effects. For a 200 Å-thick depletion layer, the number of atoms under the gate is only 40. The fluctuation, taken to be $\sqrt{N}/N$ is approximately 20%.

In the beginning of this chapter, ion implantation was defined in terms of avoiding surface effects. This requires kinetic energy. When it comes to making the shallowest possible junctions, it is not at all obvious that low-energy ($\sim$3-keV) implantation will be more useful or practical than thermal diffusion. The past problems of thermal diffusion (cleanliness, surface oxides, and control of doping concentration) are not fundamental. The shallowest junctions are probably obtainable from the use of thermal, and not kinetic, energies. However, implantation, with its obvious advantages, will continue to play a major role in VLSI in the foreseeable future.

REFERENCES

[1] J. Lindhard, M. Scharff, and H. Schiott, "Range Concepts and Heavy Ion Ranges," *Mat.-Fys. Med. Dan. Vid. Selsk* **33**, No. 14, 1 (1963).

[1a] [1a] "Proc. International Conference on Atomic Collisions (Chalk River)," *Can. J. Phys.*, **46**, 449 (1968).

[2] J. W. Mayer, L. Eriksson, and J. A. Davies, *Ion Implantation in Semiconductors*, Academic, New York, 1970, Chapter 4.

[3] J. W. Corbett, "Radiation Damage in Silicon and Germanium," in Ref. 5a, p. 1.

[4] R. W. Bower and H. G. Dill, Proc. International Electron Device Meeting, 1966, paper 16.6 (unpublished).

[5a] F. Eisen and L. Chadderton, Eds., *First International Conference on Ion Implantation, Thousand Oaks*, Gordon and Breach, New York, 1971.

[5b] I. Ruge and J. Graul, Eds., *Second International Conference on Ion Implantation, Garmish*, Springer-Verlag, Berlin, 1972.

[5c] B. L. Crowder, Ed., *Third International Conference on Ion Implantation, Yorktown Heights*, Plenum, New York, 1973.

[5d] S. Namba, Ed., *Fourth International Conference on Ion Implantation, Osaka*, Plenum, New York, 1975.

[5e] F. Chernow, J. Borders, and D. Bruce, Eds., *Fifth International Conference on Ion Implantation, Boulder*, Plenum, New York, 1976.

[6] J. F. Gibbons, "Ion Implantation in Semiconductors—Part I: Range Distribution Theory and Experiments," *Proc. IEEE*, **56**, 295 (1968).

[7] J. F. Gibbons, "Ion Implantation in Semiconductors—Part II: Damage Production and Annealing," *Proc. IEEE*, **66**, 9 (1972).

[8] K. A. Pickar, "Ion Implantation in Silicon—Physics, Processing and Microelectronic Devices," in R. Wolfe, Ed., *Applied Solid State Science*, Academic, New York, 1975, Vol. 5.

[9] H. Rupprecht, "New Advances in Semiconductor Implantation," *J. Vac. Sci. Technol.*, **15**, 1669 (1978).

[10] W. K. Hofker and J. Politiek, "Ion Implantation in Semiconductors," *Philips Tech. Rev.* **39**, 1 (1980).

[11] M. I. Current and K. A. Pickar, "Ion Implantation Processing," Electrochemical Society Fall Meeting, Montreal, May 1982, Vol. 82-1, (unpublished extended abstracts).

[12] J. Gyulai, Ed., *First International Conference on Ion Beam Modification of Materials, Budapest (1978)*, Central Res. Inst. for Phys., H-1525 Budapest 114, POB49, Hungary, 1978.

[13] R. E. Benenson, E. N. Kaufman, G. L. Miller, and W. W. Scholz, Eds., *Second International Conference on Ion Beam Modification of Materials, Albany (1980)*, North-Holland, New York, 1981.

[14] M. Wittmer and G. A. Rozgonyi, "Laser Annealing of Semiconductors: Mechanisms and Applications to Microelectronics," in E. Kaldis, Ed., *Current Topics in Materials Science*, North-Holland, New York, 1981.

[15] R. G. Wilson and G. R. Brewer, *Ion Beams with Applications to Ion Implantation*, Wiley, New York, 1973; R. G. Wilson and G. R. Brewer, *Ion Beams: With Application to Ion Implantation*, Kriegor, Huntington, New York, 1979; and J. F. Ziegler, *Ion Particle Accelerators—Applications*, Plenum, New York, 1975.

[16] A. B. Wittkower, "Calibration of Ion Implantation Systems," *Solid State Technol.*, **61**, p. 61, Nov. 1981.

[17] M. Y. Tsai, B. G. Streetman, R. J. Blattner, and C. A. Evans, "Study of Surface Contamination Produced During High Dose Ion Implantation," *J. Electrochem. Soc.*, **126**, 98 (1979).

[18] G. D. Alton, "Aspects of the Physics, Chemistry, and Technology of High Intensity Heavy Ion Sources," *Vac. Instrum. Methods*, **189**, 15 (1981).

[19] Y. Ota, "Silicon Molecular Beam Epitaxy with Simultaneous Ion Implant Doping," *J. Appl. Phys.*, **51**, 1102 (1980), and J. C. Bean "Growth of Doped Silicon Layers by Molecular Beam Epitaxy," in F. F. Y. Wang, Ed. *Material Processing Theory and Practice*, Vol. 2, North Holland, (1981), p. 175.

[20] J. F. Gibbons, W. S. Johnson, and S. W. Mylroie, in Dowden, Hutchinson, and Ross, Eds., *Projected Range in Semiconductors*, Academic, New York, 1975, Vol. 2.

[21] W. K. Hofker, "Implantation of Boron in Silicon," *Philips Res. Repts. Suppl.*, No. 8 (1975).

[22] S. Furukawa, H. Matsumura, and H. Ishiwara "Lateral Distribution Theory of Implanted Ions," in S. Namba, Ed., *Ion Implantation in Semiconductors*, Japanese Society for Promotion of Science, Kyoto, 1972, p. 73.

[23] P. D. Townsend, J. C. Kelly, and N. E. W. Hartly, *Ion Implantation, Sputtering and Their Applications*, Academic, New York, 1976.

[24] B. Smith, *Ion Implantation Range Data for Silicon and Germanium Device Technologies*, Research Studies, Forest Grove, Oregon, 1977.

[25] W. Fichtner (unpublished).

[26] D. V. Morgan, Ed., *Channeling: Theory, Observation and Applications*, Wiley, New York, 1973.

[27] P. Blood, G. Dearnaley, and M. A. Wilkins, "The Origin of Non-Gaussian Profiles in Phosphorus Implanted Silicon," *J. Appl. Phys.*, **45**, 5123 (1974).

[28] T. E. Seidel, "Channeling of Implanted Phosphorus Through Polycrystalline Silicon, *Appl. Phys. Lett.*, **36**, 447 (1980).

[29] T. Hirao, G. Fuse, K. Inoue, S. Takayanagi, Y. Yaegashi, and S. Ichikawa, "Electrical Properties of Si Implanted with As through SiO_2 Films," *J. Appl. Phys.*, **51**, 262 (1980).

[30] D. K. Brice, "Recoil Contribution to Ion Implantation Energy Deposition Distributions," *J. Appl. Phys.*, **46**, 3385 (1975).

[31] G. H. Kinchin and R. S. Pease, "The Displacement of Atoms in Solids by Radiation," *Rep. Prog. Phys.*, **18**, 1 (1955).

[32] H. J. Stein, F. L. Vook, D. K. Brice, J. A. Borders, and S. T. Picreaux, "Infrared Studies of the Crystallinity of Ion Implanted Silicon," in Ref. 5a, p. 17.

[33] L. A. Christel, J. F. Gibbons, and S. Mylroie, "An Application of the Boltzmann Transport Equation to Ion Range and Damage Distributions in Multilayered Targets," *J. Appl. Phys.*, **51**, 6176 (1980).

[34] F. F. Morehead and B. L. Crowder, "A Model for the Formation of Amorphous Si by Ion Implantation," in Ref. 5a, p. 25.

[35] T. E. Seidel, G. A. Pasteur, and J. C. C. Tsai, "Visible Interference Effects in Silicon Caused by High-Current High-Dose Implantation," *Appl. Phys. Lett.*, **29**, 648 (1976).

[36] D. K. Sadana, M. Stratham, J. Washburn, and G. R. Booker, "Transmission Electron Microscopy and Rutherford Backscattering Studies of Different Damage Structure in P^+ Implanted Si," *J. Appl. Phys.*, **51**, 5718 (1980).

[37] R. L. Petritz, "Theory of an Experiment for Measuring Mobility and Density of Carriers . . . ," *Phys. Rev.*, **110**, 1254 (1958).

[38] J. C. North and W. N. Gibson, "Channeling Study of Boron Implanted Silicon," in Ref. 5a. p. 143.

[39] R. W. Bicknell and R. M. Allen, "Correlation of Electron Microscope Studies with the Electrical Properties of Boron Implanted Silicon," in Ref. 5a, p. 63.

[40] T. E. Seidel and A. U. MacRae, "The Isothermal Annealing of Boron Implanted Silicon," in Ref. 5a, p. 149.

[41] B. L. Crowder and F. F. Morehead, Jr., "Annealing Characteristics of n-type Dopants in Ion Implanted Silicon," *Appl. Phys. Lett.*, **14**, 313 (1969).

[42] N. Yoshihiro, T. Ikeda, M. Tamura, T. Tokuyama, and T. Tsuchimoto, "Reverse Annealing of Boron and Phosphorus Implanted Silicon," in S. Namba, Ed., *Ion Implantation in Semiconductors*, Japanese Society for Promotion of Science, Kyoto, 1972, p. 33.

[43] H. Foell, T. Y. Tan, and W. Krakow, "Undissociated Dislocations and Intermediate Defects," in J. Narayan and T. Y. Tan, Eds., *Defects in Semiconductors*, North-Holland, New York, 1981, Vol. 2, p. 173.

[44] J. M. Fairfield and B. J. Masters, "Self-Diffusion in Intrinsic and Extrinsic Silicon," *J. Appl. Phys.*, **38**, 3148 (1967).

[45] A. Chu and J. F. Gibbons, "A Theoretical Approach to the Calculation of Impurity Profiles for Annealed Ion Implanted B in Si," in Ref. 5e, p. 711.

[46] L. Csepregi, J. W. Mayer, and T. W. Sigmon, *Appl. Phys. Lett.*, **29**, 92 (1976); and S. T. Picreaux, "Ion Channeling Analysis of Disorder," in J. Narayan and T. Y. Tan, Eds., *Defects in Semiconductors*, North-Holland, New York, 1981, Vol. 2, p. 135.

[47] L. Csepregi, W. K. Chu, H. Muellor, and J. W. Mayer, "Influence of Thermal History on the Residual Disorder in Implanted⟨111⟩ Silicon," *Radiat Eff.*, **28**, 277 (1976).

[48] E. I. Alessandrini, W. K. Chu, and M. R. Poponiak, "TEM Study of the Two Step Annealing of Arsenic-Implanted⟨100⟩ Silicon," *J. Vac. Sci. Technol.*, **16**, 342 (1979).

[49] T. E. Seidel and A. U. MacRae, "Some Properties of Ion Implanted Boron in Silicon," *Trans. Metall. Soc. AIME*, **245**, 491 (1969).

[50] E. C. Douglas and A. G. F. Dingwall, "Ion Implantation for Threshold Control in COSMOS Circuits," *IEEE Trans. Electron Devices*, **ED-21**, 324 (1974).

[51] B. R. Appleton and G. K. Aller, Eds., *Laser and Electron Beam Interactions with Solids*, North-Holland, New York, 1982.

[52] A. Gat et al., "Physical and Electrical Properties of Laser-Annealed Ion Implanted Silicon," *Appl. Phys. Lett.*, **32**, 276 (1978), and A. Gat, "Heat Pulse Annealing of Arsenic-Implanted Silicon with a CW Arc Lamp," *IEEE Electron Device Lett.*, **EDL-2**, p. 85 (1981), and T. O. Sedgwick, "Short Time Annealing," in C. J. Dell'Oca and W. M. Bullis, Eds., *VLSI Science and Technology*, Vol. 82-7, The Electrochemical Soc., Pennington, NJ 1982 p. 130.

[53] S. Prussin, "Role of Sequential Annealing, Oxidation and Diffusion Upon Defect Generation in Ion-Implanted Silicon Surfaces," *J. Appl. Phys.*, **45**, 1635 (1974).

[54] McD. Robinson, G. A. Rozgonyi, T. E. Seidel, and M. H. Read, "Orientation and Implantation Effects on Stacking Faults During Silicon Buried Layer Processing," *J. Electrochem. Soc.*, **128**, 926 (1981).

[55] M. Y. Tsai and B. G. Streetman, "Recrystallization of Implanted Amorphous Silicon Layers, I. Electrical Properties of Silicon Implanted with BF_2^+ or $Si^+ + B^+$," *J. Appl. Phys.*, **50**, 183 (1979).

[56] B. L. Crowder, "Influence of Amorphous Phase on Ion Distributions and Annealing Behavior of Group III and Group V Ions Implanted into Silicon," *J. Electrochem. Soc.*, **118**, 943 (1971).

[57] L. C. Kimmerling, "Defect Characterization by Junction Spectroscopy," in J. Narayan and T. Y. Tan, Eds., *Defects in Semiconductors*, North-Holland, New York, 1981, Vol. 2, p. 85.

[58] K. A. Pickar and J. V. Dalton, "Lifetime Effects in Ion Implanted Silicon," in Ref. 5a, p. 125.

[59] T. E. Seidel and R. L. Meek, "Ion Implantation Gettering and Phosphorus Diffusion Gettering of Cu and Au in Silicon," in Ref. 5c, p. 305.

[60] D. Lecrosnier, J. Paugam, F. Richow, G. Pelous, and F. Beniere, "Influence of Phosphorus-Induced Point Defects on a Gold-Gettering Mechanism in Silicon," *J. Appl. Phys.*, **51**, 1036 (1980).

[61] T. E. Seidel, R. L. Meek, and A. G. Cullis, "Direct Comparison of Ion Damage Gettering and Phosphorus-Diffusion Gettering of Au in Si," *J. Appl. Phys.*, **46**, 600 (1975).

[62] A. G. Cullis, T. E. Seidel, and R. L. Meek, "Comparative Study of Annealed Neon, Argon, and Krypton Ion Implantation Damage in Silicon," *J. Appl. Phys.*, **49**, 5188 (1978).

[63] H. J. Geipel and W. K. Tice, "Reduction of Leakage by Implantation Gettering in VLSI Circuit," *IBM J. Res. Dev.*, **24**, 310 (1980).

[64] H. Ryssel and I. Ruge, "New Applications of Ion Implantation in Semiconductor Technology," in W. A. Kaiser and W. E. Proebster, Eds., *Electronics to Microelectronics*, North-Holland, New York, 1980, p. 63.

[65] T. Y. Tan, E. E. Gardner, and W. K. Tice, "Intrinsic Gettering by Oxide Precipitate Induced Dislocations in Czochralski Si," *Appl. Phys. Lett.*, **30**, 175 (1977).

[65a] H. Shiraki, "Stacking Fault Generation Suppression and Grown-In Defect Elimination in Dislocation Free Silicon Wafers by HCl Oxidation," *Jap. Jour. Appl. Phys.* **15**, 1 (1976).

[66] W. J. M. J. Josquin, "The Oxidation Characteristics of Nitrogen Implanted Silicon," in Ref. 12, p.

1433; and J. F. Götzlich, et al., "Dopant Dependence of the Oxidation Rate of Ion Implanted Silicon," in Ref. 12, p. 1419.

[67] R. F. DeKeersmaecher and D. J. DiMaria, "Electron Trapping and Detrapping Characteristics of Arsenic-Implanted SiO_2 Layers," *J. Appl. Phys.*, **51**, 1085 (1980).

[68] S. W. Chiang, T. P. Chow, R. F. Reihl, and K. L. Wang, "The Effect of Phosphorus Ion Implantations on Molybdenum/Silicon Contacts," *J. Appl. Phys.*, **52**, 4027 (1981).

[69] M. Wittmer and T. E. Seidel, "The Redistribution of Implanted Dopants After Metal-Silicides Formation," *J. Appl. Phys.*, **49**, 5826 (1978).

[70] J. M. Shannon, "Shallow Implanted Layers in Advanced Silicon Devices," in Ref. 13, p. 545.

Additional reading

See J. L. Stone and J. C. Plunkett, "Ion Implantation Processes in Silicon" (Chapter 2) and H. Maes, W. Vandervorst, and R. van Overstraeten, "Impurity Profile of Implanted Ions in Silicon" (Chapter 8), in F. F. Y. Wang, Ed., *Material Processing Theory and Practices, Vol. 2: Impurity Doping Processes in Silicon*, North-Holland, New York, 1981.

PROBLEMS

1 A 10-μA ion beam has a $10°$ half-angle divergence as it passes through a square aperture (8 cm $\times$ 8 cm), placed 6 cm away from the target. Using a current meter, how much time is needed to implant 10^{13} atoms/cm^2 for *(a)* a singly ionized, monatomic species, *(b)* a triply ionized diatomic species? Using a charge integrator (measures It) calibrated for a singly ionized monatomic species, *(c)* what dose should be "set" to obtain 10^{13} atom/cm^2 for the triply ionized diatomic species?

2 The drift-space vacuum between a mass-separation magnet and the target is approximately 10^{-6} Torr. Consider the possibility of a neutralizing charge exchange reaction

$$I^+ + N_2 \rightarrow I^0 + N_2^+$$

with a cross section of 10^{-16} cm^2/atom. What percent of the ions are charge exchanged in a distance of 1 m? Take the (probability) fraction of unreacted particles to be exp $(-x/\lambda)$ where λ is the mean free path.

3 *(a)* Identify the three minor species peaks in Fig. 3. Use the relative magnetic-field values of 7.7, 9.9, and 13.3. F_2^+ is at $B = 8.9$, and As^+ at $B = 12.5$. Make a list of the ion species.

    (b) Can AsF_4^{++} and F_4^+ be resolved from As^+?

    (c) Discuss beam purity.

    (d) Discuss solutions that would improve beam purity.

4 An existing accelerator has a 10-μ A beam current with a 1-cm^2 area and deflection plates $(x, y$ scan) that are separated by 2 cm and have a 2-kV saw-tooth sweep voltage. Consider 10-keV As^{75} as the ion of interest. The ion beam's charge density can cause a drop in the sweeping electric field.

    (a) What is the magnitude of the drop in the sweeping electric field at the center of the beam?

    (b) Should this machine be retrofitted with a 1.0-mA beam source? Assume no geometry changes.

    (c) Discuss alternatives for scanning high beam currents.

5 *(a)* Using Fig. 7, approximate (dE/dx) total as a constant and calculate the range R. Compare at 30 and 300 keV for As, P, and B with R_p values from Fig. 8.

    (b) Calculate the sheet resistance for 30-keV 10^{15} As^{75} atoms/cm^2 and 10^{15} BF_2^{49} atoms/cm^2. Assume a fully active Gaussian nondiffused dopant profile. The profile can be approximated by equally spaced strips of constant doping and mobility.

6 *(a)* Plot the vertical and lateral dopant profiles at $x = R_p$, for 10^{15} As atoms/cm^2 and 60 keV. Use Gaussian and erfc distributions, respectively. Assume a very thick sharp vertical mask edge.
(b) Show that the vertical junction depth is

$$x_j = R_p + \Delta R_p \sqrt{2 \ln \left[\frac{\phi}{\sqrt{2\pi} \, \Delta R_p \, n_B} \right]}$$

where n_B is the background dopant concentration; assume $n_B = 10^{16}/\text{cm}^3$ in this problem.
   (c) Plot the vertical profiles for an anneal of 850 and 1000°C for 30 minutes. Use $D = 5 \times 10^{-16}$ cm^2/s and 8×10^{-15} cm^2/s, respectively, and assume Eq. (26) is valid.
   (d) Show that a mask of thickness d has a transmission factor

$$T = \frac{\phi_2}{\phi} = \frac{1}{2} \, \text{erfc} \left[\frac{d - R_p}{\sqrt{2} \, \Delta R_p} \right]$$

where ϕ_2 is the number of ions/cm^2 that penetrate the mask. How thick must an amorphous polysilicon mask be to give $T = 10^{-4}$ for 150-keV boron?
   (e) Assume a 150-keV B ion beam is perfectly aligned with a $\langle 100 \rangle$ grain of a crystallized polysilicon mask. How thick must the polycrystal be to give $T = 10^{-4}$? Use $\Delta R_p \approx 800$ Å and Fig. 11b.

7 TEM studies show that a single ion damage track has a 30 Å diameter. Using the range and $(dE/dx)_{\text{nuclear}}$ values of 30-keV As^{75} ion comment on the likelihood that the ion's damage is amorphous.

8 Assume 100-keV 10^{15} P^{31} atoms/cm^2 are uniformly implanted across a nonuniform thermally clamped silicon target. After implantation the colder end is amorphous, and exhibits higher reflectance, and colors appear between the amorphous and the other end. Approximately what is the electrically activity fraction after 30-minute anneals at 600, 900, and 1100°C at both the cold and hot ends.

9 *(a)* We are interested in producing shallow, defect-free junctions. Discuss the following: channeling tails, solid phase epitaxy, thermal cycles, and ambients.
   (b) We are interested in producing shallow junctions with no defects in the space charge region but extended defects near the surface. Discuss the same items listed in 9(a). Recommend two processes each for (a) and (b).

SEVEN

LITHOGRAPHY

D. A. McGILLIS

7.1 INTRODUCTION

Lithography, as used in the manufacture of ICs, is the process of transferring geometric shapes on a mask to the surface of a silicon wafer. These shapes make up the parts of the circuit, such as gate electrodes, contact windows, metal interconnections, and so on. Although most lithography techniques used today were developed in the past 20 years, the process was actually invented in 1798; in this first process, the pattern, or image, was transferred from a stone plate (lithos).[1]

After a test circuit or computer simulation is completed, the first step in fabricating an IC is to generate the pattern of geometric shapes. A composite drawing of the circuit is broken into levels for subsequent IC processing: gate electrodes on one level, contact windows on another, and so on. These are called *masking levels*. Interactive graphic displays and digitizers convert the geometrical layout to digital data, which is used to drive a computer-controlled pattern generator. The pattern generator is often an electron beam machine. It can transfer the design features directly to the surface of a silicon wafer, but more often it transfers the features to photosensitized glass plates called *photomasks*, or *masks*.

The final IC is made by sequentially transferring the features from each mask, level by level, to the surface of the silicon wafer. For example, between each successive image transfer an ion implant, drive-in, oxidation, or metallization operation may take place.

In the IC lithographic process, a photosensitive polymer film is applied to the silicon wafer, dried, and then exposed with the proper geometrical patterns through a photomask to ultraviolet (UV) light or other radiation. After exposure, the wafer is soaked in a solution that develops the images in the photosensitive material. Depending on the type of polymer used, either exposed or nonexposed areas of film are removed in the developing process. The wafer is then placed in an ambient that etches surface areas not protected by polymer patterns. Because the polymeric materials resist the etching process, they are called *resists*; if light is used to expose the IC pattern, they are called *photoresists*. Resists are made that are sensitive to UV light, electron beams, x-rays, or ion beams. The type of resist used in VLSI lithography depends on the type of exposure tool used to expose the silicon wafer.

Exposure tools do several jobs. First, they rigidly hold the wafer and mask in place after the mask pattern is aligned to a previous pattern already processed into the wafer. Since they provide the mechanical motion needed to make this alignment, exposure tools are sometimes called *aligners*, as are the people who operate them. Second, they provide a source of exposing radiation for the resist. Some exposure tools, such as the e-beam machine, provide a third function; they allow the silicon wafer to be exposed directly without requiring a mask. Exposure tool performance can be evaluated by three parameters: resolution, registration, and throughput. *Resolution* is defined in terms of the minimum feature that can be repeatedly exposed and developed in at least 1 μm of resist.[2] *Registration* is a measure of how closely successive mask levels can be overlaid, and *throughput* is defined as the number of silicon wafers that can be exposed per hour.

The majority of VLSI exposure tools used in IC production are optical systems that use UV light. They are capable of approximately 1-μm resolution, ±0.5-μm (3σ) registration, and up to 100 exposures per hour. Electron-beam exposure systems can produce IC features with resolution less than approximately 0.5 μm with ±0.2-μm (3σ) registration. The e-beam systems are primarily used to produce photomasks; relatively few are dedicated to direct wafer exposure. X-ray lithographic systems have approximately 0.5-μm resolution and ±0.5-μm (3σ) registration but are not yet used to produce ICs in volume.

7.2 THE LITHOGRAPHIC PROCESS

7.2.1 Masks

The first step in generating masks for IC fabrication is to draw a large-scale composite of the set of masks, typically 100× to 2000× the final size.[3] The composite layout is then converted into a set of oversized artwork with a drawing for each masking level. The artwork is photographically reduced to a 10× glass reticle. The final mask is made from the 10× reticle using another photoreduction system that reduces the image to 1×. This system exposes a site on the final photosensitive glass mask, mechanically moves to an adjacent site, exposes the mask again, and so on in step-and-repeat fashion. Each site contains a complete circuit pattern for that masking level. As many identical IC chips, or dice, are put on the mask as will ultimately fit on the silicon wafer. Figure 1 shows a mask on which IC patterns have been arrayed. The mask contains a few secondary chip sites which will produce test circuits that can be used to monitor the complete IC fabrication process or to test primary circuit design modifications.

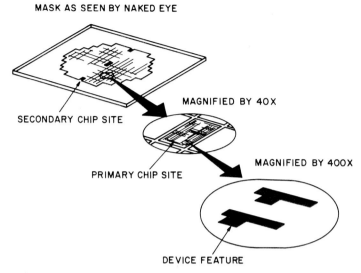

MASK AS SEEN BY NAKED EYE

MAGNIFIED BY 40X

SECONDARY CHIP SITE

PRIMARY CHIP SITE

MAGNIFIED BY 400X

DEVICE FEATURE

Fig. 1 A glass IC photomask.

The oversized artwork approach to mask making is relatively simple, but not practical when applied to VLSI circuits. Considerable effort has been invested in the development of interactive graphics systems with which designers can completely describe the circuit layout electrically. These are called computer-aided design (CAD) systems. Geometric patterns are displayed on a cathode ray tube (CRT) and positioned on the screen by using a light pen or joystick to form the desired circuit shapes. The output of the CAD system is digital data, stored on magnetic tape, which is used to drive a $1\times$ or $10\times$ pattern generator.

Masks are made from glass emulsion plates like the Kodak high-resolution plate (HRP), or glass covered with a hard surface material. Emulsion masks are the least expensive, but they are usually only used with feature sizes in the 5-μm region. All e-beam generated masks are made with hard-surface materials such as chromium, chromium oxide, iron oxide, or silicon. These masks are more expensive than emulsion but features in the 1-μm region can be defined on them.

7.2.2 The Transfer Process

The purpose of the lithographic process is to transfer the mask features to the surface of the silicon wafer (Fig. 2). Figure 3 shows an overview of a typical transfer process.[4] The silicon wafer is first oxidized to form a SiO_2 layer on the surface; the layer is usually 1000 to 10,000 Å thick. Resist is then applied to form a uniform film about 1-μm thick. After coating and drying, the resist is exposed to UV light through a photomask and developed in a solution that, in this case, dissolves the resist that was not exposed. The wafer is then put in an ambient that etches the exposed SiO_2 but does not attack the resist. Buffered hydrofluoric acid (BHF) is a typical SiO_2 etchant.

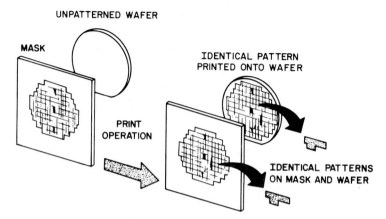

UNPATTERNED WAFER

MASK

IDENTICAL PATTERN
PRINTED ONTO WAFER

PRINT
OPERATION

IDENTICAL PATTERNS
ON MASK AND WAFER

Fig. 2 The transfer of IC patterns from a mask to a silicon wafer.

Finally the resist is stripped, leaving behind a SiO_2 image which then becomes a mask for subsequent processing. For example, an ion implant would dope the exposed silicon, but not the silicon covered by oxide. After the SiO_2 is stripped, the silicon surface is left with a dopant pattern that duplicates the design pattern on the photomask. The complete circuit is built up by aligning the next photomask in the sequence to the pattern in the silicon and repeating the lithographic transfer process. VLSI circuits may require from 5 to 11 separate masks and lithographic transfer steps to fabricate a functional device.

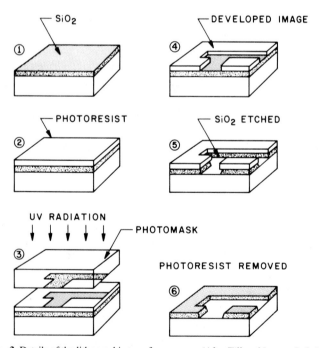

SiO₂

①

DEVELOPED IMAGE

④

PHOTORESIST

②

SiO₂ ETCHED

⑤

UV RADIATION

③

PHOTOMASK

PHOTORESIST REMOVED

⑥

Fig. 3 Details of the lithographic transfer process. *(After Till and Luxon, Ref. 4.)*

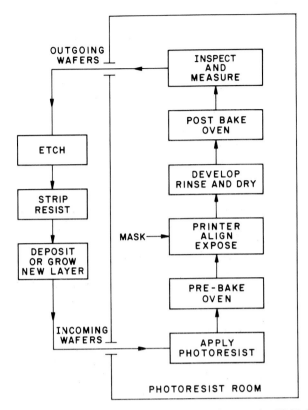

Fig. 4 Flow of silicon wafers through the lithographic processing area of an IC fabrication facility.

The complete lithographic process must be as free of defects as possible. If 10% of the chip sites become defective during each transfer process (a 90% yield), after 11 lithographic operations only 31% of the chips would work. Since defects can be introduced at all the other processing steps as well, the chip yield can easily fall to zero unless close attention is paid to limiting defects.[5]

Figure 4 shows the steps in a photolithographic process and traces the flow of silicon wafers into and out of a lithographic processing area. The photoresist room is typically illuminated with yellow light since photoresists are not sensitive to wavelengths greater than about 5000 Å. The first step is to apply the photoresist. The resist is usually spun on the wafer. The wafer is held on a vacuum spindle, and a few drops of the liquid resist are spread over its surface. The wafer is then accelerated up to a constant rotational speed, which is held for about 30 s. The thickness of the resulting resist film is proportional to the percent solids in the resist and inversely proportional to the square root of the spinspeed.[3] After spinning, the wafer is given a preexposure bake to remove the resist solvent and increase the resist adhesion to the wafer. The wafer and the appropriate mask pattern are then exposed to UV light. Before exposure the mask pattern must first be aligned to existing patterns previously etched into the wafer. The images are developed, rinsed of developer solution, and dried. A postdevelopment bake may be required to give the remaining resist images the adhesion necessary to withstand the subsequent etching process.[6] The wafers are then inspected for quality and the resist images are measured. If the quality is poor or the feature sizes are not within a specified range, the resist may be stripped and the complete photoresist process repeated. Acceptable wafers go on to be etched, resist-stripped, and cleaned, and then given further IC processing. The entire photoresist process may take a few hours to complete and is automated as much as possible.

7.2.3 Resists

Resists may be either negative or positive. Negative resists become less soluble in developer when they are exposed to radiation (as in Fig. 3), and positive resists become more soluble after exposure. Figure 5a shows typical negative and positive resist exposure response curves. At low-exposure energies the negative resist remains completely soluble in the developer solution. As the exposure is increased above a threshold energy E_T, more of the resist film remains after development. At exposures two or three times the threshold energy, very little of the resist film is dissolved. For positive resists, the resist solubility in its developer is finite even at zero-exposure energy. The solubility gradually increases until, at some threshold, it becomes completely soluble. Response curves such as these are affected by all the resist-processing variables: initial resist thickness, spectral distribution of the exposure radiation, prebake conditions, developer chemistry, developing time, and so on. These

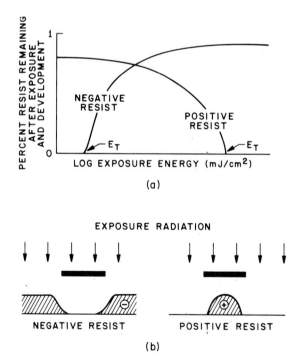

Fig. 5 (a) Positive and negative resist exposure characteristics. (b) Resist images after development.

curves can therefore be used to characterize the complete photoresist process.[7] As shown in Fig. 5a, positive resists usually require more exposure energy (longer exposure times) than negative resists to form resist images. Exposure tool throughput is therefore less when positive resists are used.

In Fig. 5b, typical resist image cross sections are drawn, showing the relationship between the edges of a photomask image and the corresponding edges of the resist images after development. Overexposure tends to reduce the resist image size relative to the mask size in both cases, but in an opposite sense. The area free of resist and consequently the area that will be etched into the silicon wafer decreases as negative resist is exposed longer, but the area free of resist increases as positive resist is given a longer exposure. This effect in optical lithography is explained by the leakage of light under the opaque mask features caused by light diffraction (see Sec. 7.3.3 for a discussion). The object of the lithographic process is to faithfully replicate the mask feature dimension in the corresponding resist images and to ultimately transfer those images into the silicon. One of the major challenges faced by lithographic engineers is to control images to the tight tolerances required in VLSI lithography, typically less than about 10% of the nominal linewidth (e.g., ± 0.2 μm for 2-μm lines).

7.2.4 Tolerances

Features on successive masking levels bear a spatial relationship to each other: metallization patterns should fully cover contact windows, emitters should lie wholly within base features, and so on. Figure 6 gives an example where a device feature on masking level 2 is designed to nest into a feature on masking level 1 with the restriction that an edge of level 1 should never touch an edge of level 2. In the circuit layout, a nesting tolerance must be included between the edges of level 1 and level 2 features. This tolerance is one of the design rules used to lay out the circuit.

The magnitude of the nesting tolerance is dictated by three factors. First, the location of device feature edges on the silicon wafer may not be exactly as specified in the original circuit layout. The size of mask features can vary from chip to chip on

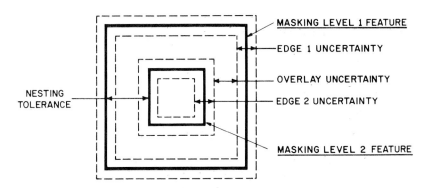

Fig. 6 Components of the nesting tolerance required between two mask levels that are registered to one another.

the mask because of improper exposure, and other factors. An absolute size variation of ±0.2 μm across a 125-mm square mask is not uncommon. When these variable size features are lithographically transferred to a silicon wafer more deviation from the original layout may occur. The resist image can deviate from the mask image because of variations in any or all of the lithographic processing variables, such as resist thickness, baking temperature, exposure, and development conditions. The etching process that finally transfers the resist image into, for example, a SiO_2 layer can also vary the etched image size from wafer to wafer and from day to day. An absolute variation of the final etched image of ±0.4 μm over a year of production is easily possible.

The second component of the nesting tolerance is the uncertainty involved in aligning the images on mask 2 to the previously etched images from mask 1. The mask-making equipment may not produce a set of masks that perfectly overlay, and the exposure machines used to align the mask patterns to the wafer patterns may have limited registration capability. The human factor involved in manually aligning one pattern to another can easily lead to ±0.5-μm uncertainties. Automatic alignment reduces the error, but does not eliminate it. The third factor is the broadening of the dopant profiles in the silicon caused by, for example, lateral diffusion.

An estimate of the nesting tolerance T can be made if the distributions of etched feature sizes (σ_{f1} for level 1, σ_{f2} for level 2) and registration (σ_r) are known (ignoring the profile factor). Assuming that σ_{f1}, σ_{f2}, and σ_r are independent random variables and have normal distributions,

$$T = 3\left[\left[\frac{\sigma_{f1}}{2}\right]^2 + \left[\frac{\sigma_{f2}}{2}\right]^2 + \sigma_r^2\right]^{1/2} \qquad (1)$$

With this tolerance the probability that the edge of an etched feature from level 1 will touch a feature from masking level 2 is approximately only 0.1%. Typical values for a well-controlled VLSI production lithographic process are $\sigma_{f1} = \sigma_{f2} = \pm0.15$ μm and $\sigma_r = \pm0.15$ μm. Using these in Eq. 1, we find T $\cong$ ± 0.6 μm. Both the minimum size feature that can be lithographically transferred and the nesting tolerance determine how tightly devices can be packed on a VLSI circuit.

An often forgotten part of lithography is the measurement technique used to determine that the size of the feature transferred to the silicon wafer is really the size that the circuit designer wanted. Current research at the National Bureau of Standards is directed toward the characterization of feature sizes for both photomasks and IC devices using optical and scanning electron microscope (SEM) techniques.[8, 9] It is unfortunately quite common to find that two measurements of the same IC device feature in two different fabrication facilities may differ by as much as 0.5 μm.

7.3 OPTICAL LITHOGRAPHY

7.3.1 Types of Optical Lithography

The three primary optical exposure methods are contact, proximity, and projection. They are illustrated in Fig. 7.

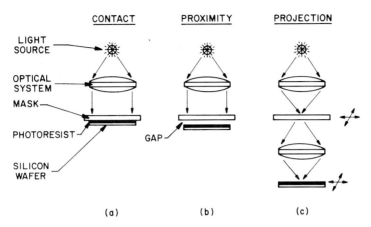

Fig. 7 Schematics of three optical lithographic techniques. (a) Contact. (b) Proximity. (c) Projection in which the mask and wafer are moved synchronously.

In contact printing, shown in Fig. 7a, a resist-coated silicon wafer is brought into physical contact with the glass photomask. The wafer is held on a vacuum chuck, and the whole assembly rises until the wafer and mask contact each other with a few kilograms of force. To align the photomask pattern to a previously etched silicon pattern, the mask and wafer are separated by about 25 μm, and a high-powered pair of objectives are brought in behind the mask to view both the mask and wafer patterns at two positions simultaneously. The objectives are connected to a split-field microscope so that the right eye sees a spot on the right side of the mask and wafer, and the left eye sees a spot on the left. The mask and wafer are aligned by mechanically translating and rotating the vacuum chuck assembly until the patterns on the mask and wafer are aligned. At this point, the wafer is brought into contact with the mask and reexamined for alignment. When the expose button on the machine is pushed, the split-field microscope is automatically withdrawn and a collimated beam of UV light illuminates the entire mask for a fixed exposure time. The exposure intensity (in mW/cm^2) at the wafer surface times the exposure time (in seconds) gives the exposure energy (mJ/cm^2), or dose, received by the resist.

Because of the intimate contact between resist and mask, very high resolution is possible in contact printing. Printing 1-μm features in 0.5 μm of positive resist is relatively easy. The problem in contact printing is dirt. A piece of dirt, such as a speck of Si dust, on the silicon wafer can damage the mask surface when the mask is forced into contact with the wafer. This damaged site then prints as a defective pattern on all subsequent wafers used with that mask. Each additional wafer may add its own damage to the mask as well. If the IC fabrication process or environment is not scrupulously clean, very few defect-free IC chips will be printed. The defect density (number of defects per centimeter squared) must be much less than one for each lithographic transfer process to realize high VLSI chip yields.[5]

The proximity exposure method is very similar to contact printing except that a small gap, 10 to 25 μm wide, is maintained between the wafer and mask during exposure. This gap minimizes (but may not eliminate) mask damage. Proximity printers operate in the Fresnel diffraction region, where resolution is proportional to $(\lambda g)^{1/2}$, where λ is the exposure wavelength and g is the gap between the mask and the wafer.[7] Approximately 2- to 4-μm resolution is possible with proximity printing.

The third exposure method, projection printing, avoids mask damage entirely. An image of the patterns on the mask is projected onto the resist-coated wafer, which is many centimeters away. To achieve high resolution, only a small portion of the mask is imaged. This small image field is scanned or stepped over the surface of the wafer. In scanning projection printers, the mask and wafer are moved synchronously. This technique achieves resolution of about 1.5-μm lines and spaces. Projection printers that step the mask image over the wafer surface are called direct-step-on-wafer or step-and-repeat systems. With these printers, the mask contains the pattern of one large chip or a group of small chips which are enlarged up to 10$\times$. The image of this pattern, or reticle, is demagnified and projected onto the wafer. After the exposure of one chip site, the wafer is moved or stepped on an interferometrically controlled *XY* table to the next chip site, and the process is repeated. Step-and-repeat reduction projection printers are capable of approximately 1-μm resolution.[10]

The optical elements in most modern projection printers are so perfect that their imaging characteristics are dominated by diffraction effects and not by lens aberrations. These printers are said to be diffraction limited systems. The resolution of a *diffraction-limited* projection printer is roughly 0.5 (λ/NA), where NA is the numerical aperture of the projection optics and λ is the exposure wavelength.[17] Projection printers have a limited depth of focus over which image quality is not degraded. The depth of focus is approximately $\pm\lambda/2(NA)^2$. High resolution (large NA) is achieved at the expense of depth of focus. For example, a projection system with NA = 0.17 and an exposure wavelength of 4000 Å will have a resolution limit of about 1.2 μm and a depth of focus approximately ±7 μm, about the thickness of a red blood cell.

7.3.2 Optical Resists

Negative resist is a cyclized polyisoprene polymer material combined with a photosensitive compound.[11] The sensitizer, or photoinitiator, becomes activated by the absorption of energy in the 2000- to 4500-Å range. Once activated the sensitizer transfers energy to the polymer molecules, which promotes crosslinking. The resultant molecular weight increase leads to insolubility in the developer system. Numerous insolubilizing reactions occur for each photon absorbed. Oxygen tends to interfere with the polymerization reactions, and so nitrogen is often directed at the negative resist surface during exposure. During development of the negative resist, the film swells, and the unexposed low molecular weight material is dissolved and rinsed away. It is this swelling action that limits the resolution of negative resists. As a rule of thumb, the minimum resolvable feature is about three times the negative resist film thickness.

Optical positive resist systems also contain a base resin material and a photosensitizer, but are totally different from negative resists in their response to exposure radiation. The sensitizer is insoluble in the aqueous developer solution and therefore prevents the dissolution of the base resin. In the exposed pattern areas, however, the sensitizer absorbs radiation and becomes soluble in an aqueous base.[12] The solubility differential leads to the development of images in positive resist. Unlike negative resist, the developer does not permeate the whole resist film; the film does not swell. Consequently, positive resists exhibit higher resolution capability.

Negative resists usually have poorer resolution capability than positive resists, but they are very sensitive and permit a large number of wafers to be exposed in an hour. This throughput can significantly reduce the cost of the ICs being made. Positive resists can be many times slower, resulting in lower throughputs and higher costs, but they offer higher resolution. Therefore, there is a tradeoff between resolution and throughput.

7.3.3 Diffraction

When exposure radiation passes through a photomask close to the edge of an opaque mask feature, the propagation is not rectilinear. Fringes are observed near the edge of the geometric shadow, and some light penetrates into the shadow region. Phenomena of this type are called *diffraction*. The theory involved in deriving the intensity distribution in the diffraction pattern can be found in several references.[13] Figure 8 shows typical diffraction patterns for contact, proximity, and projection printing. Since the energy distribution incident on the photoresist film equals the intensity distribution times the exposure time, the edge of the resist image is defined by the edges of the diffraction pattern at the position where the exposure energy equals the threshold energy for the resist (see Fig. 5). By changing either the exposure time or the diffraction pattern, the resist image can be made to grow or shrink with respect to the corresponding mask image. These changes are often not intentional.

True contact printing is performed in the geometric shadow region of the mask, which extends to a gap distance less than the wavelength of light (λ) used for the

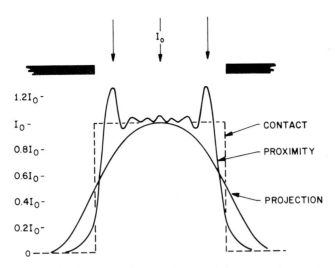

Fig. 8 Typical optical diffraction patterns from a mask feature in contact, proximity, and projection lithography. *(After Skinner, Ref. 14.)*

exposure. The contact between mask and wafer is rarely close enough for the wafer to actually be in this region. Proximity printing is performed in the Fresnel or near-field diffraction region, which extends out to about W^2/λ micrometers from the mask, where W is the mask feature width.[14] Variations in the distance between mask and wafer cause the near-field diffraction patterns of the mask images to change significantly. This in turn causes wide variations in resist image size.

Projection printing is carried out in the Fraunhoffer or far-field diffraction region. The intensity distributions in projection printing diffraction patterns may be altered by changing the system focus by as little as ± 2 µm.[15] Since silicon wafers can easily have a surface ripple that is greater than 6 µm peak to valley, most step-and-repeat projection systems have automatic focusing at each chip site.

7.3.4 Modulation Transfer Function

Optical lithographic exposure systems are characterized by their modulation transfer function (MTF). The quality of the image presented to the resist with respect to the mask image is determined by the MTF of the exposure tool. In principle, this MTF is measured by imaging sinusoidal grating masks characterized by spatial frequencies ν, defined as the inverse of the grating pitch. The modulation of the mask is a function of ν and is defined as

$$M_{mask} = \frac{I_{max} - I_{min}}{I_{max} + I_{min}} \tag{2}$$

where I_{max} and I_{min} are the local maximum and minimum intensities emerging from the mask. If the corresponding modulation of the image presented to the resist film is also measured, the MTF of the exposure tool is

$$\text{MTF}(\nu) = \frac{M_{image}(\nu)}{M_{mask}(\nu)} \tag{3}$$

The ratio I_{max}/I_{min} is called the contrast C. By plotting the MTF as a function of the spatial frequency, the lithographic performance of the exposure tool can be characterized.[16]

The degree of coherence of the light illuminating the photomask influences the image transfer capabilities of the exposure tool. The degree of optical coherence for 1:1 scanning projection systems can be measured by the ratio[17]

$$s = \frac{\text{numerical aperture of illuminator optics}}{\text{numerical aperture of projection optics}}$$

A small value of s indicates that the angular range of lightwaves incident on the mask is small so that the illumination is highly *coherent*. A large value of s indicates a large angular range of incident waves which overfill the projection optics. These waves provide what is called *incoherent* illumination.

The MTF curves in Fig. 9 represent an idealized optical exposure system that is in perfect focus. The curves closely agree with the values measured on existing optical lithographic tools. The MTF values are plotted for various degrees of illumination

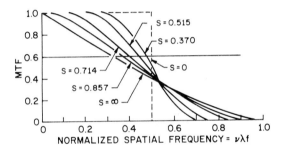

Fig. 9 MTF of an ideal imaging system as a function of illumination coherence. *(After King, Ref. 16.)*

coherence: from $s = \infty$ for completely incoherent illumination, to $s = 0$ for completely coherent illumination. The abscissa is in normalized spatial frequency units, $\nu \lambda f$, where $1/2f$ equals the numerical aperture of the projection optics. A coherent optical system images all sinusoidal grating masks equally well until the pitch becomes less than $2\lambda f$. After that point, no image is formed at all. The figure shows that completely incoherent optical systems can image gratings that are half the pitch of gratings imaged by the coherent system, but the image contrast falls monotonically as the spatial frequency is increased. To form useful resist images, most optical resists require a contrast corresponding to an MTF that is approximately 0.6. For this reason, optical exposure tools use partially coherent illumination, $0 < s < \infty$, to increase the useful image resolution while avoiding the image "ringing" that occurs with completely coherent illumination. Within the range $0.5 < s < 0.9$, tradeoffs can be made between feature size control and image sharpness. Photolithography is optimum when the size of the developed resist images are equal to or slightly larger than the corresponding mask images.[7, 17]

7.3.5 Standing Waves

In addition to diffraction effects and exposure tool MTF, light wave constructive and destructive interference within the photoresist film is another optical effect that significantly influences photoresist images.[17] This interference is illustrated in Fig. 10. Figure 10a shows monochromatic light with wavelength λ entering a photoresist film from the left (ray 1), passing through the resist and the underlying SiO_2 film (ray 2), and being reflected from the silicon substrate (ray 3). The reflected light (ray 3) passes through the resist again and exits into the air. A small percentage of the light (ray 4) is reflected at the resist-air interface and the process is repeated. Figure 10b shows the amplitudes of the incident wave $\mathcal{E}_2$ and the reflected wave $\mathcal{E}_3$. A phase change of π is assumed during the reflection at the silicon surface. Adding waves $\mathcal{E}_2$ and $\mathcal{E}_3$, the result is a standing wave of light intensity in the resist film as shown in Fig. 10c. The standing wave contains antinodes of maximum intensity and nodes of minimum intensity occurring periodically throughout the film. These standing waves can play an important role in determining the size of a developed photoresist image.

The solubility of positive resist in developer is a function of the amount of sensitizer in the resist. The sensitizer's rate of destruction is proportional to the local intensity distribution in the resist.[18] If the intensity distribution is similar to that

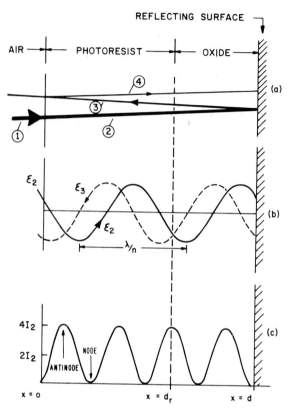

Fig. 10 Standing light waves in a resist film caused by interference between the incident and reflected light. *(After Cuthbert, Ref. 17.)*

shown in Fig. 10c, we would expect positive resist development rates to speed up and slow down as regions of maximum and minimum intensity are reached in the film as the development process proceeds. This is the case shown in Fig. 11, which plots the thickness of a positive photoresist film as a function of time in the developer solution. The resist develops slowly in regions of low exposure and rapidly in regions of high exposure, creating the steplike effect seen in the figure.

7.3.6 Summary

Optical lithographic processes and equipment exist today that will produce VLSI circuits with minimum features in the 1- to 1.5-μm range. The major problem areas being addressed by exposure tool manufactures are level-to-level registration and machine throughput. Photoresist suppliers are developing resist systems with increased photospeed and the ruggedness to withstand today's plasma etching environments. Research and development laboratories are devising multilevel resist schemes (see Sec. 7.6.2) which may push practical optical lithography to the 0.5-μm level. All of this activity will probably keep optical lithography the dominant technology of the 1980s for defining VLSI patterns on a production scale.

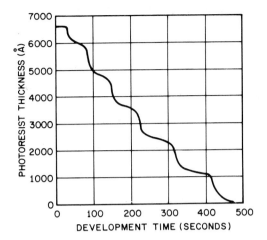

Fig. 11 Measured thickness versus time, during development of a positive resist film on a silicon wafer. *(After Konnerth and Dill, Ref. 18.)*

7.4 ELECTRON BEAM LITHOGRAPHY

7.4.1 Overview

Electron-beam fabrication of ICs offers several advantages for lithographic pattern transfer: resist geometries smaller than 1 μm can be generated, wafers can be patterned directly without a mask, and the technique can be highly automated. In addition, an electron beam has a much greater depth of focus than an optical lithographic system. An electron beam can be used to detect features on a silicon wafer. This capability can lead to extremely accurate level-to-level registration. The problem with e-beam lithographic machines is that they are slow. Their throughput is approximately only five wafers per hour at less than 1-μm resolution. These throughputs do not economically compete with optical machine throughputs of 40 wafers per hour at 1.5-μm resolution.

To write submicrometer patterns into a resist, the e-beam must be focused to a diameter of 0.01 to 0.5 μm. The current density in the focused spot should also be high, to minimize resist exposure times. Most thermionic electron guns have current densities of a few amperes per centimeter squared from a cathode that is 10 to 100 μm in diameter.[19] Therefore, electron-optical demagnifying lenses are required to reduce the e-beam diameter by as much as 10^4. The focused beam must be capable of being directed to any point in the scan field under the control of pattern generator data. This requires beam deflection and blanking systems that can operate at megahertz rates under computer control. Figure 12 gives a schematic of an e-beam lithography machine. Since the beam scan is restricted by lens aberrations to usually less than 1 cm, an interferometrically controlled *XY* table is used to position the substrate to be patterned under the e-beam. Registration to a previously defined pattern may be accomplished at each chip site by scanning the e-beam across reference marks etched in the substrate and detecting the secondary and backscattered electrons. These sig-

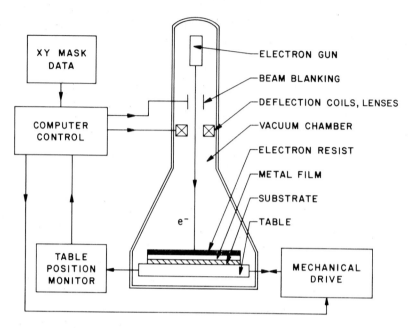

Fig. 12 Schematic of an electron-beam machine.

nals are used to automatically position the substrate under the beam. Alignment accuracy of ± 0.2 μm (3σ) is reported.[20]

Electron-beam lithography machines are usually designed for optimum performance in research and development, in the production of photomasks, or in the direct writing of silicon wafers. Machines used in research and development must provide the smallest possible focused spot so that the highest resolution can be obtained. Beam diameters as small as 5 Å have been used to etch 13-Å wide lines in NaCl crystals.[21] Device throughput in these machines is not important. A machine intended for the production of photomasks or reticles with features of 2 to 4 μm can have a relatively large beam diameter (0.25 to 1 μm) and modest throughput. Satisfactory throughput may be one mask per hour. However, a machine designed for the production of IC devices must have the highest possible throughput and, therefore, the largest beam diameter consistent with the minimum device dimensions. As a rule of thumb, the minimum device feature is about 4× the beam diameter, and the field that can be directly accessed by the e-beam without *XY* stage motion is about 2000× the minimum device feature. In other words, the smaller the device feature, the more *XY* stage motion required. More stage motion, of course, slows down production. Once again a tradeoff must be made: smaller features for wafer throughput.

7.4.2 Electron Resists

A radiation sensitive resist is one in which chemical or physical changes are induced by ionizing radiation, which allows the resist to be patterned.[22] A molecule of a polymer electron resist consists primarily of monomer units that have been polymerized into a backbone chain. Irradiation with electrons leads to two generic types of interactions: chemical bond breaking and radiation-induced polymer cross-linking.

In chemical bond breaking, or chain scission, the molecular weight is reduced in the irradiated area. If the average molecular weight is reduced enough, the irradiated material can be dissolved in a solvent that does not attack high molecular weight material. Polymers that undergo chain scission are called *positive electron resists*. Common positive resists are poly(methyl methacrylate), called PMMA, and poly(butene-1 sulfone), called PBS. A typical developer is a 1:1 mixture of methyl isobutyl ketone (MIBK) with isopropyl alcohol.

The second polymer-electron interaction is radiation-induced polymer cross-linking. The cross-linking events cause new bonds to form between adjacent chains, which creates a complex three-dimensional structure with higher molecular weight than the surrounding nonirradiated area. Polymers in which cross-linking events dominate are called *negative electron resists*. Again, development proceeds by the dissolution of the low molecular weight material. COP, poly (glycidylmethacrylate-co-ethyl acrylate), is a common negative electron resist. Swelling during development limits most negative electron resists to resolutions of about 1 μm. Positive resists have resolutions that are less than 0.1 μm.

Figure 13 shows the characteristic response curves of typical positive and negative electron resists. The curves that represent the remaining-resist thickness as a function of exposure dose are similar to curves representing optical resist characteristics (Fig. 5). The electron resist sensitivity S for positive and negative resists are defined as the electron dose required per centimeter squared to ensure complete positive resist development or to correspond to a 50% remaining thickness in the case of negative resist.[23] This definition makes the sensitivity a function of all the resist processing variables, such as thickness, developer strength, and so on. The figure shows that the positive resist PMMA is about three orders of magnitude less sensitive than the negative resist COP and would therefore require an exposure time about 1000 times longer to form useful resist images. Generally, slow resists have higher resolution than fast resists. This sensitivity-resolution tradeoff can be outlined as follows.[23]

First, imagine that the substrate to be exposed by the electron beam is subdivided into a grid of addressable locations. Each element in this grid is called a *pixel*. A pixel represents the minimum resolution element that can be defined by the complete presence or absence of charge. Pixels are combined to form pattern shapes. The minimum discernible pattern is one pixel exposed and one pixel not exposed. If the

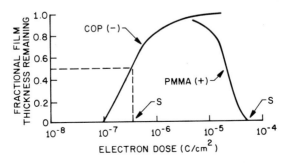

Fig. 13 Typical exposure characteristics curves for a positive and negative electron-beam resist. *(After Greeneich, Ref. 23.)*

side of a pixel is 0.5 μm, a 125-mm diameter silicon wafer would contain about 5×10^{10} pixels. To form a useful image in the resist some minimum total number of electrons N_m must strike each exposed pixel. For a given resist sensitivity S, this minimum is

$$N_m = \frac{SL_p^2}{q} \tag{4}$$

where L_p is the minimum pixel dimension, q is the electron charge, and S is the required dose in coulombs per centimeter squared.

Electron emission from the cathode of an electron gun is a random process, and the number of electrons striking a given pixel element in a time T varies statistically. One can show[23] that the probability of a pixel not receiving N_m electrons is approximately 10^{-12} if $N_m = 200$ electrons; this probability is sufficiently small that 5×10^{10} pixels can probably be exposed without error. With $N_m = 200$, Eq. 4 becomes

$$L_p = \left[\frac{200q}{S} \right]^{1/2} \tag{5}$$

which is plotted in Fig. 14. The shaded area contains combinations of pixel size and resist sensitivity that produce unacceptably high probabilities of exposure error. Equation 4 shows the basic resist-sensitivity resolution tradeoff; the product of sensitivity and pixel size is a constant fixed by N_m.

Figure 14 compares data on sensitivity and resolution for several electron resists. The data is representative of the best combination of resolution and sensitivity for the indicated resists, as the result of a single line scan under typical IC exposure conditions. The broken line represents the present state of the art in electron resists.[23]

7.4.3 Electron Scattering and Proximity Effects

When an electron beam penetrates both a resist and the IC substrate beneath it, the electrons scatter elastically and inelastically. Inelastic collisions with resist and sub-

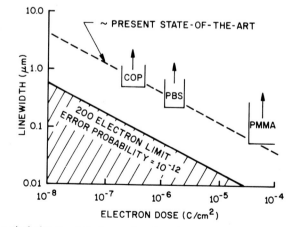

Fig. 14 Minimum pixel size (resolution) as a function of resist sensitivity. Typical resist sensitivity-resolution data shows current state of the art for e-beam lithography. (*After Greeneich, Ref. 23.*)

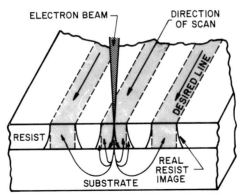

Fig. 15 Electron scattering effects in a resist-coated substrate. *(After Greeneich, Ref. 23.)*

strate atoms result in energy loss; elastic collisions cause a change in the direction of the electrons. Consequently, the incident electrons spread out as they penetrate a resist-coated wafer until either all their energy is lost or they leave the material as a result of backscattering collisions.

Electrons that are backscattered from the substrate and return to the resist deposit energy several micrometers from the center of the exposing beam. Since the resist integrates the energy contributions from all surrounding areas, the exposure dose in one pixel is affected by the exposure in neighboring pixels. This behavior is called the *proximity effect*. Figure 15 gives a specific example. The line patterns indicated by the shaded areas are to be written by the incident electron beam that is scanned along the length of the three lines. As the electrons penetrate the resist, scattering broadens the incident energy distribution. Consequently, the developed resist images are wider than would be expected from the size of the incident beam. Scattering places a limit on the minimum resist linewidth. Since backscattered electrons may travel relatively large distances before reentering the resist film, a fraction of them contribute to the exposure of resist patterns lying adjacent to the one being written. In other words, the total energy absorbed depends on the proximity of neighboring exposed areas.

In the center of a large exposed area, such as at point A in Fig. 16, there are exposure contributions from all the surrounding incident electrons. Point B, however, receives only half the energy of point A, and point C at the feature corner receives only one-fourth the dose of point A. The resist image is usually developed to a point where the width of the feature corresponds to the design width, that is, point B. The shaded area in the figure represents the developed image. Because of proximity effects, the corners are not developed out to their design location. This phenomenon is called the *intraproximity effect*. The intraproximity effect also causes large and small features to print differently. The long narrow line in the figure is smaller than its design value because the exposure dose and development conditions were optimized to produce the required edge at point B. Cooperative effects can also be seen; backscattered electrons travel large distances so patterns relatively close to each other are affected by the neighboring exposure. These are called *interproximity effects*.

To correct for proximity effects, patterns can be divided into smaller shapes. The

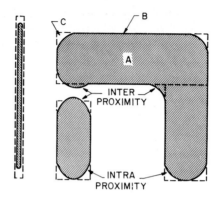

Fig. 16 Inter and intra proximity effects in e-beam exposure caused by electron scattering. *(After Greeneich, Ref. 23.)*

incident dose in each subshape is then adjusted so that the average dose in each pattern is correct.[24] A drawback is that this procedure may decrease the e-beam machine throughput because of the increased computer time required to partition and print the subdivided resist patterns.

7.4.4 Operating Strategies

To obtain a minimum-size resist image with good size control (better than $\pm 10\%$) usually requires several passes of the electron beam. Typically, four or more passes of a Gaussian-shaped beam spaced at the beam half-width are used to write a minimum size resist feature. Two major strategies are used to write e-beam patterns: vector and raster scan.[19]

In a typical vector-scan system, the digital data that specifies the feature size and location is used to direct the e-beam to the proper circuit location, turn the beam on, fill in the pattern shape by rastering the beam back and forth within the feature shape, turn the beam off and vector the beam to the next feature location, and then repeat the process. When the available scan field has been written, an *XY* table moves a new scan field under the beam. This method is particularly attractive when only a few patterns must be written and they are all the same size, for example, at a contact window level; otherwise, it may take several hours to expose a 125-mm silicon wafer.

In a raster-scan system, the electron beam scans continuously back and forth over a small field of view (typically 256 μm) while the *XY* table scans at a right angle to the beam scan. The beam is turned on and off to write the pattern. After the first stripe of a circuit pattern is written over the whole substrate, the *XY* table returns to the beginning and scans the next stripe. Systems of this type can use less complicated electron optics than vector scan machines but the *XY* table control must be precise. The raster-scan system is used primarily to make photomasks and can write a 125-mm mask in about one hour.

A third approach to writing an electron-beam pattern uses a variable-shaped electron beam in a vector scanning mode. A shaped aperture is illuminated by the electron beam and imaged onto a second shaped aperture. By deflecting the image of the first with respect to the second, a variable shaped beam can be formed and vectored to the proper circuit locations. Machine throughput can be increased substantially by this technique; a 125-mm wafer can be written in several minutes.

7.4.5 Limitations and Trends

Direct writing electron beam lithography is attractive because an e-beam system is capable of submicrometer resolution and has the best level-to-level registration capability of the major lithographic techniques. An e-beam system has the advantage of flexibility. Customized VLSI designs can be fabricated without first going through a mask-making process that is prone to errors and defects. The challenge of e-beam lithography is using the existing submicron capability at an economically justified throughput. Proximity effects can be corrected, but often at the cost of lower throughput resulting from increased computation time. High resolution is possible, but at the cost of lower resist sensitivity and lower throughput. To achieve both high throughput and high resolution, brighter, higher current sources must be developed.

7.5 X-RAY LITHOGRAPHY

7.5.1 General Principles

X-ray lithography is an extension of optical proximity printing in which the exposing wavelength is in the 4- to 50-Å range. The short wavelength of x-rays reduces diffraction effects while still using a noncontact exposure system. Because x-ray optical elements are not yet available, x-ray lithography is limited to shadow printing. An x-ray lithography system is illustrated in Fig. 17. In this system[25] a 25-kV, 4- to 6-kW electron beam generated by a ring electron gun is focused on a water-cooled Pd target. As a result, 4.37-Å x-rays are emitted and pass through a beryllium window into an exposure chamber filled with helium. (The helium prevents air from absorbing the x-rays.) A mask, with x-ray absorbing patterns, and a wafer coated with an x-ray sensitive resist are mounted on a movable stage that contains a vacuum chuck to hold the wafer flat. The mask and wafer are separated by about 40 μm. After the mask is aligned with the wafer, the stage is moved into the exposure position where the x-rays cast a shadow of the mask patterns onto the x-ray resist. The full wafer is exposed in about 1 min.

The primary reason for developing x-ray lithography is the possibility of achieving high resolution and high throughput at the same time. There are other benefits as well. The low energy of soft x-rays reduces scattering effects in both the resist and substrate; no proximity corrections have to be made. Since x-rays are not appreciably absorbed by dirt with low atomic number, dirt on the mask does not print as a defective pattern in the resist. And finally, because of the low absorption in x-ray resists, a thick resist can be uniformly exposed throughout the entire thickness, resulting in straight-walled resist images exactly replicating the mask patterns.

Geometrical effects, however, can limit the resolution of x-ray lithography.[26]

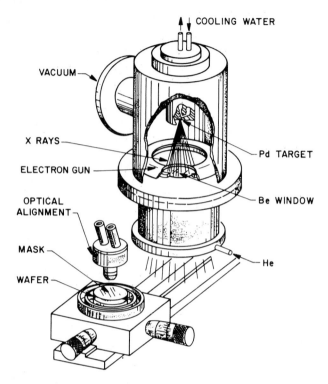

Fig. 17 Schematic of an x-ray exposure system. *(After Maydan, Ref. 25.)*

Figure 18 shows the general outline of an x-ray exposure system. A point source of x-rays of diameter ϕ is at a distance L from the x-ray mask, which in turn is separated a distance g from the resist-coated wafer. The extended point source introduces a penumbral blur δ on the position of the resist image edge

$$\delta = \phi \frac{g}{L} \qquad (6)$$

For typical values of $\phi = 3$ mm, $g = 40$ μm, and $L = 50$ cm, the penumbral blur can be on the order of 0.2 μm.

Another geometrical effect shown in Fig. 18 is the lateral magnification error, which is caused by the x-ray divergence from the point source and the finite mask to wafer separation. The projected images of the mask are shifted laterally by an amount d, given by

$$d = r \frac{g}{L} \qquad (7)$$

where r is the radial distance measured from the center of the wafer. The error is zero at the center of the wafer, but it increases linearly across the wafer. This run-out error can be as large as 5 μm at the edge of 125-mm wafer using the values $g = 40$ μm and $L = 50$ cm. In principle the error can be compensated for during the mask-making

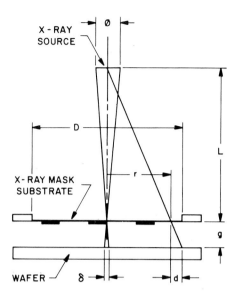

Fig. 18 Geometrical effects in x-ray lithography. *(After Fay, Ref. 26.)*

process. Variations in the mask-to-wafer gap, however, either across the wafer or from mask level to mask level, can introduce significant run-out error. Gap adjustment before each exposure may be required.

7.5.2 X-ray Resists

X-rays with wavelengths between 1 and 50 Å (photon energies between 10 and 0.25 keV) suffer negligible scattering as they go through resist materials. An x-ray moves in a straight line until it is captured by an atom, which ejects a photoelectron. The energy of the photoelectron equals the x-ray photon energy minus the few electron volts of binding energy necessary to remove the electron from its atomic shell out to infinity. The photoelectron's most probable direction is normal to the x-ray photon direction, that is, in the plane of the resist.[22] The excited atom returns to its ground state by emitting a fluorescent x-ray or an Auger electron. The x-ray fluorescence is absorbed by another atom, and the process repeats. Since all the processes end with the emission of electrons, x-ray absorption in the resist material can be thought of as releasing a swarm of secondary electrons. These electrons expose the resist by either inducing chain scission or cross-linking, depending on the type of resist. All e-beam resists are also x-ray resists.

Table 1 summarizes x-ray resist sensitivities, as previously defined in Fig. 13, for the electron beam resists COP, PBS, and PMMA. Since the flux incident on the resist from a point source x-ray exposure tool may only be between 1 and 10 mJ/cm^{2}min, the sensitivity of these resists is not adequate to achieve the goal of high resolution and high throughput.[27]

Table 1 Properties of a few x-ray resists†

Resist	Tone	Major Abs elements	λ	Sensitivity (mJ/cm^2)	Resolution (μm)
COP	$(-)$	0	4.37 Å $Pd_{L\alpha}$	175	1.0
PBS	$(+)$	S	4.37 Å $Pd_{L\alpha}$	94	0.5
PMMA	$(+)$	0	8.34 Å $Al_{k\alpha}$	600−1000	<0.1

†After Taylor, Ref. 27.

One way higher resist sensitivity can be accomplished is by increasing the x-ray absorption in the resist. The absorption of x-rays can be described by the equation

$$I = I_0 \exp(-\alpha t) \qquad (8)$$

where t is the thickness of the resist, α is the linear absorption coefficient, and I_0 and I are the intensities before and after absorption, respectively. Figure 19 shows absorption coefficients for a few selected materials[28] in the wavelength range of interest. The absorption can be increased by increasing the absorption coefficient, which is related to the x-ray atomic absorption cross sections of the elements in the resist. The cross section for x-ray capture by electrons in a given atomic shell varies with the x-ray wavelength and shows large increases at certain critical wavelengths. The critical wavelengths correspond to x-ray energies that are just sufficient to remove electrons from their atomic shells, K, L_1, and so on. For example, wavelengths slightly longer than λ_K can no longer be captured by K-shell electrons; therefore the cross section drops abruptly at that point. Materials are most transparent

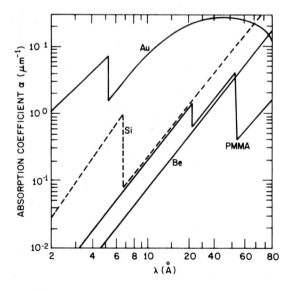

Fig. 19 Absorption coefficients of PMMA, Be, Si, and Au in the x-ray wavelength range used in x-ray lithography. *(After Spiller and Feder, Ref. 28.)*

to x-rays whose wavelength is just slightly longer than a critical wavelength, and materials have the greatest absorption when the x-ray wavelength is slightly shorter than a critical wavelength.[22] Therefore, x-ray resist sensitivity can be increased by including in the resist elements whose absorption edges are in near resonance with the exposure wavelength. Chlorine has a K edge at 4.40 Å and consequently strongly absorbs the 4.37-Å $Pd_{L\alpha}$ wavelength. Negative x-ray resists with sensitivities below 10 mJ/cm^2 have been made by incorporating Cl into the resist polymer.[27]

Although sensitive and capable of high throughputs, negative resists have limited resolution because they swell and contract during the wet chemical development process. Dry development by plasma processing avoids the swelling problem. Several types of plasma developable negative x-ray resists have been described recently. All contain an absorbing polymer host and a polymerizable monomer guest, which is locked into the host by incident x-ray radiation. Negative x-ray resists have been made by incorporating silicon-containing organometallic monomers with a chlorinated polymer absorber. These resists can be fully exposed in about 1 min and can be developed in an O_2 plasma. The resist resolution is less than 0.5 µm.[29] Figure 20 shows a plasma-developed x-ray resist process. The resist consists of a host polymer

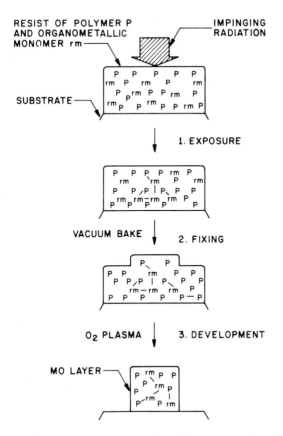

Fig. 20 Schematic of plasma-developed x-ray resist. *(After Taylor, Wolf, and Moran, Ref. 29.)*

P and an organometallic monomer rm, where r is the organic component and m is the metal component of the monomer guest. Radiation incident on the resist polymerizes the monomer and host polymer, locking in the organometallic. After exposure, the resist film is baked in a vacuum to drive off the unpolymerized monomer. The resist images are then developed out using an O_2 plasma. The underlying assumption is that the organometallic monomer is converted to a metallic oxide MO, which protects the remaining resist from attack by the O_2 plasma. This protective layer then enhances the difference in plasma removal rates between the exposed and unexposed resist, and, therefore an image is developed. These resist systems are still the subject of active research programs.

7.5.3 X-ray Masks

An x-ray mask consists of a patterned metal x-ray absorber on a thin membrane that transmits x-rays. The thickness of the absorbing material is determined by the x-ray wavelength of interest, the absorption coefficient of the material, and the contrast required by the resist to form an image. Gold is currently the most widely used absorbing material. The mask patterns are usually generated using e-beam lithography combined with dry etching techniques. To maintain high resolution and good feature size control, vertical walls are required on the absorbing gold patterns. These properties are most easily achieved with thin gold, which can be used at the longer exposure wavelengths.

The membrane that forms the mask substrate must be highly transparent to x-rays so that exposure times are minimized. It should be dimensionally stable, rugged enough to be handled frequently in production use, and, if optical alignment techniques are to be used, transparent to visible light. Many membrane materials,[30] such as polyimide, Si, SiC, Si_3N_4, Al_2O_3, and sandwich structures of $Si_3N_4/SiO_2/Si_3N_4$, have been used.

Figure 21 shows an x-ray mask structure that has been used successfully to make IC devices. It is a sandwich of boron nitride and polyimide, with 0.6-μm thick gold patterns that absorb the x-rays. The exposure wavelength used with this mask is the 4.37-Å $Pd_{L\alpha}$ characteristic line. The mask is made by first depositing a 6-μm film of boron nitride on a silicon wafer. After deposition, a 6-μm polyimide film is spun on top of the boron nitride to give additional strength. After the polyimide layer is

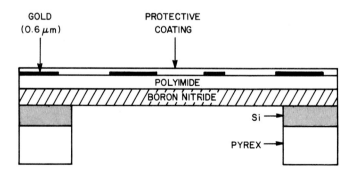

Fig. 21 An x-ray mask. *(After Maydan, Ref. 25.)*

cured, a thin Ta layer is deposited on the membrane, followed by 0.6 μm of gold which is then covered by another thin Ta layer. Electron beam resist is applied to the structure and patterned using e-beam lithography. The resist images are transferred to the top Ta layer, which subsequently acts as a masking layer for the gold etching process. After gold patterning, the Ta films are stripped and another polyimide protective coating is applied. The patterned wafer is then bonded to a pyrex ring and the silicon is etched from the back, leaving the membrane structure shown in Fig. 21.

X-ray mask making is not yet a fully developed technology. The following problems remain to be solved: improving the long-term dimensional stability of the mask, eliminating the resolution degrading effects of sloped pattern edges, and reducing the mask defect density. The viability of submicrometer x-ray lithography depends on solving these problems.[31]

7.5.4 X-ray Sources

The simplest x-ray source[26, 32] is the x-ray tube. This device focuses electrons in the keV energy range on a metal target. Here they excite an x-ray spectrum of discrete lines characteristic of the target metal and a continuous background spectrum of much lower intensity. The efficiency is usually less than one percent. Most of the e-beam energy is dissipated as heat within the target. For reasons already discussed, an extended point source of x-rays can cause considerable image edge blur due to penumbra effects. To minimize the blur the electron beam is focused to a spot a few square millimeters in diameter. Even with forced cooling of the target, the maximum-allowable thermal load is only on the order of 2 kW/mm^2, so that the x-ray flux incident on a resist-coated wafer is low, usually less than 0.1 mW/cm^2.

Another source of x-rays, however, provides an almost collimated beam (so there are no geometrical effects), a wide continuum of x-ray wavelengths, and flux densities in excess of 100 mW/cm^2 at the wafer plane. This x-ray source, called *synchrotron radiation*, is the electromagnetic radiation emitted by electrons in response to the radial acceleration that keeps them in orbit in storage rings or synchrotrons. In a synchrotron, bunches of electrons are continuously injected into a ring, raised in energy, and then removed, usually at a 50- to 60-Hz rate. In a storage ring, a single bunch of electrons is injected, raised in energy, and kept stable for several hours. Synchrotron radiation is rich in the long x-ray wavelengths between 10 and 50 Å, which are strongly absorbed by thin absorber patterns on masks and are therefore ideal for high-resolution x-ray lithography. The obvious drawback is cost.

Other x-ray sources rely on the generation of a dense plasma in a small volume to provide bursts of intense x-rays. These sources include laser focus, plasma focus, and vacuum spark techniques. It is too early to tell if any of these methods of x-ray production will find practical applications.

7.5.5 Summary

In principle, x-ray lithography offers the best conditions for achieving submicrometer resolution with high wafer throughput. Full wafers can be exposed in about 1 min using existing resists and x-ray sources, with resolution better than 0.5 μm. Step-

and-repeat exposure methods, using the intense collimated x-rays from a storage ring, may be feasible in the future. Automatic alignment techniques must be perfected, however, and mask fabrication must be improved before x-ray lithography becomes a production process.

7.6 OTHER LITHOGRAPHIC TECHNIQUES

7.6.1 Deep-UV Lithography

Standard photolithography is normally carried out in the 3100- to 4500-Å spectral region, with practical resolution about 1 to 1.5 μm. Resolution can be increased by reducing the wavelength of the exposure radiation to the 2000- to 3000-Å spectral region, called "deep UV."[13] Using conventional optical lithographic equipment that has been modified to operate at shorter wavelengths, and using mask substrates made of quartz instead of glass, resist images on the order of 0.5 μm have been printed.[13] The major advantage of the technique lies in the use of established e-beam mask-making technology.

Commercial deep-UV exposure sources are available. The xenon-mercury lamp is rich in deep UV output but has lower intensity than a standard mercury lamp. Mercury arc lamps doped with zinc or cadmium, or deuterium lamps can also be used as exposure sources.

Whether deep-UV lithography can be practical depends on the availability of a suitable photoresist. The match between the output spectrum of the exposure tool and the absorption spectrum of the resist determines the throughput capability of the technique. To achieve straight-walled resist image profiles, the resist must absorb only a small percentage of the incident radiation, usually less than 20%. On the other hand, too little absorption significantly increases exposure time. In general, any e-beam resist is a candidate for a deep UV-resist. Figure 22 shows the spectral transmission of 0.8 μm of PMMA for unexposed resist and at 2-min exposure intervals.[13] The resist exhibits a photo-dyeing effect, that is, the absorption increases with exposure. Because of the low absorption, PMMA forms straight-walled resist images. However, since the match is poor between the PMMA absorption spectrum and the output of a xenon-mercury lamp, for example, exposure times with this source are high—on the order of 10 min.

Given the availability of deep-UV exposure tools and more sensitive resists, deep UV optical lithography may soon become the dominant technology for VLSI production in the 1-μm design-rule region.[33]

7.6.2 Multilevel Resists

To develop a high-resolution straight-walled resist image, the resist must receive a uniform exposure dose throughout the depth of the resist film. Using thin resist films, usually less than 0.3 μm, greatly increases the useful resolution of an exposure tool and significantly improves feature size control.[34] However, a resist film must be thick

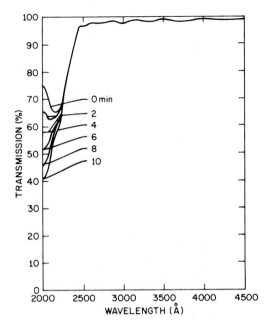

Fig. 22 Transmission spectrum of 0.8 μm of PMMA in the deep UV exposure region. *(After Lin et al., Ref. 13.)*

enough to cover the previously patterned device topography on a silicon wafer. Oxide or metal steps that are approximately 1 μm high on a VLSI device are not unusual. To adequately cover such a step the resist should be at least 1 μm thick, preferably thicker. Once such a step is covered, the resist film is not only much thicker than desired for high resolution but is also nonuniform in thickness across the step. The realization that very thin resist films lead to improved resolution, but that thick films are required for IC fabrication, led to the development of resist systems composed of multiple layers.

Multiple layer systems can be divided into two categories; those in which at least two layers are used as resists and both are exposed and developed, and those in which only the top layer is used as a resist and the other layers are removed using the top resist as a mask. In both categories, the bottom layer is usually very thick, typically two to four times the maximum step height on the IC device. This layer, if thick enough, will not only cover all the device topography, but will also form a flat surface. A very thin resist imaging layer is then spun on top of the planarizing layer. Figure 23 shows a schematic of a multilevel resist system. Device steps are covered and planarized by a thick layer of polymer, which may or may not be photosensitive. In the top figure a thin layer of isolation material such as SiO_2 or Si_3N_4 separates the thin imaging resist from the bottom material. Images are formed in the resist using optical, e-beam, or x-ray lithography. Since the resist is thin, the highest resolution capability of the exposure technique can be achieved. The top masking layer conforms to the bottom layer and is portable with the wafer, so it is called a *portable con-*

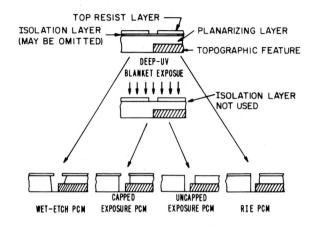

Fig. 23 Schematic of a multilevel resist system. *(After Lin, et al., Ref. 35.)*

formable mask (PCM).[35] The resist image is etched into the isolation layer and then removed (see Fig. 23, top). The isolation layer acts as a mask to transfer the image into the thick bottom layer using wet chemical isotropic etching or dry anisotropic plasma reactive-ion etching (RIE). The image transferred to the thick material then acts as a mask to pattern the IC device. When three levels are involved in the pattern transfer process, it is called a *trilevel resist process*. In the middle of Fig. 23 the intermediate isolation layer has been omitted. The thin resist is applied directly on top of the thick layer. If the bottom thick material is deep-UV sensitive, for example PMMA, and the top resist is an optical positive resist that strongly absorbs in the deep UV, the top resist can act as a PCM for the exposure of the PMMA. The top resist layer may or may not be developed off during the development of the PMMA, depending on the choice of developing conditions.[35]

Processing a multilevel resist system is much more complex than a single-layer image-transfer process. The resolution and feature size control given by a multilevel process is, however, far superior. In optical lithography, standing-wave effects are eliminated by using a multilevel resist, and in e-beam lithography, backscattering from the substrate is minimal.

7.6.3 Inorganic Resists

Germanium selenide (GeSe) glass films can act as photoresists.[36] These glasses dissolve easily in an alkaline solution but when doped with silver, they become almost insoluble. By coating a GeSe glass film with silver and exposing the film to UV light from an optical exposure tool, light-induced silver migration called *photodoping* occurs in the exposed regions. Figure 24 illustrates the process of forming an image in the GeSe inorganic resist. A thin (approximately 0.2-μm) film of GeSe is evaporated onto a substrate and then dipped into an aqueous $AgNO_3$ solution to form a layer of silver on the surface a few hundred angstroms thick. After exposure to UV light through a photomask, the unphotodoped silver in the unexposed regions is

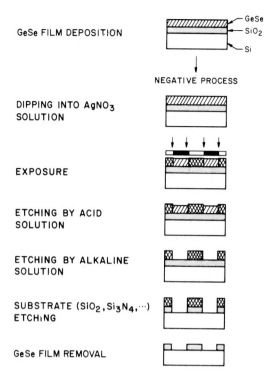

GeSe FILM DEPOSITION

NEGATIVE PROCESS

DIPPING INTO AgNO₃
SOLUTION

EXPOSURE

ETCHING BY ACID
SOLUTION

ETCHING BY ALKALINE
SOLUTION

SUBSTRATE (SIO₂, Si₃N₄,···)
ETCHING

GeSe FILM REMOVAL

Fig. 24 *(After Yoshikawa et al., Ref. 36.)*

removed in a $HNO_3 - HCl - H_2O$ solution. The GeSe not made insoluble by photo-doping is etched in an alkaline aqueous solution of NH_4OH, KOH, or NaOH.

Most polymer optical resists require an image contrast corresponding to a modulation transfer function (MTF) of approximately 0.6 to form useful images. The GeSe inorganic resist requires only an MTF of approximately 0.2 to form an image.[37] The consequence of this low-contrast requirement can be seen by referring to Fig. 9; as the MTF threshold is lowered, a greater spatial frequency can be resolved. This means that optical-projection exposure equipment that can, for example, resolve 1-μm features in a standard polymer photoresist, can resolve much smaller features in a GeSe resist. In fact, 0.5-μm lines and spaces have been printed in GeSe using commercial step-and-repeat optical exposure equipment.[37]

Additional research is needed before inorganic resists become practical. If these resists can be used in a multilevel resist system, however, existing optical lithographic equipment may be able to produce submicrometer VLSI devices.

7.6.4 Ion Beam Lithography

Ions, because of their mass, scatter much less than electrons. Ion-beam lithography[38] inherently has higher resolution capabilities than e-beam lithography because of the absence of proximity effects. Ion beams, like e-beams, can be used as a focused beam for direct writing in resists (see Chap. 10). Conventional resists can be used in ion-beam lithography, but new possibilities exist. Virtually any polymer can be used as a negative resist by implanting ions such that after reactive-ion etching the resist in a suitable plasma, the implanted ions form nonvolatile compounds. The unimplanted regions are etched away. GeSe inorganic resist can also be used with ion beams.

Ion-beam lithography is still in its infancy. Direct writing appears to be too slow to be economically attractive, but it may be useful in the future for making step-and-repeat 1:1 reticles for x-ray lithography.

7.7 SUMMARY AND FUTURE TRENDS

All of the major lithographic technologies (optical, e-beam, and x-ray) are capable of producing the 1- to 2-μm feature sizes required for state-of-the-art VLSI device fabrication. Only optical and e-beam lithographic processes are used in VLSI production today; and the overwhelming majority of the processes are optical. Each technique has its limitations: diffraction effects in optical lithography, proximity effects in e-beam lithography, and mask fabrication in x-ray lithography. Electron-beam exposure systems are capable of defining submicrometer geometries today but with very low wafer throughput. Multilevel and inorganic resist systems may push optical lithography into the submicrometer region in the near future. Any one of the three lithographies can be used to do research in submicrometer devices.

Three criteria dictate the viability of a production lithographic process: resolution, registration capability, and throughput. Adequate resolution is available for VLSI ICs, but the required overlay registration can only be achieved at the expense of throughput. Figure 25 shows the estimated resolution for all modern exposure systems as a function of 125-mm wafer throughput.[39] The outlined areas represent the maximum resolution expected. The "usable" resolution is defined as 2.5 times the machine-overlay registration capability (3σ). Optical step-and-repeat lithography and 1:1 projection lithography is expected to dominate VLSI production in the 1980s. Electron beam systems, with their excellent overlay capability but poor throughput, will continue to be important for specialized direct-write applications. As step-and-repeat x-ray systems become available and mask fabrication improves, x-ray lithography may fill the gap between e-beam and optical lithography.

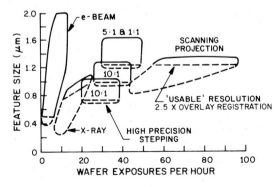

Fig. 25 Resolution capability versus throughput of 125-mm wafers for e-beam, x-ray, and optical projection exposure equipment. *(After Eklund and Landrum, Ref. 39.)*

REFERENCES

[1] M. Hepher, "The Photoresist Story," *J. Photog. Sci.*, **12**, 181 (1964).

[2] M. Hatzakis, "Lithographic Processes in VLSI Circuit Fabrication," in *Scanning Electron Microscopy Meeting*, 1979, Washington, D.C., pt 1, pp. 275-284.

[3] R. A. Colclaser, *Microelectronics: Processing and Device Design*, Wiley, New York, 1980.

[4] W. C. Till and J. T. Luxon, *Integrated Circuits: Materials, Devices, and Fabrication*, Prentice-Hall, Englewood Cliffs, N.J., 1982.

[5] For a review of IC defect analysis see, for example, A. B. Glaser and G. E. Subak-Sharpe, *Integrated Circuit Engineering, Design, Fabrication, and Applications*, Addison-Wesley, Reading, Mass., 1979.

[6] For a discussion of resist processes see, for example, D. J. Elliott, *Integrated Circuit Fabrication Technology*, McGraw-Hill, New York, 1982.

[7] D. A. McGillis and D. L. Fehrs, "Photolithographic Linewidth Control," *IEEE Trans. Electron Devices*, **ED-22**, 471 (1975); D. L. Fehrs, "An Empirical Approach to Projection Lithography," *Proc. Kodak Interface '79*, 135 (1979).

[8] W. M. Bullis and D. Nyyssonen, "Optical Linewidth Measurements on Photomasks and Wafers," in N. Einsprusch (ed.), *Microstructure Science and Engineering*, vol. 2, Academic, New York, 1981.

[9] S. Jensen, G. Hembree, J. Marchiando, and D. Swyt, "Quantitative Sub-Micrometer Linewidth Determination Using Electron Microscopy," *SPIE Semicon. Microlithog.*, **275**, 100 (1981).

[10] R. K. Watts and J. H. Bruning, "A Review of Fine-Line Lithographic Techniques: Present and Future," *Solid State Technol.*, p. 99, May 1981.

[11] M. Long and C. Walker, "Stress Factors in Positive Photoresist," *Proc. Kodak Interface '79*, 125 (1979).

[12] For other reaction possibilities see also D. W. Frey, J. R. Guild, and E. B. Hryhorenko, "Edge Profile and Dimensional Control for Positive Photoresist," *Proc. Kodak Interface '81* (1981).

[13] See, for example, B. T. Lin, "Optical Methods for Fine Line Lithography," in R. Newman (ed.), *Fine Line Lithography*, North-Holland, Amsterdam, 1980.

[14] J. G. Skinner, "Some Relative Merits of Contact, Near-Contact, and Projection Printing," *Proc. Kodak Interface '73*, 53 (1973).

[15] C. N. Ahlquist, W. G. Oldham, and P. Schoen, "A Study of a High-Performance Projection Stepper Lens," *Proc. Kodak Interface '79*, 94 (1979).

[16] M. C. King, "Principles of Optical Lithography," in N. G. Einspruch (ed.), *VLSI Electronics Microstructure Science*, vol. 1, Academic, New York, 1981.

[17] J. D. Cuthbert, "Optical Projection Printing," *Solid State Technol.*, p. 59, Aug. 1977.

[18] K. L. Konnerth and F. H. Dill, "In-Situ Measurement of Dielectric Thickness During Etching or Developing Processes," *IEEE Trans. Electron Devices*, **ED-22**, 452 (1975). See, also, the classic series of papers by F. H. Dill et al., *IEEE Trans. Electron Devices*, **ED-22**, 440-464 (1975).

[19] D. R. Herriott, "Electron-Beam Lithography Machines," in G. R. Brewer (ed.), *Electron-Beam Technology in Microelectronic Fabrication*, Academic, New York, 1980.

[20] P. Shaw, G. Pollack, R. Miller, G. Varnell, W. Lee, R. Loue, S. Wood, and R. Robbins, "E-beam fabrication of 1.25-μm 4K Static Memory," *J. Vac. Sci. Technol.*, **19**, 905 (1981).

[21] M. Isaacson and A. Murray, "In-situ Vaporization of Very Low Molecular Weight Resists Using 1/2 nm Diameter Electron Beams," *J. Vac. Sci. Technol.*, **19**, 1117 (1981).

[22] N. D. Wittels, "Fundamentals of Electron and X-Ray Lithography," in R. Newman (ed.), *Fine Line Lithography*, North-Holland, Amsterdam, 1980.

[23] J. S. Greeneich, "Electron-Beam Processes" in G. R. Brewer (ed.), *Electron-Beam Technology in Microelectronic Fabrication*, Academic, New York, 1980.

[24] E. Kratschmer, "Verification of a Proximity Effect Correction Program in Electron-Beam Lithography," *J. Vac. Sci. Technol.*, **19**, 1264 (1981).

[25] D. Maydan, "X-Ray Lithography for Microfabrication," *J. Vac. Sci. Technol.*, **17**, 1164 (1980).

[26] B. Fay, "X-Ray Techniques and Registration Methods (Micro-Lithography)," in H. Ahmed and W. C. Nixon (eds.), *Microcircuit Engineering*, Cambridge University Press, London, 1980.

[27] G. N. Taylor, "X-Ray Resist Materials," *Solid State Technol.*, p. 73, May 1980.

[28] E. Spiller and R. Feder, "X-Ray Lithography," in H. J. Queisser (ed.), *X-Ray Optics*, Springer-Verlag, New York, 1977.

[29] G. N. Taylor, T. M. Wolf, and J. M. Moran, "Organosilicon Monomers for Plasma-Developed X-Ray Resists," *J. Vac. Sci. Technol.*, **19**, 872 (1981).

[30] R. K. Watts, "X-Ray Lithography," *Solid State Technol.*, p. 68, May 1979.

[31] See, for example, the series of papers by W. D. Buckley et al., *J. Electrochem. Soc.*, **128**, 1106-1120 (1981).

[32] A. Heuberger, H. Betz, and S. Pongratz, "Present Status and Problems of X-Ray Lithography," in J. Truesch (ed.), *Advances in Solid State Physics*, Plenary Lectures of the German Physical Society, March, 1980.

[33] E. Chandross, E. Reichmanis, C. Wilkins, Jr., and R. Hartless, "Photoresists for Deep-UV Lithography," *Solid State Technol.*, p. 81, Aug. 1981.

[34] M. Hatzakis, "Multilayer Resist Systems for Lithography," *Solid State Technol.*, p. 74, Aug. 1981.

[35] B. J. Lin, E. Bassous, V. Chao, and K. Petrillo, "Practicing the Novolac Deep-UV Portable Conformable Masking Technique," *J. Vac. Sci. Technol.*, 19, 1313 (1981).

[36] A. Yoshikawa, O. Ochi, H. Nagai, and Y. Mizushima, "A Novel Inorganic Photoresist Utilizing Ag Photodoping in SeGe Glass Film," *Appl. Phys. Lett.*, **29**, 677 (1976).

[37] K. L. Tai, R. Vadimsky, C. Kemmerer, J. Wagner, V. Lamberti, and A. Timko, "Submicron Optical Lithography Using an Inorganic Resist/Polymer Bilevel Scheme," *J. Vac. Sci. Technol.*, **17**, 1169 (1980).

[38] W. L. Brown, T. Venkatesan, and A. Wagner, "Ion Beam Lithography," *Solid State Technol.*, p. 60, Aug. 1981.

[39] M. H. Eklund and G. Landrum, "1982 Forecast on Processing," *Semiconductor Int.*, p. 43, Jan. 1982.

PROBLEMS

1 Suppose that you are required to specify the resist thickness that will be used in a production lithographic process. The following data is available:

- 1.5-μm minimum features must be printed. Resolution is adequate when the resist thickness t is in the range 0.5 to 2.0 μm but feature size control is better for thinner resists.
- Each wafer has 150 chip sites; each chip has a 0.2-cm^2 active area.
- 5 mask levels are required to complete the device.
- 2000 finished wafers must be produced each day (20 h per day = 3 shifts).
- The resist defect density D_0 increases as the resist is made thinner, where D_0 is the number of defects per square centimeter, and is approximated by $D_0 = 1.4t^{-3}$; t is in micrometers.
- The chip yield (percentage good) can be approximated at each mask level by $y = (1 + qD_0a)^{-1}$, where q is the fraction of defects that render a chip inoperable (fatality rate) and a is the active area of the chip.
- On average, 50% of the defects are fatal defects.
- More time is needed to expose thick resist than to expose thin resist. The exposure tool throughput in wafers/h is approximated by $125 - 50t$ for $(0.5 \leq t \leq 2.0 \ \mu m)$.

    (*a*) Specify the resist thickness to be used and justify your recommendation with tabular and graphical data.

    (*b*) If exposure tools cost $350,000 each, what is the difference in equipment cost for a process using 1 μm and 1.5 μm of resist.

2 Referring to Fig. 10, assume that the amplitude of the incident light wave E_2 is given by

$$E_x(x) = E_2 \sin (wt - kx + \phi)$$

and the amplitude of the reflected wave E_3 is given by

$$E_3(x) = E_2 \sin [wt - k(2d - x) + \phi + \pi]$$

where $k = 2\pi n/\lambda$, n is the real part of the film dielectric constant and is assumed equal for photoresist and SiO_2, and λ is the exposing wavelength.

(a) Referring to Fig. 10c, derive an expression for the standing wave intensity attributable to the interference between E_2 and E_3.

(b) Derive equations that predict the positions of the intensity minima and maxima with respect to the reflecting surface.

(c) Consider a positive photoresist film on 1250 Å of SiO_2 over a silicon substrate. Discuss the effect on the resist image that might result from a SiO_2 thickness change of ±250 Å. Assume $n = 1.6$ and $\lambda = 3200$ Å.

3 In electron beam lithography the term *Gaussian beam diameter* (d_G) describes the diameter of an electron beam in the absence of system aberrations, that is, a beam distorted only by the thermal velocities of the electrons. The current density in a Gaussian beam is given by $J = J_p \exp[-(r/\sigma)^2]$, where J_p is the peak current density, r is the radius from the center of the beam, and σ is the standard deviation of electron distribution in the beam. Defining $d_G = 2\sigma$, derive an expression relating d_G to the peak current density J_p and the total current in the electron beam I.

Answer: $I = (\pi/4)J_p d_G^2$

4 The maximum current density J_m that can be focused toward a spot with a convergence half-angle α is limited by the transverse thermal emission velocities of the electrons in a Gaussian electron beam. J_m is given by the *Langmuir limit equation*

$$J_m = J_c \left[1 + \frac{eV_0}{kT_c}\right] \sin^2 \alpha$$

where J_c is the cathode (source) current density, T_c is the temperature corresponding to the electron energy, k is Boltzmann's constant (1.38×10^{-23} J/°K), and e is the electronic charge (1.6×10^{-19} C). For small convergence angles α, derive an expression that relates the Gaussian beam diameter d_G to the electron source parameters J_c, T_c, and V_0.

Answer: $d_G^2 \geq IkT_c / [(\pi/4)J_c eV_0\alpha^2]$

5 (a) The brightness B of a source of electrons is defined as the current density J emitted per unit solid angle Ω, that is, $B = J/\Omega$. The units of B are amperes per square centimeter per steradian. Assume that the current is emitted from (or converges toward) a small area through a cone of included half-angle α and that α is small. Derive an expression relating the maximum source brightness to the source parameters J_c, T_c, and V_0.

Answer: $B \cong J_c eV_0 / \pi kT_c$

(b) Assuming that brightness is conserved in the electron beam column, show that the Gaussian beam diameter d_G is related to the source brightness.

Answer: $d_G \cong (2/\pi)(1/\alpha)(I/B)^{1/2}$

6 Suppose that an x-ray resist must see a mask modulation greater than or equal to 0.6 in order to form useful resist images. What is the minimum gold thickness required on an x-ray mask to satisfy this requirement if the exposure wavelength is 4 Å?

Answer: $t \geq 0.31$ μm

EIGHT

DRY ETCHING

C. J. MOGAB

8.1 INTRODUCTION

Resist patterns defined by the lithographic techniques described in Chapter 7 are not permanent elements of the final device but only replicas of circuit features. To produce circuit features, these resist patterns must be transferred into the layers comprising the device. One method of transferring the patterns is to selectively remove unmasked portions of a layer, a process generally known as etching.

As the title of this chapter suggests, "dry etching" methods are particularly suitable for VLSI processing. Dry etching is synonymous with plasma-assisted etching[1] which denotes several techniques that use plasmas in the form of low-pressure gaseous discharges. These techniques are commonly used in VLSI processing because of their potential for very-high-fidelity transfer of resist patterns.

The earliest application of plasmas to silicon ICs dates back to the late 1960s, when oxygen plasmas were being explored for the stripping of photoresists.[2] Work on the use of plasmas for etching silicon was also initiated in the late 1960s and was signaled by a patent[3] detailing the use of CF_4-O_2 gas mixtures. At that time, there was no universal endorsement of dry methods which were largely novel replacements for existing wet chemical techniques.

This early work set the stage for an important period in the evolution of IC technology. From 1972 to 1974, workers at several major laboratories were heavily involved in the development of an inorganic passivation layer for MOS devices. The preferred passivation turned out to be a plasma-deposited silicon nitride layer. While this material exhibited many desirable characteristics, there was one immediate difficulty. No suitable wet chemical etchant could be found to etch windows in the nitride in order to expose underlying metallization for subsequent bonding. This problem was circumvented by the use of CF_4-O_2 plasma etching.[4] Concurrently, CF_4-O_2 plasma etching was developed for patterning CVD silicon nitride layers being used as junction seals.[5] These efforts marked the first significant applications of plasma etching in IC manufacture and the beginning of large-scale efforts to develop plasma etching techniques.

Not long after this, an awareness of the potential of plasma techniques for highly anisotropic etching evolved. In particular, there were many observations of a vertical etch rate that greatly exceeded the lateral etch rate when etching through a layer of material. As will become apparent, anisotropy is necessary for high-resolution pattern transfer. The significance of etch anisotropy was recognized by researchers who were hoping to achieve ever larger scales of integration by designing circuits with ever smaller features. By the mid-1970s, therefore, most major IC manufacturers had mounted substantial efforts to develop plasma-assisted etching methods. These methods were no longer seen as merely novel substitutes for wet etching, but rather as techniques having capabilities uniquely suited to meeting forseeable requirements on pattern transfer.

8.2 PATTERN TRANSFER

"Pattern transfer" refers to the transfer of a pattern, defined by a masking layer, into a film or substrate by chemical or physical methods that produce surface relief.

8.2.1 Subtractive and Additive Methods

In the *subtractive* method of pattern transfer shown in Fig. 1a, the film is deposited first, a patterned masking layer is then generated lithographically, and the unmasked portions of the film are removed by etching. In the *additive* (or lift-off) method shown in Fig. 1b, the lithographic mask is generated first, the film is then deposited over the mask and substrate, and those portions of the film over the mask are removed by selectively dissolving the masking layer in an appropriate liquid so that the overlying film is lifted off and removed.

The subtractive methods collectively known as dry etching are the preferred means for pattern transfer in VLSI processing today. The lift-off process is capable of high resolution, but is not as widely applicable as dry etching.

8.2.2 Resolution and Edge Profiles in Subtractive Pattern Transfer

The resolution of an etching process is a measure of the fidelity of pattern transfer, which can be quantified by two parameters. Bias is the difference in lateral dimension between the etched image and the mask image, defined as shown in Fig. 2. Tolerance is a measure of the statistical distribution of bias values that characterizes the lateral uniformity of etching.

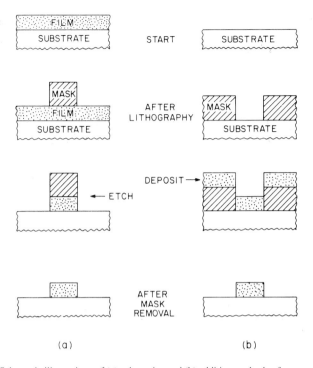

Fig. 1 Schematic illustrations of (a) subtractive and (b) additive methods of pattern transfer.

A zero-bias process produces a vertical edge profile coincident with the edge of the mask, as shown in Fig. 3a. In this case, there is no etching in the lateral direction and the pattern is transferred with perfect fidelity. This case represents the extreme of *anisotropic* etching. When the vertical and lateral etch rates are equal or, more precisely, when the etch rate is independent of direction, the edge profile appears as a quarter-circle after etching has been carried just to completion, as shown in Fig. 3b. In this case of *isotropic* etching, the bias is twice the film thickness.

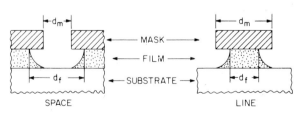

$$BIAS = B = d_f - d_m$$

Fig. 2 Etch bias is a measure of the amount by which the etched film undercuts the mask at the mask-film interface.

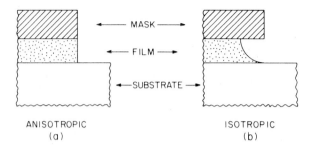

Fig. 3 Ideal etch profiles for (a) fully anisotropic ($A_f = 1$) and (b) isotropic ($A_f = 0$) etching with no mask erosion.

Any edge profile, corresponding to etching just to completion, which lies between the extremes depicted in Fig. 3a and 3b results from an etch rate that is anisotropic. We can define the degree of anisotropy A_f by

$$A_f = 1 - \frac{v_l}{v_v} \tag{1}$$

where v_l and v_v are the lateral and vertical etch rates, respectively. With reference to a feature etched just to completion, Eq. 1 can be written:

$$A_f = 1 - \frac{|B|}{2h_f} \tag{2}$$

where B is the bias and h_f is the film thickness. Thus for isotropic etching $A_f = 0$ while $1 \geq A_f > 0$ represents anisotropic etching. In practice the term "anisotropic etching" is often taken to mean the extreme case, $A_f = 1$ (Fig. 3a).

In early IC fabrication practice, etch bias was usually dealt with by introducing an appropriate amount of compensation in the masking layer. Consider, for example, the etching of a pattern consisting of lines and spaces of equal size. To simplify matters let the film features have a final dimension d_f and the mask features a dimension d_m, as shown in Fig. 4. For a nonzero-bias process, the mask pattern will not consist of lines and spaces of equal size. Instead the mask is compensated, as shown, so that the minimum feature l that must be resolved in the mask is

$$l = d_f - |B| \tag{3}$$

and substituting from Eq. 2

$$l = d_f \left[1 - \frac{2h_f}{d_f} (1 - A_f) \right] \tag{4}$$

Equation 4 shows that the minimum lithographic feature is proportional to the desired feature size with a proportionality factor determined by the degree of etch anisotropy and the aspect ratio of the etched feature. It is apparent from Eq. 4 that as d_f tends to the resolution limit of the lithographic technique employed for generating the masking

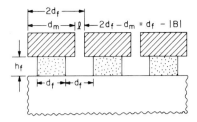

Fig. 4 Mask dimensions have been compensated for etch bias to achieve an etched pattern with equisized lines and spaces. Edge profile is assumed to be vertical for simplicity.

layer, A_f must tend to unity (except in the case $h_f \ll d_f$, which is not of practical interest for VLSI). In other words, as features become smaller, for a fixed or nearly fixed aspect ratio, the margin for compensation diminishes, and a higher degree of etch anisotropy is required. Such is the case for many of the pattern transfer operations needed in the fabrication of VLSI devices.

8.2.3 Selectivity and Feature Size Control

In the previous section we have focused on the etching of the film, and have implicitly treated both the mask and substrate as unetchable. This ideal situation occurs rarely in actual practice, particularly with dry etching. More often, all of the materials exposed to the etchant have a finite etch rate. Thus, a parameter of considerable importance in pattern transfer for VLSI is the *selectivity* of an etching process. Selectivity is defined as the ratio of etch rates between different materials.

Selectivity with respect to the resist mask has an impact on feature size control. Selectivity with respect to the substrate affects performance and yield. The substrate may be the silicon substrate or a film grown or deposited in the fabrication of a previous level of the device.

The selectivity required with respect to the mask is determined by the uniformity of etch rate for both film and mask, the film thickness uniformity, the extent of overetching, the mask edge profile, the anisotropy of etch rate for the mask, and the maximum permissible loss of linewidth in the etched feature. We can quantify these contributions with reference to Fig. 5.

Consider the etching of a film with mean thickness h_f and with a uniformity specified by a dimensionless parameter δ, such that $h_f(1 + \delta)$ is the maximum thickness, and $h_f(1 - \delta)$ is the minimum thickness and $0 \leq \delta \leq 1$. Suppose that the mean etch rate is v_f and the uniformity of etching is such that the etch rate varies spatially over the range $v_f(1 \pm \phi_f)$ where ϕ_f is a dimensionless parameter $(0 \leq \phi_f \leq 1)$. Taking worst-case conditions is the most conservative approach to deriving the selectivity necessary to assure a loss of linewidth due to resist erosion (etching) within permissible limits on any portion of any wafer being etched. This corresponds to using a maximum etch rate for the mask, and assuming that the film etches at the slowest rate where it is thickest. (The etch rate is defined as the vertical

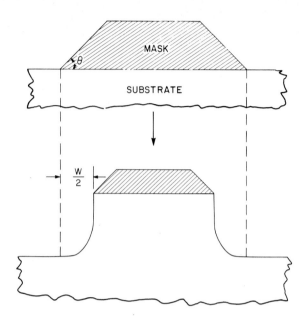

Fig. 5 The evolution of an etched feature when the mask has a finite etch rate. The difference between the intended pattern width and the actual linewidth is W.

depth of etching divided by the time of etching.) In this region of the film the time to complete etching t_c is:

$$t_c = \frac{h_f \ (1 + \delta)}{v_f \ (1 - \phi_f)} \tag{5}$$

If Δ is the fractional overetch time, the time to completion is extended to $t_c (1 + \Delta)$, so that the total etch time t_t is:

$$t_t = \frac{h_f}{v_f} \ \frac{(1 + \delta) \ (1 + \Delta)}{(1 - \phi_f)} \tag{6}$$

During this time, the mask is eroded by etching as shown in Fig. 5. If the mask has *maximum* vertical and lateral etch velocities v_v and v_l, respectively, then the edge of the mask recedes by a maximum amount $W / 2$ given by:

$$\frac{W}{2} = [v_v \ \cot \theta + v_l \ ]t_t \tag{7}$$

where θ is shown in Fig. 5. Substituting for t_t from Eq. 6 we find after rearrangement:

$$W = 2 \ \frac{v_v}{v_f} \ h_f \ \frac{(1 + \delta) \ (1 + \Delta)}{(1 - \phi_f)} \ \left[\cot \theta + \frac{v_l}{v_v} \right] \tag{8}$$

The etch rate of the mask is defined by the vertical etch velocity. In the present case

v_v has been taken as a maximum value, thus providing the most conservative estimate of the selectivity required for a given value of W.

We can define the mask etch rate in terms of a uniformity parameter ϕ_m such that $v_v = v_m(1 + \phi_m)$ where v_m is the mean mask etch rate. Then noting that $v_f / v_v = S_{fm}$ is the desired selectivity of the film with respect to the mask, and $v_l / v_v = 1 - A_m$ where A_m is the degree of etch anisotropy for the mask, Eq. 8 can be rearranged to yield:

$$S_{fm} = 2\frac{h_f}{W} U_{fm}[\cot \theta + (1 - A_m)] \qquad (9)$$

where $U_{fm} = [(1 + \delta)(1 + \Delta)(1 + \phi_m)]/(1 - \phi_f)$ is the "uniformity" factor that accounts for a worst-case coincidence of the various nonuniformities.

It is instructive to consider an example that illustrates the application of Eq. 9. Suppose that etching is carried out using a process that is fully anisotropic for the film ($A_f = 1$). In this case the only linewidth loss results from resist erosion. Further, let us assume that the etch rate uniformity for both film and mask is 10%, that the film thickness uniformity is 5%, and that a 20% overetch is used. Then we have: $\phi_f = \phi_m = 0.1$, $\delta = 0.05$, and $\Delta = 0.2$. Substituting these values in Eq. 9 we find,

$$S_{fm} = 3.08 [\cot \theta + (1 - A_m)] \frac{h_f}{W} \qquad (10)$$

Figure 6 shows a plot of this expression for the particular cases $\theta = 60°$ and $90°$, and for isotropic and fully anisotropic etching of the masking layer. For photoresist masks, the angle θ is determined by the lithographic method (Section 7.3.1) and can be influenced by post-exposure processing. An angle of $60°$ is typical for scanning-type projection printers whereas θ of about $90°$ can be achieved with contact printing. Vertical walled masks ($\theta = 90°$) are also typical of multilevel resist systems (Section 7.6.3). Note that the most favorable case for linewidth control corresponds to aniso-tropic etching of a vertical-wall mask. This ideal is approached when multilevel resists are used in conjunction with reactive ion etching processes (Section 8.4).

The selectivity required with respect to etching of the substrate material can be determined by an approach analogous to the one just used for the mask. Again, taking the conservative, worst-case view that the fastest etching and thinnest portion of the film overlays the fastest etching portion of the substrate, we find:

$$S_{fs} = \frac{h_f}{h_s} U_{fs} \qquad (11)$$

where h_s is the maximum permissible depth of penetration into the substrate and

$$U_{fs} = \left[\frac{\phi_f(2 + \Delta + \Delta\delta) + \delta(2 + \Delta) + \Delta}{(1 - \phi_f^2)} \right] \qquad (12)$$

with ϕ_f, Δ, and δ defined as before.

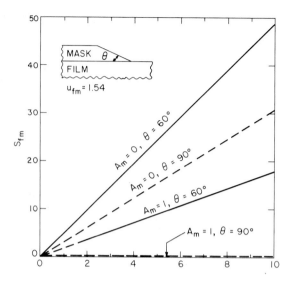

Fig. 6 The selectivity S_{fm} needed with respect to the mask is plotted as a function of the ratio of film thickness to loss of linewidth for various mask profiles, and for the extremes of isotropic and anisotropic etching of the mask.

Obviously, if the film thickness and etch rate were perfectly uniform ($\delta = \phi_f = 0$) and no overetching was required ($\Delta = 0$), selectivity with respect to the substrate would not be a concern. In actuality, this ideal is rarely encountered in VLSI. This is true not only for the obvious reason that perfect uniformity is highly unlikely, but more importantly because (even with perfect uniformity) overetching is required whenever anisotropic etching is coupled with stepped topography. Figure 7 illustrates the need for overetching. If the etching is anisotropic, then overetching (i.e., etching beyond the "endpoint" where the slowest etching region of film has been cleared from the planar surface) is necessary to clear the residual film. From Fig. 7 it can be seen that for fully anisotropic etching ($A_f = 1$), $\Delta = h_1 / h_2$ and this is the minimum possible value of U_{fs}.

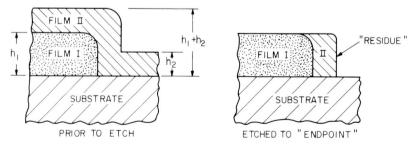

Fig. 7 If etching is anisotropic, overetching is needed to remove residual material at steps. $A_f = 1$ in the example shown.

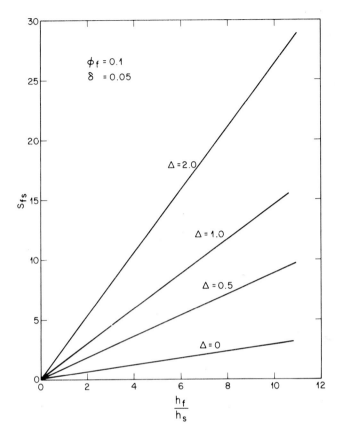

Fig. 8 The selectivity needed with respect to the substrate S_{fs} is plotted as a function of the ratio of film thickness to the amount of substrate removed for various amounts of overetching.

As an example, pertinent to MOSFET fabrication, consider etching a 0.3-μm polysilicon layer that passes over a 0.6-μm field oxide step with a 0.05-μm sublayer gate oxide. For these conditions $\Delta = 2$ (200% overetch!) and the *minimum* selectivity required for anisotropic etching is 2(0.3/0.05) = 12, if the polysilicon film is uniform in thickness, etches uniformly, and etching is complete at the instant when all of the gate oxide has been etched away. The concern with etching beyond this point should be clear. Because the etch is designed to remove polysilicon, presumably relatively rapidly, continued etching after the gate oxide has been removed results in substantial etching of the silicon substrate, causing irreparable damage to the device. Figure 8 illustrates the impact of overetching on selectivity for the particular case $\phi_f = 0.1$ and $\delta = 0.05$.

We have seen that selectivity with respect to the mask is needed to enable feature size control with projection printed resist masks ($\theta < 90°$) and/or when the mask has a finite etch rate in the lateral dimension. Selectivity with respect to the "substrate" is needed to prevent unwanted removal of previously processed portions of the device. Anisotropy is favored for etching fine features, because very little etch bias can be tolerated. However, anisotropic etching in the presence of stepped topography necessitates overetching, which increases the selectivity required.

8.3 LOW-PRESSURE GAS DISCHARGES

Plasma-assisted pattern transfer techniques rely on partially ionized gases consisting of ions, electrons, and neutrons produced by low-pressure ($\sim 10^{-4}$- to 10^{+1}-torr) electric discharges. The generic term "plasma-assisted etching" includes ion milling, sputter etching, reactive ion beam etching, reactive ion etching (also known as reactive sputter etching), and plasma etching. These techniques, which are described in Section 8.4, differ in the specifics of discharge conditions, type of gas, and apparatus; the common thread is the discharge, often referred to simply as the plasma.

8.3.1 Self-Sustained Discharges

When an electric field of sufficient magnitude is applied to a gas, the gas breaks down. The process begins with the release of an electron by some means such as photoionization or field emission. The released electron is accelerated by the applied field and gains kinetic energy, but in the course of its travel through the gas, it loses energy in collisions with gas molecules. There are two types of collisions, elastic and inelastic. Elastic collisions deplete very little of the electron's energy (fractional loss $\sim 10^{-5}$), because of the great mass difference between electrons and molecules. Ultimately the electron energy becomes high enough to excite or ionize a molecule by inelastic collisions. In ionizing collisions the electron loses essentially all of its energy. Ionization frees another electron which is accelerated by the field, and so the process continues. If the applied voltage exceeds the breakdown potential, the gas rapidly becomes ionized throughout its volume.

Electrons released in ionizing collisions and by secondary processes (which will be discussed later) are lost from the plasma by drift and diffusion to the boundaries, by recombination with positive ions, and, in certain electronegative gases, by attachment to neutral molecules to form negative ions. The discharge reaches a self-sustained steady state when electron generation and loss processes balance each other.

Nonionizing, inelastic collisions between electrons and gas molecules or atoms also occur. Two important types of nonionizing collisions are electronic excitation of molecules (or atoms) and molecular fragmentation. Electronically excited molecules and atoms account for much of the luminous glow of the plasma by emitting photons as they relax to lower-lying electronic states. Molecular fragments are often highly reactive atoms and free radicals. A free radical is a molecular fragment having an unpaired electron.

The electron density for the plasmas of interest ranges from 10^9 to 10^{12} cm^{-3}. Considering that the density of gas molecules at 1 torr is about 10^{16} cm^{-3} it can be seen that these discharges are weakly ionized. This results in a gas temperature near ambient, despite a mean electron temperature of about 10^4 to 10^5 K. The relatively low gas temperature permits the use of thermally sensitive materials, such as organic resists, for etch masks.

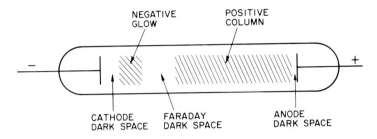

Fig. 9 Schematic view of a dc glow discharge showing the most prominent regions of the discharge.

In summary, the application of an electric field to a gas results in the conversion of electrical energy to potential energy of activated gaseous species such as ions, atoms, and free radicals which can be used to produce etching by physical and chemical interactions with solid surfaces. The energy is transferred by free electrons colliding inelastically with gas molecules.

8.3.2 Methods for Plasma Production

DC discharge The simplest discharge to produce is the glow discharge, in which a dc potential is applied between two metal electrodes in a partially evacuated enclosure. The discharge is visibly nonuniform between the electrodes, and is composed of a series of luminous light and dark zones, shown schematically in Fig. 9.

Positive ions are accelerated toward the negative electrode (cathode) and, on impact, cause ejection of secondary electrons. Additionally, if the ions have sufficient energy, they can produce atom displacement in the cathode as well as sputtering[6]; that is, the ejection of cathode atoms. Secondary electrons are rapidly accelerated away from the cathode causing a space charge of less mobile positive ions to form in the region known as the cathode dark space. The dark space has a relatively low conductivity, because it is depleted of the more mobile electrons, and consequently most of the applied voltage drops across it. When the secondary electrons have been accelerated to a high enough energy, ionization takes place; the point where ionization begins marks the leading edge of the negative glow. The width of the negative glow zone reflects the distance over which the accelerated electrons dissipate their energy through inelastic collisions. Upon leaving this zone, most of the electrons have energies too small to cause further ionization and another relatively dark region (Faraday dark space) is established. Finally, the positive column is reached where electrons and ions have equal densities.

Typically, dc glow discharges operate at pressures exceeding 30×10^{-3} torr and applied voltages exceeding a few hundred volts. A useful variant of this arrangement uses a cathode that is heated to produce copious thermionic emission. This ensures an ample supply of electrons to sustain the plasma and allows operation at lower pressure. At still lower pressures ($\sim 10^{-3}$ torr), the mean free path for electrons exceeds typical dimensions of discharge chambers, and the probability of ionizing collisions is too small to maintain the discharge unless the electrons are confined by an external magnetic field.[7]

The dc glow illustrates three common characteristics of gaseous discharges:

1. Because electrons are much more mobile than ions, positive space charge tends to form adjacent to the negative electrode. In fact, the disparity in mobilities also causes these ''ion sheaths'' to form at any surface immersed in the plasma.
2. The ion sheath is a poor conductor compared to regions of higher electron density; consequently, the largest voltage drops occur across the ion sheaths.
3. The mean electron energy is increased as pressure is reduced or more precisely as the parameter $\mathcal{E}/p$ is increased, where $\mathcal{E}$ is the electric field and p is the pressure. Since the electron mean free path is inversely proportional to p, $\mathcal{E}/p$ is a measure of the energy imparted to an electron by the field between collisions.

AC discharges If a low-frequency alternating field is applied across the electrodes in Fig. 9, their polarity changes every half-cycle so that each electrode alternates as the cathode. The ions and the electrons can both follow the field and establish a glow discharge identical to that of Fig. 9, except for periodic polarity reversal. As the frequency of the applied field is increased, a point is reached where the ions created during breakdown cannot be fully extracted from the gap prior to field reversal. As the frequency is increased further, a large fraction of the electrons have insufficient time to drift to the positive electrode during a half-cycle. These electrons then oscillate in the interelectrode gap and undergo collisions with gas molecules. The lower limit of frequency for oscillations depends on the electron mobility, the electrode spacing, and the amplitude of the applied field. The frequency limit is typically in the rf range.

Three advantages are realized with rf discharges, which make their use widespread. First, electrons can pick up sufficient energy during their oscillations in the gap to cause ionization. The discharge can thus be sustained independent of the yield of secondary electrons from the walls and electrodes. Second, the probability of ionizing collisions is enhanced by electron oscillations allowing operation at pressures as low as $\sim 10^{-3}$ torr. The third advantage is that electrodes within the discharge can be covered with insulating material. This permits sputter etching and reactive sputter etching of insulators, and also eliminates problems due to the build-up of insulating material on metal electrodes that can occur when reactive gases are employed in plasma etching. The mechanism of insulator sputtering in an rf discharge has been discussed at length in the literature.[8]

The potentials that develop at various points in the rf discharge are important in determining the energies of ions incident on surfaces in the plasma.[9] Three potentials pertinent to various etching techniques are labeled in Fig. 10. V_t is the potential at the surface of the rf-powered electrode measured with respect to ground. V_p is the plasma potential with respect to ground, and V_f is the potential (relative to ground) of an electrically floating surface, such as an insulating wall or a substrate isolated from ground by an insulating film. The potentials across the ion sheaths are: $V_p - V_t$ at the rf-powered electrode, $V_p - V_f$ at the floating surface, and V_p at a grounded surface. The potential of the surface with respect to the plasma determines the maximum possible energy of ions bombarding that surface.

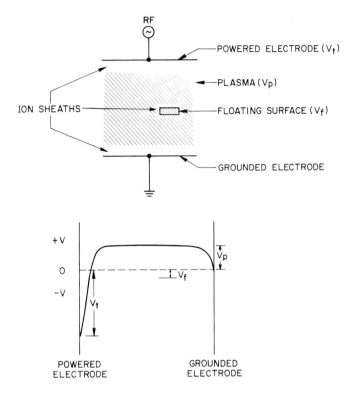

Fig. 10 Schematic view of rf discharge. The potential is shown as a function of position in the discharge for the case where the area of the powered electrode is much less than the area of all grounded surfaces in contact with the discharge.

To a first approximation, the rf coupling across the ion sheaths is capacitive with the area and thickness of a sheath determining the capacitance. For this reason, the ratio R of the area of the rf-powered electrode to the area of all grounded surfaces in contact with the plasma is a key parameter in determining how the applied voltage is distributed among the ion sheaths.[8-11] The potential $V_p - V_t$ increases as R decreases. As a practical consequence, this relationship means that sputter etching, which requires relatively large $V_p - V_t$, is most efficient when R is small and the substrate forms or is attached to the rf-powered electrode (the target).

Under the same conditions the ground-electrode ion sheath has a comparatively small potential drop V_p across it so little or no sputtering occurs there. In a typical diode sputtering system, R is about 0.05 and $V_p - V_t$ can be in the kilovolt range when V_p is less than 100 V. Plasma-etching systems tend to be more symmetric ($R \sim 0.5$) and are operated at higher pressure (usually in the 0.1- to 1.0-torr range). Hence the potentials across the ion sheaths, including the powered electrode, are on the order of V_p. The floating potential, V_f, is usually only a few volts below ground. Therefore, ions bombarding a floating surface do not usually have energies much greater than V_p.

8.3.3 Physical and Chemical Phenomena in Gas Discharges

In dry etching, the plasma serves as a source of species that produce or in some manner catalyze etching. The steady-state constitution of any discharge is governed by the rates of production and loss of the various species.

Production of ions, atoms, and radicals As already noted, electron impact is the primary mechanism of ion production in noble and molecular gas discharges. In molecular gases, ionization may be concurrent with fragmentation, in which case dissociative ionization is said to occur. As examples consider:

Simple ionization:

$$Ar + e \rightarrow Ar^+ + 2e$$
$$O_2 + e \rightarrow O_2^+ + 2e \tag{13}$$

Dissociative ionization:

$$CF_4 + e \rightarrow CF_3^+ + F + 2e \tag{14}$$

Dissociative ionization with attachment:

$$CF_4 + e \rightarrow CF_3^+ + F^- + e \tag{15}$$

Electron impact can also result in molecular dissociation (fragmentation) without ionization, which generally requires less energetic electrons. Most atoms, radicals, and in some cases negative ions are produced by these impact events. As examples,

$$O_2 + e \rightarrow 2O + e$$
$$\rightarrow O + O^- \tag{16}$$
$$CF_3Cl + e \rightarrow CF_3 + Cl + e \tag{17}$$
$$C_2F_6 + e \rightarrow 2CF_3 + e \tag{18}$$

The production of atoms and radicals in molecular gas discharges is essential to etching, because the feed gases themselves are almost always virtually unreactive. As an example, CF_4 is a relatively inert gas that does not react with Si at any temperature up to the melting point ($1412°C$). However, when a discharge is initiated in CF_4, one of the by-products is atomic fluorine which reacts spontaneously with Si at room temperature to form volatile SiF_4. Similarly, O_2 does not attack photoresists significantly at or near room temperature, but the atomic oxygen produced in an O_2 discharge rapidly converts resist to volatile by-products such as CO, CO_2, and H_2O. The rate of production of ions, atoms, and radicals depends on discharge parameters such as pressure, power density, frequency, and feed-gas flow rate. However, exact relationships between discharge parameters and production rates for various species are generally not known.

Loss mechanisms Electrons are lost from a discharge in ways that have already been noted: drift, diffusion, recombination, and attachment. In molecular gases the recombination and attachment events can be dissociative. For example,

Dissociative recombination:

$$e + O_2^+ \rightarrow 2O \tag{19}$$

Dissociative attachment:

$$e + CF_4 \rightarrow CF_3 + F^- \tag{20}$$

Ions can also drift to the electrodes or diffuse to the walls and be lost.

Atoms and radicals can be lost either by homogeneous reactions or by heterogeneous reactions. Homogeneous reactions occur entirely in the gas phase. Heterogeneous reactions take place on surfaces. Which type of loss reaction will dominate for any species depends on many factors such as the pressure, the type of surfaces present (rough, smooth, reactive, nonreactive, etc.), the surface area to volume ratio of the discharge, and the particular gas. For example, two oxygen atoms cannot recombine directly because the energy evolved cannot be dissipated. However, a third body (e.g., an O_2 molecule) can provide the needed energy sink. The rates of such reactions depend strongly on pressure.

Surfaces can serve as reaction sites regardless of the pressure. All surfaces are not equally effective with respect to recombination of reactive species, however. As an example, the recombination of F atoms proceeds much more rapidly on a copper surface than on an oxidized aluminum surface. The materials used in constructing reactive etching systems must be chosen carefully to avoid unwanted heterogeneous reactions.

8.4 PLASMA-ASSISTED ETCHING TECHNIQUES

Plasma-assisted etching can take several different forms. The ion etching techniques, which include sputter etching and ion milling, produce etching solely by physical sputtering (Section 8.3.2). The reactive techniques, which include plasma etching, reactive ion etching, and reactive ion beam etching, rely, to various degrees, on both chemical reactions that form volatile or quasi-volatile compounds and physical effects such as ion bombardment.

The term "plasma etching" is often taken to represent the pure case of chemical reaction, where the plasma serves merely as a source of reactive, neutral species that combine with a solid surface to form a volatile product. There are examples of plasma etching in VLSI technology where this description is essentially accurate. However, physical effects such as ion bombardment often play an important role in plasma etching, much as they do in reactive ion etching. Thus one must be cautious about the implicit assignment of a "mechanism" to a given etch process based on the terminology used to describe that process.

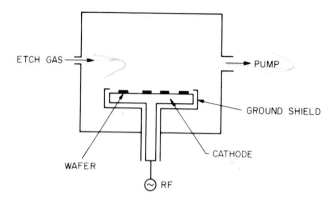

Fig. 11 An rf diode system for reactive ion etching. The cathode is the powered electrode, while internal grounded parts of the system serve as the anode. Note that the area of the cathode is much smaller than the area of the anode. The plasma is unconfined and fills the entire chamber. The ground shield prevents sputtering of the enclosed portions of the powered electrode.

It is preferable to distinguish these techniques on an operational rather than a mechanistic basis. That is, each can be said to occupy a different portion of the operating parameter space.

8.4.1 Sputter Etching and Ion Beam Milling

Both sputter etching and ion-beam milling use high-energy ($\gtrsim$ 500-eV) noble gas ions, such as Ar^+, derived from a discharge. Sputter etching is accomplished most simply in an rf diode system shown schematically in Fig. 11. The material to be etched is clamped to the powered electrode and bombarded by ions drawn from the plasma. Recall from Section 8.3.2 that if the ratio of cathode surface to grounded surface is small enough, most of the voltage drop occurs across the ion sheath at the cathode. The direction of the electric field in the sheath region is normal to the cathode surface so that, at typical operating pressures ($\sim 10^{-2}$ to 10^{-1} torr), ions arrive predominantly at normal incidence and the degree of etch anisotropy is inherently high.

In ion beam milling, the ion source is usually a magnetically confined dc discharge that is physically separated from the substrate by a set of grids. The grids are biased so as to extract an ion beam (typically Ar^+) from the source as shown in Fig. 12. Ion voltages (energies) exceeding 500 V are required[7] for practical beam current densities ($\lesssim$ 1 mA/cm^2). Usually the beam is well collimated, so that the angle of incidence can be controlled by tilting the substrate holder. A hot filament emitter is placed in the beam path to provide low energy electrons for beam neutralization.

Although both sputter etching and ion beam milling have the potential for high resolution, they are not used to any significant extent in VLSI technology. The main reason for this is that selectivity is insufficient.

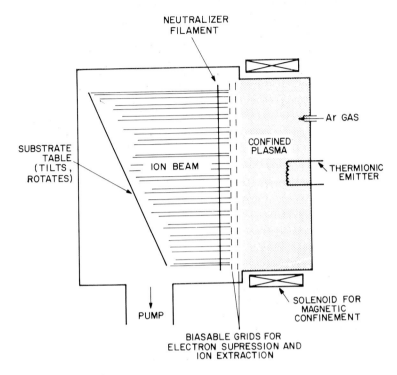

Fig. 12 An ion milling system.

8.4.2 Plasma Etching

Molecular gases containing one or more halogen atoms are used for plasma etching silicon, silicon compounds, and certain metals. These gases are selected because the fragments they produce in a plasma react with the materials of interest to form volatile compounds at temperatures low enough to be appropriate for pattern transfer.

Parallel-plate systems, such as shown in Fig. 13, are used for high-resolution etching.[12] Such systems have several distinguishing characteristics. First, the electrodes are nearly symmetric (ratio of powered to grounded surfaces tends to be much nearer to unity than for sputter etching or reactive ion etching systems). The degree of plasma confinement is relatively high, brought about by electrodes which are closely spaced and have lateral dimensions nearly equal to those of the vacuum enclosure. Plasma confinement tends to increase the plasma potential. The other distinguishing characteristics are that the material to be etched is placed on the grounded electrode and the operating pressure is relatively high, ranging from 10^{-1} to 10^{+1} torr.

The possibility of a high plasma potential in plasma etching systems must not be overlooked.[9] Since the substrates are either grounded or floating, the energy of ions incident on them can be as high or slightly higher than the plasma potential, and can reach several hundred volts under certain conditions, despite the high operating pressures. When high plasma potentials prevail, the surface reactions involved in the

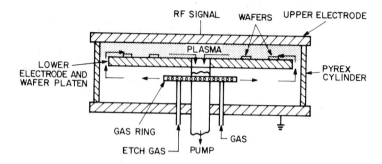

Fig. 13 A parallel-plate plasma etching system. The plasma is largely confined to the region between the powered electrode and the grounded wafer platen. Confinement tends to increase the plasma potential.

etching can be strongly influenced by ion bombardment as discussed in Section 8.5.1. Plasma etching then becomes mechanistically indistinguishable from reactive ion etching (Section 8.4.3).

Generally only a mechanical pump is needed for plasma etching. Two-stage, oil-sealed rotary pumps are common, with pumping speeds ranging up to 1500 L/min. Corrosive and/or toxic gases can be formed in the discharge (e.g., CO, COF_2, $COCL_2$, F_2, and Cl_2) even with relatively inert feed gases, so good safety practice must be adhered to in venting the pump, changing the pump oil, and routine pump maintenance.[13]

Pressure and feed-gas flow rate should be controlled independently; hence a throttle valve is required to regulate pumping speed. Flow rate typically ranges from 50 to 500 sccm (standard cm^3/min, i.e., cm^3/min at standard temperature and pressure).

RF power is most often delivered to the plasma through an impedance matching network at 13.56 MHz. (This "ISM" frequency is allotted by the Federal Communications Commission for industrial, scientific, and medical use.) Recent work on the influence of frequency, however, has revealed the importance of this parameter in determining ion energy and has prompted some departures from this standard operating point.[14] The matching network is used to match the plasma impedance to the output impedance of the rf generator, thus assuring efficient power transfer.

8.4.3 Reactive Ion Etching and Reactive Ion Beam Etching

Reactive ion etching (RIE), also known as reactive sputter etching (RSE), employs apparatus similar to that for sputter etching (Fig. 11). However, in RIE the noble gas plasma is replaced by a molecular gas discharge generated in gases identical to those used for plasma etching. The distinguishing operating conditions are: (1) asymmetric electrodes (i.e., ratio of cathode area to grounded surface area much less than 1); (2) substrates placed on the powered electrode; and (3) relatively low operating pressures ranging from about 10^{-3} to 10^{-1} torr. Each condition contributes to providing relatively high-energy ions at the substrate surface during etching (Section 8.3).

The lower operating pressures used in RIE necessitate the use of more complex vacuum pumps, and lower feed-gas flow rates (~10 to 100 sccm). In other respects these systems are similar to parallel-plate plasma etchers.

Reactive ion beam etching is the newest of the reactive plasma techniques.[15] The equipment and operating parameters are similar to those used in ion milling (Section 8.4.1 and Fig. 12). However, molecular gases identical to those used in plasma and reactive ion etching replace the noble gases in the ion source.

Although initial results indicate a very high degree of etch anisotropy ($A_f \cong 1$) is obtainable, reported selectivities are poor. In light of the considerably greater equipment complexity and potential drawbacks in connection with ion source maintenance, reactive ion beam etching is unlikely to become a preferred method for VLSI pattern transfer in the near future.

8.5 CONTROL OF ETCH RATE AND SELECTIVITY

The importance of adequate selectivity in dry etching is discussed in Section 8.2.4. The etch rate for a given process must be sufficiently reproducible and high enough to assure its utility for VLSI manufacturing practice. In this section we consider the major factors governing etch rate and selectivity.

8.5.1 Ion Energy and Angle of Incidence

The influence of ion energy and angle of incidence on sputtering yield, defined as the number of ejected atoms per incident ion, is of interest, because sputtering and related effects take place in the reactive plasmas most often used in VLSI. The ion energy must exceed a threshold value of about 20 eV for sputtering to occur at all, and should be much higher than this (several hundred eV) to obtain practical sputter etch rates. The sputtering yield for most materials increases monotonically with ion energy in the energy range characteristic of dry etching (ion energy $\leq$ 2 keV), although for energies exceeding ~300 eV the rate of increase diminishes. Typical sputtering yields for VLSI materials with 500-eV Ar^+ range from ~0.5 to 1.5. Consequently, selectivity is inherently poor for ion etching.[1, 16]

Sputtering yield is sensitive to the angle at which ions impinge on the surface. Ions arriving with oblique angles of incidence have a higher probability of producing a substrate atom with velocity vectored away from the surface. In addition, these ions tend to transfer more of their energy to atoms near the surface which have a higher probability of escape.

Ions from the plasma collide with surfaces in both plasma and reactive ion etching. Sputtering can and does result, but under usual operating conditions produces only a small contribution to the etch rate. Of much greater importance is the effect that impacting ions can have on chemical reactions occurring at the surface. These ion-assisted reactions are currently under intensive study. A growing body of experimental evidence indicates that ion-assisted reactions between neutral etchant species derived from the plasma and solid surfaces play a dominant role in many of the dry etching processes developed for VLSI.

Figure 14 shows an example of an ion-assisted reaction. In this case separate beams of Ar^+(450 eV) and XeF_2 were incident on a Si surface. The etch rates for

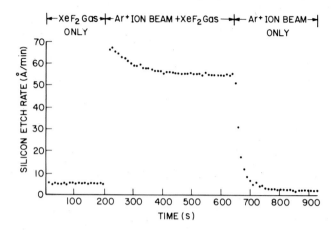

Fig. 14 An ion-enhanced reaction. The rate of reaction between XeF$_2$ and Si is increased dramatically when a 450-eV Ar$^+$ beam irradiates the Si surface. The Ar$^+$ beam alone sputter etches the Si at a much lower rate. *(After Coburn and Winters, Ref. 18.)*

each beam were measured separately and found to be relatively low. The Ar$^+$ produces etching by physical sputtering; the XeF$_2$ molecules cause etching by dissociating on the surface to Xe, which simply desorbs, and two F atoms. The F atoms then react spontaneously with Si to form volatile silicon fluorides. The measured etch rate for both beams incident simultaneously was much higher (about eight times) than the sum of the individual rates, indicating a synergistic effect. [17]

Figure 15 shows another example of an ion-assisted reaction, Ar$^+$ and Cl$_2$ on Si.[17] Unlike F atoms, Cl$_2$ does not spontaneously etch Si, yet when an Ar$^+$ beam is simultaneously incident on the surface, Si is etched with a gas, SiCl$_4$, as the byproduct. The etch rate measurements shown in Figs. 14 and 15 were made by detecting the change in mass of a Si film with a very sensitive quartz crystal microbalance. The brief transient seen in Fig. 15 when the Cl$_2$ gas is admitted (at ~220 s) corresponds to an increase in mass due to initial adsorption of chlorine on Si.

A number of mechanisms could account for the influence of ion bombardment on reaction rates[18]:

1. Ion bombardment creates damage or defects on the surface which catalyze chemisorption or reaction.
2. Ion bombardment directly dissociates reactant molecules (e.g., XeF$_2$ or Cl$_2$).
3. Ion bombardment removes involatile residues that would otherwise retard etching.

The relative importance of these mechanisms and alternatives are still the subject of study, speculation, and some controversy. For our purposes, it is sufficient to realize that bombardment by energetic ions causes physical processes such as lattice damage, thermal spikes, and molecular dissociation that can greatly enhance or even enable chemical reactions between neutral etchant species and solid surfaces. In the first case (XeF$_2$ + Si) the solid can be etched spontaneously (i.e., in the absence of

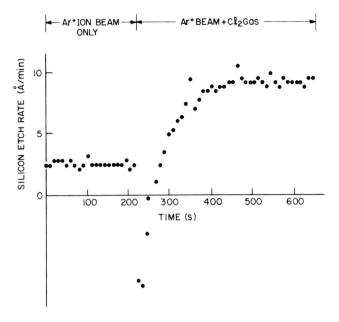

Fig. 15 An ion-induced reaction. Cl_2 does not spontaneously etch Si, but etching occurs at a rate much greater than can be accounted for by sputter etching when the Si surface is irradiated with a 450-eV Ar^+ beam during exposure to Cl_2. *(After Coburn and Winters, Ref. 18.)*

ion bombardment) by the etchant species (F atoms), but the overall rate of reaction is accelerated by energetic ion bombardment. In the second case (Cl_2 + Si) no etching occurs unless there is energetic ion bombardment. We shall refer to the former case as an *ion-enhanced* reaction, and the latter as an *ion-induced* reaction.

The examples given show that the effect of the ion beam is related to physical processes, since no mechanism exists for a chemical contribution by noble gas ions. For 1-keV ions Ar^+ is more effective than Ne^+ which is more effective than He^+ in assisting the XeF_2 + Si and Cl_2 + Si reactions.[19, 20] The effectiveness of heavier ions also suggests that physical processes related to momentum transfer are important. But what about etching in plasmas such as CF_4 and related gases used in dry etching? In these instances the ions themselves contain potential reactants (e.g., CF_3^+). Work on etching Si with XeF_2 under simultaneous ion bombardment reveals essentially no change in the etch rate when CF_3^+ replaces Ar^+ as the bombarding ion.[20] Thus energetic ions can enhance or induce reactions through physical processes irrespective of the chemical identity of the ions. In fact, the high etch rates often obtained in dry etching would be difficult to reconcile with the relatively low ion fluxes arriving at surfaces immersed in these low-ion-density plasmas, if the etching were attributed mainly to reaction with ions.

The picture that emerges for etching in reactive plasmas is that reactants, mainly neutrals, are generated in the plasma, adsorb on the surface, and react to form products that subsequently desorb, with the overall reaction possibly being initiated and/or accelerated by energetic ions extracted from the plasma. Of course, the extent to which the ions increase reaction rates depends on the specific gases, materials, and operating parameters chosen.

8.5.2 Feed-Gas Composition

Gas composition is a dominant factor in determining etch rate and selectivity for plasma and reactive ion etching. Table 1 lists some representative gases together with materials reported to be etched by plasmas generated in these gases. Halogen-containing gases have been used almost exclusively for etching in VLSI except for photoresist removal and the patterning of organic layers where O_2 plasmas have been employed. The choice of these gases reflects the fact that the formation of volatile or quasi-volatile halide compounds from the inorganic materials used in VLSI is both thermodynamically and kinetically possible at or near room temperature. The preponderance of halocarbons seen in Table 1 results because they are relatively easy to handle and have minimal operating hazards.

Multicomponent mixtures are frequently used for reactive etching. These mixtures usually take the form of a major component plus one or more additives, which are introduced to produce a desired effect in connection with etch rate, selectivity, uniformity, or edge profile.[21] An example of additive effects on etch rate is the plasma etching of Si and SiO_2 with CF_4-containing mixtures.

The etch rates of Si and SiO_2 in a CF_4 plasma are relatively low. If O_2 is added to the feed gas, the etch rates of Si and SiO_2 increase dramatically, as seen in Fig. 16. A maximum etch rate[22] is reached at about 12% O_2 for Si and 20% O_2 for SiO_2. The etch rates decrease with continued addition of O_2, more rapidly for Si than for SiO_2 (Fig. 16). These effects can be explained by considering the plasma and surface chemistry involved. F atoms are formed by electron impact dissociation of CF_4 and consumed by combination with CF_x radicals ($x \leqslant 3$). The rates of these processes in a pure CF_4 plasma are such that the steady-state concentration of F atoms is low, and since F atoms are the etchant species the etch rates are also low. Added oxygen results in depletion of CF_x radicals by formation of COF_2, CO, and CO_2, which reduces the consumption of F atoms. The net result is an increase in the F atom concentration, up to about 23% O_2, and a corresponding increase in etch rates. Ultimately the F atom concentration decreases because of dilution.

Table 1 Some gases used in dry etching for VLSI

Material	Gases
Si	CF_4, CF_4+O_2, SF_6, SF_6+O_2, NF_3 Cl_2, CCl_4, CCl_3F, CCl_2F_2, $CClF_3$
SiO_2, Si_3N_4	CF_4, CF_4+H_2, C_2F_6, C_3F_8, CHF_3
Al, Al−Si, Al−Cu	CCl_4, CCl_4+Cl_2, $SiCl_4$, BCl_3, BCl_3+Cl_2

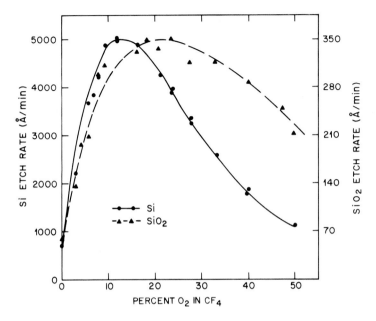

Fig. 16 The addition of O_2 to a CF_4 plasma affects the rate of etching of both Si and SiO_2.

If oxygen additions affected only the plasma chemistry, one would expect the maximum etch rates for both Si and SiO_2 to occur at the O_2 concentration which produces a maximum F atom density. However, as can be seen in Fig. 16, the maxima for Si and SiO_2 are not coincident because oxygen is also involved in the surface chemistry. In the case of Si etching, oxygen tends to chemisorb on the surface, thereby partially blocking direct access by F atoms. Since this effect increases as more oxygen is added, the maximum etch rate for Si occurs at an oxygen concentration much less than 23%. A similar effect is absent for etching SiO_2, because the surface is, in effect, covered with oxygen to begin with. Thus the maximum etch rate for SiO_2 occurs near the oxygen addition producing a maximum F atom concentration. Oxygen chemisorption also accounts for the more rapid decrease of etch rate for Si beyond the maximum in Fig. 16. F atoms react much more rapidly with Si than with SiO_2, so CF_4-O_2 plasmas offer high selectivity of Si over SiO_2.

If H_2 is added to a CF_4 plasma, quite different effects are noted.[23, 24] In reactive ion etching, at relatively low pressure, the etch rate of SiO_2 is nearly constant for H_2 additions up to about 40% while the etch rate for Si decreases monotonically to a value near zero at $\geq 40\%$ H_2 as seen in Fig. 17. H_2 in amounts exceeding $\sim40\%$ causes unwanted polymer formation on the SiO_2. (See Section 8.7.1 for a discussion of polymer deposition.) Selectivities for SiO_2:Si of 40:1 are possible with CF_4-H_2 reactive ion etching. In plasma etching, at higher pressure (~1 torr), the addition of H_2 can both increase the etch rate of SiO_2 and decrease the etch rate of Si. Again, the selectivity for SiO_2:Si can be controlled by adjusting the H_2 content of the feed gas.

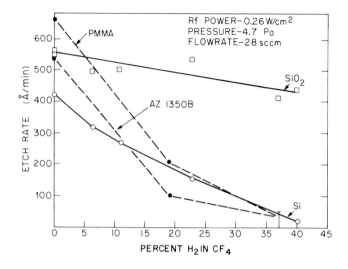

Fig. 17 The addition of H_2 to CF_4 results in a rapid decrease of the etch rate of Si relative to SiO_2. Additions exceeding $\approx 40\%$ cause polymer deposition and cessation of etching. *(After Ephrath, Ref. 23).*

A generally accepted explanation for these observations is as follows. CF_x ($x \leq 3$) radicals etch SiO_2 by an ion-induced reaction, probably involving dissociative chemisorption, which ultimately results in formation of SiF_4. The C derived from these radicals is removed from the surface by combining with oxygen from the SiO_2 to form CO, CO_2, and possibly COF_2 gases. A similar reaction path with Si is unavailable, because there is no way to remove the adsorbed C, which blocks etching (i.e., blocks access to surface sites for F). The role of hydrogen then is twofold. It combines with F atoms to form stable HF thus removing a potential Si etchant, and, particularly at higher pressures, it changes the plasma chemistry so that higher concentrations of etchant CF_x are produced. The overall effect can be depicted schematically as:

$$CF_4 + e \rightarrow CF_x + F + e \tag{21}$$

$$\frac{1}{2} H_2 + F \rightarrow HF \tag{22}$$

and

$$CF_x + SiO_2 \rightarrow SiF_4 + (CO, CO_2, COF_2) \tag{23}$$

$$CF_x + Si \rightarrow C \text{ adsorbed on Si} \tag{24}$$

To summarize, etch rates and selectivities for Si and SiO_2 can be controlled in CF_4 reactive plasmas by the addition of oxidizing or reducing components to the feed gas. With oxidant additions, etching of Si is favored relative to SiO_2, while reducing agents favor the inverse selectivity.

8.5.3 Pressure, Power Density, and Frequency

Pressure, power density, and frequency are independent parameters, but in practice, the individual contributions each makes to an etching process are sometimes difficult to unravel or predict. However, certain general trends are evident.

Lowering pressure and/or frequency, and increasing power density, increases the mean electron energy and the energy of ions incident on surfaces. An increase in power also increases the density of radicals and ions in the plasma. Thus, if etching is ion-assisted, a decrease in pressure or frequency or an increase in power favors etch rate anisotropy.

In general, etch rates increase monotonically with power, although at a diminishing rate. Essentially all the applied power is ultimately dissipated as heat, so that at very high power densities, substrates require heat sinking to avoid deleterious effects such as photoresist flow and charring or loss of selectivity. Recently, there has been much interest in very-high-rate etching for single-wafer etching systems[25] that operate at relatively high pressure and very high power density (several W/cm^2).

Very few studies of the variation of etch rate with frequency have been reported, but it is clear that a main influence of frequency is in its effect on ion energy.[14, 26] This is most apparent from observations of etch rate anisotropy that are discussed in Section 8.6. Frequency has been explored in the range from about 10 kHz to 30 MHz.

8.5.4 Flow Rate

The flow rate of the feed gas determines the maximum possible supply of reactant. The actual supply depends on the balance between generation and loss of active species in the plasma, as already discussed. One mechanism by which etchant species are lost is convective flow. The rate of loss is inversely proportional to the residence time t_r given by:

$$t_r = \frac{Vp}{760F} \qquad (25)$$

where p is the pressure in torr and V and F are the plasma volume and flow rate, respectively, in consistent units. Residence time is a measure of the mean time a molecule spends in the plasma.

Under usual operating conditions, flow rate has only a small influence on etch rate. More pronounced effects are seen at the extremes, where either flow rate is so small that etch rate is limited by the available supply of reactant or flow rate is so high that convection becomes a major pathway for loss of active species. Whether or not convective losses are observed depends on the available pumping speed, the particular gas, and the materials within the reactor. If the active species have an inherently short lifetime due to other loss processes, then flow rate effects may not be encountered. This is usually the case when the etchant species is atomic chlorine, for example. Etch rates can be affected when species with longer lifetimes, such as F atoms, are the etchant. Figure 18 shows[27] the reciprocal of etch rate plotted against flow rate for CF_4-O_2 etching of SiO_2 and Si_3N_4. The linear dependence shown is consistent with the inverse dependence of residence time on flow rate.

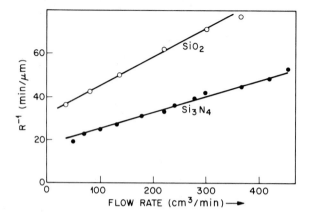

Fig. 18 The reciprocal of etch rate R is linear in flow rate for etching in a $CF_4-4\%$ O_2 plasma at 1.0 torr. This dependence indicates that the lifetime of the active species is determined by convective losses. *(After Kalter and Van deVen, Ref. 27.)*

8.5.5 Temperature

Temperature influences etch rate in reactive etching primarily through its effect on the rates of chemical reactions. An Arrhenius dependence [etch rate $\sim \exp(-Q/kT)$ where Q is the activation energy, T is the absolute temperature of the substrate, and k is Boltzmann's constant] usually prevails with relatively small values of activation energy ($Q \leq 0.5$ eV/mole), although exceptions have been noted where the rate decreases with temperature.[28] The rate decrease may be due to an increase in the rate of thermal desorption of etchant species from the surface. Selectivity can also be affected by temperature because the activation energy is material-dependent.

Some means of controlling substrate temperature is desirable for obtaining uniform and reproducible etch rates. Heating by the plasma is a major source of temperature rise in thermally isolated wafers. In addition, the heat generated from exothermic reactions that produce etching can be appreciable. Figure 19 illustrates the effect of the heat of reaction on wafer surface temperature during the stripping of photoresist in a barrel reactor,[1] where wafers are relatively isolated thermally. The maxima in the curves correspond to the times when resist stripping is completed. The shift in the positions of the maxima with wafer load is due to the loading effect discussed in the next section.

8.5.6 The Loading Effect

In reactive etching the etch rate is sometimes found to decrease as the amount of etchable surface area is increased. This phenomenon is known as the loading effect. Loading occurs when the active etch species reacts rapidly with the material being etched, but the species has a long lifetime in the absence of etchable material. Etching is then the primary loss mechanism for the species, so the greater the area of material, the more etchant species are consumed. The generation rate of the active

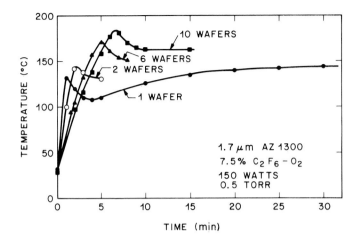

Fig. 19 The variation of wafer temperature with time during plasma stripping of a photoresist layer. The four sets of data points are for separate runs with only the number of wafers varied. The maxima correspond closely to the endpoints of stripping.

species is fixed by operating parameters (pressure, power, frequency, etc.) and is largely independent of the amount of etchable material present. Thus, the average concentration of active species, as determined by the difference between the rates of generation and loss, decreases as the etchable surface area increases.

The dependence of etch rate R on etchable surface area Φ in the simplest case of a single etchant species takes the form[29]

$$R = \frac{\beta \tau G}{1 + K\beta\tau\Phi} \tag{26}$$

where β is a reaction rate constant, τ is the lifetime of the active species in the absence of etchable material, G is the generation rate of active species, and K is a constant for a given material and reactor geometry.

Equation 26 indicates that no noticeable loading effect will occur so long as $K\beta\tau\Phi \ll 1$. This condition can be met by employing plasmas in which the inherent lifetime (τ) of the active species is very small; that is, where loss mechanisms other than etching dominate.

Figure 20 shows an example[30] of the loading effect for etching polysilicon in $CF_4 - O_2$. Notice that the etch rate is independent of area over the range studied at 40°C, but that a strong effect is seen at 140°C. This can be attributed to the Arrhenius temperature dependence of the rate constant β in Eq. 26. The effect would appear at 40°C, of course, if still larger areas were exposed.

The most serious concern caused by the loading effect is with feature size control when lateral etching occurs. As the endpoint of etching is reached, the surface area decreases very rapidly and any overetching is carried out at a higher rate than nominal. This makes linewidth control extremely difficult, since, in effect, accelerated lateral etching occurs on clearing.

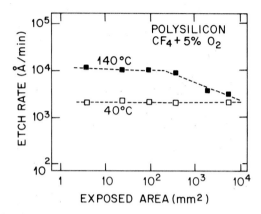

Fig. 20 The loading effect. The etch rate of Si in a $CF_4-5\%$ O_2 plasma decreases when the surface area of Si is increased sufficiently and/or the reactivity is increased sufficiently by increasing temperature. For reference, a single 100 mm-diameter wafer has an area of approximately 7.8×10^3 mm². *(After Enomoto et al., Ref. 30.)*

The loading effect is macroscopic in the sense that the presence of one wafer in a reactor influences the etch rate at a second wafer in another part of the reactor. This implies that transport processes in the plasma are rapid enough that no appreciable concentration gradients can exist for the etchant species within the bulk of the plasma.

Microscopic loading effects have also been observed, wherein the size and density of features being etched can influence the etch rate.[31] These effects result from localized concentration gradients of etchant species, which are caused by differing rates of reaction with mask and substrate materials. For example, the material near the edge of a masking feature may etch more or less rapidly than the same material further removed from the edge.

8.6 CONTROL OF EDGE PROFILE

Considerations relevant to feature size control were covered in Sections 8.2.2 and 8.2.3. This section deals with some effects that influence etch profiles and methods for controlling them.

8.6.1 Mechanisms for Anisotropy in Reactive Etching

When etching occurs by an ion-assisted reaction, etch rate anisotropy can be expected, because ions are incident normal to the wafer surface. (Normal incidence occurs in the usual operating modes for reactive etching, with wafers having surface topography with vertical dimensions much less than typical ion sheath thicknesses.) Consequently, the bottom surface of an etching feature receives a much greater flux of energetic ions than the sidewalls, as shown in Fig. 21.

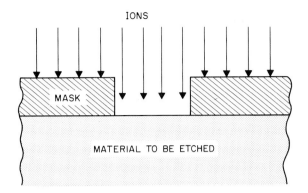

Fig. 21 The sidewalls of etched features are not subject to energetic ion bombardment under typical conditions, since the ions arrive predominantly at normal incidence.

If the etching reaction is ion-induced (Section 8.5.1), there will be no lateral etching; whereas if it is ion-enhanced, the mask will be undercut by an amount determined by the spontaneous reaction rate. Figure 22 shows a hypothetical example in which the degree of anisotropy obtained for the ion-enhanced reaction depends on ion energy. For a given gas, material, power density, and frequency the reactive ion etching mode usually provides higher ion energies than the plasma etching mode so that a higher degree of anisotropy results. Etching of polysilicon in Cl_2 plasmas illustrates the effect of ion energy. When this is done in the RIE mode, nearly ideal anisotropy $(A_f = 1)$ results,[28] whereas plasma etching results in considerable undercutting.[31]

A useful approach to minimizing lateral etching with ion-enhanced reactions is to incorporate a gas additive that provides a recombinant species.[31] The function of the recombinant is either to combine with etchant species on surfaces to produce a volatile product, or to serve as a precursor for the formation of a passivating film. The detailed mechanisms underlying this approach are not well understood. However, it is suspected that ion bombardment not only enhances the rate of reaction between etchant and substrate, but also stimulates desorption of recombinant species, thereby reducing their concentration on ion-bombarded surfaces. An example will illustrate these ideas.

When Si is plasma-etched with Cl_2, undercutting occurs. However, as C_2F_6 is added to the feed gas, the degree of undercut diminishes until at sufficiently high C_2F_6 concentrations ($\geq 85\%$), lateral etching is virtually absent. This effect can be accounted for by the reactions:

Generation of etchant species:

$$e + Cl_2 \quad \rightarrow \quad 2Cl + e \qquad (27)$$

Ion-enhanced reaction:

$$Si + xCl \quad \rightarrow \quad SiCl_x \qquad (28)$$

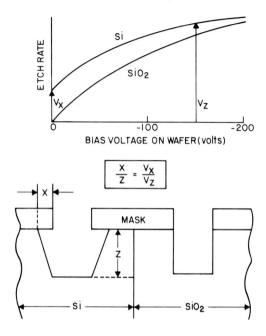

Fig. 22 A hypothetical example illustrates the variation of etch rate and etch profile for ion-enhanced (the curve labeled Si) and ion-induced (the curve labeled SiO_2) reactions. The ion energy is assumed to correspond to the value indicated by the arrow labeled V_z. V_x and V_z are, respectively, the lateral and vertical etch velocities. *(After Coburn and Winters, Ref. 18.)*

Generation of recombinant species:

$$e + C_2F_6 \quad \rightarrow \quad 2CF_3 + e \tag{29}$$

Recombination:

$$CF_3 + Cl \quad \rightarrow \quad CF_3Cl \tag{30}$$

The last reaction is likely to be suppressed by ion bombardment because of dissociation or desorption of CF_3 and/or dissociation of CF_3Cl. Thus, if the gas composition is adjusted properly, one can obtain a situation where the rate of etching exceeds the rate of recombination on ion-bombarded surfaces, whereas the reverse is true on sidewalls where ion bombardment is minimal. The degree of anisotropy can be controlled simply by adjusting the feed-gas composition.[31]

8.6.2 Other Effects Influencing Edge Profile

Faceting, trenching, and redeposition are three effects that arise from physical sputtering and can influence edge profiles in reactive etching. The extent of their influence depends on sputter yield and ion flux so they often can be completely suppressed. These effects tend to be more prevalent with reactive ion etching than with plasma etching, because of the higher ion energies involved.

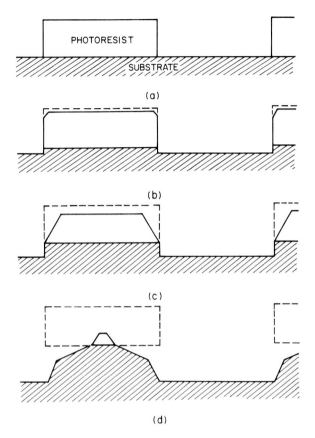

Fig. 23 Faceting results from the dependence of sputter etch rate on the angle of incidence of ions striking the surface. (a) Prior to etching, (b) initiation of the facet, (c) facet intersects substrate surface, (d) substrate is exposed and forms its own facet.

Figure 23 shows the phenomenon of faceting. The sputter etch rate depends on the angle of incidence that arriving ions make with the surface and, for most materials, is a maximum for angles off normal incidence. The facet is inclined to the incident ions at an angle corresponding to the maximum etch rate. The facet does not affect the edge profile unless etching proceeds long enough for it to intersect the surface (Fig. 23d).

Trenching, depicted in Fig. 24, results mainly from an enhanced ion flux at the base of a step due to ion reflection off the side of the step. The etch rate resulting both from physical sputtering and any ion-assisted reactions increases at the location of the trench because of the greater ion flux there.

Physically sputtered material that has not been converted to volatile products condenses on any surface it encounters. Sputtered material is ejected from the surface with approximately a cosine distribution, and therefore a significant fraction can redeposit on the walls of adjacent masking features. This redeposition changes the edge profile and the linewidth. Redeposition is not usually a problem in reactive

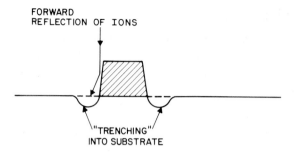

Fig. 24 Trench formation arises from an ''excess'' flux of ions resulting from reflection off the sidewall.

etching, because it can be avoided by choosing feed gas, plasma parameters, and masking materials so that only volatile products form.

8.6.3 Endpoint Determination

When lateral etching occurs, linewidth and edge profile can be controlled, to a certain extent, by minimizing the amount of overetching. As noted in Section 8.2.4, overetching is almost always required to compensate for nonuniformities, and for patterning stepped surfaces when $A_f > 0$. Various methods of detecting the endpoint of etching have been used.[32, 33] They include: (1) direct visual observation of the etched layer; (2) monitoring of optical reflections from the etched layer; (3) detection of changes in the concentration of etchant species in the plasma by emission spectroscopy; (4) detection of etch products by emission spectroscopy or mass spectrometry; and (5) detection of changes in plasma impedance. Methods (1) and (2) are independent of the area of material being etched, but are not suited to dealing with nonuniform batch etching. Methods (3), (4), and (5) require a minimum area of material, determined by etch rate and detector sensitivity, and tend to average over nonuniformities. In addition, method (3) requires a loading effect.

Endpoint detection, used with etching processes that have little or no anisotropy, is seldom a suitable substitute for highly anisotropic and selective etching in VLSI applications with stringent requirements for linewidth control. However, endpoint detection is a useful adjunct to any etching process for overall process control and process diagnostics. Endpoint detection permits compensation for variations in etch rate that result from fluctuations in material composition or thickness, or from changes in operating parameters.

8.7 SIDE EFFECTS

Etching with reactive plasmas is not without side effects, which are mostly unwanted. Several of the more important ones are discussed briefly in this section.

8.7.1 Polymer Deposition

Discharges in halocarbon gases produce unsaturated (halogen-deficient) fragments that can react rapidly on surfaces to produce polymeric films. An example is the reaction of CF_2 radicals to produce fluorocarbon films. Obviously, such films impede etching if they form on the material to be etched, and so are undesirable. On the other hand, if polymer films can be made to form selectively on the mask or substrate, then very high selectivity is possible.

An excess of unsaturates, low ion energy, and reducing conditions generally favor film deposition. Thus, for certain gases such as CHF_3, films may form on grounded and floating surfaces, but not form simultaneously on rf-powered surfaces that are subject to higher-energy ion bombardment. Similarly, films may form on a Si surface but not on an SiO_2 surface, because the oxygen released during etching of the latter surface reacts with the unsaturated species to form volatile products.

Polymer film deposited on reactor surfaces can cause problems with adsorption of atmospheric contaminants, particularly water vapor, and with release of gaseous species during subsequent plasma operations. For example, when an oxidizing plasma is run in a system coated with fluorocarbon film, a substantial quantity of F atoms is released in the plasma.

8.7.2 Radiation Damage

The variety of energetic particles (ions, electrons, and photons) present in a plasma creates a potentially hostile environment for processing VLSI devices. The gate oxide and the SiO_2-Si interface are particularly susceptible to damage by irradiation with these particles.[34, 35]

The damage can take several forms: (1) atomic displacement resulting from energetic ion impact; for reactive etching this is usually limited to a region no more than 100 Å below the exposed surface; (2) primary ionization where Si−O bonds are broken and electron-hole pairs formed; this process is caused mainly by deep UV photons and soft x-rays; and (3) secondary ionization where electrons created by atom displacement or primary ionization interact with defects in the Si−O network.

Each of these forms of damage produces similar electronic defects—trapped positive charge and neutral traps. The former defect can cause shifts in threshold and flat-band voltages, while the latter tends to trap energetic electrons.

If a gate oxide is exposed directly to a reactive etching plasma with energetic ions (~400 eV), atom displacement damage is not observed, probably because the damaged layer is continuously removed by etching. However, photon damage is manifested as trapped holes and neutral traps. The trapped holes can be removed by annealing at 400°C, whereas removal of the neutral traps requires annealing at 600°C or more.

When gate oxides are exposed directly to non-reactive plasmas, atom displacement damage is observed. Removing this damage requires annealing at 1000°C.

Fortunately, in actual MOS device fabrication the sensitive gate oxide region of the device is protected by the gate metallization, typically polysilicon, during plasma exposure. Most of the radiation is not energetic enough to penetrate the gate electrode, so damage is confined to the periphery of that electrode. In addition, the processing usually includes subsequent high-temperature steps that ensure annealing of the damage.

The primary concern is with the creation of neutral traps after aluminum metallurgy is in place, for then the required annealing is precluded. Care must be taken to keep the maximum voltages in a reactive etching system below the thresholds that correspond to unannealable damage. These voltages depend on the specific device structure and mask level.

8.7.3 Impurity Contamination

All of the internal surfaces of a reactive etching system are subject to ion bombardment and can be sputtered. Unless the construction materials are properly chosen and voltages carefully controlled, the sputtered material can deposit on wafer surfaces and be incorporated in the device being etched.[36] Heavy metal contamination, which severely degrades minority-carrier lifetime, has been observed, especially in stainless steel systems.[37]

Sputter deposition of nonvolatile materials onto the etching surface impedes or completely blocks etching. This is one cause of etch "residues." When highly anisotropic etching is done, even very localized contamination of this sort presents a problem. Polymer films, sometimes only a few monolayers thick, can also cause device contamination. Usually dry etching must be followed by a wet chemical cleaning procedure to remove various contaminants, particularly following etching of small contact windows.

8.9 DRY ETCHING PROCESSES FOR VLSI TECHNOLOGY

8.8.1 Silicon Dioxide

Dry etching of SiO_2 is used mainly for opening contact windows, which usually are the smallest feature with the highest aspect ratio (film thickness to feature size) on the device. An aspect ratio of 1:2 is typical. Patterning of contact windows is a severe test, because the degree of anisotropy and the selectivity required are high.

The oxide is usually a deposited form of SiO_2 such as phosphosilicate glass (PSG), which isolates two conductor levels. In etching contact windows to both a first-level conductor and the Si substrate, the conductor thickness and/or substrate junction depths determine the required selectivity. Extensive efforts have been made to develop processes with high selectivity and anisotropy for etching PSG over Si and polysilicon that can be applied to MOS devices.

Following the suggestion that CF_3 radicals might react with SiO_2 in preference to Si,[38] gases such as CF_4, C_2F_6, C_3F_8, and CHF_3 have been used individually or in combination with H_2 (or hydrocarbons such as CH_4, C_2H_4, and C_2H_2) to maximize the CF_3 radical concentration and minimize the F atom concentration (see Section 8.5.2). In fact, CF_3 has never been conclusively identified as the etchant species, but

Table 2 Typical etch rates and selectivity for some dry etching processes for VLSI

Etched Material (M)	Gas	Etch rate (Å/min)	Selectivity		
			M/resist	M/Si	M/SiO$_2$
Al, Al−Si, Al−Cu	BCl$_3$ + Cl$_2$	500	5−8	3−5	20−25
Polysilicon	Cl$_2$	500−800	5	. . .	25−30
SiO$_2$	CF$_4$ + H$_2$	500	5	20	
PSG	CF$_4$ + H$_2$	800	8	32	

the weight of evidence suggests that CF$_x$, with $x \leqslant 3$, is the likely etchant and that the reaction is ion-induced. Anisotropy is relatively easily obtained both in RIE and parallel-plate plasma etching.

The selectivity over Si generally improves as operating parameters, especially gas composition, are altered to increase the concentration of unsaturated (fluorine-deficient) species in the plasma. As unsaturates also increase the tendency to polymer deposition (Section 8.7.1), parameters must be carefully controlled when working at maximum selectivity. Near the maximum in selectivity, polymer forms on grounded and floating surfaces, while etching occurs at the cathode in RIE. Additionally, since the release of oxygen discourages polymerization, SiO$_2$ can be etched even while polymer deposits on adjacent Si. In this extreme, very high selectivity is obtained (see Fig. 17).

Another means of eliminating F atoms from the discharge to maximize selectivity is to introduce a large surface area of a material which reacts rapidly with F atoms such as Si or C. The material can be used as the support electrode for the wafers being etched.[39]

Oxide etch rate and selectivity over Si are generally increased by an increase in rf power, consistent with an ion-induced reaction. Typical values for etch rate and selectivity are shown in Table 2. At high power densities some heat sinking of substrates is usually required to minimize resist degradation. With good heat sinking the selectivity with respect to conventional photoresists is excellent.

The SEM micrograph in Fig. 25a illustrates a contact window etched in a C$_2$F$_6$−CHF$_3$ plasma. The high degree of anisotropy apparent in the figure becomes increasingly important as design rules shrink, but creates a step coverage problem for subsequently deposited metallization. Various methods for achieving tapered windows with dry etching have been proposed, but all of them involve tradeoffs either in selectivity, dimensional control, or process complexity.[40, 41]

8.8.2 Silicon Nitride

Two more or less distinct types of nitride films are used in VLSI processing. Films deposited by low-pressure or atmospheric CVD are used as an oxidation and/or diffu-

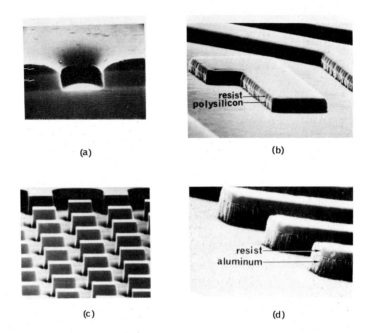

(a) (b)

(c) (d)

Fig. 25 SEM micrographs illustrate the results of highly anisotropic etching in reactive plasmas for several materials used in VLSI technology. (a) A plasma etched contact window in a 2-μm-thick phosphorous-doped SiO_2 layer; the substrate is Si. (b) Plasma etched pattern in a 1-μm-thick phosphorous-doped polycrystalline Si film; the substrate is SiO_2. (c) 1-μm-wide features created in single-crystal Si by RIE; note the trenching at the base of the features. (d) Plasma-etched pattern in a 1.5-μm-thick Al$-$0.7% Cu film; the substrate is an SiO_2 film.

sion mask and do not become a permanent part of the device. As an example in n-channel MOS fabrication a thin ($\sim$1000 Å) nitride film is deposited over a thin oxide ($\sim$250 Å) on the Si substrate and patterned. A thick oxide ($\sim$5000 Å) is then grown in unmasked regions by high-temperature oxidation. Because the nitride is thin and lateral dimensions relatively coarse, a high degree of anisotropy is not required for patterning the nitride. CF_4-O_2 and other plasmas that produce F atoms have been used in this application. The selectivity with respect to SiO_2 must be sufficient to avoid complete loss of the underlying oxide, or else the Si substrate will be substantially etched, since the selectivity of CVD silicon nitride relative to Si is about 1:8 for F atom plasmas.

Nitride films deposited from SiH_4-NH_3 or SiH_4-N_2 plasmas at low substrate temperatures ($\leq$ 350°C) are used for passivation and sometimes as intermediate dielectrics. In the former application the films are thick ($\leq$ 1.5 μm), but only coarse features must be etched in them to expose underlying metallization for bonding. Isotropic etching in CF_4-O_2 or other F atom source gases is usually employed for patterning. Plasma-deposited nitride, which is really a polymer-like Si$-$N$-$H material, etches much faster than CVD silicon nitride in plasmas containing F atoms. The etch rate is similar to that for Si.

When plasma-deposited silicon nitride film is used as an intermediate dielectric, the same considerations for etching windows apply as with SiO_2, and the same gases and process conditions are employed, although with somewhat lower etch rates than for SiO_2.

8.8.3 Polysilicon and Refractory-Metal Silicides

A high degree of anisotropy is required for etching polysilicon or polycide[42] gates. (A polycide is a composite layer consisting of a layer of metal silicide over a layer of polysilicon.) The gate length is a critical, fine line dimension that fixes the device channel length in the self-aligned gate technology. For example, if the etch profile is tapered rather than vertical, then portions of the gate will not be thick enough to be effective in masking the source and drain dopant implant. The resultant substrate doping profile depends on the amount of taper which, if uncontrolled, results in a variable channel length.

Similarly, the selectivity with respect to SiO_2 must be high, because the thin gate oxide (250 to 500 Å) exposed at the completion of etching overlays shallow (~2500-Å) source and drain junctions in the Si substrate. Also, if polysilicon or polycide runners cross field oxide steps, the runners will have a greater vertical thickness at the steps. This additional thickness requires overetching if the etching is anisotropic (Section 8.2.4).

Gases and gas mixtures containing chlorine have predominated for anisotropic etching of polysilicon. Cl_2 and Cl_2-Ar plasmas have been used for reactive ion etching with $A_f \cong 1$ for undoped material. Heavily doped ($\gtrsim 10^{20}$ cm^{-3}) n-type material is undercut for identical conditions.[28] For plasma etching conditions, both doped and undoped materials etch laterally in Cl_2 plasma, and the etch rate for heavily doped n-type poly-Si is more than an order of magnitude higher than for undoped or p-type polysilicon. The influence of doping has been explained on the assumption that Cl atoms are the etchant species, and that n-type doping, by raising the Fermi level, reduces the energy barrier for electron transfer to a bound Cl atom.[31] Etching can be made highly anisotropic ($A_f \cong 1$) by use of a gas additive containing a recombinant, as discussed in Section 8.6.1. Examples of additives that have been used are CCl_4 in RIE and C_2F_6 in plasma etching.

Bromine containing analogs of the chlorinated gases can also be used for anisotropic etching of polysilicon. The selectivity with respect to SiO_2 for both Cl and Br containing plasmas is generally good. Loading effects also tend to be minimal or absent with Cl and Br containing gases. Figure 25b and c illustrates anisotropic etching of patterns in polysilicon and single-crystal silicon, respectively.

Fluorinated gas plasmas generally etch polysilicon isotropically, with a strong loading effect. However, gases such as CF_4, CF_4-O_2, and SF_6 have been reported to etch polysilicon anisotropically under conditions that produce high-energy ion bombardment of the surface, such as low frequency and low pressures.[43-45]

Comparatively little information has been published on dry etching processes for refractory-metal silicides. Both isotropic and anisotropic etching have been reported for CF_4-O_2 plasmas.[46, 47] Isotropic profiles have been observed with plasma etching, and anisotropic profiles have been observed with reactive ion etching. Reactive ion etching with SF_6 has also been reported to provide a high degree of etch anisotropy, but with low selectivity ($\leq 4:1$) relative to SiO_2.[45] Much better selectivity is required for two-level polysilicon or polycide structures, when the layers pass over field oxide steps.

A single-step process for anisotropic dry etching of polycides has not been reported. Three problems are encountered in this application. (1) The etch rate anisotropy is insufficient for one or both layers. (2) Polysilicon etches faster than the silicide, leading to undercutting that causes loss of adherence of the silicide or subsequent step coverage problems. (3) Selectivity with respect to the underlying gate oxide is insufficient.

Multi-step processes have been designed to circumvent these problems.[46, 48] For example, a two-step process, for 1-μm MOSFETs,[46] consists of a reactive ion etch step to define the silicide and part of the polysilicon layer. This step is followed by an isotropic plasma etch with good selectivity over oxide to clear "extra" material from vertical steps (see Fig. 7).

8.8.4 Aluminum (Al−Si, Al−Cu)

Chlorine-containing gases, such as CCl_4, BCl_3 and $SiCl_4$, or mixtures of these gases with Cl_2 have been favored for etching aluminum alloys (Al−Si, Al−Cu) used in VLSI.[49] A freshly exposed aluminum surface, uncovered by Al_2O_3, reacts spontaneously with Cl or Cl_2 to form quasi-volatile $AlCl_3$ even in the absence of a plasma. However, aluminum is usually covered by a thin ($\sim$30-Å), layer of native oxide that does not react with Cl or Cl_2.

The native oxide must be removed by sputtering or chemical reduction before etching can proceed. Gases such as BCl_3 and CCl_4, when dissociated in a plasma, produce fragments capable of reducing the thin oxide layer. This step in the etching is observable as an induction period. Part of the irreproducibility initially reported for aluminum etching is related to the deleterious effect of residual gases, particularly water vapor, on the duration of the induction period. Water vapor can prolong initiation of etching by reacting with or scavenging oxide-reducing species and by reacting with aluminum to reform the oxide.

Further difficulty with residual water vapor is related to the quasi-volatility of the etch product ($AlCl_3$), which can redeposit on etch system walls and adsorb considerable moisture on exposure to atmosphere. Upon initiation of a plasma, desorption of water vapor interferes with etching. Some commercial etching systems have vacuum load-locks to avoid this problem.

Anisotropic etching can be achieved in either the RIE or plasma etching mode of operation, particularly if a recombinant gas mixture such as CCl_4-Cl_2 or BCl_3-Cl_2 is used. BCl_3 offers an advantage over CCl_4, in that polymer deposition does not occur over a wide range of operating parameters.

Evidence indicates that the reaction of Cl atoms with aluminum is unaffected by ion bombardment.[21] If this is the case, anisotropy is attributable to the influence of ions on rates of recombination-type reactions. There is some evidence that with CCl_4

a non-reactive layer forms on sidewalls and blocks etching there, but is removed by ion bombardment elsewhere.[14]

Al−Si alloys, containing up to several percent Si, are readily etched in chlorine containing gases, as silicon forms volatile chlorides. However, Al−Cu alloys ($\leq 4\%$ Cu), used to suppress electromigration, present a more difficult problem, because copper forms no volatile halides. Cu containing residues often result after reactive etching of these alloys, unless ion energies are sufficient to remove them by sputtering. Wet chemical procedures have been used to remove such residues.

Selectivity with respect to SiO_2 is sufficient for VLSI devices even when aluminum passes over steps. However, selectivity with respect to silicon (or polysilicon) is generally poor with chlorine containing gases. Consequently, conductors must overlap contact windows and this restricts the density of conductor lines. Table 2 indicates etch rates and selectivity typical of reactive etching of aluminum, and Fig. 25d illustrates a typical edge profile.

Another problem that has plagued the development of aluminum dry etching is post-etch corrosion, which results when atmospheric moisture hydrolyzes chlorine containing residues on the wafer to form HCl. Much of the residue is associated with the photoresist, making it desirable to remove this material as soon after etching as possible, preferably in-situ. However even this precaution may be insufficient to protect fine line patterns. A more expedient approach is to follow etching by exposure to a fluorocarbon plasma, which converts chloride residues to unreactive fluorides. Al−Cu alloys are especially susceptible to post-etch corrosion, because the enrichment of the involatile Cu component in the surface region causes copper chlorides to form.[50]

8.9 SUMMARY AND FUTURE TRENDS

VLSI processing requires methods for transferring circuit features from resist masks, defined by lithography, into active circuit materials, with a high degree of dimensional accuracy. Dry etching techniques that use low-pressure gas discharges (plasmas) can produce highly directional etching to meet the requirements for dimensional accuracy.

Etching processes used for pattern transfer in VLSI must be highly selective. Ideally, neither the resist mask nor previously processed portions of a circuit should be removed during etching. The requirements for selectivity are best met by plasma-assisted etching techniques that use gases containing reactive constituents. Fragmentation of these gases in a plasma produces species that can chemically combine with the material to be etched to form a volatile product.

Reactive ion etching and plasma etching in parallel-plate systems are the dominant techniques for VLSI dry etching. The superior control of circuit dimensions achieved with these methods results from etch rate anisotropy. Anistropy occurs when chemical reactions between neutral species, generated in the plasma, and the surface being etched are influenced by directional energetic particle bombardment. The energetic particles are usually positive ions drawn from the plasma by an imposed electric field.

Etch rate, selectivity, and anisotropy are determined by various parameters including gas composition, gas pressure, wafer temperature, and the operating frequency and power density of the plasma. These parameters must be carefully controlled to avoid unwanted effects such as polymer deposition and radiation damage.

Plasma processes have been developed for etching most of the materials used in VLSI technology. However, improvements in selectivity, edge profile control, reproducibility, overall process control, automation, and throughput are needed.

Better selectivity and profile control will be particularly important to efforts to further reduce the size of circuit features. Future work will lead to increased etch rates. Faster etching will shift the focus in dry etching equipment to single-wafer systems, with cassette to cassette, fully automated wafer processing.

The future will also see dry etching extended to new materials and devices, and significant advances in our understanding of the complex physics and chemistry underlying dry etching processes.

REFERENCES

[1] C. M. Melliar-Smith and C. J. Mogab, "Plasma-Assisted Etching Techniques for Pattern Delineation," in J. L. Vossen and W. Kern, Eds., *Thin Film Processes*, Academic, New York, 1979.

[2] S. M. Irving, "A Plasma Oxidation Process for Removing Photoresist Films," *Solid State Technol.*, 14(6), 47 (1971).

[3] S. M. Irving, K. E. Lemons, and G. E. Bobos, "Gas Plasma Vapor Etching Process," U.S. Patent 3,615,956. (Filed March 27, 1969; patented Oct. 26, 1971.)

[4] T. C. Penn, "Forecast of VLSI Processing—A Historical Review of the First Dry-Processed IC", *IEEE Trans. Electron Devices*, ED-26, 640 (1979).

[5] D. L. Tolliver, "Plasma Processing in Microelectronics—Past, Present, and Future," *Solid State Technol.*, 25, 99 (1980).

[6] G. K. Wehner and G. S. Anderson, "The Nature of Physical Sputtering," in L. I. Maissel and R. Glang, Eds., *Handbook of Thin Film Technology*, McGraw-Hill, New York, 1970.

[7] D. Bollinger and R. Fink, "A New Production Technique: Ion Milling," *Solid State Technol.*, 25, 79 (1980).

[8] B. Chapman, *Glow Discharge Processes*, Wiley, New York, 1980.

[9] J. L. Vossen, "Glow Discharge Phenomena in Plasma Etching and Deposition," *J. Electrochem. Soc.*, 126, 319 (1979).

[10] H. R. Koenig and L. I. Maissel, "Applications of RF Discharges to Sputtering," *IBM J. Res. Dev.*, 14, 168 (1970).

[11] J. W. Coburn and E. Kay, "Positive-Ion Bombardment of Substrates in RF Diode Glow Discharge Sputtering," *J. Appl. Phys.*, 43, 4965 (1972).

[12] A. R. Reinberg, "Dry Processing for Fabrication of VLSI Devices," in N. G. Einspruch, Ed., *VLSI Electronics*, Academic, New York, 1981, Vol. 2.

[13] D. B. Fraser et al, *Pumping Hazardous Gases*, American Vacuum Society, New York, 1980.

[14] R. H. Bruce, "Anisotropy Control in Dry Etching," *Solid State Technol.*, 24, 64 (1981).

[15] D. F. Downey, W. R. Bottoms, and P. R. Hanley, "Introduction to Reactive Ion Beam Etching," *Solid State Technol.*, 26, 121 (1981).

[16] C. M. Melliar-Smith, "Ion Etching for Pattern Delineation," *J. Vac. Sci. Technol.*, 13, 1008 (1976).

[17] J. W. Coburn and H. F. Winters, "Ion- and Electron-Assisted Gas-Surface Chemistry—An Important Effect in Plasma Etching," *J. Appl. Phys.*, 50, 3189 (1979).

[18] J. W. Coburn and H. F. Winters, "Plasma Etching—A Discussion of Mechanisms," *J. Vac. Sci. Technol.*, 16, 391 (1979).

[19] U. Gerlach-Meyer, "Ion Enhanced Gas-Surface Reactions: a Kinetic Model for the Etching Mechanism," *Surf. Sci.*, 103, 524 (1981).

[20] U. Gerlach-Meyer, J. W. Coburn, and E. Kay, "Ion-Enhanced Gas-Surface Chemistry: The Influence of the Mass of the Incident Ion," *Surf. Sci.*, 103, 177 (1981).

[21] D. L. Flamm and V. M. Donnelly, "The Design of Plasma Etchants," *Plasma Chemistry and Plasma Processing*, 1, 317 (1981).

[22] C. J. Mogab, A. C. Adams, and D. L. Flamm, "Plasma Etching of Si and SiO$_2$—The Effect of Oxygen Additions to a CF$_4$ Plasma," *J. Appl. Phys.*, 49, 3769 (1978).

[23] L. M. Ephrath, "Selective Etching of Silicon Dioxide Using Reactive Ion Etching with CF$_4$/H$_2$," *J. Electrochem. Soc.*, 126, 1419 (1979).

[24] K. Hirata, Y. Ozaki, M. Oda, and M. Kimizuka, "Dry Etching Technology for 1-μm VLSI Fabrication," *IEEE Trans. Electron Devices*, ED-28, 1323 (1981).

[25] R. F. Reichelderfer, "Single Wafer Plasma Etching," *Solid State Technol.*, 25 160 (1982).

[26] J. Taillet, "Plasma Physics: Ion Energy in RF Plasma Etching," *J. Physique*, 11, L-223 (1979).

[27] H. Kalter and E. P. G. T. Van deVen, "Plasma Etching in IC Technology," *Philips Tech. Rev.*, 38, 200 (1978/79).

[28] G. C. Schwartz and P. M. Schaible, "Reactive Ion Etching of Silicon," *J. Vac. Sci. Technol.*, 16, 410 (1979).

[29] C. J. Mogab, "The Loading Effect in Plasma Etching," *J. Electrochem. Soc.*, 124, 1262 (1977).

[30] T. Enomoto, M. Denda, A. Yasuoka, and H. Nakato, "Loading Effect and Temperature Dependence of Etch Rate in CF$_4$ Plasma," *Jpn. J. Appl. Phys.*, 18, 155 (1979).

[31] C. J. Mogab and H. J. Levinstein, "Anisotropic Plasma Etching of Polycrystalline Silicon," *J. Vac. Sci. Technol.*, 17, 721 (1980).

[32] J. E. Greene, "Optical Spectroscopy for Diagnostics and Process Control During Glow Discharge Etching and Sputter Deposition," *J. Vac. Sci. Technol.*, 15, 1718 (1978).

[33] P. J. Marcoux and P. D. Foo, "Methods of End Point Determination for Plasma Etching," *Solid State Technol.*, 24, 115 (1981).

[34] R. A. Gdula, "The Effects of Processing on Radiation Damage in SiO$_2$," *IEEE Trans. Electron Devices*, ED-26, 644 (1979).

[35] L. M. Ephrath and D. J. DiMaria, "Review of RIE Induced Radiation Damage in Silicon Dioxide," *Solid State Technol.*, 24, 182 (1981).

[36] S. P. Murarka and C. J. Mogab, "Contamination of Silicon and Oxidized Silicon Wafers During Plasma Etching," *J. Electron. Mater.*, 8, 763 (1979).

[37] K. Hirata, Y. Ozaki, M. Oda, and M. Kimizuka, "Dry Etching Technology for 1-μm VLSI Fabrication," *IEEE Trans. Electron Devices*, ED-28, 1323 (1981).

[38] R. A. H. Heinecke, "Control of Relative Etch Rates of SiO$_2$ and Si in Plasma Etching," *Solid State Electron.*, 19, 1039 (1976).

[39] S. Matsuo, "Selective Etching of SiO$_2$ Relative to Si by Plasma Reactive Sputter Etching," *J. Vac. Sci. Technol.*, 17, 587 (1980).

[40] J. A. Bondur and R. G. Frieser, "Shaping of Profiles in SiO$_2$ by Plasma Etching," in R. G. Frieser and C. J. Mogab, Eds., *Plasma Processing*, Electrochem. Soc., Pennington, New Jersey, 1981.

[41] L. B. Rothman, J. L. Mauer IV, G. C. Schwartz, and J. S. Logan, "Process for Forming Tapered Vias in SiO$_2$ by Reactive Ion Etching," in R. G. Frieser and C. J. Mogab, Eds., *Plasma Processing*, Electrochem. Soc., Pennington, New Jersey, 1981.

[42] S. P. Murarka, "Refractory Silicides for Integrated Circuits," *J. Vac. Sci. Technol.*, 17, 775 1(1980).

[43] P. D. Parry and A. F. Rodde, "Anisotropic Plasma Etching of Semiconductor Materials," *Solid State Technol.*, 22, 125 (1979).

[44] H. Mader, "Anisotropic Plasma Etching of Polysilicon with CF$_4$," in R. G. Frieser and C. J. Mogab, Eds., *Plasma Processing*, Electrochem. Soc., Pennington, New Jersey, 1981.

[45] W. Beinvogl and B. Hasler, "Anisotropic Etching of Polysilicon and Metal Silicides in Fluorine Containing Plasma," *Proc. 4th Int. Symp. on Si Materials and Technol.*, 81−5, 648 (1981).

[46] B. L. Crowder and S. Zirinsky, "1μm MOSFET VLSI Technology: Part VII—Metal Silicide Interconnection Technology—A Future Perspective," *IEEE Trans. Electron Devices*, ED-26, 369 (1979).

[47] T. P. Chow and A. J. Steckl, "Plasma Etching Characteristics of Sputtered MoSi$_2$ Films," *Appl. Phys. Lett.*, 37, 466 (1980).

[48] E. C. Whitcomb and A. B. Jones, "Reactive Ion Etching of Submicron MoSi$_2$/Poly-Si Gates for CMOS/SOS Devices," *Solid State Technol.*, 25, 121 (1981).

[49] D. W. Hess, "Plasma Etching of Aluminum," *Solid State Technol.*, 24, 189 (1981).

[50] W. Y. Lee, J. M. Eldridge, and G. C. Schwartz, "Reactive Ion Etching Induced Corrosion of Al and Al-Cu Films," *J. Appl. Phys.*, 52, 2994 (1981).

PROBLEMS

1 Assuming a mask that cannot erode, sketch the edge profile of an isotropically etched feature in a film of thickness h_f on an unetchable substrate for (*a*) etching just to completion, (*b*) 100% overetch, and (*c*) 200% overetch. What shape does the profile tend toward as overetching proceeds? Comment on the advisability of estimating the degree of anisotropy of etching from scanning electron micrographs of edge profiles taken after removal of the masking layer.

2 By tracing the trajectory during etching of a point on the beveled edge of the mask in Fig. 5, arrive at Eq. 7.

3 Show that Eq. 12 results when the thinnest and fastest etching portion of the film is assumed to be over the fastest etching portion of the substrate.

4 Polysilicon lines 0.5 μm thick pass over a field oxide step 1.1 μm high and across a gate oxide 0.05 μm thick. Calculate the selectivities required with respect to mask and gate oxide if the polysilicon is etched with a process having 10% etch rate uniformity, $A_f = 1.0$, and $A_m = 0.5$. Assume 5% polysilicon thickness uniformity, 5% mask etch rate uniformity, a mask edge profile of 60°, and that linewidth must be controlled to 0.2 μm.

5 Consider a discharge in CF$_4$. Assume that electrons and gas molecules can be treated as hard spheres with masses m and M, respectively. Calculate the maximum fractional loss of kinetic energy for an electron striking a CF$_4$ molecule. Consider the CF$_4$ molecule initially at rest, and the collision to be elastic. Repeat the calculation for an inelastic collision where the potential energy of the CF$_4$ molecule increases by the maximum amount possible.

6 A CF$_4$–O$_2$ rf plasma is operated at 300 W, and 0.5 torr, with a feed-gas flow rate of 100 cm^3/min (STP). The plasma occupies a volume of 4000 cm^3. Under these conditions atomic fluorine is generated at a rate of 10^{16} cm^{-3} – s^{-1} in the plasma. The combined effect of loss mechanisms for F atoms results in a rate of loss proportional to the F atom concentration. In steady state, this concentration is measured as 3×10^{15} cm^{-3}. What is the mean lifetime of F atoms for these conditions? How does it compare to the residence time of an average molecule in the plasma? How would the etch rate of Si be affected if the flow rate were increased tenfold while holding other parameters constant?

7 What is the function of the ground shield shown in Fig. 11? What limitation is there on the spacing between the ground shield and the powered electrode?

8 What are the major distinctions between reactive ion etching and parallel-plate plasma etching? Compare the advantages and limitations of these techniques.

9 Would you expect the rate of an ion-assisted reaction between neutral species and a solid surface to be independent of the angle of incidence of the ion beam? Why? What do you expect would happen to the reaction rate as the ion energy was increased continuously beyond 5 keV?

10 The enthalpy of the exothermic reaction

$$Si + 4F \rightarrow SiF_4$$

is 370 kcal/g-mole at 25°C. At what rate is heat generated when a 100-mm-diameter, 0.5-mm-thick Si wafer is etched on one face at a rate of 1.0 μm/min in an F generating plasma? Suppose that the wafer is thermally isolated during etching. By how much will its temperature rise if 5.0 μm are etched away?

11 Explain why the peaks of the curves in Fig. 19 are shifted in time relative to each other. Plot the peak positions (time coordinate) against the number of wafers. What functional form do you get? Show that this form is expected when a loading effect exists and the activation energy for the reaction is negligible. What conclusion can be drawn from the fact that peak wafer temperature depends on the number of wafers?

12 Explain why endpoint detection by monitoring of reactant species requires a loading effect.

METALLIZATION

D. B. FRASER

9.1 INTRODUCTION

Metallization is perhaps best defined operationally by giving an example. Figure 1 shows a schematic view of a conventional MOSFET with an n^+ source and drain implanted in a p-type substrate. The source and drain are contacted through windows by metal (e.g., Al) and connected to a power supply. Current flows between the source and drain when a threshold voltage V_T is applied to the gate electrode. This voltage creates a field across the gate oxide, which causes the adjacent p substrate to invert to n-type, thus creating a conductive n channel between the source and drain. The gate electrode, usually conductive polysilicon, is connected by metal to a signal voltage. Thus, the metallization requires low-resistance interconnections, and the formation of low-resistance contacts to n^+, p^+, and polysilicon layers. Also, the structures should be stable under use—that is, metal adherence, electromigration (material transport in conductors carrying high currents), and corrosion should not significantly reduce reliability—and, finally, the structure should be easily patterned by a straightforward process.[1]

Most silicon MOS and bipolar integrated circuits now manufactured are metallized with Al or one of its alloys. Because Al has a low-room-temperature resistivity of approximately 2.7 $\mu\Omega$-cm, and that of its alloys may be 30% greater, these metals satisfy the requirements of low resistance. Al and its alloys adhere well to thermally grown SiO_2 and to deposited silicate glasses (because the heat of formation of Al_2O_3 is higher than that of SiO_2). Despite these advantages, the use of Al in VLSI applications where junctions are shallow often encounters problems with electromigration and corrosion. However, as will be shown, viable solutions exist. Electromigration can be reduced by control of the deposited film characteristics and corrosion can be minimized by careful processing and packaging techniques.

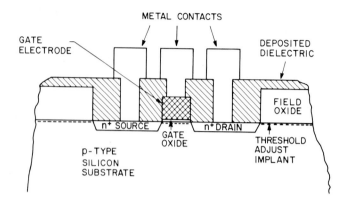

Fig. 1 Schematic view of a MOSFET cross section. The areas of concern (in this chapter) are the gate electrode and metal regions.

Other metallization structures have been used but their complex processing makes them undesirable for VLSI. Among these structures[2] are Ti-Pd-Au and Ti-Pt-Au. The Ti-Pt combination is currently used as the first-level conductor in a two-level metal device structure for LSI applications with Ti-Pt-Au as the second-level metal layer.[3]

The MOSFET gate electrode and interconnect structures are another category of metallization structures. The reason for the metallization designation is that the refractory metals and refractory metal silicides that are used to augment or replace the polysilicon are generally deposited by physical vapor deposition processes. These processes are similar to those used to deposit Al and its alloys. These refractory materials are necessary, since the nominal 500-$\mu\Omega$-cm resistivity of n^+ polysilicon is too high for VLSI applications when the device channel lengths are 1.5 μm or less and where a single chip may have more than 100,000 devices.

9.1.1 Contacts

In general the contact between the metallization and the substrate may be characterized as rectifying or ohmic. When devices require diode characteristics in the contact, a barrier must be used. Otherwise, the goal is to achieve low resistance. A figure of merit, the specific contact resistance R_c, is useful in characterizing ohmic contacts:[4]

$$R_c = \left[\frac{\partial J}{\partial V} \right]_{V=0}^{-1} \qquad (1)$$

For low doping in the semiconductor, the metal specific contact resistance can be represented by

$$R_c = \frac{k}{qA^*T} \exp\left[\frac{q\phi_B}{kT} \right] \qquad (2)$$

In Eq. 2 A^* is the Richardson's constant given by $A^* = 4\pi qm^* k^2/h^3$ where q is the charge, m^* is the effective mass of the charge carrier (e.g., electron), k is Boltzmann's constant, h is Planck's constant, and ϕ_B is the barrier height. Equation 2 leads to the conclusion that low-resistance contacts are obtained with low barrier heights, since thermionic emission over the barrier dominates the charge transport. At higher doping levels the barrier width decreases, tunneling becomes important, and the specific contact resistance may be written as

$$R_c \approx \exp\left[\frac{4\pi}{h} \sqrt{\epsilon_s m^*} \left[\frac{\phi_B}{\sqrt{N_D}} \right] \right] \tag{3}$$

where ϵ_s is the dielectric permittivity of silicon and N_D is the doping concentration. Roughly, for $N_D \geq 10^{19}$ cm^{-3}, R_c will be dominated by tunneling and will decrease rapidly as the doping level goes above 10^{19} cm^{-3}. For $N_D \leq 10^{17}$ cm^{-3}, thermionic emission dominates R_c which is then independent of the doping level. Calculated and experimental values for R_c are shown in Fig. 2. For low-barrier contacts on highly

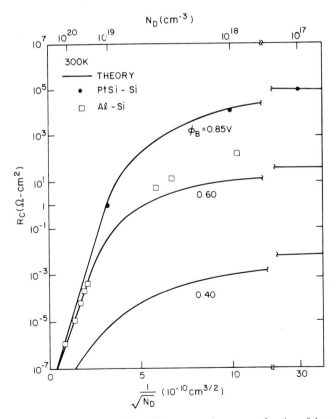

Fig. 2 Theoretical and measured values of specific contact resistance as a function of donor concentration and barrier height. *(After Sze, Ref. 4.)*

Table 1 Schottky barrier height ϕ_B

Contact material	ϕ_B (V) for n-type Si	ϕ_B (V) for p-type Si
Al	0.72	0.58
Cr	0.61	0.50
Mo	0.68	0.42
Ni	0.61	0.51
Pt	0.90	
Ti	0.50	0.61
W	0.67	0.45
CoSi	0.68	
CoSi$_2$	0.64	
IrSi	0.93	
Ni$_2$Si	0.7−0.75	
NiSi	0.66−0.75	
NiSi$_2$	0.7	
PtSi	0.84	
Pd$_2$Si	0.72−0.75	
TaSi$_2$	0.59	
TiSi$_2$	0.60	
WSi$_2$	0.65	

doped substrates, $R_c \approx 10^{-7}$ Ω-cm^2 may be used as a target value. Table 1 lists a number of ϕ_B values.[4]

9.1.2 Fundamentals of Physical Vapor Deposition

VLSI metallization is currently done in vacuum chambers.[5] Figure 3 shows a schematic view of a system. The chamber shown is a bell jar, a stainless-steel cylindrical vessel closed at the top and sealed at the base by a gasket. Beginning at atmospheric pressure the chamber is evacuated by a roughing pump, such as a mechanical rotary-vane pump or a combination mechanical pump and liquid-nitrogen-cooled molecular sieve system. The rotary-vane pump can reduce the system pressure to about 20 Pa, and the combination pump system can achieve about 0.5 Pa. At the appropriate pressure, the chamber is opened to a high-vacuum pumping system that continues to reduce the pressure of the process chamber. The high-vacuum pumping system may consist of a liquid-nitrogen-cooled trap and an oil diffusion pump, a trap and a turbomolecular pump, or a trap and a closed cycle helium refrigerator cryopump. In a low-throughput system, a trap, a titanium sublimation pump, and an ion pump could also be used. The choice of pumping system depends on required pumping speed, ultimate pressure attainable (in a reasonable time), desired film quality, method of film deposition, and expense. Traditional systems have used oil diffusion pumps but fear of contaminating the films with oil has led to the use of turbomolecular and cryopumps. The cryopump acts as a trap and must be

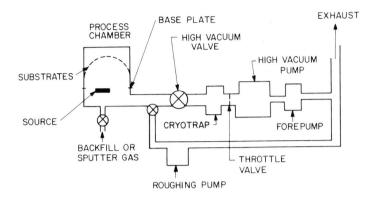

Fig. 3 Schematic view of a high-vacuum chamber with substrates mounted in a planetary substrate support above the source. Gages are not shown for simplicity.

regenerated periodically; the turbomolecular and diffusion pumps act as transfer pumps, expelling their gas to a forepump.

The high-vacuum pumping system brings the chamber to a low pressure which is tolerable for the deposition process. This low pressure is considered the "working" or base pressure. As an example, the desired base pressure may be 6.6×10^{-5} Pa (5×10^{-7} torr) for an aluminum evaporation system, but when auxiliary heaters are turned on the chamber pressure may rise by an order of magnitude or more. The pressure may rise still further when the evaporation source is heated. To reduce the time required for the deposition process including the pump-down period, system cleanliness is an absolute necessity on several levels. All components in the chamber are chemically cleaned and dried. Generally, warm water circulates in the coolant channels of the vacuum chamber to reduce the adsorption of water vapor on the freshly coated interior when it is opened to the atmosphere. Any interior film buildup is removed frequently to avoid a major source of trapped atmospheric gas. Freedom from sodium contamination is vital when coating MOS devices.[6] This requirement involves cleaning the substrates to be coated in HF solutions, avoiding skin contact with any interior portion of the coating system, and using pure-metal sources.

Sputter deposition demands similar precautions. The system operates with about 1 Pa of argon pressure during the film deposition. Despite the relatively high system pressure, sputtering is as demanding a process as evaporation because other gases, such as water vapor and oxygen, may be detrimental to film quality if present at background pressures of about 10^{-2} Pa. The purity of the argon sputtering gas is also a factor. Thus, to maintain purity, the lines connecting the gas source to the sputter chamber should be clean and vacuum tight. For sputtering, the throttle valve should be placed between the trap and the high-vacuum pumping system. The argon gas pressure can then be maintained by reducing the effective pumping speed of the high-vacuum pump, while the full pumping speed of the trap for water vapor is utilized.

Assuming a vacuum station of volume V has no leaks, and is equipped with pumps of adequate capacity and that the ultimate limitation is set by outgassing of water vapor, the chamber pressure P at any time t after pump down has been initiated is given by the approximate relation[5]

$$P = P_0 \exp\left[\frac{-St}{V}\right] + \frac{Q}{S} \qquad (4)$$

where P_0 is the initial pressure, S is the pumping speed, and Q is the rate of outgassing within the system. After the first hour of pumping the second term dominates and $P = Q_1/S$ where Q_1 is the outgassing rate after approximately 1 hour. Note that Q is a slowly varying function of time, since the source of outgassing, in principle, will eventually be depleted. This characteristic of vacuum systems has led to the introduction of "load-lock" systems, where the substrates are introduced into the process chamber through a lock chamber that cycles between atmospheric pressure and some reduced pressure. At the reduced pressure the substrates are transferred from the lock into the process chamber, and only the substrates have to be outgassed rather than the whole chamber interior. After completion of the process the substrates are transferred through the same or another lock and brought out of the system. Use of such systems in production facilities is growing because the number of wafers processed per day can exceed what can be processed in a simple chamber (when silicon wafers 100 mm and larger in diameter are used). We used the term "process" in the above description because, in addition to film deposition, reactive sputter etching and plasma etching are also performed in load-lock systems.

9.1.3 Thickness Measurement and Monitoring

In VLSI applications control of conductive film thickness is essential, because a film thinner than desired can cause excess current density and failure during operation. Conversely, excessive thickness can lead to difficulties in etching. The use of thickness monitors is common in evaporation deposition, and in magnetron sputter deposition where planetary systems support the substrates. In some magnetron deposition processes, the film is deposited without monitoring during the deposition, but is checked after the deposition.

The most common thickness monitor is a resonator plate made from a quartz crystal. The plate is oriented relative to the major crystal axes, so that its resonance frequency is relatively insensitive to small temperature changes.[7] The acoustic impedance and the additional mass of any film deposited on the resonator cause a frequency change that can be measured accurately. After calibrating the monitor in the deposition system, it may be used to control the deposition rate as well as the final thickness of the deposited film. The resonator crystal has a finite useful life and must be replaced; however, no recalibration is necessary if the deposition system has not been modified. The resonator has a finite useful life because $\Delta f \propto \Delta M$ holds true for $\Delta f / f_0 \leq 0.05$, where Δf is the resonator frequency change, ΔM is the additional deposited mass, and f_0 is the initial resonator frequency.

We can calibrate such systems and measure unmonitored film thickness in at least two different-ways. The simplest is to use a microbalance and weigh the substrate before and after film deposition. The film is assumed to have bulk density ρ_D, so that the increase in mass Δm is related to the film thickness t by

$$\text{Volume} = \frac{\Delta m}{\rho_D} = At \tag{5}$$

and

$$t = \frac{\Delta m}{\rho_D A} \tag{5a}$$

where A is the area of the film.

Another technique uses a surface profile measuring device. A fine stylus, usually diamond, is drawn over the surface of the substrate and encounters a step where the film has been etched or masked during deposition. The entire height of the step is detected by differential capacitance or inductance measurements. Calibration is maintained by using standard film samples which can be checked periodically. Films as thin as 100 Å or less can be measured by such devices. Other techniques for measuring conductor film thickness include optical interference techniques and eddy current measurements.

9.1.4 Application of Kinetic Theory of Gases

The kinetic theory of gases yields two concepts that are useful in physical vapor deposition. The first is the concept of rate of bombardment by gas molecules of the exposed surfaces in the chamber.

$$N = (2\pi mkT)^{-1/2}p \tag{6}$$

where N is the bombardment rate in molecules $cm^{-2}\text{-}s^{-1}$ for a gas of molecular mass m at temperature T in kelvins and pressure p. Equation 6 may be rewritten as

$$N = 6.4 \times 10^{19}\,(MT)^{-1/2}p \tag{7}$$

where M is the gram-molecular mass and p is in Pa. Possible effects of residual gas or gas added intentionally during the deposition of films can be estimated using the bombardment rate. The second useful concept is that of mean free path λ where

$$\lambda = \frac{kT}{p\,\pi\sigma^2\sqrt{2}} \tag{8}$$

and where σ is the diameter of a gas molecule. For residual gases, such as He, O_2, N_2, and H_2O, found frequently in vacuum chambers, the value of σ ranges[8] from 2 to 5 Å. Thus for air at constant pressure the product

$$\lambda p = \text{constant} \approx 0.7 \text{ cm-Pa} \tag{9}$$

Because the collision process is statistical, the fraction of total molecules n_0 not suffering a collision while traveling a distance d is

$$n/n_0 = \exp(-d/\lambda) \tag{10}$$

For a distance $d = \lambda$, only 37% of the n_0 molecules do not undergo collision. These concepts are useful in sputtering applications where Ar pressures of about 1 Pa are frequently used, and $\lambda \approx 0.7$ cm may be expected. The implication for sputter deposition is that the sputtered vapor may undergo considerable kinetic scatter prior to reaching the substrate. In contrast, in evaporation processes with chamber pressure $P \leqslant 10^{-2}$ Pa, residual gas molecules would have $\lambda \approx 1$ m, thus validating the assumption that evaporated vapor travels in straight lines from the source to the substrate. The assumption of straight-line travel is basic to treatments of film step coverage on substrates.

9.2 METHODS OF PHYSICAL VAPOR DEPOSITION

The rate of evaporation of metal from a melt is estimated by use of the Hertz-Knudsen[9] equation:

$$N_e = (2\pi \, mkT)^{-\frac{1}{2}} p_e \tag{11}$$

where N_e is the number of molecules per unit area per time, m is the molecular (atomic) mass, k is Boltzmann's constant, T is the surface temperature in kelvins, and p_e is the equilibrium vapor pressure of the evaporant. This vapor pressure may be written as a rate of mass loss per unit area from the source

$$R = 4.43 \times 10^{-4} \left[\frac{M}{T} \right]^{\frac{1}{2}} p_e \qquad \text{g/cm}^2\text{-s} \tag{12}$$

where M is the gram-molecular mass and p_e is in Pa. For example, p_e (Al) ≈ 1.5 Pa at 1500 K.

The total loss R_T per unit time from the source may be found by integrating over the source area:

$$R_T = \int R \, dA_s \tag{13}$$

The flux of material to the receiving substrate is dependent on the cosine of the angle, ϕ, between the normal to the source surface and the direction of the receiving surface a distance r away. If θ is an angle between the receiving surface normal and the direction back to the source, then

$$D = \frac{R_T}{\pi r^2} \cos \phi \cos \theta \tag{14}$$

where D is the deposition rate in g/cm²-s.

The deposition rate at various points on a substrate plane above a point or area source may be found from Eq. 14. For example,

$$\frac{D}{D_0} = \left[1 + \left[\frac{L}{H} \right]^2 \right]^{-2} \tag{15}$$

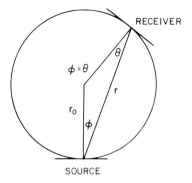

Fig. 4 Idealized view of vapor source and film gathering surface mounted on a sphere of radius r_0.

for a small area source and

$$\frac{D}{D_0} = \left[1 + \left[\frac{L}{H} \right]^2 \right]^{-3/2} \tag{16}$$

for a point source where D_0 is the rate directly above and H away from the source and D is the rate at a point L away from the center of the substrate plane.

When the receiving surface is spherical and has a radius r_0, and the source is on the surface (see Fig. 4), then

$$\cos \theta = \cos \phi = \frac{r}{2r_0} \tag{17}$$

and Eq. 14 is written

$$D = \frac{R_T}{4\pi r_0^2} \tag{18}$$

Therefore the deposition rate is the same for all points on the spherical surface, which is the principle behind the planetary (rotating spherical sections) substrate supporting systems used in deposition chambers.

In chronological order, the method of deposition of Al and (to some extent) its alloys has proceeded through: (1) resistance-heated evaporation, (2) electron-beam evaporation, (3) inductively heated evaporation, (4) magnetron sputter deposition, and (5) chemical vapor deposition. Each technique has advantages and disadvantages that must be carefully considered before deciding which may be used in a given application.

9.2.1 Resistance-Heated Evaporation

Figure 5 shows a refractory metal filament (e.g., W) with small pieces of the Al to be evaporated shown suspended from the filament coils. Other, more complex source structures may also be formed from sheets of the refractory metal. The resistance-heated approach is attractive because it is simple, inexpensive, and produces no ioniz-

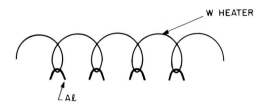

Fig. 5 Refractory wire coil W acting as a support and heat source for the vacuum evaporation of Al.

ing radiation. Its disadvantages are the possibility of contamination from the heater, the small charge which limits ultimate film thickness, short filament life, and, unless flash evaporation techniques[10] are used, the difficulty in preserving alloy composition in the film. Flash evaporation employs a heated surface onto which particles of the alloy to be vaporized are dropped. The heating is rapid and all constituents are transported to the substrate. Despite its disadvantages, filament evaporation continues to be used for deposition of Al electrodes that are used in the evaluation of test capacitors for furnace qualification and in experiments. A large array of charged filaments can be used to approximate a large area source. The large area source gives better metal film step coverage than a single source. When a number of substrates are to be coated simultaneously, a planetary system may be used to provide uniform film thickness. Other metals, such as Au and Pd, may be conveniently evaporated by this method.

9.2.2 Electron-Beam Evaporation

Figure 6 shows a schematic view of an e-beam evaporation source. A hot filament supplies current of the order of 1 A to the beam and the electrons are accelerated through (typically) 10 kV, and strike the surface to be evaporated. Using a magnetic field to curve the path of the e-beam permits screening of the hot filament so that impurities from the filament cannot reach the substrates. Scanning the e-beam over the surface of the melt prevents the nonuniform deposition that would otherwise occur by the formation of a cavity in the molten source. By using a large source, thick-film deposition may be performed without breaking the vacuum and recharging the source. The large source also permits moving the substrates farther away from the source such as in a planetary system. With a number of sources in one chamber, the system can deposit films sequentially without breaking the vacuum. Coevaporation to form alloy films may also be performed with multiple sources. Because of the high power available in the e-beam very high film deposition rates can be attained. Depending on source-to-substrate distance, rates as high as 0.5 µm/min are common. The use of excessive power, however, can lead to deposition on the substrates of metal droplets that have been blown out of the source by expanding metal vapor.

In addition to Al and its alloys, other elements (Si, Pd, Au, Ti, Mo, Pt, and W) and dielectrics such as Al_2O_3 may be evaporated by the e-beam process. Generally Al and its alloys are evaporated from a charge sitting directly in the water-cooled

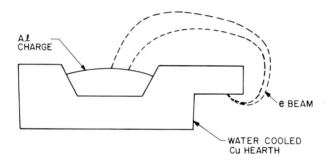

Fig. 6 Electron-beam evaporation system. Note that a magnetic field causes the electrons to follow a curved path so that the substrates are protected from the hot filament.

copper hearth of the e-beam source. Heat transfer to the cooling water is reduced if a crucible liner (e.g., boron nitride) is used to contain the source. However, the liner may contaminate the deposited films. Contamination from Cu may also occur when no liner is used, if the Al melt wets the copper hearth and begins dissolving it.

At voltages of the order of 10 kV, the characteristic Al K-shell x-rays, along with a continuum, are generated by the e-beam. This ionizing radiation penetrates the surface layers of the silicon substrates and causes "damage" which changes the MOS capacitor characteristics. The silicon then requires subsequent annealing.

9.2.3 Inductively Heated Sources

Figure 7 schematically shows an evaporation source that is heated by rf induction. The crucible is generally made of BN. This process also achieves high deposition rates. Its advantage over the e-beam source is the absence of ionizing radiation. Like e-beam evaporation excessive material heating by rf induction can also cause molten drops to be transported to the substrates. Another disadvantage of this process is the mandatory use of the crucible. As in the e-beam method, dilute Al alloys may be deposited, as well as other metals compatible with the crucible. Note a lower temperature sinter can be used to form the contacts of the Al film to the substrate, because of the absence of ionizing radiation during the deposition process. This point will be discussed later when contact problems are described (Section 9.4.1).

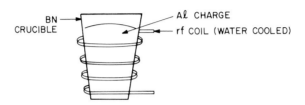

Fig. 7 Inductively heated evaporation source. The molten Al charge is contained by the dielectric BN crucible.

9.2.4 Sputter Deposition

Conventional sputtering[11] has found wide application in IC processing. Metals such as Ti, Pt, Au, Mo, W, Ni, and Co are readily sputtered using either a dc or rf discharge in a diode system. Sputtering is a physical phenomenon involving the acceleration of ions, usually Ar^+, through a potential gradient, and the bombardment by these ions of a "target" or cathode. Through momentum transfer, atoms near the surface of the target material become volatile and are transported as a vapor to the substrates. At the substrates, the film grows through deposition. The sputtering of dielectrics, such as Al_2O_3 or SiO_2 requires the use of an rf power source, while conductors may be sputtered with either power source. Al is difficult to sputter by conventional means because residual oxidants form a stable oxide on its surface during sputtering. A high electron density is required in the discharge to increase the ion current density at the target surface and thus prevent the oxide from forming. The high density may be achieved by introducing an auxiliary discharge, as in triode sputtering,[11] or through the use of magnetic fields to capture the electrons and increase their ionizing efficiency, as is done in magnetron sputtering.[12, 13, 14]

Ion beam sputtering[15] has also been used to sputter both metals and insulators. The flux of energy to the target can be modified through independent variation of ion current and energy. Furthermore, the target is in a lower pressure chamber than in other sputter processes, so that more of the sputtered material is transferred to the substrate and less background gas is incorporated in the deposited film. No ion beam sputter deposition system has yet been developed for the metallization of large numbers of silicon wafers.

Some characteristics of sputter deposition are: (1) the ability to deposit alloy films with composition similar to that of the target, (2) the incorporation of Ar ($\sim$2%) and background gas ($\sim$1%) in the film, and (3) in conventional diode systems, considerable heating of the substrates ($\sim$350°C) by the secondary electrons emitted by the target. Often rf energy may be applied to the substrates which causes them to be bombarded by ions. If the rf is applied prior to metal deposition, the process is termed "sputter etching."[16] Sputter etching may clear residual films from window areas and enhance the contact between the metal and exposed areas. If the rf energy is applied during film deposition, it is a bias-sputter deposition,[17] and may enhance the step coverage of the film or reduce the severity of the surface topography. Bias-sputter deposition of SiO_2 produces planar Si wafer surfaces prior to metal deposition.[18]

9.2.5 Magnetron Sputter Deposition

When magnetron sputter deposition[12, 13, 14] was introduced, high-rate sputter deposition of Al and its alloys found practical application. The reason appears to be the much higher current density that occurs at the magnetron target surface during operation. Introducing a magnetic field converts the sputtering device from a high-impedance to a low-impedance structure. Two versions are available in large-scale film coating machines.

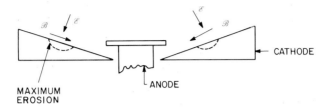

Fig. 8 Cross section of a conical magnetron. The magnetic field, $\mathcal{B}$, is provided by permanent magnets and is perpendicular to the electric field, $\mathcal{E}$, near the cathode. The anode is usually biased positively (20 to 40 V) relative to ground.

The first version, shown in Fig. 8, is the conical magnetron or S-Gun. The incorporation of a concentric anode and the circular symmetry are unique to this structure. The conical magnetron has a sputtered flux that is less than that found for the cosine distribution, and, if many substrates are to be coated simultaneously, planetary systems similar to those used with evaporation sources may be used.

Figure 9 shows the other source, a planar magnetron. It can be made in varying lengths so that large substrate areas can be coated. Usually, this magnetron is used with substrates translated in a plane before the magnetron. The magnetron also can be mounted in systems equipped with planetary substrate holders.

Both magnetrons operate at voltages an order of magnitude or more below the e-beam source voltage, and thus generate less penetrating radiation. Deposition rates depend on the source-to-substrate distance, and can be as high as 1 μm/min for Al and its alloys.

9.2.6 Chemical Vapor Deposition

The attractiveness of chemical vapor deposition (CVD)[19, 20] for metallization stems from the conformal nature of the coating (i.e., good step coverage), the ability to coat large numbers of substrates at a time, and the relatively simple equipment required.

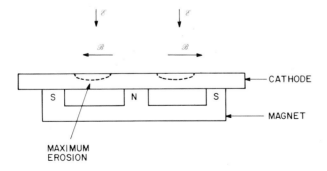

Fig. 9 Cross section of a planar magnetron. The magnets may be permanent or electromagnets. The anode is a separate entity, usually nearby, and biased positively relative to ground.

Unlike physical vapor deposition, which suffers from shadowing effects and imperfect step coverage, low-pressure CVD can yield conformal film coverage over a wide range of step profiles, and often yields lower bulk electrical resistivity.

The major effort in depositing metal for ICs by CVD has been in W deposition. Processes for W deposition have been developed with preferential deposition on silicon but not oxide.[21] The W is attractive because of its low electrical resistivity (5.3 $\mu\Omega$-cm) and its refractory nature. Both pyrolytic and reduction reactions have been used. For example, WF_6 may produce W films by:

$$WF_6 + \text{thermal energy} \rightarrow W + 3F_2 \tag{19}$$

or

$$WF_6 + 3H_2 \rightarrow W + 6HF \tag{20}$$

or

$$WF_6 + \text{plasma or optical energy} \rightarrow W + 3F_2 \tag{21}$$

Temperatures may range from 60 to 800°C in the various reactors. The use of WF_6 may cause loss of oxide during deposition, and WCl_6 may find application where the fluoride is not suitable, although the chloride requires higher temperatures.

Figure 10 shows a schematic view of a CVD reactor. The reactor tube is surrounded by a furnace and is considered a "hot wall" system. Using rf induction heating of a susceptor on which the substrates are positioned and cooling the reactor vessel walls constitutes a "cold wall" system. Arguments for either system have been based on efficiency of gas usage and particulate contamination.

In addition to W, other metals such as Mo, Ta, Ti, and Al are of interest for VLSI applications. Reactions such as[22]

$$2MoCl_5 + 5H_2 \xrightarrow{800°C} 2Mo + 10HCl \tag{22}$$

$$2TaCl_5 + 5H_2 \xrightarrow{600°C} 2Ta + 10HCl \tag{23}$$

$$2TiCl_5 + 5H_2 \xrightarrow{600°C} 2Ti + 10HCl \tag{24}$$

and using metal organic compounds such as tri-isobutyl aluminum

$$[(CH_3)_2CH-CH_2]_3Al \xrightarrow{150°C} [(CH_3)_2CH-CH_2]_2AlH$$
$$+ (CH_3)_2C=CH_2 \tag{25}$$

followed by

$$[(CH_3)_2CH-CH_2]_2AlH \xrightarrow{250°C} Al + \tfrac{3}{2}H_2$$
$$+ 2(CH_3)_2C=CH_2 \tag{26}$$

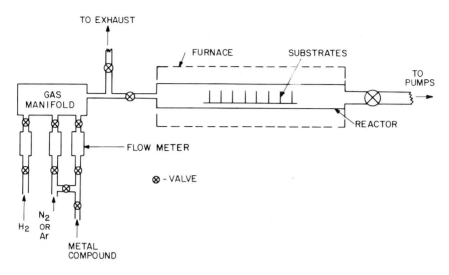

Fig. 10 Highly simplified view of a low-pressure CVD reactor system. To obtain enhanced reactions, the furnace could be augmented by a plasma source, intense light source, or other energy source.

are typical. The CVD deposition of Al films suitable for VLSI has yet to be demonstrated, although the other metals have been successfully deposited. The deposition of the refractory metal may be a preliminary step in forming the silicide, as demonstrated by WSi_2 films formed on polysilicon.[23]

9.3 PROBLEMS ENCOUNTERED IN METALLIZATION

Assume that an Al-based metallization is to be used. Questions remain about which alloy, what method of deposition, and what etching process should be employed. The answers are not simple since device performance, economics, and reliability will be factors that must be considered in each case.

9.3.1 Description

Step coverage presents a problem in metallization, because metal is deposited well into the process sequence. At this point the wafer has already had many steps generated in it. Metal alloy composition is another problem, because an excess of a constituent may cause device malfunction. A related problem is obtaining a low contact resistance. Particulates generated within the deposition chamber can severely limit the yields in the narrow linewidths associated with VLSI. Hillock (small, elevated areas) formation dependent on alloy composition and thermal history can change the specular nature of the film reflectivity and introduce difficulties in lithography and subsequent film coverage. Etching the metal layer has been a problem, because conventional wet etching cannot be used in VLSI. Figure 11 schematically shows an

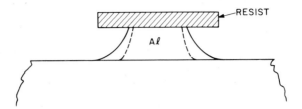

Fig. 11 Schematic view of a cross section of wet-etched Al beneath a resist mask. The solid profile is the expected boundary if etching is sufficient just to clear the surface. The broken line is the boundary for an overetch of 15 to 20%.

example of the results of isotropic etching. Because metal is attacked beneath the etch mask, compensation must be made for the linewidth lost in transferring the lithographic pattern to the etched metal. As lateral dimensions decrease and lines get closer together, compensation becomes physically impossible; anisotropic etching of the metal is therefore necessary.

9.3.2 Solutions

Solutions to the step coverage problem have been approached in several ways. First, raising the temperature of the substrate during film deposition (~300°C) creates greater surface mobility of the deposited material, thus reducing the severity of cracks that exist in corner regions. Next, orientation of the substrate relative to the source can be optimized.[24, 25] Optimization is especially important since shadowing occurs in the deposition process when using a point source such as an e-beam or an inductively heated melt. Computer simulation has been useful in modifying the supporting planetary system.

Since most planetary systems do not use rotation of the individual substrate about its own axis, orientation within the planet is significant in reducing step coverage problems.[24] Step edges that are parallel to the planet radius are coated symmetrically. Steps with edges placed perpendicular to the planet radius tend to be coated asymmetrically, and also tend more to exhibit cracks (Fig. 12).

If small contact windows are to be coated, the course of action may be different than outlined above. For VLSI, a plane surface may be approximated by depositing the interlevel dielectric by bias-sputter deposition (see Section 9.2.4) or by using planarization.[26] Planarization is a low-temperature process that reduces surface features. A thick resist layer is applied to the dielectric, and a plasma-etch process is used that attacks the dielectric and the resist at equal rates. To accommodate this process a thicker than normal (usually by a factor of 2) intermediate dielectric layer is needed. The extensive heat treatment that normally would be used to make the dielectric flow, thus reducing the severity of the step contours, cannot be tolerated in VLSI where implanted dopants are not permitted to diffuse extensively. Contact-window step coverage remains a problem even on planar surfaces, because extensive taper etching of the window edge would consume excessive area.

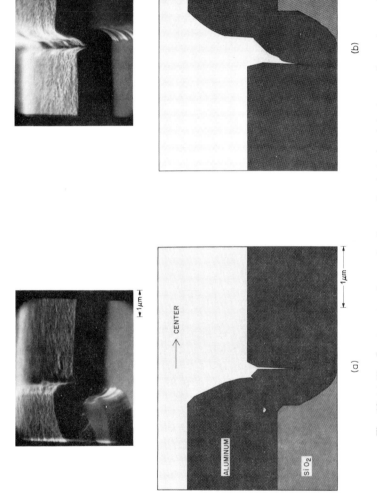

Fig. 12 (a) The figure on top shows what is actually obtained and the bottom figure shows what is predicted by a computer model. This step is perpendicular to the planet radius and the arrow points to the planet center. (b) The figure on top shows what is actually obtained and the bottom figure shows what is predicted by the computer model. The step is parallel to the planet radius. *(After Blech, Fraser, and Haszko, Ref. 24.)*

363

The use of sources that have larger areas than point sources, such as magnetrons, relieves many of the step coverage problems. If the substrates are relatively distant (20 to 30 cm) from the source, such as planetary-mounted wafers, the directionality of the sputtered metal vapor becomes more random. Randomness occurs because at pressures of about 0.5 Pa the mean-free path of the Ar atoms is of the order of 1 cm. Thus the metal vapor incident on the planetary-mounted substrate during magnetron sputtering is more random in direction than evaporated vapor but the vapor is "colder" because it transfers energy to the Ar gas. The vapor's lower energy, which is characteristic of the incident vapor, leads to less movement of the deposited species on the substrate surface. Decreased movement can limit the grain growth and the development of ordered (fiber texture) structures. The substrates can be relatively close and stationary, or they can move slowly before a large-area magnetron. This proximity to the source permits high deposition rates with material that has undergone an order of magnitude less travel through the Ar. Significantly more heating of the substrate can be achieved, resulting in improved step coverage. Sidewall and flat-surface film thickness ratios ranging from 50 to 100% have been obtained on steps. In windows this ratio is dependent on the aspect ratio (depth/width).

Alloy films may be deposited by single- or multiple-source evaporation. Electron-beam evaporation from Al-2% Cu, for example, yields a deposited film of Al-0.5% Cu. Silicon is usually added by co-evaporation, and thus requires control of evaporation from more than one source. The degree of control of alloy composition is critical, because the commonly used postmetallization 450°C sinter of Al alloys can remove Si by dissolution (see Figs. 13 and 14) from the substrate, if the alloy lacks sufficient Si. Redeposition of previously dissolved Si in windows occurs upon cooling if excess Si is present. The magnetron sputter sources offer the opportunity to use complex alloy sources for film deposition. In some early studies all constituents of commercial alloy targets were found in approximately the same concentration in the deposited films. The choice of alloy composition may be directed by the need to preserve a specular, hillock-free surface.

Particulate contamination during the metallization process can create defects. When the chamber containing the substrates is evacuated the gas flow may be turbulent,[5] and particles may be stirred up and deposited on the substrates. This condition can be minimized by using controlled throttling during the pump down. Also during venting, when substrates are to be removed, turbulence should be avoided in the chamber. Moving components which may have previously deposited layers that can flake are also another source of particulates. System cleanliness and minimizing film buildup on rolling or sliding surfaces are essential.

VISI requires anisotropic etching of metal layers. Both plasma etching and reactive sputter etching have been developed enough that commercial equipment is available. Another approach, which may be attractive if an unusual alloy is used, is lift-off.[27, 28] In lift-off the inverse pattern is formed by lithography and metal is deposited on the masked substrate. Then the desired pattern is revealed by lifting the mask and undesired metal (Fig. 15). The lifting is accomplished by using solvents that attack the lithographic pattern, thus undercutting the overlayer of metal. When the metal is deposited through apertures in the lithographic mask, it sits directly on the substrate,

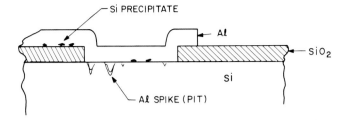

Fig. 13 A schematic view of an Al film contacting a large window ($\sim 10 \times 10~\mu m^2$). Note the Al spikes (pits) in the silicon and the precipitated silicon.

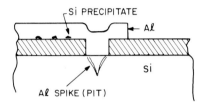

Fig. 14 A schematic view of a VLSI Al contact to a window ($\sim 1.5 \times 1.5~\mu m^2$). Note that the pit in the silicon can fill the window and that it may have a thin covering of precipitated silicon.

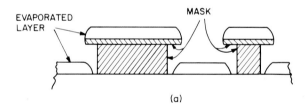

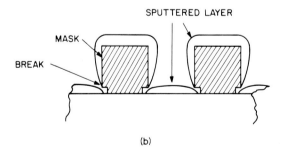

Fig. 15 Views of lift-off cross sections for (a) evaporated metal and (b) sputtered metal. Note the high-shadowing features used in evaporation and the undercut masks used in sputtering. These different features are necessary because the two deposition processes have different step coverages.

and thus remains after lifting. Unless thermally stable materials are used to form the mask layer, constraints on substrate temperature during film deposition may be imposed which would limit the usefulness of the metal layer. The masking layer should also withstand predeposition cleaning.

As well as being affected by the purity of the source material, the microstructure and purity of the film deposited in a vacuum chamber may be affected by how the system is pumped, the base pressure, and the rate of deposition. For example, if a chamber is evacuated to a base pressure of 10^{-2} Pa (7.5×10^{-5} torr), the residual gas (if there are no air leaks) is primarily water vapor. From the kinetic theory of gases

$$N = 6.4 \times 10^{19} (T)^{-1/2}P \tag{27}$$

where N is the bombardment rate of H_2O molecules (in cm^{-2}-s^{-1}), T is the absolute temperature, and P is the pressure in Pa. At $T = 300$ K and 10^{-2} Pa, N is significant, because it approximates the rate of arrival of Al at the substrate for a deposition rate of 50 Å/s. Al films normally have less than 50% O content (usually $\leq$ 0.1%); thus the probability of the H_2O molecule relinquishing the O atom to the metal film is significantly less than 1. Nevertheless, a low-background pressure prior to initiating film deposition minimizes incorporation of oxygen.

Similarly, the base pressure prior to sputter deposition should be low. The addition of Ar, along with any impurities it contains, to the ambient for sputter deposition also increases the impurity content of the film. The type of pumps used to evacuate the deposition chamber and the traps are important considerations. Oil contamination from mechanical and diffusion pumps may be minimized by controlling the pump down sequence and using cryotraps. Closed-cycle helium cryopumps and turbomolecular pumps are frequently used in evaporation and sputter deposition equipment. These pumps are primarily used to avoid oil contamination, and also to reduce the operating costs incurred by continuous use of liquid nitrogen in cold traps.[5]

Of course, the Si substrates should be cleaned before being placed in the metallization chamber. Most common cleaning techniques involve the use of buffered HF or HF solutions. These solutions remove thin residual oxides from Si and polycrystalline Si, as well as remove some oxide from the intermediate dielectric. Surface contaminants containing sodium are removed along with the surface layers. Extensive deionized water rinses follow the aggressive cleaning to remove the fluoride. Spin drying in a warm, dry nitrogen flow removes the water. The substrates are loaded shortly after drying to avoid recontamination. The small amount of SiO_2 ($\leq$ 20 Å) resulting from the deionized water rinse and exposure to air offers no significant barrier to the Al metal film when sintering is performed at 300°C and above. No barrier exists because of the high energy of formation of the Al_2O_3 (400 kcal/mol)[29] relative to that of SiO_2 (205 kcal/mol).[29] These energy values permit the contacting Al to reduce the thin SiO_2 layer.

The quality of the metal films deposited in a system should be checked frequently by evaluating the C-V characteristics of a gate oxide-type SiO_2 layer (which comes from a furnace known to produce clean oxides). These checks should be made if the system is cleaned, a new source is installed, the system has had questionable test results, or unusual substrate material has been processed.

9.4 METALLIZATION FAILURE

As device structures diminish in size, the migration of Al into the Si substrate at contact windows will cause failure. This migration may occur during the fabrication process or in subsequent device operation. In addition, migration of Al in the metal lines during device operation may result in failure through open circuits. As VLSI emerges the pursuit of solutions to these failure mechanisms is an area of intense activity.

9.4.1 Junction Spiking

Junction spiking is a penetration of a p-n junction interface by a conductive projection. Although the problem of penetration of the Si substrate by Al "spikes," caused by the local dissolution of Si, can occur generally in IC processing, the problem is compounded in VLSI (see Figs. 13 and 14). In VLSI the junctions are shallow, typically of the order of 0.3 μm deep, and the contact windows are small. This combination is formidable since the Si that satisfies the solubility requirements of the Al is only accessed through small area contacts, which increase the depth of the spike.

The problem of junction spiking may be solved by depositing Al with Si added. The amount of Si required should be determined by the maximum process temperature and the Al-Si phase diagram (see Fig. 16). For example, heating the substrate to 450°C should require that the Al contain 0.5 wt. % Si. However, in practice slightly more than 1 wt. % Si is required. If the Al contacts are all to p^+ Si, this method of

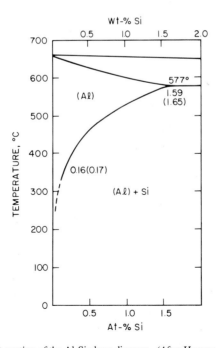

Fig. 16 A portion of the Al-Si phase diagram. *(After Hansen, Ref. 58.)*

solving the spiking problem is acceptable. However, another problem may be evident if n^+ Si must be contacted. Because of the excess Si present in the Al, some precipitation of Si occurs in the contact window and will form a nonohmic contact to n^+ Si, because the recrystallized Si precipitate contains Al (which is a p dopant).[26] Another method of satisfying the Si requirements of the Al film is to deposit the film on a layer of polysilicon. The polysilicon may be doped p^+ or n^+ by in-situ or post-deposition doping. If the polysilicon is deposited by CVD methods, a conformal coating exists in the windows. The solubility requirements of the Al are satisfied locally since the polysilicon is present beneath the Al at all points.

A structure that will work equally well for both n^+ and p^+ contacts is required An elegant structure involving multiple layers has been proposed.[30] Figure 17 shows this structure. The bottom layer in the window is a silicide formed by reacting a noble or near-noble metal with the substrate. Covering the silicide is a barrier layer that prevents the top layer (Al) from reacting with the silicide. The contact structure may be formed by depositing a single layer consisting of a mixture of a refractory and a noble metal. By controlling the deposition temperature and restricting the process temperature, only the noble metal silicide forms at the contact to the silicon, while above the contact the remaining film acts as a barrier to the Al layer. The combination of refractory and noble metals in a single layer makes etching a formidable task so that lift-off patterning becomes a viable solution to the problem. These combination structures have been evaluated for $Al-Pd_{.80}W_{.20}-Si$ and $Al-Pt_{.10}Cr_{.90}-Si$, and found to be stable[30] for normal contact sintering at 450°C.

Another method for forming shallow contacts is to deposit a mixture of metal and Si such that some Si is consumed in the contact window, although not as much as if the metal alone were deposited. These structures have been used for Schottky barrier contacts where Pd-Si and Pt-Si mixtures were deposited and subsequently reacted to form Pd_2Si and PtSi contacts.[30] Similar shallow silicide contacts should be feasible using other silicide systems for non-Schottky contacts to VLSI structures. Of course, the need for a barrier on the silicides still exists if reactions with Al are possible.

Epitaxial Si layers in contact windows may be able to supply the needed barrier. One approach is to use molecular beam epitaxy (MBE), a low-temperature process, and grow a Si plug in the window. This Si plug would raise the contact interface away from the junction, and also alleviate metal step coverage at the window.

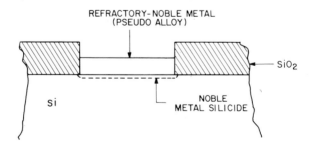

Fig. 17 Idealized cross-sectional view of a barrier-noble silicide contact to silicon.

Another technique is to use solid phase epitaxy where an upward diffusing silicide ($TiSi_2$) layer[31] passes through a covering polysilicon layer and converts it to an epitaxial crystalline plug. Again this results in the Al contact being moved away from the junction.

Diffusion barriers may be considered essential to incorporation of stable contacts in VLSI structures. In the preceding paragraphs, reference was made to a barrier layer (refractory metal) on a silicide layer. Such layers are generally polycrystalline and therefore have different diffusion characteristics than bulk materials have at low temperature. The rapid diffusion of material through grain boundary regions requires that impurities be incorporated to passivate the grain boundaries. As an example, Mo or Ti-W films may be improved by the addition of O or N at levels of 10^{-3} wt.% or less. These are examples of "stuffed barriers,"[32] since they inhibit rapid diffusion along grain boundaries.

Passive compounds such as nitrides are also attractive as barriers, because they may be deposited by reactive sputter deposition using metal targets. However, because of possible reactions between Al and TiN at $\sim500°C$, the use of another layer such as Ti between the nitride and the Al is necessary if high-temperature processing will be used.[33] Sufficient Ti must be present to satisfy the Al in forming Al_3Ti, otherwise Al will get through the TiN and react with the Si substrate. For structures that will be heated to 450°C, Ta may be a preferable metal since it too forms a nitride, TaN. It has been observed that Ta metal between Al and TaN has been found superior to the Ti-based structure.[33] Using one metal to form both the nitride and metal layer is attractive, because the layers can be deposited sequentially.

Refractory metals deposited directly on the Si exposed in the windows may be considered. Ta reacts with Si only at temperatures above 600°C, and may form a Al_3Ta layer with Al above 450°C. Provided sufficient Ta is deposited, Al spiking should be prevented. Sputter-deposited Ti-W (10 to 90 wt.%) has been evaluated as a barrier layer between Al and Si, as well as between Al and Pt-Si.[34] Contacts were stable after heating to 550°C. The use of the Ti-W beneath the Al caused a 10% increase in the resistivity, because either Ti or W diffused into the Al layer.[34]

9.4.2 Electromigration

A prime consideration in device reliability is the electromigration resistance of the metallization. Electromigration is observed as a material transport of the conductive material. It occurs by the transfer of momentum from the electrons, moving under the influence of the electric field applied along the conductor, to the positive metal ions.[35] Hence, after conductor failure, a void or break in the conductor is observed and nearby a hillock or other evidence of material accumulation in the direction of the anode (Figs. 18 and 19) is found.[36]

Figure 18 shows SEM views of S-Gun sputter-deposited Al-0.5 wt. % Cu and In-Source-evaporated Al-0.5 wt. % Cu failures. Melting is evident in both cases but view (b) also clearly displays hillock formation in the direction of the electron flow. Figure 19 shows an SEM view of failure of e-beam-evaporated Al on polysilicon. In

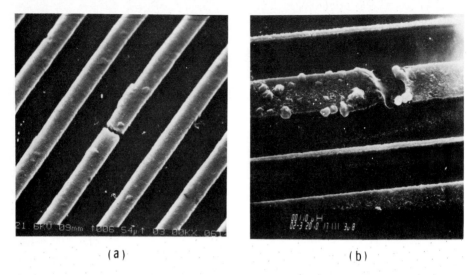

(a)   (b)

Fig. 18 SEM micrographs of electromigration failure. (a) S-Gun magnetron-deposited Al-0.5 wt. % Cu and (b) In-Source-evaporated Al-0.5 wt. % Cu. *(After Vaidya, Fraser, and Sinha, Ref. 36.)*

view (a) catastrophic melting and balling of the Al is evident. View (b) shows the area after etching the Al to reveal the polysilicon runners and the Si precipitates. The arrows indicate the direction of the electron flow.

Electromigration resistance of Al film conductors can be increased by several techniques. These techniques include alloying with Cu, incorporation of discrete layers such as Ti, encapsulating the conductor in a dielectric, or incorporating oxygen during film deposition.[37] The mean-time-to-failure (MTF) of the conductor can be related to the current density J in the conductor and an activation energy Q by

$$\text{MTF} \propto J^{-2} \exp[Q / kT] \tag{28}$$

for $10^5 \leq J \leq 2 \times 10^6$ (A/cm^2). Experimentally, a value of $Q \sim 0.5$ eV is obtained, and taken to indicate that low-temperature grain boundary diffusion is the

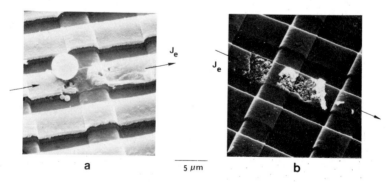

a   5 µm   b

Fig. 19 SEM view of an electromigration failure of e-beam-evaporated Al on polysilicon. (a) Metal is on the polysilicon. (b) Al is etched to reveal the polysilicon *(After Vaidya, Fraser, and Sinha, Ref. 36.)*

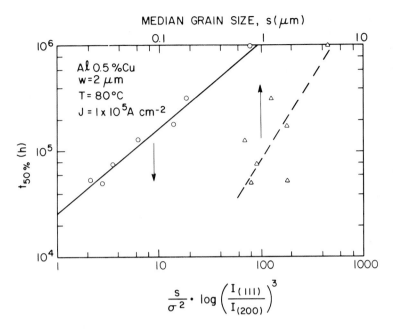

MEDIAN GRAIN SIZE, s(μm)

$$\frac{s}{\sigma^2} \cdot \log \left(\frac{I_{(111)}}{I_{(200)}} \right)^3$$

Fig. 20 Plot of median conductor lifetime, $t_{50\%}$, versus median grain size of Al-0.5 wt. % Cu (denoted by triangles) and $t_{50\%}$ versus an empirical parameter (denoted by circles). *(After Vaidya, Fraser, and Sinha, Ref. 36.)*

primary vehicle of material transport, since $Q \approx 1.4$ eV would characterize the self-diffusion of Al in bulk crystalline material. Experiments have also related the MTF to grain size in the metal film, distribution of grain size, and the degree to which the conductor exhibits fiber texture ($\langle 111 \rangle$). Figure 20 shows the relationship between two parameters and the time for 50% ($t_{50\%}$) of the conductors to fail.[36] One parameter is the median grain size s and the other is $s/\sigma^2 \times \log \left[I_{(111)}/I_{(200)} \right]^3$. Here σ^2 is a measure of the distribution in grain sizes, $I_{(111)}$ is the intensity of the $\langle 111 \rangle$ reflection, and $I_{(200)}$ is the intensity of the $\langle 200 \rangle$ reflection obtained from x-ray diffractometer measurements of the films. The latter parameter is strongly dependent on the method of film deposition. Figure 21 shows how e-beam-evaporated films have demonstrated superior lifetimes compared to In-Source and S-Gun magnetron-sputtered films. Inferior lifetimes may be due both to the 2 wt. % Si and to the lower surface mobility of the incident metal vapor atoms during sputter deposition when the substrates are many free path lengths from the sputter source (see Section 9.1.4). Sputter deposition is to be contrasted with an evaporation deposition where the vapor travels in straight lines to the substrates and loses virtually no energy in transit. A recent discovery,[36] significant to VLSI, is that decreasing linewidth (below 2 μm) results in increased MTF for e-beam-deposited Al-0.5% Cu (Fig. 21).[36] This discovery is related to the fact that when linewidth shrinks sufficiently, the metal line is composed of single-crystal segments.

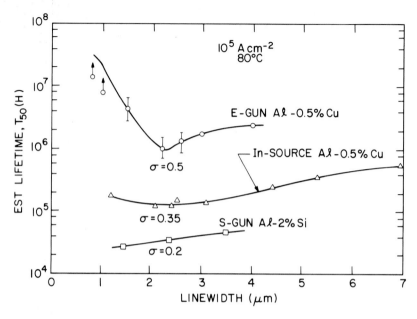

Fig. 21 Median conductor lifetime versus linewidth for Al alloys deposited in three different ways *(After Vaidya, Fraser, and Sinha, Ref. 36.)*

Attempts have been made to eliminate contact spiking by combining Al metal films with n^+ doped polysilicon. Polysilicon was used for two reasons: first, to provide Si that satisfies the solubility requirements of the Al, and second to provide a conformal conductive layer beneath the Al at steps. The first requirement was satisfied, but the second requirement was not completely met. Si was found to move in the Al grain boundaries and failure due to electromigration occurred predominantly at steps where the evaporated Al was thinned.[38] Failure rapidly followed the loss of continuity in the metal film, due to electromigration of the Si resulting in local heating and melting (see Fig. 19).

Note also that merely using an Al-(1−3%) Si alloy does not necessarily protect the junctions.[39] For example, a circuit under test and locally operating below 250°C will show Si electromigrating within the metal and pits forming at the cathode, while Si (p^+) will be deposited at the anode. Perhaps, the use of refractory layers sandwiched between Al and the underlying polysilicon would perform the task of isolating the interactive layers while also providing a shunt if the Al were to form an open circuit.

9.5 SILICIDES FOR GATES AND INTERCONNECTIONS

9.5.1 Application Requirements

The general requirements for gates and interconnects are that the film material have $\rho \lesssim 60\ \mu\Omega$-cm, be stable throughout the remaining process steps, and be reliable. Refractory metals such as W and Mo and the silicides $TiSi_2$, WSi_2, $MoSi_2$, and $TaSi_2$

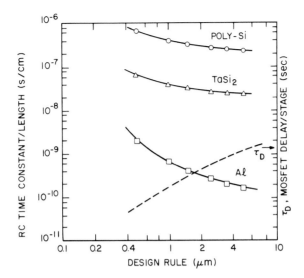

Fig. 22 *RC* time constant per unit length for three conductive materials as a function of feature size. Also shown is delay per stage of ring oscillators as a function feature size. *(After Sinha, Ref. 47.)*

have been proposed and used as MOSFET gate electrode materials either alone or with doped polysilicon.[40-46] These disilicides are stable in contact with the polysilicon, and, as will be seen, the presence of the polysilicon helps to stabilize the structures in oxidizing ambients. However, the metals (notably W and Mo), if used directly on gate oxides, are not stable in oxidizing ambients.

To appreciate the need for higher-conductivity gate and interconnect materials, consider the fact that *RC* delay time is a key factor in VLSI or high-speed circuits.[47] Figure 22 compares the delay per unit length versus linewidth for polysilicon (30 $\Omega/\square$), TaSi$_2$ (1.25 $\Omega/\square$), and Al (.025 $\Omega/\square$). The conductive layers are assumed to be 1 μm thick and sandwiched between two 1.5-μm-thick SiO$_2$ layers. These facts imply that for a given maximum tolerable delay, a conductor may be more than an order of magnitude longer if a silicide is used instead of polysilicon. Of course, the use of an additional metal layer to interconnect short lengths of polysilicon is an option, but the added process complexity and cost make it less attractive than using a single-level conductor.

9.5.2 Deposition Techniques

Silicides may be formed in several ways. A metal film may be deposited on polysilicon, and the structure sintered to obtain the silicide. (1) The silicide film may be deposited by co-deposition by sputtering[42] or evaporating[44] simultaneously from metal and silicon sources either onto oxide or polysilicon; (2) sputtering from a single source, such as a composite or sintered target, onto oxide or polysilicon; or (3) chemical vapor deposition (either thermally or plasma enhanced) of the silicide on oxide or polysilicon. The most widely used techniques have been sputter deposition and e-beam evaporation. Co-deposition by either process permits control of the ratio of

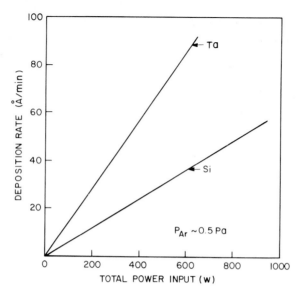

Fig. 23 S-Gun magnetron deposition rates versus power for Ta and Si. The anode was at +40 V relative to ground.

metal to Si atoms in the deposited layer. As an example, Fig. 23 shows the sputter deposition rates for Ta and Si, respectively, in a system equipped with 45-cm planetary and S-Gun magnetrons. Figure 24 shows the symmetric placement of sources for co-deposition. Because of the stability of the sources, co-deposition may be performed by maintaining each source at a predetermined power dissipation. Electron-beam evaporation may be performed in a similar manner using two independent sources. However, co-evaporation is likely to be used less often than co-sputtering in production, because the desired material constituent ratio is more difficult to maintain in the deposited film.

9.5.3 Properties

Room temperature resistivities of various silicides[48] on n^+ poly Si are given in Table 2. The lowest resistivity is obtained in $TiSi_2$ formed by sintering a metal layer on polysilicon and should be contrasted with a value 1.5 to 2 times larger obtained when

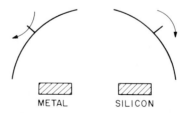

Fig. 24 Schematic view of magnetrons and a planetary system for co-sputter deposition. Note the symmetric placement of the sources, which is necessary to get uniform films.

Table 2 Silicide resistivities (300 K)

Material	Starting form	Resistivity ($\mu\Omega$-cm)
TiSi$_2$	Metal/polysilicon	13−16
	Co-sputtered	25
ZrSi$_2$	Metal/polysilicon	35−40
HfSi$_2$	Metal/polysilicon	45−50
TaSi$_2$	Metal/polysilicon	35−45
	Co-sputtered	50−55
MoSi$_2$	Co-sputtered	100
WSi$_2$	Co-sputtered	70
CoSi$_2$	Metal/polysilicon	17−20
	Co-sputtered	25
NiSi$_2$	Metal/polysilicon	50
	Co-sputtered	50−60
PtSi	Metal/polysilicon	28−35
Pd$_2$Si	Metal/polysilicon	30−35

TiSi$_2$ is obtained by co-sputter deposition. This difference is believed to be due to electron mobility, which may be higher in the silicide formed by sintering the metal since larger crystals of the silicide are obtained. Figure 25 shows that the reflectivity from the silicide surface formed by reacting the metal with polysilicon is quite diffuse. Such films are difficult to work with in photolithography and may even disturb automatic alignment devices. In contrast the co-deposited film reflectivity is more like that of a metal, and the film merely replicates the underlying polysilicon surface. Resistivity and reflectivity, however, are not the only parameters that determine which silicide to use. The stability of the desired phase can be extremely significant. For example, the existence of eutectics would limit the maximum temperature of the silicide in contact with Si. Thus, Pd$_2$Si is limited to about 700°C, PtSi to about 800°C, and NiSi$_2$ to about 900°C. The other silicides in Table 2 should be stable to temperatures above 1000°C. Stability in an oxidizing ambient is also important.

Stress, because of its magnitude, is a significant parameter in silicide layers formed on Si wafers. To a first approximation, the source of the tensile stress observed in sintered silicide layers appears to be caused by the net volume loss occurring when the volumes of the metal and the Si are combined to form the silicide. However, temperature-dependent measurements of stress have shown that the tensile stress in TaSi$_2$ decreases as temperature is increased and that the coefficient of thermal expansion of TaSi$_2$ is approximately $9 \times 10^{-6}/\text{C}°$, while that of Si is approximately $3 \times 10^{-6}/\text{C}°$.[49] The major portion of the room temperature stress is due to thermal expansion differences and the relatively high temperature at which the silicide is formed by sintering. The stress levels may be reduced in the silicides by forming them at lower temperatures through the introduction of other sources of energy. The

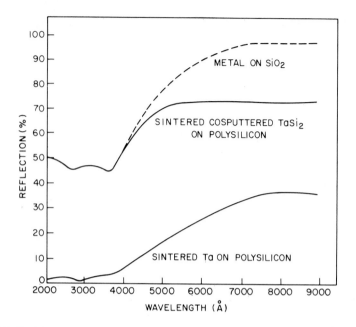

Fig. 25 Reflectivity of three surfaces versus wavelength. Note the low reflectivity of the bottom curve.

significance of stress may be appreciated by examining Fig. 26, where the room temperature stress of a 2500-Å co-sputter-deposited $TaSi_2$ film is shown as a function of various MOS process steps. Should the value of stress rise appreciably above $\sim 2 \times 10^{10}$ dyn/cm^2, the adherence of the layer would be uncertain and the useful thickness would also be limited. The preceding discussion illustrates the need for a silicide layer compatible with the process sequence.

Similarly, Fig. 26 shows the variation in the room temperature sheet resistance of the same (stress) specimen measured by a four-point probe as a function of the same process steps. In addition, the silicide may be exposed to chemicals, such as $NH_4 F/HF$ solutions that attack silicides. Thus $TiSi_2$ is attractive because of its resistivity, but unattractive because of its susceptibility to attack by HF solutions.

The next requirement is the determination of the work function ϕ_M of the MOS electrode material by obtaining ϕ_{MS} $(= \phi_M - \phi_{Si})$, where $\phi_{Si} \approx 4.35$ V. The value of ϕ_{MS} is obtained by making a series of capacitance measurements on oxides of different thickness and plotting the flatband voltage V_{FB} versus oxide thickness.

$$V_{FB} = \phi_{MS} - \left[\frac{Q_f + Q_m + Q_{ot}}{C} \right] \qquad (29)$$

where Q_f is the fixed charge per unit area, Q_m is the mobile charge (e.g., Na^+) per unit area, Q_{ot} is the trapped charge per unit area in the oxide, and C is the capacitance. To keep the parameters in the second term common to all the capacitors, a thick oxide is grown and selectively etched to provide four or more dielectric thicknesses. With these different thicknesses, one substrate may yield four or more

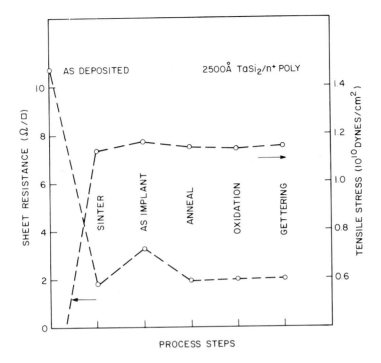

Fig. 26 Room temperature stress and sheet resistance of TaSi$_2$ on polysilicon as a function of process steps.

experimental values. As an example the plots for TiSi$_2$/n$^+$ polysilicon, TaSi$_2$/n$^+$ polysilicon, and Al/n$^+$ polysilicon are shown in Fig. 27.[47] The values of ϕ_M for the two silicides are 3.30 $\pm$ 0.05 V and are very similar to ϕ_M for Al/n$^+$ polysilicon, which is 3.25 V. Should the values for the silicides differ from the "standard" Al/n$^+$ polysilicon, modifications would have to be made in a process sequence (e.g., implants to adjust threshold voltage) to incorporate the silicides and still have device parameters meet specifications.

Also test capacitors must be subjected to temperature-voltage bias stressing in order to evaluate the stability of the capacitors. Since many of the metals used to form silicides are not as free from impurities as polysilicon, mobile charge and slow trapping should be monitored. This is conveniently done by making test capacitors that have been subjected to the full range of the process sequence. Figure 28 shows an example of such a test capacitor.[47] In this case a TaSi$_2$/n$^+$ polysilicon electrode on a 526-Å-thick oxide. The tests involved $\pm$10 V bias at 250°C which did not shift the C-V curve and is indicative of no Q_m contamination. The C-V curve also gives a fixed charge of $Q_f = 2 \times 10^{10}$/cm^2 which is acceptable.

The refractory metals may exhibit oxidation at lower temperatures than their silicides do. The formation of volatile oxides such as occur with W or Mo can lead to a reaction that ruptures the film. In Fig. 29 films of co-sputtered W-Si and WSi$_2$ deposited directly on oxide are shown after exposure to air at 1000°C.[47] However, if additional Si is available, such as in an underlying layer of polysilicon, the oxidation of

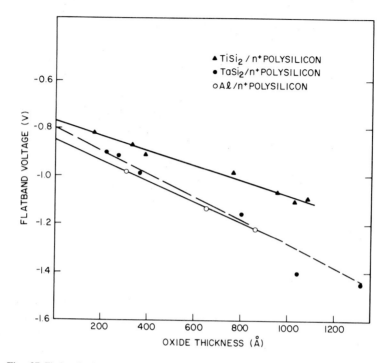

Fig. 27 Flatband voltage versus oxide thickness for three electrodes. *(After Sinha, Ref. 47.)*

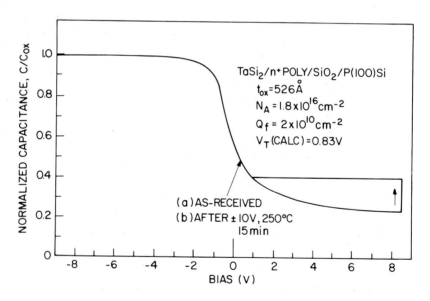

Fig. 28 C-V measurements of TaSi$_2$/n$^+$ polysilicon test capacitors on a fully processed device wafer. *(After Sinha, Ref. 47.)*

PEST REACTION UPON AIR OXIDATION

WSi/SiO₂/Si

(a)

WSi₂/SiO₂/Si

(b)

Fig. 29 SEM views of adherence failure of co-sputtered (a) W-Si and (b) WSi₂ films deposited on oxidized silicon. *(After Sinha, Ref. 47.)*

the WSi_2/polysilicon structure proceeds by the diffusion of Si to the silicide surface to form SiO_2. This diffusion of Si leaves the silicide layer intact until the polysilicon has been consumed. The same process has been observed for $TaSi_2$/polysilicon and is demonstrated by the change in resistance of such a layer through steam oxidation.[50] The stress of the $TaSi_2$ did not change until the polysilicon had been consumed. The (covering) SiO_2 that results from oxidizing a silicide/polysilicon structure is not expected to be very different from the oxide found on polysilicon, and this has been observed for $TaSi_2/n^+$ polysilicon[50] and WSi_2/n^+ polysilicon.[44] For structures involving more than one layer of silicide/n^+ polysilicon the oxide isolation should be similar to that obtained with multilevel polysilicon structures.

9.5.4 Device Performance

Various integrated circuits, both CMOS and NMOS, have been made from refractory gate materials other than polysilicon. The only reported discrepancies have been between $TaSi_2/n^+$ polysilicon and n^+ polysilicon controls in CMOS enhancement mode transistors, where the threshold voltages V_T differed by 0.3 V for the n- and p-channel devices, respectively.[51] Also, the use of $MoSi_2$ directly on gate oxide leads to a work function difference of 0.5 V compared to n^+ polysilicon.[52] Increases in circuit operational speed have been observed where high-conductivity gates have been used both in normal silicon and silicon-on-sapphire devices.[53] Compatibility with existing processing has been demonstrated. The requirements for gate and interconnect level sheet resistances of less than 3 $\Omega/\square$ can be met by the silicide/polysilicon structures presently used.

Where will the future devices lead technology? The use of shallow junctions of approximately 0.1-μm depth (e.g., in the source and drain areas) will require the use of a conductive film to reduce the sheet resistance of about 100 $\Omega/\square$ by an order of magnitude. A method to accomplish this reduction has already been demonstrated in an MOS structure utilizing PtSi Schottky contacts in the source and drain areas.[54] The use of materials other than PtSi and high doping in the substrate leads to low-resistance (non-Schottky) contacts. Lower processing temperatures and the possible use of presently unsuitable materials will be required, along with the shallow junctions. The use of other refractory materials is possible as well, since the resistivity of TiB_2 has been reported to be about 10 $\mu\Omega$-cm.[55]

Incorporation of a silicide layer at the gate and interconnect level also raises the question of stability of the primary metallization (Al) contact to the silicide. The major concern of course is the extent of the reaction that may occur in sintering the Al (450°C for e-beam and 300°C for Al deposition processes without ionizing radiation). Al can apparently penetrate $TaSi_{2.2}$ when heated to 500°C for 60 minutes. This may be due to precipitated excess Si in the $TaSi_{2.2}$ film acting as a soluble defect, and opening the layer beneath to the Al. However, no test device failure could be related to a failure of the Al-$TaSi_2$ contact when sintering was performed at 450°C for 30 min. In NMOS structures, if a layer of n^+ polysilicon is deposited over the windows under the Al, no interaction between the Al and the $TaSi_2$ occurs for sintering at 450°C. Al has been found to interact with PtSi, Pd_2Si, $CoSi_2$, and $MoSi_2$, and then penetrate into the underlying Si substrate when sinter temperatures range from 200 to 550°C.

9.6 CORROSION AND BONDING

Once the device structure has been completed, a passivation layer may be applied over the final metal layer. Leads which connect the chip to the outside world are bonded to the metal through windows etched in the passivating film. The layer used for passivation may be a low-temperature phosphosilicate glass, a plasma CVD dielectric (either oxide or nitride), a spin-on layer of glass-containing suspension, or a spin-on organic layer. In general, the passivating layer protects the metal pattern from being scratched during handling prior to bonding.

The bonding wires may be either Al-1% Si or Au. With Al alloy wire bonds, failure may occur in the wire just beyond the bond, due to thinning or fracture. Gold wire bonds can be made easily, because of the ductility of the wire; however, intermetallics may form and weaken the structure. The so-called "purple plague," $AuAl_2$, is an indication of intermetallic formation.[56] To prevent formation of the intermetallic, the length of time that the Au and Al alloy are in contact at high temperature must be carefully controlled.

Metallization corrodes significantly in a high-humidity environment.[57] One approach is to utilize hermetically sealed packages that can prevent the corrosion. If the structures are not sealed, then residuals, such as Cl that may be present after plasma or reactive sputter etching, react with moisture to attack the Al, even without an imposed electric field

$$Al + 3HCl \rightarrow AlCl_3 + {}^3\!/_2 H_2 \tag{30}$$

$$AlCl_3 + 3H_2O \rightarrow Al(OH)_3 + 3HCl \tag{31}$$

Note that the Cl is not bound after the $Al(OH)_3$ is formed, leading to further attack of exposed Al. The problem is compounded by placing metal lines close together and imposing an electric field between them, such as would occur in VLSI structures. Passivation by removing residual Cl is common in most Al dry-etching processes. This residual Cl may be removed by a CF_4-O_2 or O_2 plasma treatment immediately after etching and before exposing the wafers to the atmosphere. Further stability may be gained by thermally oxidizing the metal.[57] Excessive P in phosphosilicate glass may cause formation of HPO_3 on the dielectric surface, which may in turn lead to attack of the Al alloy structure. Maintaining a maximum of 6% P content in the dielectric minimizes this source of corrosion. Corrosive environments where the reactants are present in the atmosphere require not only cleaning, but passivation and encapsulation.

9.7 FUTURE TRENDS

Multi-level metallization may be necessary in order to keep VLSI chip areas to a size compatible with reasonable yield. Cross-unders in the bulk silicon substrate may be possible by using epitaxially grown silicon on high-conductivity silicides. These buried high-conductivity patterns may be an alternative to additional metal layers above the substrate surface. With the use of lower processing temperatures will come the use of materials currently not considered (i.e., materials that are thermodynamically unstable in present process sequences), such as PtSi. A major stumbling block to an all-low-temperature process is the required high-temperature treatment to getter impurities during the final stages of the process. Photochemical deposition may be useful in maintaining low-temperature processing. Increased use of thermal CVD techniques is also likely, since the more conformal nature of the films is attractive in metal step coverage. Development of a conductor plug to remove the metal step coverage problem at contact windows is another possibility. Much of the burden for achieving practical VLSI technologies will fall on metallization, and, for the foreseeable future, metallization will remain an active area for the introduction of new materials and processes.

REFERENCES

[1] J. L. Vossen, Ed., *Bibliography on Metallization Materials and Techniques for Silicon Devices*, Thin Film Division, American Vacuum Society, New York, 1974−1982.
[2] M. P. Lepselter, "Beam Lead Technology," *Proc. Second Symposium on the Deposition of Thin Films by Sputtering*, University of Rochester, June 1967, p. 48.
[3] W. D. Ryden and E. F. Labuda, "A Metallization Providing Two Levels of Interconnect for Beam-Leaded Silicon Integrated Circuits," *IEEE J. Solid-State Circuits*, SC-12, 376 (1977).
[4] S. M. Sze, *Physics of Semiconductor Devices*, 2d ed., Wiley, New York, 1981, p. 304.
[5] J. F. O'Hanlon, *A User's Guide to Vacuum Technology*, Wiley, New York, 1980.
[6] A. S. Grove, *Physics and Technology of Semiconductor Devices*, Wiley, New York, 1967.
[7] R. Glang, "Vacuum Evaporation," in L. I. Maissel and R. Glang, Eds., *Handbook of Thin Film Technology*, McGraw-Hill, New York, 1970, Chapter 1, p. 1-107.
[8] S. Dushman, *Scientific Foundations of Vacuum Technique*, Wiley New York, 1962, p. 21.

[9] R. Glang, "Vacuum Evaporation," in L. I. Maissel and R. Glang, Eds., *Handbook of Thin Film Technology*, McGraw-Hill, New York, 1970, Chapter 1, p. 1-26.

[10] R. Glang, "Vacuum Evaporation," in L. I. Maissel and R. Glang, Eds., *Handbook of Thin Film Technology*, McGraw-Hill, New York, 1970, Chapter 1, p. 1-92.

[11] J. L. Vossen and J. J. Cuomo, "Glow Discharge Sputter Deposition," in J. L. Vossen and W. Kern, Eds., *Thin Film Processes*, Academic, New York, 1978, p. 12.

[12] J. A. Thornton and A. S. Penfold, "Cylindrical Magnetron Sputtering," in J. L. Vossen and W. Kern, Eds., *Thin Film Processes*, Academic, New York, 1978, p. 76.

[13] D. B. Fraser, "The Sputter and S-Gun Magnetrons," in J. L. Vossen and W. Kern, Eds., *Thin Film Processes*, Academic, New York, 1978, p. 115.

[14] R. K. Waits, "Planar Magnetron Sputtering," in J. L. Vossen and W. Kern, Eds., *Thin Film Processes*, Academic, New York, 1978, p. 131.

[15] J. M. E. Harper, "Ion Beam Deposition," in J. L. Vossen and W. Kern, Eds., *Thin Film Processes*, Academic, New York, 1978, p. 175.

[16] C. M. Melliar Smith and C. J. Mogab, "Plasma Assisted Etching Techniques for Pattern Delineation," in J. L. Vossen and W. Kern, Eds., *Thin Film Processes*, Academic, New York, 1978, p. 497.

[17] J. L. Vossen and J. J. Cuomo, "Glow Discharge Sputter Deposition," in J. L. Vossen and W. Kern, Eds., *Thin Film Processes*, Academic, New York, 1978, p. 50.

[18] T. N. Kennedy, "Sputtered Insulator Film Contouring over Substrate Topography", *J. Vac. Sci. Technol.*, *13*, 1135 (1976).

[19] W. Kern and V. S. Ban, "Chemical Vapor Deposition of Inorganic Thin Films," in J. L. Vossen and W. Kern, Eds., *Thin Film Processes*, Academic, New York, 1978, p. 258.

[20] A. C. Adams, Chapter 3, this volume.

[21] J. M. Shaw and J. A. Amick "Vapor Deposited Tungsten For Devices," *RCA Review*, *31*, 306 (1970).

[22] C. F. Powell, "Chemically Deposited Metals," in C. F. Powell, J. H. Oxley, and J. M. Blocher, Eds., *Vapor Deposition*, Wiley, New York, 1966, p. 277.

[23] N. E. Miller and I. Beinglass, "Hot-Wall CVD Tungsten for VLSI," *Solid State Technol.*, *23*, 79 (1980).

[24] I. A. Blech, D. B. Fraser, and S. E. Haszko, "Optimization of Al Step Coverage Through Computer Simulation and Scanning Electron Microscopy," *J. Vac. Sci. Technol.*, *15*, 13 (1978); Errata, *J. Vac. Sci. Technol.*, *15*, 1856 (1978).

[25] W. Fichtner, Chapter 10, this volume.

[26] A. C. Adams, "Plasma Planarization," *Solid State Technol.*, *24*, 178 (1981).

[27] T. Sakurai and T. Serikawa, "Lift-Off Metallization of Sputtered Al Alloy Films," *J. Electrochem. Soc.*, *126*, 1257 (1979).

[28] T. Batchelder, "A Simple Metal Lift-Off Process," *Solid State Technol.*, *25*, 111 (1982).

[29] *Handbook of Chemistry and Physics*, Chemical Rubber Co., Cleveland, 1970.

[30] K. N. Tu, "Shallow And Parallel Silicide Contacts," *J. Vac. Sci. Technol.*, *19*, 766 (1981).

[31] A. K. Sinha, W. S. Lindenberger, D. B. Fraser, S. P. Murarka, and E. N. Fuls, "MOS Capability of High Conductivity $TaSi_2/n^+$ Poly Si Gate MOSFET", *IEEE Trans. Electron Devices*, *ED-27*, 1425 (1980).

[32] M.-A. Nicolet and M. Bartur, "Diffusion Barriers in Layered Contact Structures," *J. Vac. Sci. Technol.*, *19*, 786 (1981).

[33] M. Wittmer, "High-Temperature Contact Structures for Silicon Semiconductor Devices," *Appl. Phys. Lett.*, *37*, 540 (1980).

[34] P. B. Ghate, J. C. Blair, C. R. Fuller, and G. E. McGuire, "Application of Ti:W Barrier Metallization For Integrated Circuits," *Thin Solid Films*, *53*, 117 (1978).

[35] J. Black, "Physics of Electromigration," *Proc. 12th Reliability Physics Symposium*, IEEE, New York, 1974, p. 142.

[36] S. Vaidya, D. B. Fraser, and A. K. Sinha, "Electromigration Resistance of Fine Line Al," *Proc. 18th Reliability Physics Symposium*, IEEE, New York, 1980, p. 165.

[37] A. Gangulee, P. S. Ho, and K. N. Tu, Eds., *Low Temperature Diffusion and Application to Thin Films*, Elsevier, New York, 1975.

[38] S. Vaidya, "Electromigration in Aluminum/Poly-silicon Composites," *Appl. Phys. Lett.*, *39*, 900 (1981).

[39] P. B. Ghate, J. C. Blair, and C. R. Fuller, "Metallization In Microelectronics," *Thin Solid Films*, *45*, 69 (1977).

[40] R. C. Henderson, R. F. W. Pease, A. M. Voschenkov, R. P. Helm, and R. Wadsack, "A High Speed P-channel Random Access 1024 Bit Memory Made with Electron Lithography," *IEEE Solid-State Circuits*, *SC-10*, 92 (1975).

[41] D. M. Brown, W. E. Engler, M. Garfinkel, and P. V. Gray, "Self-Registered Molybdenum-Gate MOSFET," *J. Electrochem. Soc.*, *115*, 874 (1966).

[42] S. P. Murarka and D. B. Fraser, "Silicide Formation in Thin Cosputtered (Titanium + Silicon) Films on Polycrystalline Silicon and SiO_4," *J. Appl. Phys.*, *51*, 350 (1980).

[43] K. L. Wong, T. C. Holloway, R. F. Pinizotto, Z. P. Sobczak, W. R. Hunter, and A. F. Tasch, "Composite $TiSi_2/n^+$ Poly-Si Low Resistivity Gate Electrode and Interconnect for VLSI Device Technology," *IEEE Trans. Electron Devices*, *ED-29*, 547 (1982).

[44] B. L. Crowder and S. Zirinsky, "1-μm MOSFET VLSI Technology: Part VII-Metal Silicide Interconnection Technology—A Future Perspective," *IEEE J. Solid State Circuits*, *SC-14*, 291 (1979).

[45] T. Mochizuki, K. Shibata, T. Inoue, and K. Ohuchi, "A New MOS Process Using $MoSi_2$ As A Gate Material," *Jpn. J. Appl. Phys.*, *17* (suppl. 17-1) 37 (1977).

[46] S. P. Murarka, D. B. Fraser, A. K. Sinha, and H. J. Levinstein, "Refractory Silicides of Titanium and Tantalum for Low Resistivity Gates and Interconnects," *IEEE Trans. Electron Devices*, *ED-27*, 1409 (1980).

[47] A. K. Sinha, "Refractory Metal Silicides for VLSI Applications," *J. Vac. Sci. Technol.*, *19*, 778 (1981).

[48] S. P. Murarka, "Refractory Silicides for Integrated Circuits," *J. Vac. Sci. Technol.*, *17*, 775 (1980).

[49] T. F. Retajczyk and A. K. Sinha, ' Elastic Stiffness and Thermal Expansion Coefficients of Various Refractory Silicides and Silicon Nitride Films," *Thin Solid Films*, *70*, 241 (1980).

[50] S. P. Murarka, D. B. Fraser, W. S. Lindenberger, and A. K. Sinha, "Oxidation of Tantalum Disilicide on Polycrystalline Silicon," *J. Appl. Phys.*, *51*, 3241 (1980).

[51] D. B. Fraser, S. P. Murarka, A. R. Tretola, and A. K. Sinha, "Tantalum Silicide/Polycrystalline Silicon-High Conductivity Gates for CMOS LSI Applications," *J. Vac. Sci. Technol.*, *18*, 345 (1981).

[52] T. Mochizuki, T. Tsujimaru, M. Kashiwagi, and Y. Nishi, "Film Properties of $MoSi_2$ and Their Application to Self-Aligned $MoSi_2$ Gate MOSFET," *IEEE Trans. Electron Devices*, *ED-27*, 1431 (1980).

[53] B. C. Leung and J. S. Maa, "Refractory Metal Silicide/N^+ Polysilicon in CMOS/SOS," *Tech. Digest Int. Elect.*, 827 (1981).

[54] C. J. Koeneke, S. M. Sze, R. M. Levin, and E. Kinsbron, "Schottky MOSFET for VLSI," paper 15.6, IEEE Electron Device Meeting, Washington, D. C., Dec. 1981.

[55] R. Kieffer and F. Benesovsky, *Hartstoffe*, Springer, Vienna, 1963.

[56] D.-Y. Shih and P. J. Ficabora, "The Reduction of Au-Al Intermetallic Formation and Electromigration in Hydrogen Environments," *IEEE Trans. Electron Devices*, *ED-26*, 27 (1979).

[57] J. M. Eldridge, "Corrosion Problems of Metal Conductor Lines in Integrated Circuits," *Solid State Devices*, Inst. Phys. Conf. Ser. No. 57, 1980, p. 211.

[58] M. Hansen, *Constitution of Binary Alloys*, McGraw-Hill, New York, 1958.

PROBLEMS

1 If the residual water vapor pressure is 5×10^{-5} Pa in an Al evaporation station at 300 K, what O content does a deposited film have if Al is deposited at 50 Å/s? Assume the reaction results in incorporating Al_2O_3 in the film and that each H_2O molecule has a reaction probability 10^{-3}.

2 Assume that you have no way of co-depositing Al-Si while maintaining the compositional ratio throughout the film thickness. However, individual discrete films may be deposited with accurate thickness. A sandwich of 1 μm of Al on a Si film must be deposited on a Si contact. It must be stable so that sintering at 450°C can be performed without attacking the substrate. Assuming equilibrium, how thick must the Si layer be?

3 Pulsed currents of density 10^3 A/cm^2 must be passed by a metal silicide-to-semiconductor ohmic contact ($\phi_B \approx 0.4$ V) in the emission range (light doping). What voltage drop is associated with the contact?

4 Metallization requires an electromigration-resistant Al conductor that must make contact through small-diameter (~ 1.25-μm) windows. Which film deposition method would you choose? Give reasons for your choice.

5 A MOS test capacitor is formed by depositing e-beam-evaporated Al on an oxidized silicon wafer and patterning the metal film. The flatband voltage of the capacitor is shifted by 1 V relative to a measured value obtained before a 450°C hydrogen heat treatment. What causes the shift (see Eq. 29)? If the Al severely penetrates the silicon at 450°C, what alternatives might be used?

6 In the preceding problem what charge will correspond to the voltage shift if the capacitor is formed on a 1000-Å-thick oxide? How can the shift be attributed to radiation-induced changes in the oxide rather than to other sources?

7 A circuit's design requires a maximum permissible current density of 5×10^5 A/cm^2 through a conductor 1 mm long, 1μm wide, and nominally 0.5 μm thick. Assume that 10% of the conductor length passes over steps and is 50% of the nominal metal film thickness. What maximum voltage may be used across the conductor if the sheet resistance is 5.6×10^{-2} $\Omega/\square$? (Neglecting the thinner cross sections at steps can lead to reliability problems.)

8 Given the various Al film parameters that influence the electromigration resistance of Al conductors, describe how you would deposit Al films to ensure a maximum service lifetime. How would you test such films?

TEN

PROCESS SIMULATION

W. FICHTNER

10.1 INTRODUCTION

Numerical simulation has emerged recently as an important aid to process and device developments. In fact, process and device simulations are now as common as circuit simulation for two major reasons: Computer simulations are less expensive and much faster than experimental approaches. For example, suppose a semiconductor manufacturer plans to develop a new CMOS process with 1.5-μm design rules. This new process may involve nine lithography steps, six ion implantations, and several diffusion, annealing, and oxidation steps. Using available software and a medium-size computer (e.g., VAX* 11/780), one can simulate all critical process steps (e.g., channel-stop condition, threshold adjustment, etc.) in a matter of minutes or hours. A real experiment, on the other hand, usually takes from several days to a few weeks. By using computer simulations, we can save enough time to obtain results on process sensitivity by also modeling variations on the process. Figure 1 summarizes process and device simulation steps that will be treated in this chapter. For VLSI devices and circuits, process conditions are tightly coupled to the performance of finished devices. Therefore, process simulation cannot be a stand-alone field but has to be closely coupled to device simulation. Device design is only possible when both fields are considered together.

10.2 EPITAXY

This section describes a model that simulates epitaxial doping profiles in a variety of growth conditions.[1-4]

*Trademark of Digital Equipment Corporation.

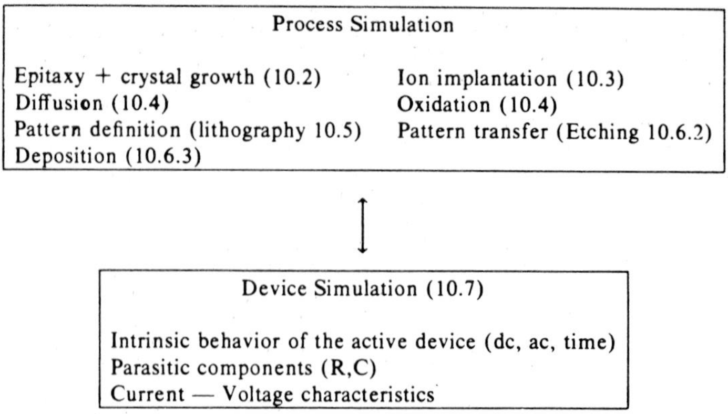

```
┌─────────────────────────────────────────────────────────────────┐
│                        Process Simulation                         │
│                                                                   │
│  Epitaxy + crystal growth (10.2)      Ion implantation (10.3)     │
│  Diffusion (10.4)                     Oxidation (10.4)            │
│  Pattern definition (lithography 10.5)  Pattern transfer (Etching 10.6.2) │
│  Deposition (10.6.3)                                              │
└─────────────────────────────────────────────────────────────────┘
```

$$\Big\uparrow$$

```
┌─────────────────────────────────────────────────────────────────┐
│                     Device Simulation (10.7)                      │
│                                                                   │
│  Intrinsic behavior of the active device (dc, ac, time)          │
│  Parasitic components (R,C)                                       │
│  Current — Voltage characteristics                                │
└─────────────────────────────────────────────────────────────────┘
```

Fig. 1 Schematic showing the coupling between process and device simulation.

10.2.1 Epitaxial Doping Model

In this model arsine (AsH_3) is the doping species considered and silane (SiH_4) is used to grow silicon in a hydrogen ambient in an atmospheric pressure reactor. Fick's second law is applied throughout the silicon to account for the thermal redistribution of impurities during epitaxial growth.[5] Thus,

$$\frac{\partial C(z,t)}{\partial t} = \frac{\partial}{\partial z}\left[D\frac{\partial C}{\partial z}\right] \qquad \infty > z > z_f \qquad (1)$$

has to be solved in a region as shown in Fig. 2. C is the dopant concentration in the silicon, D is the diffusion coefficient, and z and t are the space and time variables, respectively.

The solution of Eq. 1 must satisfy the following initial and boundary conditions:

$$C(z, 0) = f_1(z) \qquad (2)$$

$$D\left.\frac{\partial C}{\partial z}\right|_{z\to\infty} = 0 \qquad (3)$$

$$D\left.\frac{\partial C}{\partial z}\right|_{z=z_f} = f_2(t) \qquad (4)$$

where $f_1(z)$ represents the impurity diffusion just before the epitaxial deposition, and Eq. 3 states that the flux of impurities deep inside the silicon is zero. Equation 4 accounts for the fact that, during epitaxial growth, the diffusion flux of impurities in the solid at the gas-solid interface is a function of time. An expression for $f_2(t)$ can be derived from a mathematical description of the mechanisms that control the incorporation of the impurities into the silicon host lattice during the growth process.

Figure 9 in Chap. 2 shows schematically the step sequences that occur in the gas phase of an epitaxial reactor. Three main sequences are indicated in the figure:

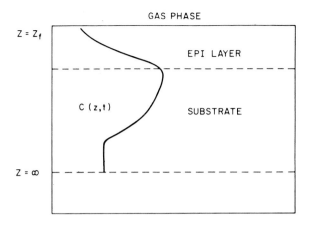

GAS PHASE

$Z = Z_f$

EPI LAYER

$C(z,t)$

SUBSTRATE

$Z = \infty$

Fig. 2 Schematic cross section of a silicon wafer for the purpose of solving Fick's second law. *(After Reif and Dutton, Ref. 3.)*

Step 1. Forced-convection mass transport of As_2H_3 from the reactor tube entrance to the deposition region.

Step 2. Boundary-layer mass transport of As_2H_3 from the well-mixed main gas stream through the boundary layer to the surface.

Step 3. Dissociation of As_2H_3 through gas-phase chemical reactions into several As containing species.

The description of mechanisms at the growing surface is based on the terrace-ledge-kink model[6] which divides the surface into adsorption (or terrace) sites, step (or ledge) sites, and kink sites. Figure 9 in Chapter 2 also illustrates the sequence of steps occurring at the surface.

Step 4. Adsorption of the As-containing species at a terrace site on the growing surface.

Step 5. Chemical dissociation into As and H in the adsorbed layer. Different species (AsH_3, As, H, etc.) occupy terrace sites and are able to move at the surface.

Step 6. Surface diffusion and incorporation of adsorbed As at step and kink sites.

Step 7. The incorporated As is buried by subsequently arriving Si atoms during epitaxial growth.

Step 8. Desorption of hydrogen from the surface.

Based on this step sequence, Eq. 4 can be written as[3]

$$D \left. \frac{\partial C}{\partial z} \right|_{z=z_f} = f_2(t) = -k_{mf}\left[P_D^0 - \frac{C(z_f)}{K_p} \right]$$

$$+ gC(z_f) + K_A \frac{\partial C(z_f)}{\partial t} \qquad (5)$$

The first term describes the flux of dopant species leaving the boundary layer by adsorbing at the surface (steps 4 through 6). The variable k_{mf} is a kinetic coefficient associated with the mechanism dominating the dopant-incorporation process. P_D^0 is the input partial pressure, $C(z_f)$ is the dopant concentration at the interface, and K_p is a segregation coefficient relating the epitaxial dopant concentration to the concentration of dopant species in the gas phase. The second term $gC(z_f)$ represents the rate at which the adsorbed layer decreases its concentration of dopant species due to the silicon covering step (step 7). The last term represents diffusion of dopant atoms between the adsorbed layer and the bulk silicon. The variable K_A relates the epitaxial dopant concentration to the concentration of the dopant species in the adsorbed layer.

Fick's second law (Eq. 1) can now be solved subject to the boundary and initial conditions, specified in Eqs. 2 to 5.

10.2.2 Computer Implementation and Results

Figure 3 shows schematically a finite difference discretization of the simulation region. The silicon region is partitioned into discrete spatial cells with a constant dopant concentration within each cell [that is, $C = C_{i+1}$ for $(z_i + z_{i+1})/2 \leqslant z \leqslant (z_{i+1} + z_{i+2})/2$].

At the initial time $t = 0$, the doping profile is given by Eq. 2. The simulation starts by adding a new cell z_{i-1} (Fig. 3b). The dopant concentration C_{i-1} of this new cell is computed from Eq. 5 by setting the left-hand side to zero

$$0 = k_{mf}\left[P_D^0 - \frac{C_{i-1}}{K_p}\right] - gC_{i-1} - K_A\frac{\partial C_{i-1}}{\partial t} \tag{6}$$

This equation accounts only for dopant introduction into the added cell and neglects the simultaneous impurity redistribution in the silicon. Now we calculate the impurity redistribution that occurs during the growth of all z_{i-1} cells by solving Fick's law (Eq. 1). This is illustrated in Fig. 3c. No flux of impurities enters the silicon at this time. This concludes the calculation at one time step Δt, and we arrive at $t = t_0 + \Delta t$, where we start the cycle over until the total epitaxial deposition time is over.

Figure 4 compares a doping profile simulated by using this model with a profile measured by the spreading resistance technique. For the comparison shown in the figure, two consecutive independent arsenic-doped epitaxial films were deposited as indicated in the inset. The arsine flows corresponding to the first and second layers were adjusted to produce epitaxial doping levels of approximately 10^{17} and 10^{15} cm^{-3}, respectively.

Between the end of the first deposition cycle and the beginning of the second, the reactor was purged with H_2 for 8 min at 1050°C. The transition between the high and low doping levels is typical if a lightly doped layer is grown epitaxially on top of a heavily doped substrate or buried layer. This transition is at first abrupt and then becomes gradual. This graded transition is a result of the autodoping phenomenon.

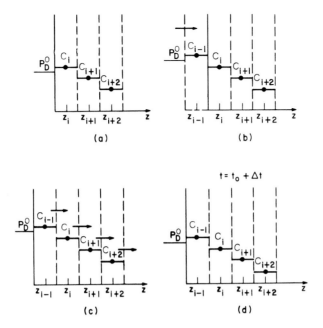

Fig. 3 Implementation of the numerical technique used to solve Fick's second law with the surface boundary condition dictated by the epitaxial deposition process. *(After Reif and Dutton, Ref. 3.)*

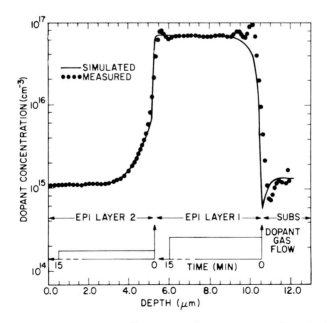

Fig. 4 Measured and simulated doping profiles corresponding to two consecutive epitaxial depositions. The growth rate is 0.35 μm/min at $T = 1050°C$. *(After Reif and Dutton, Ref. 3.)*

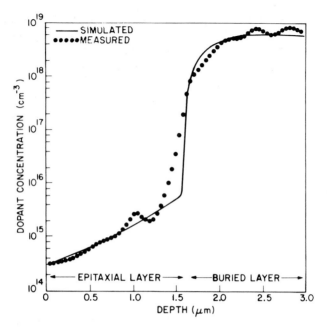

Fig. 5 Measured and simulated doping profiles corresponding to a typical autodoping situation. The growth rate is 0.27 μm/min at $T = 1050°C$. *(After Reif and Dutton, Ref. 3.)*

Figure 5 compares a more typical experimental autodoping result with its simulation. Arsenic was implanted $(3 \times 10^{15}$ cm^{-2}, 100 keV) into a boron-doped, 10 Ω-cm, (100) silicon wafer, and then redistributed for 2 h at 1250°C. The substrate was vapor etched with HCl (0.5% by volume, 2 min, 1200°C) and then baked in hydrogen (32 min, 1200°C) before the epitaxial-deposition step. The epitaxial layer was intended to be intrinsic (i.e., no arsine flow entered into the reactor). The epitaxial growth rate was approximately 0.27 μm/min, and the total deposition time was 6 min. The results in Figs. 4 and 5 show the excellent agreement between simulation and experiment.

10.3 ION IMPLANTATION

Successful application of ion implantation depends strongly on the ability to predict and control electrical and mechanical effects for given implant conditions. In the past, the basic theory of ion stopping in solids has been the LSS theory, named after its developers Lindhard, Scharff, and Schiott (see Chap. 6). This theory has been used widely to predict primary ion range and damage distributions in amorphous, semi-infinite substrates. According to the LSS theory, the ion distribution has a Gaussian shape with a projected range R_p and a standard deviation ΔR_p. Range data for different ion-target combinations have been derived on the basis of LSS and are available in the literature.[7, 8]

In VLSI processing, however, it is quite common to implant into a substrate that is covered by one or more thin layers of different materials. Typical examples are threshold—adjust implants, chanstop and source/drain implants into gate—and field-oxide regions which may be covered by Si_3N_4. Furthermore, implantations may be performed through thin layers of heavy metals (e.g., Ta or $TaSi_2$). The existence of multilayered structures results in implant profile discontinuities at the interfaces between layers. Additionally, atoms from surface layers may be knocked into deeper layers by impinging ions. This *recoil* effect might degrade the electrical performance of the finished device.

The basic assumptions of the LSS theory do not allow its application to multilayered structures. In the following sections, we apply results from Chap. 6 that are pertinent to the theory of ion collisions in solids. The Boltzmann transport equation (BTE) and Monte Carlo (MC) methods are widely used to simulate ion implantation phenomena in solids. We introduce these two approaches and compare theoretical results with experimental data.

10.3.1 The Boltzmann Transport Equation Approach and Monte Carlo Methods

Let us consider the case of a 100-keV $^{75}As^+$ implant through a double layer of Si_3N_4 and SiO_2 into silicon[9] (Fig. 6). An arsenic atom entering the system may be scattered not only from silicon atoms but also from nitrogen atoms in Si_3N_4 and oxygen atoms in SiO_2. Furthermore, if the transferred energy E_T is high enough, the target atoms, or "recoils," are set into motion, possibly creating recoils themselves until they come to rest. Particles at each position z are described by their energy E and the direction θ in which they are traveling with respect to the z axis.

For each particle of interest, a momentum distribution $F(E, \theta, z)$ can be defined. The number of particles with energies and angles in the two-dimensional interval

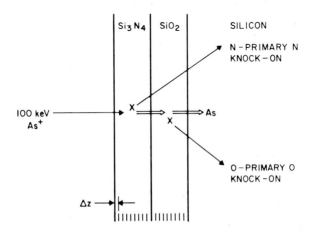

Fig. 6 Arsenic implantation through a $Si_3N_4-SiO_2$ double layer. *(After Smith and Gibbons, Ref. 9.)*

$dE \; d\theta$ that go through a unit-area element at depth z normal to the surface is given by $F(E, \theta, z) \; dE \; dz$.

The spatial evolution of this momentum distribution is determined by a Boltzmann equation for each different species k:

$$
\frac{\partial F_k(E, \theta, z)}{\partial z} = N \int \left[\frac{F_k(E', \theta') \; d\sigma(E' \to E, \theta' \to \theta)}{\cos \theta'} \right.
$$

$$
\left. - \frac{F_k(E, \theta) \; d\sigma(E \to E', \theta \to \theta')}{\cos\theta} \right] + Q_k(E, \theta, z) \qquad (7)
$$

Ions can be scattered from an energy E' and angle θ' into a final state (E, θ) or they can be scattered out of (E, θ) into (E', θ') (second rhs term). For recoil distributions, Q_k describes the creation of recoil particles from rest.

The distribution functions are assumed known in the surface plane $z = 0$. Recoil distributions are identically zero there, and the momentum distribution of the primary ions is determined by delta functions

$$
F(E, \theta, z = 0) = \Phi_0 \delta(E - E_0) \delta(\theta - 0) \qquad (8)
$$

where Φ_0 is the total implanted dose in cm^{-2} and E_0 is the incident beam energy.

The coupled set of transport equations (Eq. 7) is numerically integrated to obtain the distributions for all depths $z > 0$. This step requires that the motion of each particle be confined to a finite number of discrete momentum states. Each state F_{ij} is defined[10] by an energy E_i ($0 \leqslant E_i \leqslant E_0$) and an angle θ_j ($0 \leqslant \theta_j \leqslant \pi/2$).

Reasonable computation times restrict the number of elements for F_{ij}. Fifteen equally spaced energy intervals and ten angular intervals have been found sufficient for 5 to 10% accuracy in the range distributions.[10]

In the MC approach, ion implantation is simulated by following the history of a projectile through its successive collisions with target atoms using the binary collision approximation. Distributions for the range parameters of primary and recoiled ions and the associated damage (electronic and nuclear energy loss) can be obtained by following N histories (where N is large, $N \gtrsim 10^3$).

Each history begins with a given energy, position, and direction. The particle is assumed to change direction with each binary nuclear collision and to move in a straight, free-flight path between collisions. The energy is reduced as a result of nuclear and electronic (inelastic) energy losses. The ion will stop either when its energy drops below a prespecified value or when its position is outside the target (a reflected ion).

Monte Carlo calculations are possible for both amorphous and crystalline targets. In the amorphous model, the position of the target atoms is Poisson distributed. The ion interacts with one target at a time as indicated by the impact parameter given by

$$
p = \sqrt{\frac{R_n}{\pi N^{2/3}}} \qquad (9)
$$

where R_n is a uniformly distributed random number between zero and one and N is the target density.

The final success of Monte Carlo calculations—as measured by comparing theoretical results to experimental data—is strongly dependent on the choice of the interaction potential between the projectile ions and the target atoms. Best results are obtained with an approximation to the Thomas-Fermi potential which was described in Chap. 6 with the screening function[11]

$$\phi(R) = 0.35 \exp(-0.3R) + 0.55 \exp(-1.2R)$$
$$+ 0.1 \exp(-6R) \tag{10}$$

At this stage, we could theoretically integrate the equations of motion for the scattering angle θ, which in turn would allow us to calculate E_T. A computer program called MARLOWE[11] based on this *exact* technique is available. However, direct integration is time-consuming and can be avoided by an elegant analytical technique to evaluate the scattering angle,[12] thus allowing the calculation of E_T. The azimuthal scattering angle ϕ is randomly selected using

$$\phi = 2\pi R_n \tag{11}$$

Compared to the BTE approach, the MC technique has three major advantages. First, it is intrinsically a three-dimensional technique. In modern device processing, ions are only implanted into finite areas (e.g., windows) of a wafer, resulting in a lateral distribution of ions under the mask edge. Although Eq. 7 could be generalized to two or three dimensions, this has not yet been done. A second advantage of the MC technique over the BTE approach arises when very light ions are implanted into heavy targets ($M_1/M_2 \gg 1$), such as in the case of H^+ implantation (see Ion Beam Lithography, Sec. 10.5.3). In this case, many ions are backscattered towards the surface, which is no problem in the MC model. In the BTE approach, however, these ions scatter back into regions where the solution is supposedly already known.[10] A third advantage arises in the simulation of ion implantation into crystalline materials. In reality, of course, we know that silicon is not a random medium, but has the regular structure of the diamond lattice. No BTE results have been published accounting for lattice effects.

10.3.2 Results and Comparisons

In this subsection, we show the results of a number of representative calculations with both the BTE model and the MC technique. Basically, both models give the same results in all cases, if they use the same physical parameters (e.g., potential). The theoretical basis is the same for both methods.

Arsenic in silicon In Fig. 7, we compare BTE calculations using both the Kalbitzer[13] and Wilson[14] cross sections with a Pearson type IV distribution (see Chap. 6) generated from LSS.[15] The implant dose is 10^{16} cm^{-2} at an energy of 355 keV. All theoretical results can be compared to experimental data.[16] For these conditions, the reduced energy ϵ is 0.5, which means that nuclear stopping dominates completely. We see that, near the peak of the distribution, the BTE and LSS results using the Kalbitzer cross sections agree. The LSS profile, however, is too skewed, and both pro-

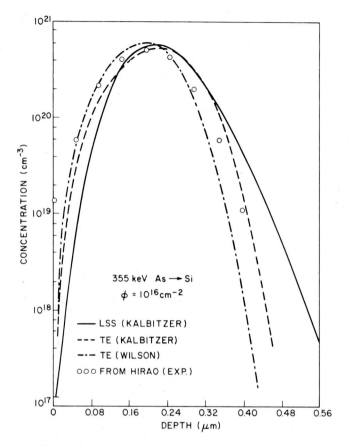

Fig. 7 Comparison of LSS and transport equation calculations and experimental results for the range profile of 355-keV arsenic implanted into silicon to a dose of 10^{16} cm^{-2}. The cross section used for each calculation is indicated in the key. *(After Christel and Gibbons, Ref. 10.)*

files are slightly deeper than the experimental result. The BTE result with the Wilson cross section is in excellent agreement with the experiment.

Boron in silicon The boron-Si ion-target combination is a good test for any simulation because excellent experimental data are available.[17] Figure 8 shows MC results obtained from simulation of 10,000 ion trajectories for the two implant energies of 50 and 100 keV. Measured electronic stopping power data are used to correct the LSS expression (that is, $k = 1.59 k_L$). Electronic stopping is the dominant energy-loss mechanism, especially in the 100-keV case. The MC results have been fitted to Pearson type IV distributions, given by the full drawn lines in Fig. 8. Agreement with SIMS data[17] is excellent, particularly for the 50-keV case. The influence of the electronic stopping parameter is indicated for the 100-keV case, which is also shown for $k = 1.50 k_L$, a value more consistent with the results in Ref. 14.

Figure 9 shows the final damage-density distribution for a 100-keV boron implant as calculated by the BTE method[13] together with an LSS calculation.[18] In the BTE

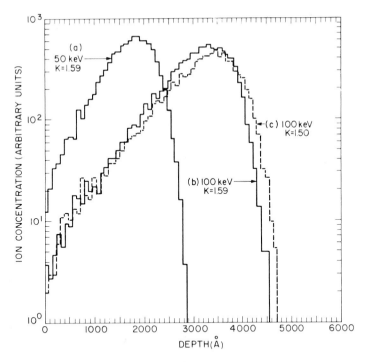

ION CONCENTRATION (ARBITRARY UNITS)

(a)
50 keV
K=1.59

(c) 100 keV
K=1.50

(b) 100 keV
K=1.59

DEPTH(Å)

Fig. 8 Results of MC calculations for boron implantation into silicon. 10,000 trials were used for each run. (a) $E = 50$ keV, $k/k_L = 1.59$. (b) $E = 100$ keV, $k/k_L = 1.59$. (c) $E = 100$ keV, $k/k_L = 1.50$.

result, only nuclear events contribute to the final damage. In reality, a minimum threshold energy E_d is required to remove a silicon atom from its lattice position. The Brice calculation includes corrections for recoil-energy loss caused by electronic processes and shows the closeness between both calculations.

Interfaces, channeling, and lateral effects The BTE and MC models both extend without any particular modification to multiple layer problems. Figure 10 shows MC results[19] obtained for a 150-keV, 2×10^{15} cm^{-2} phosphorus implant through 1500-Å SiO$_2$ into amorphous silicon. No discontinuity occurs in the MC data in agreement with experimental results. The curve represents an LSS profile matched analytically. Discontinuities occur only if the variation in the mass density between the different layers is large.[13]

Effects of crystal structure[19] and lateral effects[20] have been studied only by MC methods. The influence of the crystal structure exhibits itself in the channeling phenomenon. For the calculations, the positions of the crystal atoms are fixed according to the diamond lattice. Thermal vibrations are taken into account. The ions interact in general with all the atoms bordering the channel. Dechanneling is an integral part of the model. Shown in Fig. 11 are computed profiles of 150-keV P$^+$ through 1500-Å SiO$_2$ into crystalline silicon tilted $7.5°$ off the $\langle 110 \rangle$ axis. The effect of increasing damage during implantation has been taken into account. In the upper right corner, the computed damage distribution is shown.

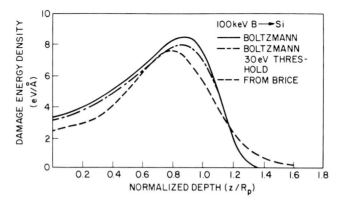

Fig. 9 As-deposited energy deposition profiles for 100-keV boron into silicon comparing the Brice and transport equation calculations. The abscissa is normalized to the projected range of the boron and the ordinate is energy density per incident particle. *(After Christel and Gibbons, Ref. 10.)*

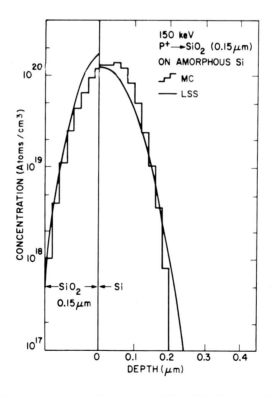

Fig. 10 Theoretical and experimental ion distributions in SiO_2–Si double-layer substrate. *(After De Salvo and Rosa, Ref. 19.)*

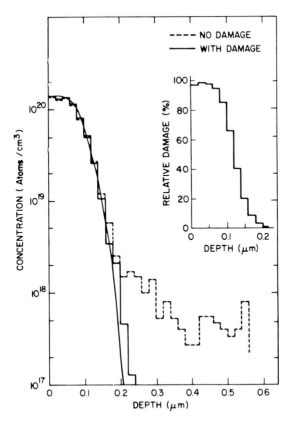

Fig. 11 Computed penetration profile of 150-keV P^+ in SiO_2 (0.15 μm) on Si crystal tilted at 7.5° off the $\langle 110 \rangle$ axis, taking into account the effect of increasing damage during implantation for a total dose of 2×10^{15} ions cm^{-2} (continuous histogram). The inset shows the damage distribution vs. depth. *(After De Salvo and Rosa, Ref. 19.)*

10.4 DIFFUSION AND OXIDATION

Solid-state diffusion is the physical mechanism that is responsible for the impurity migration within the silicon crystal during high-temperature processing. Together with ion implantation and epitaxy, solid-state diffusion is one of the key methods for controlling the type, concentration level, and distribution of impurities within specific regions of the silicon wafer. To obtain a diffused layer, impurity atoms are introduced into the surface region either by a predeposition step or by ion implantation. For VLSI devices, ion implantation is the preferred method since it allows both accurate control of the amount of dopant introduced and considerable freedom in the profile position (by a suitable choice of implant energy). A high-temperature step is usually required to activate the implanted ions and to remove the damage associated with the implant process (see Chap. 6).

As we have seen above, implantations are often performed through mask windows which make the diffusion process a two-dimensional problem.[21]

10.4.1 Impurity Diffusion and Thermal Oxidation

The basic law governing the transport of the ith impurity is given by the continuity equation[22]

$$\frac{\partial C_i}{\partial t} = \text{div } \mathbf{J}_i = \nabla \cdot (D_i \nabla C_i + Z_i \mu_i N_i \mathcal{E}) \tag{12}$$

where $C_i = C_i(x,z)$ and $\mathbf{J}_i$ are the concentration and the flux of the ith impurity, D_i is the concentration dependent diffusion coefficient, Z_i and μ_i are the charge state ($+1$ for acceptors, -1 for donors) and the mobility of the impurity, respectively, N_i is the electrically active concentration, and $\mathcal{E}$ is the electric field. Let us consider a two-dimensional problem with the lateral dimension x, the depth coordinate z, and the time t. Equation 12 has to fulfill a set of initial and boundary conditions:

Condition 1: $\qquad\qquad\qquad\qquad C_i(x,z,0) = f(x,z) \tag{13}$

Condition 2: $\qquad\qquad\qquad\qquad C_i(x, \infty, t) = 0$

or

$$C_i(x, \infty, t) = C_B \;(= \text{bulk concentration}) \tag{14}$$

Condition 3: No impurity flux is allowed along the lines of symmetry ($x = x_R$ and $x = x_L$).

$$\frac{\partial C_i}{\partial x} = 0 \quad \text{for } x = x_R \text{ and } x = x_L \tag{15}$$

Condition 4: The boundary condition at the surface depends on whether the surface is being oxidized

$$D_i \left.\frac{\partial C_i}{\partial z}\right|_{z=0} = C_i \left|\frac{1}{m} - b\right| z \tag{16}$$

or is exposed to an impurity gas source

$$D_i \left.\frac{\partial C_i}{\partial z}\right|_{z=0} = h(C_i - C_i^*) \tag{17}$$

In Eq. 16, m is the segregation coefficient given by the ratio of the dopant concentrations in silicon and SiO_2

$$m = \frac{C_i^{\text{Si}}}{C_i^{\text{SiO}_2}} \tag{18}$$

and b accounts for the volume change associated with the formation of SiO_2 (1 unit of SiO_2 consumes 0.44 units of Si). Equation 16 is valid under the assumption that the diffusion coefficient in the oxide is much smaller than in the silicon. If this is not true, Eq. 16 must be modified, and Eq. 12 has to be solved also in the oxide.

In Eq. 17, h is the mass transfer (or evaporation) coefficient and C^* is the dopant concentration in the gas phase.

When the impurity concentration is extremely high, some precipitation of impurity atoms (i.e., clusters) may occur.[23, 24] This precipitation makes some impurity atoms electrically inactive. During impurity diffusion, these clusters can "decluster" and therefore become electrically active. We can describe this phenomenon by the equations[24]

$$\frac{\partial N}{\partial t} = k_d C_c - k_c N^{\overline{m}} \tag{19}$$

$$\frac{\partial C_c}{\partial t} = -k_d C_c + k_c N^{\overline{m}} \tag{20}$$

$$C = N + C_c \tag{21}$$

which together with Eq. 12 describe the impurity flux. The variables k_c and k_d are the clustering and declustering rates, C_c is the concentration of the clustered impurity, N is the electrically active part ot the total chemical concentration C, and $\overline{m}$ is the cluster size. Numerical solutions to these equations indicate that the kinetics of arsenic clustering is especially important at annealing temperatures below 1000°C. Figure 12 shows the influence of the arsenic clustering on the carrier concentration. All samples are doped to a concentration of 2×10^{21} cm^{-3} by ion implantation. At 1000°C, equilibrium is reached with a carrier concentration $N = 2.81 \times 10^{20}$ cm^{-3} (dashed line). From the clustering, one can predict that subsequent anneals at lower temperatures will significantly decrease the carrier concentration. While the equilibrium carrier concentration monotonically decreases with temperature, the time required to reach equilibrium rapidly increases at higher temperatures.

For most practical processing situations ($T > 900°C$, $t > 20$ min) the effect of dynamic clustering and declustering is not significant. The clustering phenomenon

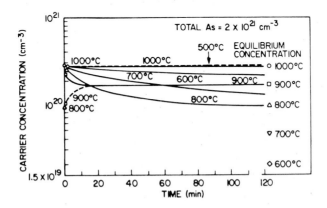

Fig. 12 Carrier concentration as a function of annealing time. *(After Tsai, Morehead, and Baglin, Ref. 24.)*

can then be modeled by an equilibrium clustering relation[4, 25]

$$C = N + \beta N^{\bar{m}}$$ (22)

where β is the temperature and impurity-dependent equilibrium constant given by

$$\beta = \frac{k_d}{k_c}$$ (23)

The diffusion coefficient D_i in Eq. 12 is, in general, a function of the concentration of the impurities for high dopant concentrations (see Chap. 5). All process simulation programs reported include the concentration dependence of D_i obtained from the vacancy-diffusion model

$$D_i = D_i^x + D_i^- f + D_i^{2-} f^2 + \frac{D_i^+}{f}$$ (24)

with $f = N/n_i$. The variables D_i^x, D_i^-, D_i^{2-}, D_i^+ are the intrinsic diffusivities of the various vacancy states in silicon, N is the electron concentration that depends on all C_i, and n_i is the intrinsic concentration at the diffusion temperature.

At low impurity concentrations, N is approximately equal to n_i and the diffusion coefficient reduces simply to the sum of the various vacancy states, independent of concentration

$$D_i = D_i^x + D_i^- + D_i^{2-} + D_i^{\lrcorner}$$ (25)

The individual diffusivities in Eq. 24 or Eq. 25 are given in Arrhenius form

$$D_i^* = D_{i0}^* \exp\left[-\frac{Q_i^*}{kT}\right]$$ (26)

with the prefactor D_{i0}^* and the activation energy Q_i^* (see Chap. 5).

The electron concentration N can be approximated by

$$N = \frac{C_{net} + \sqrt{C_{net}^2 + 4n_i^2}}{2}$$ (27)

with

$$C_{net} = -\sum_{i=1}^{n} Z_i N_i$$

Diffusion of all important group III (B) and group V (As, Sb) elements in silicon is described well by the diffusion model, Eq. 12, together with Eq. 24. The diffusion of phosphorus, however, is governed by a rather complex diffusion behavior and is modeled by the three-region model (see Chap. 5).

An accurate description of the phosphorus diffusion is especially important in emitter-push situations usually encountered in bipolar technology. A typical process would include a base boron implant followed by a series of drive-in steps. The first base drive is a dry oxidation step, followed by two wet oxidation steps and another dry oxidation. Next we form the emitter by chemical predeposition followed by drive-in steps. The resulting final doping profiles are shown[26] in Figs. 13a to c. Figure 13a is a three-dimensional surface plot, and Fig. 13b is the corresponding contour plot result. Figure 13c shows the emitter push-out phenomenon which results in the considerable deepening of the base-collector junction under the emitter as compared to the inactive base region.

Although of less importance, coupled-diffusion effects also occur in MOS fabrication. Consider an NMOS process with 1-μm design rules. After growing the gate oxide of 250-Å we implant B at high energy to adjust the threshold voltage and to prevent punch-through. Polysilicon is deposited and doped, and the source-drain regions are opened up by lithographic and etching steps. Source and drain are formed by a high-dose arsenic implant. Several drive-in steps follow until the device is finally processed. Figures 14a and b are surface plots of the total concentration and boron concentration, respectively, in the source drain regions and under the gate. Note the redistribution of boron in the source and drain junction areas caused by the emitter-dip effect.

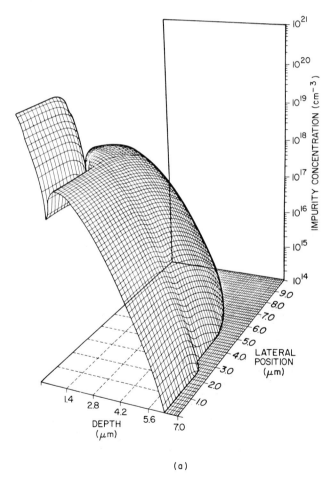

(a)

Fig. 13 Final phosphorus and boron doping profiles. (a) Surface plot of the phosphorus-boron impurity distribution.

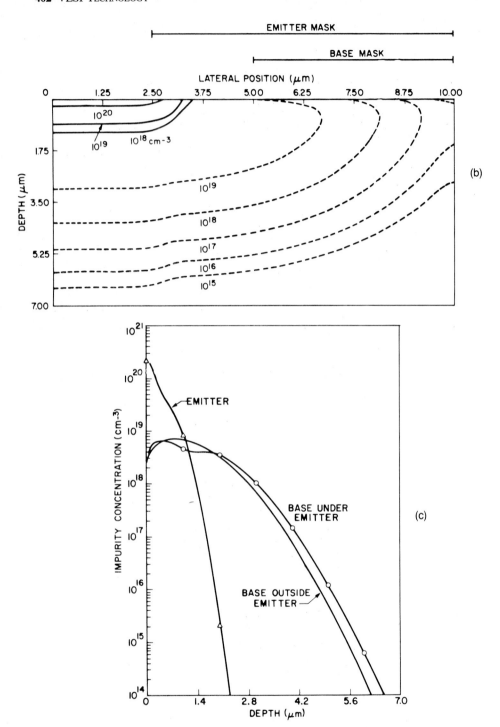

Fig. 13 (continued) (b) Contour plot of phosphorus and boron concentration. (c) Phosphorus and boron profiles in the emitter and the inactive base region. *(After Penumalli, Ref. 25.)*

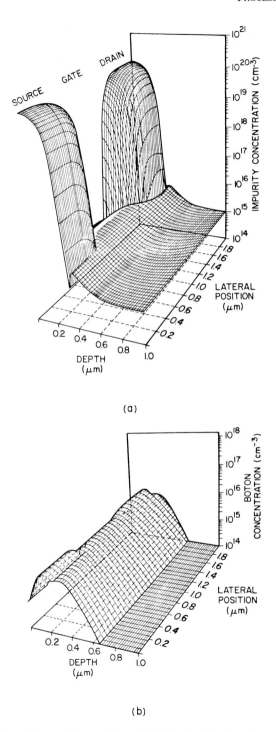

(a)

(b)

Fig. 14 Final arsenic and boron doping profiles. (a) Surface plot for the total concentration in a 1-μm-gatelength MOS device. (b) Surface plot of the boron concentration. *(After Penumalli, Ref. 25.)*

10.4.2 Thermal Oxidation

Thermal oxidation at high temperatures, which forms a layer of SiO_2 on silicon, is an integral process step in the fabrication of silicon devices. The kinetics of oxidation are fairly well understood for one-dimensional problems. According to the theory, the oxide thickness is expressed as

$$d_{ox}^2(t) + Ad_{ox}(t) = B(t + t_0) \tag{28}$$

with the oxide thickness d_{ox}, the rate constants A and B, and the correction time t_0 which accounts for the initial oxide thickness $d_{ox}(0)$ at $t = 0$:

$$t_0 = \frac{d_{ox}^2(0) + Ad_{ox}(0)}{B} \tag{29}$$

A and B are related to the linear and parabolic growth coefficients k_L and k_P and to the normalized oxygen partial pressure P_{O_2} by

$$A = \frac{P_{O_2}k_P}{k_L} \tag{30}$$

and

$$B = P_{O_2}k_P \tag{31}$$

For low dopant concentrations, k_P and k_L depend only on the oxidizing ambient and the crystal orientation. The temperature dependence of these rate constants can be expressed by one activation energy. Under high surface-concentration conditions, the oxidation rate is significantly enhanced.[27] The reason can be attributed to the generation of excess point defects at the $Si-SiO_2$ interface. In Refs. 4 and 26, the oxidation rate enhancement is included into Eq. 30 and Eq. 31 by adjusting the linear and parabolic rate coefficients to

$$k_L = k_L^i \left[1 + \gamma(C_V^T - 1) \right] \tag{32}$$

and

$$k_P = k_P^i \left[1 + \delta(C_V^T)^{0.22} \right] \tag{33}$$

The variables k_L^i and k_P^i are the intrinsic (low concentration) rate constants and is a parameter determined empirically.

$$\gamma = 2.62 \times 10^3 \exp\left[-\frac{1.10}{kT} \right] \tag{34}$$

The variable C_V^T is the normalized total vacancy concentration,

$$C_V^T = \frac{1 + C_v^+ \left[\dfrac{n_i}{N} \right] + C_v^- \left[\dfrac{N}{n_i} \right] + C_v^{2-} \left[\dfrac{N}{n_i} \right]^2}{1 + C_v^+ + C_v^- + C_v^{2-}} \tag{35}$$

with the vacancy concentrations

$$C_v^+ = \exp\left[\frac{E^+ - E_i}{kT}\right] \qquad\qquad E^+ = 0.35 \text{ eV}$$

$$C_v^- = \exp\left[\frac{E_i - E^-}{kT}\right] \qquad\qquad E^- = E_g - .57 \text{ eV} \qquad (36)$$

$$C_v^{2-} = \exp\left[\frac{2E_i - E^- - E^{2-}}{kT}\right] \qquad E^{2-} = E_g - 0.11 \text{ eV}$$

E_i is the position of the intrinsic level in the gap

$$E_i(T) \cong \frac{E_g}{2} \qquad (37)$$

In Eq. 33, δ is an empirical constant

$$\delta = 9.63 \times 10^{-16} \exp\left[\frac{2.83}{kT}\right] \qquad (38)$$

and C_T is the total dopant concentration.

At this time, a well established theory is not available that would allow a first-principles simulation of two-dimensional oxidation phenomena, such as the lateral oxidation under a Si_3N_4 mask that gives rise to the "bird's beak" geometry. Conceptually, such a theory involves a calculation of the oxygen flux to the silicon surface by solving the oxygen diffusion equation[28]

$$D_{ox}\nabla^2 C_{ox} = \frac{\partial C_{ox}}{\partial t} \approx 0 \qquad (39)$$

where D_{ox} and C_{ox} are the oxygen diffusion coefficient and concentration, respectively. The boundary conditions are the same as in the one-dimensional model. The volume expansion rate and the velocity at each volume oxide element are obtained from the oxygen flux. The motion of the outer oxide boundary is also described by this velocity as well as by the oxide boundaries and boundary conditions. At sufficiently high temperatures ($T = 900°C$), the oxide material can flow although the viscosity is extremely high. Assuming incompressibility, a simplified Navier-Stokes equation[28]

$$\mu\nabla^2 V = \nabla p \qquad (40)$$

treats this flow of oxide subject to the volumetric expansion at the interface. μ, **V**, and **p** are viscosity, velocity, and pressure in the two-dimensional oxide region.

This rigorous treatment is rather involved and will consume large amounts of CPU time. A simplified treatment of lateral oxidation uses a coordinate transformation method from the physical domain to a coordinate system in which the moving boundary remains stationary in time. With this approach the solution domain is simplified at the expense of complicating the diffusion equation (Eq. 12).

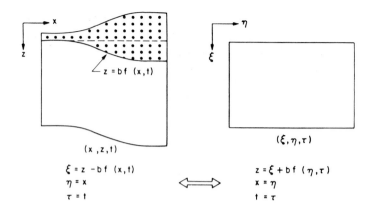

Fig. 15 Simulation regions.

Let (x, z, t) be the two spatial variables and the time in the physical coordinate system. Let (ξ, η, τ) be the corresponding variables in the transformed coordinate system. The simulation regions in both systems are illustrated in Fig. 15. The coordinate transformation is

$$\xi = z - bf(x, t) \tag{41}$$

$$\eta = x \tag{42}$$

$$\tau = t \tag{43}$$

where
$$f(x, t) = \frac{d_{ox}(t)}{2} \, \mathrm{erfc} \sqrt{\frac{2x}{k_l d_{ox}(t)}} \tag{44}$$

and where f is a function of lateral position and time, k_l is the ratio of lateral to vertical oxidation, and b is defined as in Eq. 16. Applying Eqs. 41 to 44 to Eq. 12 yields transformed equations for the (ξ, η) variables.

Local oxidation is commonly used in MOS processing. Figures 16a and b show the region and the boron profile before and after local oxidation respectively. In Fig. 16a, the as-implanted boron profile is shown. Oxidizing this profile for several hours in wet and dry atmospheres not only redistributes the boron considerably (note the different length scales), but also results in the bird's beak geometry of Fig. 16b.

10.4.3 Numerical Aspects

Depending on the number of impurities present, a set of coupled nonlinear partial differential equations, like Eq. 12, has to be solved in either one or two dimensions subject to initial and boundary conditions. Here we shall only consider the two-dimensional case, since it is more useful in simulating diffusion phenomena in VLSI devices. The spatial derivatives in Eq. 12 are discretized in the usual manner by centered differences[29] on a two-dimensional grid. This reduces Eq. 12 to a set of N non-linear ordinary differential equations, where N is the number of grid points in the

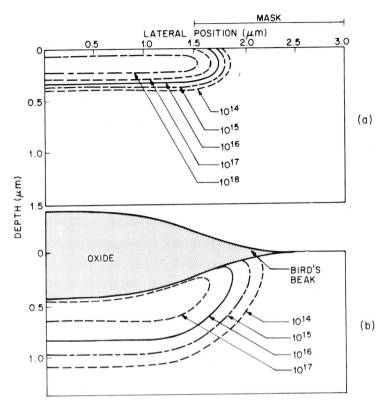

Fig. 16 Effect of local oxidation. (a) Region and boron profile before oxidation. (b) Region and boron profile after oxidation. *(After Penumalli, Ref. 25.)*

mesh. A variety of different methods is available for the time integration. Implicit[26] and explicit[30] methods are both suitable for the numerical solution of diffusion problems. In terms of operations per time step, implicit methods are more CPU time-intensive than explicit methods. However, implicit methods allow larger time steps and are usually more stable than explicit methods.

The standard technique is the first-order backward difference method[26]

$$\frac{\partial C_i}{\partial t}\bigg|_{t=t^{n+1}} \rightarrow \frac{C_i^{n+1} - C_i^n}{\Delta t^{n+1}} \tag{45}$$

where the superscripts denote the concentration levels at the time steps t^{n+1} and t^n with $\Delta t^{n+1} = t^{n+1} - t^n$.

The time and space discretization finally yield for each impurities species, a set of N nonlinear algebraic equations that can be expressed in matrix notation as

$$C_i^{n+1} = C_i^n + \mathbf{B}(C_i^{n+1})C_i^{n+1} + S(C_1, C_2, \ldots, C_n) \tag{46}$$

where $\mathbf{B}(C)$ is a matrix whose elements are functions of C, and S is a vector representing boundary conditions, etc. Equation 46 is solved by applying Newton's method. Rewriting it in the form

$$g(C) = 0 \qquad (47)$$

and applying Newton's method, we obtain

$$\frac{\partial g}{\partial C} x = Ax = -g(C) \qquad (48)$$

Solving the linearized system Eq. 48 by any conventional method[31] for x concludes one Newton iteration by updating the concentration to

$$C^{new} = C^{old} + x \qquad (49)$$

This cycle is repeated until a suitable error criterion is reached. We can now update the time step and continue the time integration. Automatic time-step selection schemes allowing a convenient solution of Eq. 46 are available.

10.5 LITHOGRAPHY

In this section we present the basic theory and simulation results for optical, electron-beam, and ion-beam lithography. Optical lithography is the standard pattern-definition process in IC fabrication, as described in Chap. 7. We shall derive the basic relations of resist exposure for positive resists which are the basis of a comprehensive computer program called SAMPLE.[32] Electron-beam lithography is the standard technique today for the fabrication of masks for optical and x-ray lithography. Furthermore, direct electron-beam writing on wafers is the only technique to obtain extremely small linewidths. Electron scattering is responsible for the formation of the final image. Ion-beam lithography, however, can achieve the smallest linewidths of all lithographic techniques. Ion-beam lithography is basically ion implantation using a focused beam.

10.5.1 Optical Lithography

A generalized optical system is shown in Fig. 17. The information to be replicated is contained on a thin optically opaque layer supported by a transparent substrate. This pattern (the mask) is transferred by the exposure system to form an aerial image, which consists of a spatially dependent light-intensity pattern in the vicinity of the wafer. Exposure of the resist-coated wafer to the aerial image makes the resist more soluble (in case of a positive resist) to a chemical developer, which allows for easy removal of the exposed sections.

The simulation of this process consists of three parts:

1. *Optical computations.* The end product of the optical computations is the two-dimensional net (incident and reflected) intensity distribution I. The necessary input information for computing I relates to the optical system, the intensity distribution pattern of the light source, and the resist and substrate parameters.

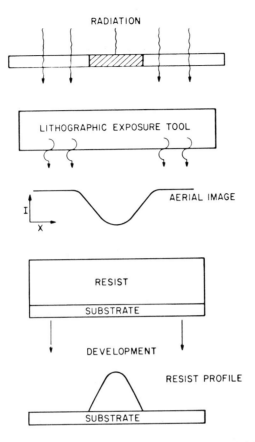

RADIATION

LITHOGRAPHIC EXPOSURE TOOL

AERIAL IMAGE

I

X

RESIST

SUBSTRATE

DEVELOPMENT

RESIST PROFILE

SUBSTRATE

Fig. 17 Idealized photolithographic system. *(After King, Ref. 34.)*

2. *Exposure computations*. The interaction of the exposing radiation I with the resist reduces the local inhibitor concentrations M. Calculation of the local instantaneous value of M requires a knowledge of specific exposure parameters that depend on the resist.

3. *Development calculations*. The development response of the resist to the developer requires a knowledge of empirical resist constants that permit computation of the development rates from M. The development rates then permit profile calculations for any particular development time.

We shall first consider the theoretical simulation of proximity printed images[33-35] (see also Chap. 7). If feature size and mask-to-wafer spacing are comparable to the wavelength λ of the exposing light, the diffraction from the mask edges is an electromagnetic diffraction problem that is given by Maxwell's equations under the appropriate boundary conditions. We have to calculate the square of the electric field $|\mathscr{E}|^2$, since photoresists react only to the intensity of $\mathscr{E}$. Assuming that the opaque material on the mask is infinitely thin and perfectly conducting, any one-

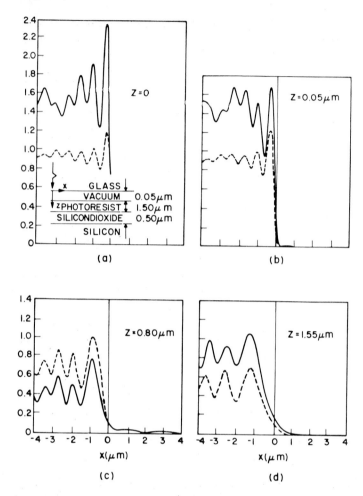

Fig. 18 Diffraction by a perfectly conducting infinitesimally thin half plane imbedded between the glass substrate of a mask and a photoresist on top of SiO₂ and a silicon substrate. The solid line represents the normalized field intensity $| \mathcal{E} |^2$ and the dotted line represents normalized power flow density $| \mathcal{E} \times \mathcal{H}^* |$. (a)–(d) show different diffraction results when z is made to vary from 0 to 1.55 μm. *(After Heitman and van der Berg, Ref. 35.)*

dimensional pattern can be synthesized with a combination of transverse electric (TE) slits and half-planes. The diffraction of a TE plane wave through a perfectly conducting half-plane of infinitesimal thickness depends strongly on the optical properties of the material beneath the mask.

Figure 18 depicts simulated results[35] for a realistic proximity printing situation with an air gap between the glass substrate of the mask and the resist. The relative permittivities for the glass, photoresist, SiO₂, and Si are 2.25, 2.56 + $i0.032$, 3.5, and 21.17 + $i0.466$, respectively. The permeabilities μ of all materials are identical to the vacuum value. The silicon layer under the resist causes a partial reflection that combines with the incoming wave to form a standing wave in the resist. The irregularities in the peak positions at the resist surface ($z = 0$) are also caused by the reflection. The SiO$_2$ layer causes a phase change of the reflected wave. The small vacuum gap (0.05 μm) between the glass and the resist simulates the imperfect contact situation.

Image formation in projection printing Most projection systems are designed to yield a diffraction-limited image over the entire image field. The systems are usually monochromatic (which allows the projection lens to be optimized for resolution, field flatness, and distortion), avoiding chromatic aberration. The quality of the aerial image relative to that of the mask pattern is determined by the modulation transfer function (MTF) of the lithographic exposure tool defined by

$$\text{MTF}(\nu) = \frac{M_{\text{image}}(\nu)}{M_{\text{mask}}(\nu)} \tag{50}$$

where M_{image} and M_{mask} are the mask and image modulation, respectively, for a spatial frequency ν (Chap. 7). The expression is valid for a mask with a sinusoidally varying transmission.

For an idealized imaging system, as in Fig.19a, the angle Ψ between the maximum pupil diameter and the image plane determines the resolution. This angle can be described by the numerical aperture NA,

$$NA = n \sin \Psi \tag{51}$$

or the effective f/number

$$f/\text{number} = \frac{1}{2NA}$$

where n is the index of refraction of the surrounding media ($n \approx 1$ for air).

The nature of the image depends on how the mask is illuminated and the wavelength λ of the light. Figure 19b shows schematically the coherent illumination of a mask with sinusoidal transmission of period P. Increasing λ or decreasing P increases the diffraction angle Φ. As long as the condition $\Phi \leq \Psi$ is fulfilled a perfect image is formed, since all the light is collected. Since the pattern of the mask consists of equal lines and spaces of spatial frequency $\nu = 1/P$, the pattern can be expressed as an infinite Fourier series[34]

$$\text{Mask pattern } (x) = a_0 + \sum_{k=1}^{\infty} a_k \sin (2\pi k \nu x) \tag{52}$$

with the Fourier coefficients a_k. The Fourier coefficients of the aerial image I can be found from those of the mask pattern by using the definition of the MTF, Eq. 50,

$$I(x) = a_0 + \sum_{k=1}^{\infty} \text{MTF}(k\nu)a_k \sin (2\pi k \nu x) \tag{53}$$

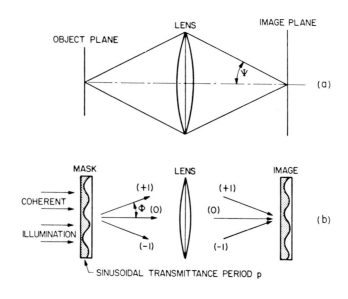

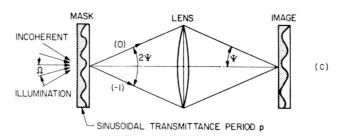

Fig. 19 (a) Simple imaging system. (b) Coherent illumination of a mask with sinusoidal transmittance of period *P*. (c) Incoherent illumination of a mask with sinusoidal transmittance of period *P*. *(After King, Ref. 34.)*

Assuming that the lithographic system operates near its limiting capability, only the fundamental spatial frequency is important and MTF $(k \nu) = 0$ for $k > 1$. Now we calculate the amplitude of the aerial image

$$A(x,z) = \frac{A_{max}}{2} \left\{ 1 + \frac{4}{\pi} \exp\left[i \phi(z)\right] \sin\left(2\pi \nu x\right) \right\} \qquad (54)$$

with the phase angle ϕ describing various aberrations of the optical system. For a perfect exposure system ϕ depends only on the focus condition, given approximately by

$$\phi(z) = \frac{\pi z \lambda}{P^2} \qquad (55)$$

where z is the distance to the focal plane. The intensity of the coherent aerial image is given by

$$
\begin{aligned}
I(x) &= |A|^2 \\
&= \frac{I_{max}}{4} \left[1 + \frac{8}{\pi} \cos \phi \, \sin (2\pi\nu x) + \left[\frac{4}{\pi} \right]^2 \sin^2 (\pi\nu x) \right]
\end{aligned}
\qquad (56)
$$

In the case of coherent illumination, Eqs. 55 and 56 describe the image formation for both projection printers and contact-proximity printers. In the latter case z describes the separation between the mask substrate and the wafer.

The other extreme of image formation occurs when the illumination conditions are similar to the situation in Fig. 19c. For an angle $\Omega > \Psi$, the system is described as incoherent. Light can be diffracted by an angle 2Ψ (compare this with Ψ in the coherent case) and still be collected by the projection optics.

Starting from Eq. 52 and applying the same approximation, we calculate the aerial image for the incoherent case

$$
I(x) = \frac{I_{max}}{2} \left[1 + M_I(\nu) \frac{4}{\pi} \sin (2\pi\nu x) \right]
\qquad (57)
$$

where M_I is an approximation to the incoherent MTF for a circular pupil

$$
M_I(\nu) \simeq 1 - \frac{4}{\pi} \sin (\nu\lambda f)
\qquad (58)
$$

In reality, all projection printers operate in a region between the two extremes of coherent and incoherent imagery because the pupil of the objective lens is partially filled, as shown[36] in Fig. 20. This condition is called *partial coherence*. It is characterized by the parameter σ; the ratio of the numerical aperture of the condenser lens,

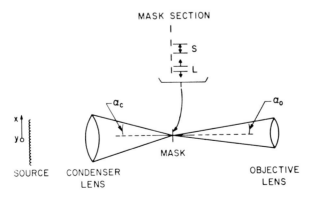

Fig. 20 Definition of symbols in a partially coherent system. *(After O'Toole and Neureuther, Ref. 36.)*

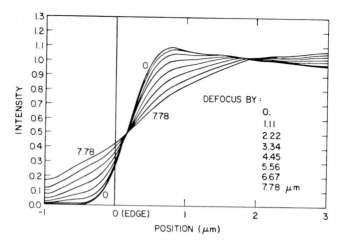

Fig. 21 Effect of focus error for $\sigma = 0.6$ on the image of a mask pattern with 2-μm lines and 6-μm spaces. $NA = 0.28$, $\lambda = 0.436$ μm. *(After O'Toole and Neureuther, Ref. 36.)*

$NA_c = \sin \alpha_c$ (remember $n \approx 1$) and the numerical aperture of the objective lens, $NA_o = \sin \alpha_o$

$$\sigma = \frac{NA_c}{NA_o} \tag{59}$$

A coherent system is characterized by $\sigma = 0$ and an incoherent system by $\sigma = \infty$. The difference between $\sigma = \infty$ and $\sigma = 1$ is small.

The basic effects of imaging with partially coherent light can be seen in Fig. 21, which shows the calculated image intensity[32] near the edge of a mask pattern consisting of 2-μm lines and 6-μm spaces. Since the mask is periodic, it is reflexive around both the $x = -1$-μm and $x = 3$-μm axis.

The numerical aperture of the lens is 0.28 and the wavelength is 0.436 μm. The focus error for the curves is taken in units of 0.4 Rayleigh units; one Rayleigh unit is 2.78 μm $(= \lambda/2NA_o^2)$. The focus error d is the distance in micrometers between the resist surface and the plane of perfect focus.

Calculation of photoresist exposure[37] Calculations of resist exposure require a knowledge of the optical constants of the substrate and any overlying layers and of the thicknesses of all the corresponding layers. The key to describing the exposure dependent optical properties of the photoresist are the exposure parameters A, B, and C. A and B describe the absorption constant α according to

$$\alpha = AM(z,t) + B \tag{60}$$

where M is the relative amount of photoactive inhibitor present at any position z and time t during exposure.

In the calculation, the complex index of refraction $\mathbf{n}$ of the photoresist is used

$$\mathbf{n} = n - ik \tag{61}$$

where n is the real part of the index and k is the extinction coefficient at the exposing wavelength λ

$$k = \frac{\alpha\lambda}{4\pi} \qquad (62)$$

The index **n** can be expressed with Eqs. 60 to 62 as

$$\mathbf{n} = n - i\,\frac{\lambda[AM(z,t) + B]}{4\pi} \qquad (63)$$

During exposure, **n** changes as the inhibitor is destroyed by the exposing light with intensity I. The optical sensitivity parameter C relates the destruction rate to the light intensity:

$$\frac{\partial M(z,t)}{\partial t} = -I(z,t)M(z,t)C \qquad (64)$$

As in the proximity printing case (see Fig. 18), the light intensity can vary appreciably within the resist film over thicknesses that are small compared to the resist thickness. Standing waves are caused by interference between the incident light and reflected components, resulting in a nonuniform inhibitor concentration and a corresponding nonuniformity in **n**.

Because the optical properties of the resist vary during exposure as a function of depth, the resist film is subdivided in layers thin enough to be treated as if they had isotropic properties.[38] Furthermore, the computation is divided into time (i.e., exposure) steps small enough to minimize changes in intensity and corresponding changes in the inhibitor concentration. If I_j and M_j denote the intensity and concentration in the jth sublayer, we calculate I_j by holding the inhibitor concentration constant. We proceed by incrementing the exposure-time variable by Δt_e and calculating new values for M_j to so that the computation of I_j can be repeated. M_j is altered by the exposure to

$$M_j \,\big|_{\,t_e + \Delta t_e} = M_j \,\big|_{\,t_e}\, \exp\,(-I_j C \Delta t_e) \qquad (65)$$

with the initial condition $M_j \,\big|_{\,t_e = 0} = 1$. Reasonable accuracy is obtained if the resist layer thickness δz_j is less than 0.03λ and the exposure time increment Δt_e is chosen so that the largest change in any M_j is 0.2 or less.

Figure 22 shows a computed result of the intensity distribution $I(z)$ with a 0.584-μm thick photoresist film on silicon and with a 600-Å SiO_2 layer at the beginning of exposure by a uniform incident illumination. After an exposure flux of 57 mJ cm^{-2}, the resulting inhibitor concentration is shown in Fig. 23. This is a typical result for a uniform exposure of photoresist on a real substrate.

Photoresist development The description of an image exposure in the photoresist is given by a two-dimensional matrix of inhibitor concentration values $M(x,z)$. The development process is modeled as a surface-controlled etching reaction which is con-

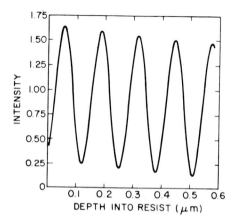

Fig. 22 Intensity of exposing light within a 0.584-μm AZ1350J photoresist film on 600 Å of oxide on silicon. *(After Dill et al., Ref. 37.)*

trolled by the local value of M. $M(x,z)$ is assumed constant in unit cells of dimensions Δx and Δz around x and z. The etch rate R is expressed as

$$R(M) = a \ \exp \left[E_1 + E_2 M + E_3 M^2 \right] \qquad (66)$$

where a gives the etch rate in μm/s, and E_1, E_2, and E_3 are experimental constants of the resist, depending on the developer, the temperature, and the processing conditions, respectively.

Development starts along the surface in contact with the developer. Cells of constant M are removed by the developer according to Eq. 66 and depending on the number of cells in contact with the developer. New cells are allowed to start etching

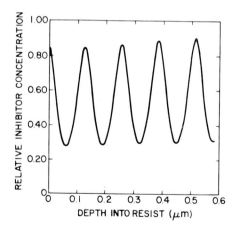

Fig. 23 Inhibitor concentration within a resist film on oxide on silicon after exposure to 57 mJ cm^{-2} at a wavelength of 0.4358 μm. *(After Dill et al., Ref. 37.)*

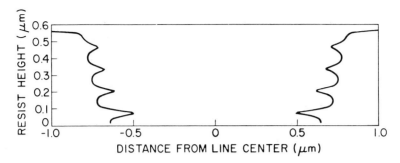

Fig. 24 Edge profile for a nominal 1-μm line in AZ1350 photoresist developed for 85 s in 1:1 AZ developer:water. *(After Dill et al., Ref. 37.)*

when old cells are removed. The time to remove a cell which has only the top side exposed is given by

$$t_r = \frac{\Delta z}{R_{ij}}$$ (67)

where R_{ij} is the etch rate of the particular cell, and Δz is the layer thickness. Similarly, if the top and one side are exposed

$$t_r = \frac{\Delta x \, \Delta z}{R_{ij} \sqrt{\Delta x^2 + \Delta y^2}}$$ (68)

Figure 24 shows the calculated resist edge profile of a nominal 1-μm line that has been exposed by a lens with $NA = 0.45$ at $\lambda = 0.4358$ μm. The development time is 85 s. The edge fingers on the line are typical for a monochromatic exposure of AZ 1350J.

10.5.2 Electron-Beam Lithography

In electron-beam lithography, finely focused electron beams are used to expose polymeric resist layers. The interaction and scattering of electrons within the resist layer and the underlying substrate depend on the beam energy, the resist type and thickness, the substrate parameters, and so on. The best resolution obtainable is not limited by the characteristics of the incident beam but rather by electron scattering. The actual process of electron scattering in solids is so complex that we have to rely on numerical models for quantitative results. The only model of practical importance is the Monte Carlo (MC) technique. With this technique we simulate a large number of individual electron trajectories to obtain the energy deposited in the resist (similar to ion implantation described in Sec. 10.3.2). Electrons undergo scattering events with the target nuclei (elastic scattering). In addition, they suffer energy loss by inelastic scattering processes with the target electrons. Elastic scattering results

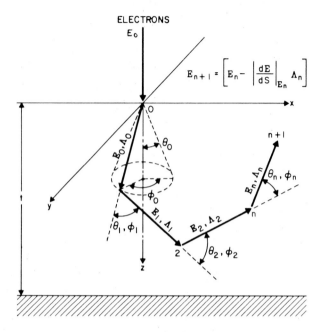

Fig. 25 Schematic diagram showing the initial Monte Carlo step lengths for electron scattering in a thin-resist film on a thick substrate. *(After Kyser and Murata, Ref. 39.)*

mainly in a change in the direction of the incoming electron. To model elastic scattering, the screened Rutherford formula is used for the differential cross section

$$\frac{d\sigma}{d\Omega} = \frac{Z^2 q^4}{16E^2} \left[\sin^4\frac{\theta}{2} + \frac{\theta_0^2}{4} \right]^{-1} \tag{69}$$

where $d\sigma/d\Omega$ is the differential cross section per unit solid angle, and θ_0 is the screening parameter

$$\theta_0 = 3.7 Z^{1/3} E^{-1/2} \tag{70}$$

The electron is assumed to travel in a succession of short straight paths between elastic scattering events, as shown[39] in Fig. 25. At each scattering point, the resulting azimuthal angle θ is determined by selecting a random number weighted with the differential cross section, Eq. 69. The path length λ_i between scattering events is selected by weighting the mean-free path between collisions by another random number in the zero to one range. The energy of the electron is reduced at each step by multiplying the path length by the Bethe energy loss rate

$$\frac{dE}{ds} = \frac{2\pi N_0 q^4 \rho Z}{AE} \ln \left[\frac{1.1658E}{I} \right] \tag{71}$$

where E is the electron energy, Z and A are the atomic number and atomic weight of the solids, respectively, N_0 is Avogadro's number, ρ is the density, and I is the mean excitation energy.

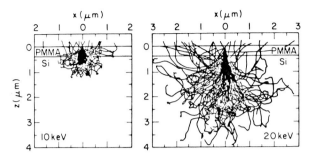

Fig. 26 Simulated trajectories of 100 electrons in PMMA. (a) Simulation for a 10-keV delta function beam. (b) Simulation for a 20-keV delta function beam. *(After Kyser and Murata, Ref. 39.)*

This process is repeated until the electron comes to rest. Depending on the angle, scattering events can be divided into two categories: forward-scattered and backscattered. Figure 26 shows 100 simulated electron trajectories for a 10-keV and a 20-keV delta function beam incident at the origin for a 0.4-μm PMMA film on a thick Si substrate. The beam is incident along the z axis and all trajectories have been projected onto the xz plane. These figures qualitatively show the degree of lateral forward scattering within the film, as well as the degree and position of backscattering. Backscattered electrons can emerge at distances far away from the origin.

Monte Carlo results for a delta line 20-keV beam on 0.4-μm PMMA on Al are shown in Fig. 27. Twenty thousand electron trajectories are simulated. The radial distribution of energy density is shown for two different depths (0.1 μm and 0.4 μm). For comparison, analytical results[40, 41] are included. At the origin the results of the MC models are consistently higher than the results of other models.

The latent image which is the absorbed energy density of the δ-function line source allows the calculation of the spatial distribution of energy density for any arbitrary beam shape by Fourier transformation. If an exposure profile is to be written with a rectangular beam, as in Fig. 28a, the profile for the absorbed energy density is obtained from the MC data by a convolution of a Gaussian distribution with itself over the square dimension.[42] The result of the convolution is

$$f(x) = \overline{K}\left[\text{erf}\left|\frac{a-x}{\sqrt{2}\sigma}\right| + \text{erf}\left|\frac{a+x}{\sqrt{2}\sigma}\right|\right] \qquad (72)$$

where the beam width FWHM (full-width half-maximum) $= 2a$, σ is the standard deviation, and $\overline{K}$ is a constant. For $a/\sigma \gg 1$, the edge slope is

$$\left|\frac{df}{dx}\right|_{x=\pm a} = \frac{2\overline{K}}{\sqrt{2\pi}\sigma} \qquad (73)$$

The edge width is given by $\sqrt{2\pi}\sigma/2$, and is defined by the tangent to $f(\pm a)$ intercepting $f(x) = 0$ and $f(x) = 2\overline{K}$ erf $(a/\sqrt{2}\sigma)$. The edge of $f(x)$ is symmetric around its half-height.

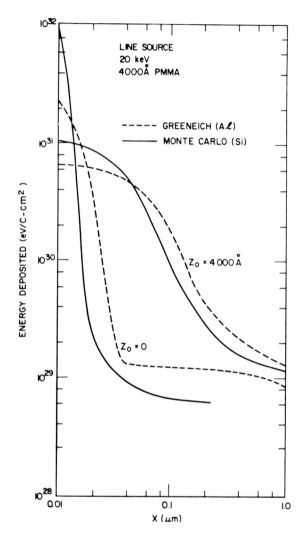

Fig. 27 Energy density profiles for a line source. *(After Kyser and Murata, Ref. 39, and Hawryluk et al., Ref. 40.)*

If the exposure pattern is to be written with lines composed of one or more Gaussian shaped beams possibly with different weights, as in Fig. 28b, the beam is described by

$$f(x) = K \exp \left[\frac{x^2}{\delta^2} \right] \tag{74}$$

Depending on the actual beam shape, either Eq. 72 or 73 is used as the envelope function for the digital convolution of the latent image from the ideal line source. This convolution assumes that superposition of electron exposure and subsequent energy deposition holds.

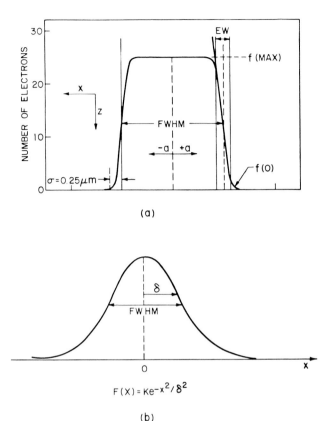

Fig. 28 Exposure patterns for arbitrary beam shapes. (a) Definition of terms. The vertical axis is the number of electrons distributed over the incident line, normalized to 10^5 electrons. *(After Kyser and Pyle, Ref. 42.)* (b) Schematic representation of a Gaussian round beam.

Figure 29a shows the simulated MC result of the energy deposited within 1.8 μm of resist at three depths for a 25-keV beam. At the surface ($z = 0$), the distribution is very narrow, but for increasing depth, it becomes broader due to backscattering contributions from the substrate. To calculate the lateral distribution of deposited energy (see Fig. 29b), the δ-function distribution is convoluted with the rectangular beam in Fig. 28a. The energy deposited varies with z. The tails in the original line response deep in the resist are a significant part of the total distribution in Fig. 29b.

As in optical lithography, we can calculate resist development. For positive resists, a general relationship between R and E is

$$R = (A + BE^n)[1 - \exp(-\alpha z)] + \epsilon(E) \qquad (75)$$

where R is the etch rate in Å/s, z is the distance below the surface, E is the local absorbed energy density in keV cm^{-3}, and A, B, and n are appropriate constants. The dependence of ϵ is modeled as

$$\epsilon(E) = \epsilon_0 + CE^m \qquad (76)$$

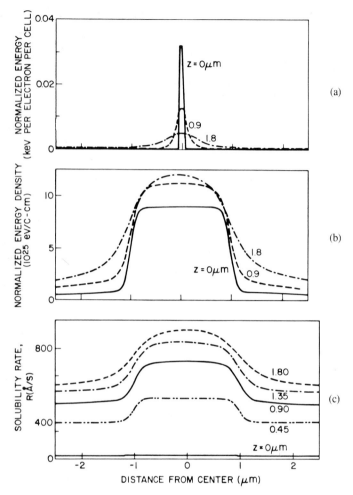

Fig. 29 Energy distributions and etch rate. (a) Lateral distribution of energy deposited in the film of a 1.8-µm polymeric resist on Si (25-keV, 2.0-µm written linewidth) using a Monte Carlo simulation for an ideal line source. (b) Lateral distribution of energy deposited within a 1.8-µm resist by 25-keV electrons for the 2-µm line in Fig. 44a. (c) Lateral distribution of etch rate for the same resist film and the latent image of $Q = 20$ µC cm^{-2}. *(After Kyser and Pyle, Ref. 42.)*

where C and m are constants and E is evaluated at $z = 0$. The form of Eq. 75 implies that for $z \ll 1/\alpha$ and vanishing incident dose Q, $R = \epsilon_0$. For $z = 1/\alpha$ and $Q = 0$, $R = A + \epsilon_0$. This type of dissolution behavior has been observed experimentally for certain positive resist materials under optical and electron-beam exposure. If α becomes large and $C = 0$, Eq. 75 reduces to the solubility rate behavior of PMMA used in SAMPLE.[32] The parameter α can be interpreted to describe the distance the solvent must diffuse into the resist before any significant development reaction starts corresponding to a diffusion distance $1/\alpha$. The correction term $\epsilon(E)$ provides the proper surface rate.

Equation 75 transforms the latent image of Fig. 29b into a solubility rate image. Figure 29c gives the lateral etch rate distribution with the constants in Eqs. 75 and 76 set to $A = 50$ Å/s, $\alpha = 1.5$ μm^{-1}, $B = 2.5 \times 10^{-18}$, $n = 1.05$, $C = 2.0 \times 10^{-30}$, $m = 1.5$, and $\epsilon_0 = 0.5$ Å/s. The development proceeds in the same manner as outlined before in the discussion of photoresist development.

Monte Carlo simulation, together with resist modeling is an extremely powerful tool for investigating proximity effects (see Chap. 7). Suppose we would like to develop a fine-line array of 0.5-μm lines and spaces in 1 μm of PMMA resist. Because of electron scattering, the various lines do not develop at the same time. Figure 30a shows the normalized energy density for two depths in the resist. Although each line receives the same dose, the outer lines receive less absorbed energy E within the film. The developed profile in Fig. 30a shows the case where the calculation is stopped when the center line just begins to open at the interface. With an adjusted dose for each line, all lines can be developed to the same size at the same time. By simply specifying the depth z at which the maximum absorbed energy density ought to be uniform, a computer program iteratively adjusts the relative line doses. For the same line array, the dose modulation (at $z = 0.5$ μm) is calculated to be 1.111 for the outer two lines and 1.041 for the inner two lines. The center line is exposed to a value of 1.000. Figure 30b shows corresponding latent images and developed profiles with this dose modulation.

The use of MC models to perform proximity-effect correction is expensive, requiring large computer programs and long computing times. The use of analytic functions facilitates the study of proximity effects.

The proximity function $f(r)$, defined in Fig. 31, can be approximated by two Gaussian distributions with standard deviations β_f, β_b, and relative areas η_E

$$f(r) = K \left[\exp \left[\frac{-r^2}{\beta_f^2} \right] + \frac{\eta_E \beta_f^2}{\beta_b^2} \exp \left[\frac{-r^2}{\beta_b^2} \right] \right] \qquad (77)$$

Monte Carlo calculations[43] and experimental techniques[44] are used to obtain the parameters in Eq. 77 for each particular resist-substrate situation and energy E.[41] In actual problems, complex patterns are decomposed into primitive figures. If all primitive figures are rectangles, a pattern comprised of N rectangles has $5N$ adjustable parameters, that is, four geometric parameters (two x values and two y values) and one exposure parameter per rectangle. The proximity function (Eq. 77) is used to calculate the dose due to exposure of the ith primitive figure with area A_i

$$D_i = \iint\limits_{A_i} f(r) \, dx \, dy \qquad (78)$$

For rectangles, Eq. 78 provides an analytic solution. Next we define a set of M numbers, whose values express the quality of the correspondence between the predicted and the desired patterns. We limit each primitive figure to only one

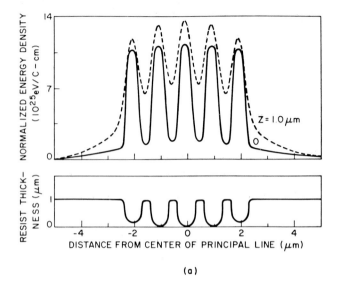

(a)

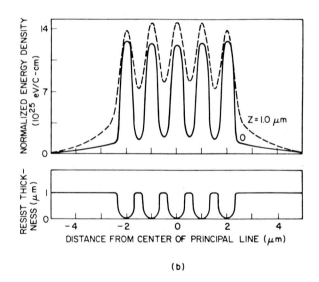

(b)

Fig. 30 Energy distributions (a) Top: Lateral distribution of energy deposited in the film by 20-keV electrons at different levels in the film. Bottom: The simulated profile for the latent image using developer parameters: $A = 1$ Å/s, $B = 8.0 \times 10^{-37}$ $(cm^3/keV)^2$, and $n = 2.0$ (appropriate for PMMA in 1:1 MIBK-IPA). Only the profile corresponding to the first line to reach the substrate (Si) surface is shown; $Q = 80$ μC cm^{-2}. (b) Top: The curves give the lateral distribtuion of energy deposited at different levels in the resist film with dose modulation of 1.111 on the outer two lines and 1.041 on the inner two lines (1.000 for the center line) of the five-line array of 0.5-μm lines and gaps. Bottom: Simulated developed profile for the latent image above. Note that all five lines now reach the substrate surface at the same time; $Q = 80$ μC cm^{-2}, developer parameters are the same as in (a). *(After Kyser and Pyle, Ref. 42.)*

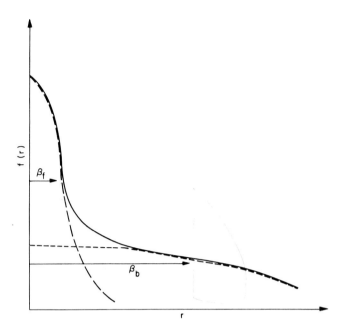

Fig. 31 Schematic of the proximity function $f(r)$ for arbitrary resist, substrate, and for incident electron energy. The forward-scattered electron distribution (— — —) has a characteristic width β_f, while the backscattered-electron distribution (— — —) has a characteristic width β_b. *(After Parikh, Ref. 45.)*

adjustable variable $(M = N)$.[45] The single parameter is the primitive figure exposure E_i; the single quantifier M_i is the average dose in the primitive figure. Using Eq. 78, we can calculate the average dose in the jth figure due to exposure of the ith figure:

$$D_{ji} = \frac{1}{A_j} \iint_{A_j} D_i \ dx \ dy \qquad (79)$$

The total average dose in the jth figure is the sum of all contributing component doses

$$D_j = \sum_i D_{ji} \qquad (80)$$

where D_{ji} is linear with respect to the exposures and is expressed in the form

$$D_{ji} = E_i K_{ji} \qquad (81)$$

and the K_{ji} are symmetric in i and j $(K_{ij} = K_{ji})$.[46]

Setting each of the D_i equal to some average dose $\overline{D}$ results in N equations

$$E_1 k_{11} + E_2 k_{12} + \cdots + E_N k_{1N} = \overline{D}$$
$$\cdots\cdots\cdots\cdots\cdots\cdots\cdots\cdots\cdots\cdots\cdots\cdots\cdots \qquad (82)$$
$$E_1 k_{N1} + E_2 k_{N2} + \cdots + E_n k_{NN} = \overline{D}$$

This system can be solved to yield values for all E_i.

The quality of correction that can be achieved with this algorithm is limited by the subdivision of the total pattern into primitive figures. The increase in the number of shapes can be controlled by partitioning a pattern only at those locations that are influenced most strongly by proximity effects.[47] The strategy for partitioning a pattern is as follows:

1. Attempt a proximity correction at a given pattern.
2. Assess pattern quality.
3. If the pattern quality test fails at certain points, subdivide such points and their associated regions.
4. Reattempt correction.

This procedure is repeated until the pattern quality is sufficient or until it becomes impossible to subdivide the pattern given the technical limitations of the electron-beam machine.

Applying this algorithm to the eight rectangles in the pattern shown in Fig. 32a

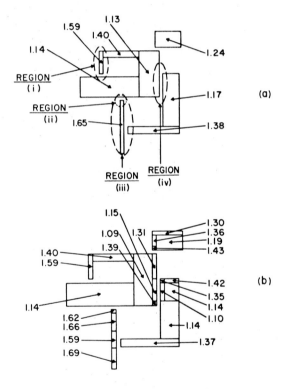

Fig. 32 Partitioned pattern (a) A pattern consisting of eight rectangles. Note regions (i) to (iv) where proximity corrections are needed to complete the dissolution of the resist as well as pattern fidelity. If this pattern is not proximity corrected, a relative exposure value of unity is given to each rectangle. If this pattern is corrected via the self-consistent algorithm, a relative exposure value (noted in the figure) is given to each rectangle. (b) Partitioned pattern with 21 rectangles that are obtained by using the algorithm described in Sec. 2. The self-consistent algorithm was used to compute the relative incident electron exposures for each of the rectangles computed. *(After Parikh and Schreiber, Ref. 47.)*

leads to the partitioned pattern in Fig. 32b. Note the regions (i) to (iv) in Fig. 32a, where proximity effects necessitate corrections for complete dissolution of the resist and for pattern quality. If this pattern is corrected by the algorithm given by Eqs. 79 through 82, a relative exposure as noted in the figures is given to each rectangle.

10.5.3 Ion Beam Lithography

Ion beam lithography can achieve higher resolution than optical, x-ray, and electron lithographic techniques because:

1. Ions have a higher mass and therefore scatter less than electrons.
2. Resists such as PMMA are more sensitive to ions than to electrons.
3. Ion beams (like electron beams) can be used both in an efficient projection printing mode and in a focused-beam direct-writing mode.

The simulation of ion-beam lithography is similar to electron-beam calculations and is based on MC calculations.[48] Figure 33 shows simulated trajectories for 50 H^+ ions implanted at 60 keV into PMMA and various substrates. The following points should be noted:

1. The spread of the ion beam at a depth of 0.4 μm is about 0.1 μm in all cases (compare with Fig. 26).
2. Backscattering is completely absent for the Si substrate.
3. The amount of backscattering is limited for the Au substrate.

The simulation proceeds analogous to the electron-beam case by convolution of the delta-line response with the real beam shape (see Eq. 72 or Eq. 74).

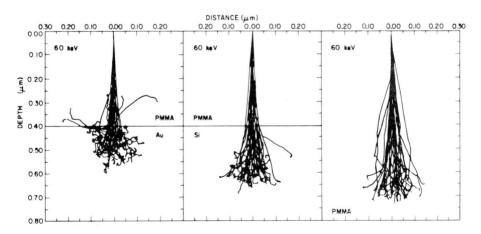

Fig. 33 Trajectories of 60-keV H^+ ions traversing through PMMA into Au, Si, and PMMA. *(After Karapiperis et al., Ref. 48.)*

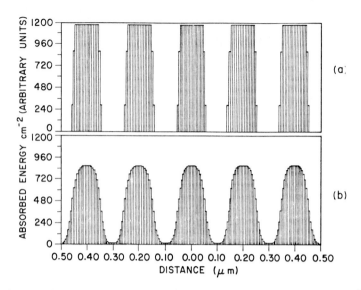

Fig. 34 Histogram of absorbed energy in the *xz* plane for five 1000-Å-wide lines of 60-keV H$^+$ ions in the *y* direction. Line spacing is 1000 Å. (a) Absorbed energy at 400 Å. (b) Absorbed energy at 4000 Å. *(After Karapiperis et al., Ref. 48.)*

Figure 34 shows histograms of absorbed energy in the *xz* plane for five 0.1-μm lines with 0.1-μm spaces for 60-keV H$^+$ implants at two depths of 0.04 μm and 0.4 μm. Note the limited degree to which the absorbed energy spreads. At the interface between the PMMA and Si, the overlap between neighboring lines at the midpoint of their separation is extremely small (1/80 of the peak value; compare this value to that in Fig. 30). Proximity effects are therefore negligible which is also reflected in simulated development profiles obtained by the same technique as in the electron-beam case. Figures 35a and b shows developed profiles for a high dose, 2×10^{-6} C cm^{-2}, and a low dose, 0.6×10^{-6} C cm^{-2}. The developed lines have vertical profiles and the shape of the walls is unaffected by the exposure of the neighboring lines.

10.6 ETCHING AND DEPOSITION

In silicon processing, etching and deposition steps become more and more important as device sizes shrink. The control of the etched profiles and the shapes of deposited layers have a direct impact on the performance of the final device and circuit. Dry etching techniques (Chap. 8) are necessary to achieve the fine linewidth in VLSI technologies. These techniques range from chemical, isotropic processes to directional, physical processes (ion milling) with mixed physico-chemical techniques, such as reactive-ion etching, in between. In most IC processes, at least two layers of interconnect are used. These layers are obtained by depositing and patterning polysilicon and Al. Low-temperature processing requirements, together with the enhanced anisotropic etching techniques results in steep edge profiles that are difficult to cover with a

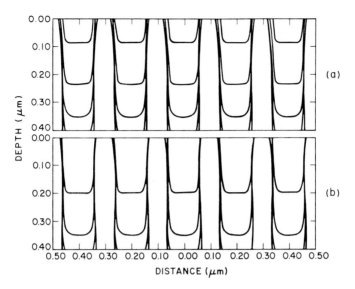

Fig. 35 Developed profiles in PMMA. Five incident 1000 Å-wide lines, 1000 Å apart, of 60-keV H^+ ions. 1:1 (MIBK:IPA) developer. (a) Dose: 0.6×10^{-6} C cm^{-2}. Profiles after 1, 3, 5, 7, and 9 min. (b) Dose: 2.0×10^{-6} C cm^{-2}. Profiles after 15, 30, 45, 60, and 75 s. *(After Karapiperis et al., Ref. 48.)*

film of uniform thickness. Simulation of the deposition is needed to produce accurate results that can help to optimize a particular source-substrate situation.

10.6.1 The String Model Algorithm

From the simulation point of view, both etching and deposition are problems which are essentially geometric in nature. In the string model, the boundary between processed and unprocessed regions (e.g., developed and undeveloped regions during etching) is approximated by a series of points joined by straight line segments.[49, 50] The resulting profile is determined by the initial profiles that move through a medium in which the speed of propagation is a function of the local variables at each point. Consider the examples in Figs. 36a and b, which illustrate the application of the string model algorithm to isotropic and anisotropic etching. A typical isotropic etching case equivalent to this case is the development of a silicon layer being plasma etched in a fluorine source such as CF_4 or SF_6. The basic reaction

$$Si + 4F \rightarrow SiF_4 \tag{83}$$

describes the chemistry of the process. If only free fluorine radicals are present, the etching proceeds isotropically wherever the absorption of fluorine has exposed the Si. Under these isotropic conditions, the etching is simulated by advancing all string points at a constant rate in the direction of the perpendicular bisector of the adjacent elements. The anisotropic etch rate is proportional to the cosine of the angle between the flux direction and the surface normal as in Fig. 36b. In the extreme case, shadowing can occur as illustrated in Fig. 36c. To incorporate the shadowing mechanism,

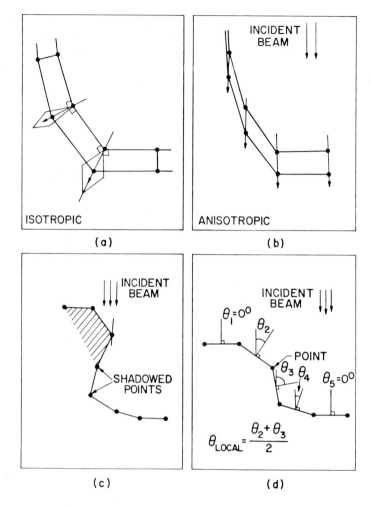

Fig. 36 Application of the string model to isotropic and anisotropic etching and the extreme ion milling situation. (a) and (b) Isotropic and anisotropic advances of points along line edge profile. (c) Shadowed points along a string. (d) Extraction of local angular orientation in ion milling model.

the positions of all points are considered with respect to a line parallel to the radiation flux. Points that are shaded by other segments are advanced according to the isotropic background rate. Ion milling or sputtering simulation[51] uses the same point advancement and shadowing algorithm as directional etching. In addition, average angular information is taken from adjacent segments and the incident ion beam. The ion milling situation is presented schematically in Fig. 36d. In SAMPLE,[32] the etch rate is modeled along these lines according to the sputtering yield

$$S(\theta) = S_0 \left[\frac{\phi}{\rho} \right] \left[A \cos \theta + B \cos^2 \theta + C \cos^4 \theta \right] \qquad (84)$$

where S_0, A, B, and C characterize the sputtering yield of the material to be ion etched, ρ is the atomic layer density, and ϕ is the current density of the ion flux.

With the string model, the growth of films through deposition is simulated by reverse etching. The advancement of each point of the line-edge profile is controlled by the deposition conditions. During the advancement, string segments that become very long or very short are adjusted automatically by adding and deleting points.

10.6.2 Etching

A number of plasma etching processes can be classified[51, 52] empirically as isotropic and anisotropic. Figure 37 clearly illustrates the nature of those two processes. A layer of polysilicon is anisotropically etched below a layer of isotropically overetched silicon dioxide. The upper figure shows the simulated result,[32, 50] which demonstrates the two separate etching components. The simulation uses the isotropic component in the etching of the SiO_2 layer first and then proceeds with the anisotropic component in the silicon. The result ought to be compared to the experimental data at the bottom of Fig. 37. By identifying isotropic and anisotropic components from experimental information, a complete dry etching process can be simulated. Figure 38 compares simulated results[32] with an SEM micrograph for a reactive-ion etched SiO_2-Si structure covered by Al_2O_3. The etch rates used in the simulation are obtained from experimental data.[53] SiO_2 etches more anisotropically, while the silicon shows a more chemical isotropic etch during the process.

An ion milling example[51] is given in Figs. 39a and b. Experimental results for a hard Ti mask on a soft Au layer or silicon are shown in Fig. 39a. A 0.4-μm thick Ti layer is deposited by a lift-off process and pure Ar is used during ion milling. Note that the Ti facets at an angle of 45°. After 4 min, this facet reaches the Au interface. For short milling times (2 min), the resulting Au profile is vertical but becomes less steep for longer times. The corresponding simulation result in Fig. 39b clearly reproduces with high accuracy all phenomena found experimentally.

10.6.3 Deposition

The simulation of deposition profiles uses the same string model algorithm that the etching model uses. The following assumptions are made in this simulation:[32, 54]

1. The mean free path of the atoms is larger than the distance between the source and the substrate.
2. The source-to-substrate distance is large compared to the step height.
3. The magnitude of the film growth rate follows the cosine distribution law; that is, the growth rate is proportional to $\cos \alpha$ where α is the angle between the vapor stream and the surface normal.
4. The growth direction is toward the vapor stream.
5. The sticking coefficient is set to one for a cold substrate.
6. At elevated temperatures, surface migration on the substrate follows a random-walk law. An increase of the substrate temperature increases the migration distance.

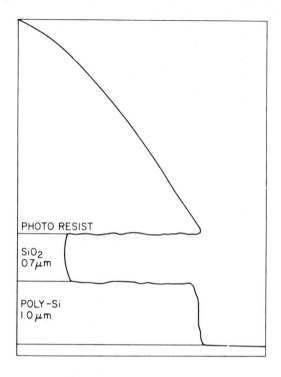

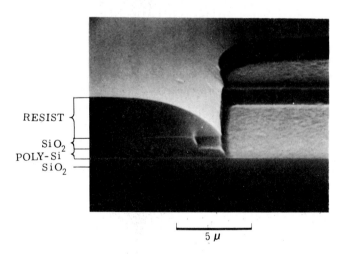

Fig. 37 Isotropic undercut followed by anisotropic etch: simulation (*after Reynolds, Neureuther, and Old-ham, Ref. 50*) and experiment (*after Mogab and Harshberger, Ref. 52*).

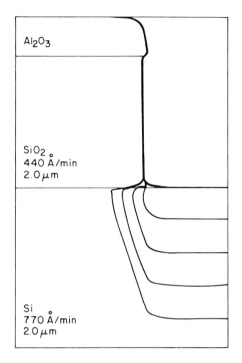

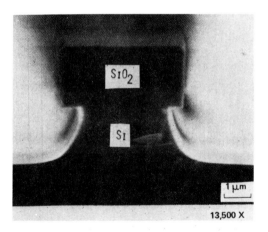

Fig. 38 Reactive ion etching of SiO_2 and Si: simulation (*after Reynolds, Neureuther, and Oldham, Ref. 50*) and experiment (*after Schwartz, Rothman, and Schopen, Ref. 53*).

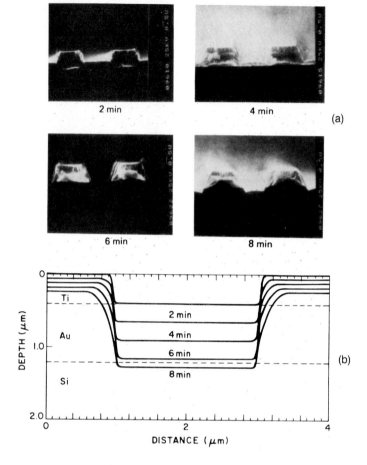

Fig. 39 Example of ion milling (a) SEM micrograph of a 0.4-μm Ti (lift-off) mask on 1 μm of evaporated Au or Si. (b) Simulation of the ion milling in (a). *(After Neureuther, Liu, and Ting, Ref. 51.)*

Deposition results depend strongly on the type of evaporation source actually used. For a unidirectional source as in Fig. 40a, the vapor stream arises at the surface in one direction only. No film growth can occur in shadowed regions. For unshadowed points, the growth rate is expressed as

$$R(x,z) = C \sin \alpha_1 \mathbf{i} + C \cos \alpha_1 \mathbf{k} \qquad (85)$$

where α_1 is the angle between the z axis and the vapor stream, $\mathbf{i}$ and $\mathbf{k}$ are the unit vectors in x and z directions, respectively, and C is the growth rate of an unshadowed surface normal to the vapor stream. In the case of dual evaporation sources (Fig. 40b), each point in an unshadowed region is exposed to two vapor streams, allowing the growth rate to be written as

$$R(x,z) = C \left[\sin \alpha_1 + \sin \alpha_2 \right] \mathbf{i} + C \left[\cos \alpha_1 + \cos \alpha_2 \right] \mathbf{k} \qquad (86)$$

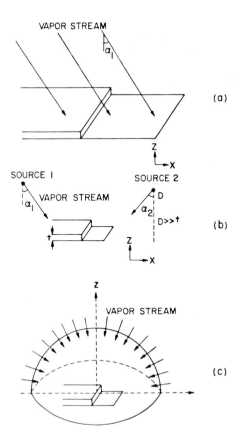

Fig. 40 Evaporation sources (a) Unidirectional; (b) dual evaporation; (c) hemispherical.

For a hemispherical source (Fig. 40c) the vapor flux is distributed in a range of directions with the growth rate

$$R(x,z) = C\left[\cos \alpha_1 - \cos \alpha_2\right]\mathbf{i} + C\left[\sin \alpha_2 - \sin \alpha_1\right]\mathbf{k} \qquad (87)$$

where α_1 and α_2 are the lower and upper bounds of the incident angles of the vapor streams.

A planetary system is shown in Fig. 41. In this configuration, the rotation of the planet along the system control axis does not affect the deposition rate. The growth rate is calculated by holding the planet stationary and rotating only the source along the planet axis. The growth rate for a deposition from all angles between α_{min} and α_{max} is[55]

$$R_x(x,z) = \int_{\alpha_{min}}^{\alpha_{max}} \Delta x_i(\alpha - \beta)\,d\alpha \qquad (88)$$

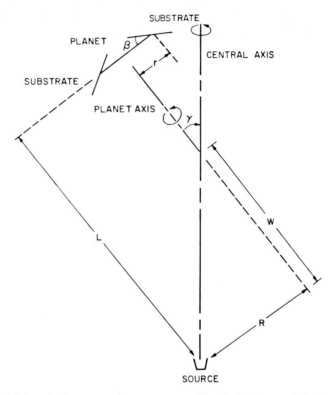

SUBSTRATE

CENTRAL AXIS

PLANET

SUBSTRATE

PLANET AXIS

γ

β

r

L

W

R

SOURCE

Fig. 41 Schematic planetary evaporator geometry. *(After Blech, Fraser, and Haszko, Ref. 55.)*

and

$$R_z(x,z) = \int_{\alpha_{min}}^{\alpha_{max}} \Delta z_i(\alpha - \beta) \, d\alpha \qquad (89)$$

The variables $\Delta x_i(\alpha)$ and $\Delta z_i(\alpha)$ are given by

$$\Delta x_i(\alpha) = I(\omega) \cos \theta'' \tan \alpha \, \frac{dw}{d\alpha} \qquad (90)$$

$$\Delta z_i(\alpha) = I(\omega) \cos \theta'' \frac{dw}{d\alpha} \qquad (91)$$

where $I(\omega) \, dw$ is the amount of material arriving per unit area of the wafer for a rotation dw.

$$I(\omega) \sim \frac{R}{L} \left\{ \frac{1 + \dfrac{WL}{R^2} - \left[\dfrac{r}{R}\right]^2 - \dfrac{rL}{R^2} \tan(\alpha - \beta)}{L^3 \sqrt{1 + \left[\dfrac{\omega}{R}\right]^2} \left[1 + \left[\dfrac{R}{L}\right]^2 - \left[\dfrac{r}{L}\right]^2 - \dfrac{2r}{L} \tan(\alpha - \beta)\right]^{3/2}} \right\} \qquad (92)$$

and θ'' is the angle between z and the vapor stream. The cone source is the special case of a planetary source with $\beta = r = 0$. Equations 89 through 92 can be evaluated analytically to

$$
R_x(x,z) = -\left[\sqrt{1 - \left[\frac{L}{R} \tan \alpha_{max}\right]^2}\right.
$$
$$
\left. - \sqrt{1 - \left[\frac{L}{R} \tan \alpha_{min}\right]^2}\right] \frac{R(R^2 + LW)}{(R + L)^2\sqrt{R^2 + W^2}} \qquad (93)
$$

$$
R_z(x,z) = \left[\arcsin\left[\frac{L}{R} \tan \alpha_{max}\right] - \arcsin\left[\frac{L}{R} \tan \alpha_{min}\right]\right]
$$
$$
\times \frac{L(R^2 + LW)}{(R^2 + L^2)\sqrt{R^2 + W^2}} \qquad (94)
$$

Figure 42 shows an example of a simulated equal-time contour for a planetary evaporator.[56] The shadow of the surface results in a thinning of the deposited material both on the face and the bottom of the step. Small cracks appear at the boundary between the shadowed and unshadowed regions and at the bottom of the vertical step. Simulated and experimental results for 1-μm lines and spaces are compared in Fig. 43. In the symmetric case (Fig. 43a), a slight depression caused by shadowing from both sides is predicted. In the asymmetrical case, the dip on the left and the partial deposition on the right show good agreement.

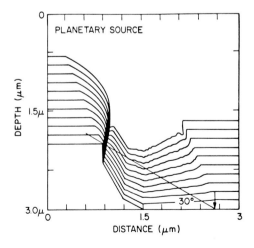

Fig. 42 Simulated time evolution of surface contours for Al deposition. *(After Ting and Neureuther, Ref. 56.)*

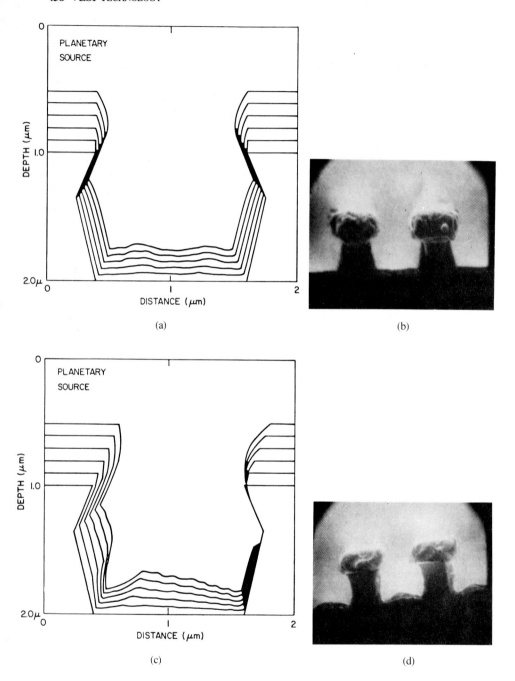

Fig. 43 Simulation and experimental comparison for 1-μm lines and spaces. 2-in wafer located in the outboard planet position of an Airco Temescal 1800 system. (a) and (b) Theory and experiment for symmetrical case; (c) and (d) asymmetrical case. *(After Ting and Neureuther, Ref. 56.)*

10.7 DEVICE SIMULATION

For the device design to be successful, process simulation has to be coupled to device modeling to account for the interrelation between processing and device behavior. Device modeling is based on the numerical solution of the coupled nonlinear partial differential equations,[57] which model the intrinsic behavior of semiconductor devices. Depending on the problem, one-dimensional[58] or higher-dimensional models[57] are required. In the following we consider a NMOS process for the fabrication of submicrometer size MOSFETs.[59] Electron-beam lithography with a novel multilevel resist structure defines the pattern. Dry etching techniques transfer the patterns. Table 1 summarizes the wafer process. At all lithographic levels a three-layer resist structure is used (see Chap. 7). At the bottom is a thick layer of HPR resist followed by a thin intermediate stencil layer of amorphous silicon and an upper layer of electron resist. Positive (PBS) and negative resists (GMC) are used at different levels. Apart from the lithographic steps, the fabrication sequences of major importance are chan-stop and threshold-adjust implants (steps 3, 5, and 7), and the source-drain formation implant (step 15).

Several low-temperature ($T = 900°C$) annealing steps occur in the whole process. All major process steps have been modeled in two dimensions. Figure 14a shows the total concentration as calculated by simulating the process in Table 1. Figure 44

Table 1 Wafer process outline

(substrate is 6 to 8 Ω-cm, B doped)

1.	Grow field oxide, 3500 Å
2.	Ion implant B, 150 keV, 2×10^{12} cm^{-2}
3.	Active-area level lithography, PBS(+) resist
4.	Field oxide etch
5.	Ion implant B, 150 keV, 0.5×10^{12} cm^{-2} (optional)
6.	Depletion level lithography, PBS(+) resist
7.	Ion implant As, 60 keV, 3×10^{12} cm^{-2}
8.	Grow gate oxide, 250 Å
9.	Deposit polysilicon, 1500 Å
10.	Polycon level lithography, PBS(+) resist
11.	Etch polysilicon; etch oxide
12.	Deposit polysilicon, 2000 Å
13.	Polysilicon level lithography, GMC(−) resist
14.	Etch polysilicon
15.	Ion implant As, 30 keV, 7×10^{15} cm^{-2}
16.	Grow thin oxide, deposit PSG, and planarize
17.	Window-level lithography, PBS(+) resist
18.	Etch oxide
19.	Form silicide
20.	Deposit polysilicon plus Al
21.	Metal level lithography, GMC(−) resist
22.	Etch Al, etch polysilicon
23.	Sinter Al, metallize backside

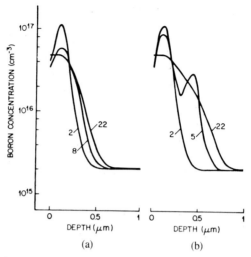

Fig. 44 Calculated doping concentration in the channel of enhancement devices for (a) single and (b) double boron implant. The numbers on the curves correspond to the process steps in Table 1. *(After Watts et al., Ref. 59.)*

gives calculated doping concentrations in the channel of enhancement devices for the single (step 3) and double (steps 3 and 5) boron implants, including all heat treatments in the process. With Fig. 14a as input to the two-dimensional device simulator, excellent agreement between measured and calculated results is obtained, as illustrated by the I/V characteristics in Fig. 45. Results of similar quality are obtained for the gate-length dependence of the threshold voltage for both enhancement and depletion devices.

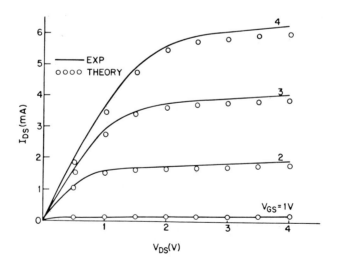

Fig. 45 Measured (full lines) and calculated (full dots) results of enhancement device with $L = 0.62$ μm, $W = 30$ μm. *(After Watts et al., Ref. 59.)*

10.8 SUMMARY AND FUTURE TRENDS

This chapter summarizes the relevant theory and provides examples of process simulation. Theoretical results have been minimized by referencing the important literature. The material presented, together with the references at the end, should enable the reader to understand the basic features and goals of process simulation.

Process simulation is a rapidly expanding field and the literature is growing. With the exception of several theoretical papers, most references cited are from the late 1970s with the bulk being published in the last few years. We can expect increased sophistication in computer programs together with an improved understanding of physical processes to take over the job of designing new processes and devices completely. Optimization of processes will become an automatic tool in the 1980s. Process and device simulation coupled with circuit simulation will be user-oriented. In the future, a total design system will allow on-line process design to predict the desired device and circuit parameter sensitivities, and to facilitate circuit design and layout with given design rules.

REFERENCES

[1] R. Reif, T. I. Kamins, and K. C. Saraswat, "A Model for Dopant Incorporation into Growing Silicon Epitaxial Films. I. Theory," *J. Electrochem. Soc.*, **126**, 644 (1979).

[2] R. Reif, T. I. Kamins, and K. C. Saraswat, "A Model for Dopant Incorporation into Growing Silicon Epitaxial Film. II. Comparison of Theory and Experiment," *J. Electrochem. Soc.*, **126**, 653 (1979).

[3] R. Reif and R. W. Dutton, "Computer Simulation in Silicon Epitaxy," *J. Electrochem. Soc.*, **128**, 909 (1981).

[4] D. A. Antoniadis, S. E. Hansen, and R. W. Dutton, "SUPREM-II-A Program for IC Process Modeling and Simulation," Stanford Electronics Laboratories Technical Report No. 5019-2, June 1978.

[5] P. H. Langer and J. I. Goldstein, "Impurity Redistribution during Silicon Epitaxial Growth and Semiconductor Device Processing," *J. Electrochem. Soc.*, **121**, 563 (1974); see also *J. Electrochem. Soc.*, **124**, 591 (1977).

[6] M. M. Faktor and I. Garrett, *Growth of Crystals from the Vapor*, Chapman and Hall, New York, 1974.

[7] J. Gibbons, W. S. Johnson, and S. Mylroie, *Projected Range Statistics*, 2d ed., Wiley, New York, 1975.

[8] B. Smith, *Ion Implantation Range Data for Silicon and Germanium Device Technologies*, Research Studies Press, Oregon, 1977.

[9] D. H. Smith and J. F. Gibbons, "Application of the Boltzmann Transport Equation to the Calculation of Range Profiles and Recoil Implantation in Multilayered Media," in F. Chernow, J. A. Borders, and D. K. Brice, eds., *Ion Implantation in Semiconductors 1976*, Plenum, New York, 1977.

[10] L. A. Christel and J. F. Gibbons, "An Application of the Boltzmann Transport Equation to Ion Range and Damage Distributions in Multilayered Targets," *J. Appl. Phys.*, **51**, 6176 (1980).

[11] M. T. Robinson and I. M. Torrens, "Computer Simulation of Atomic Displacement Cascades in Solids in the Binary Collision Approximation," *Phys. Rev.*, **B9**, 5008 (1974).

[12] J. P. Biersack and L. G. Haggmark, "A Monte Carlo Computer Program for the Transport of Energetic Ions in Amorphous Targets," *Nucl. Instrum. and Methods*, **174**, 257 (1980).

[13] S. Kalbitzer and H. Oetzmann, "Ranges and Range Theories," *Radiat. Eff.*, **47**, 57 (1980).

[14] W. D. Wilson, L. G. Haggmark, and J. P. Biersack, "Calculations of Nuclear Stopping, Ranges and Straggling in the Low Energy Region," *Phys. Rev.*, **B15**, 2458 (1977).

[15] C. Lehmann, *Interaction of Radiation with Solids and Elementary Defect Production*, North-Holland, New York, 1977.

[16] T. Hirao, K. Inoue, S. Takayanagi, and Y. Yaegashi, "The Concentration Profiles of Projectiles and Recoiled Nitrogen in Silicon after Ion Implantation Through Si_3N_4 Films," *J. Appl. Phys.*, **50**, 193 (1979).

[17] W. K. Hofker, D. P. Oosthoek, N. J. Koeman, and H. A. M. de Grefte, "Concentration Profiles of Boron Implantations in Amorphous and Polycrystalline Silicon," *Radiat. Eff.*, **24**, 223 (1975).

[18] D. K. Brice, *Ion Implantation Range and Energy Deposition Distributions*, IFI/Plenum, New York, 1975.

[19] A. DeSalvo and R. Rosa, "A Comprehensive Computer Program for Ion Penetration in Solids," *Radiat. Eff.*, **47**, 117 (1980).

[20] A. DeSalvo and R. Rosa, "Monte Carlo Calculations on Spatial Distribution of Implanted Ions in Silicon," *Radiat. Eff.*, **31**, 41 (1976).

[21] R. W. Dutton, H. G. Lee, and S. Y. Oh, "Simplified Two Dimensional Analysis for Time Dependent Carrier Transport and Impurity Redistribution," in B. T. Browne and J. J. H. Miller, eds., *Numerical Analysis of Semiconductors Devices and Integrated Circuits*, Boole Press, Dublin, 1981.

[22] S. M. Hu and S. Schmidt, "Interactions in Sequential Diffusion Processes in Semiconductors," *J. Appl. Phys.*, **39**, 4272 (1968).

[23] H. Ryssel, K. Müller, K. Haberger, R. Henkelmann, and F. Jahnel, "High Concentration Effects of Ion Implanted Boron in Silicon," *Appl. Phys.*, **22**, 35 (1980).

[24] M. Y. Tsai, F. F. Morehead, and J. E. E. Baglin, "Shallow Junctions by High-Dose As Implants in Si: Experiments and Modeling," *J. Appl. Phys.*, **51**, 3230 (1980).

[25] B. R. Penumalli, "Lateral Oxidation and Redistribution of Dopants," in B. T. Browne and J. J. H. Miller, eds., *Numerical Analysis of Semiconductor Devices and Integrated Circuits*, Boole Press, Dublin, 1981.

[26] B. R. Penumalli, "A Comprehensive Two-Dimensional VLSI Process Simulation Program—BICEPS," *IEEE Trans. Electron Devices*, **ED-36**, Sept. (1983).

[27] A. M. Lin, D. A. Antoniadis, and R. W. Dutton, "The Oxidation Rate Dependence of Oxidation-Enhanced Diffusion of Boron and Phosphorus in Silicon," *J. Electrochem. Soc.*, **128**, 1131 (1981).

[28] D. Chin, R. W. Dutton, and S. M. Hu, "Two-Dimensional Modelling of Local Oxidation," presented at the 40th Device Research Conference, Ft. Collins, Colo., June 21-23, 1982.

[29] R. S. Varga, *Matrix Iterative Analysis*, Prentice-Hall, Englewood Cliffs, N.J., 1962.

[30] D. Chin, M. Kump, and R. W. Dutton, "SUPRA-Stanford University Process Analysis Program," Stanford Electronics Laboratories Technical Report, July 1981.

[31] L. A. Hageman and D. M. Young, *Applied Iterative Methods*, Academic, New York, 1981. A program called ITPACK is available from these authors which contains all methods described in this reference.

[32] "SAMPLE 1.5 User's Guide," University of California at Berkeley, 1982.

[33] B. J. Lin, "Optical Methods for Fine Line Lithography," in R. Newman, ed., *Fine Line Lithography*, North-Holland, New York, 1980.

[34] M. C. King, "Principles of Optical Lithography," in N. G. Einspruch, ed., *VLSI Electronics—Microstructure Science*, Vol. 1, Academic, New York, 1981.

[35] W. G. Heitman and P. M. van der Berg, "Diffraction of Electromagnetic Waves by a Semi-Infinite Screen in a Layered Medium," *Can. J. Phys.*, **53**, 1305 (1975).

[36] M. M. O'Toole and A. R. Neureuther, "Influence of Partial Coherence on Projection Printing," SPIE, Vol. 174, Developments in Semiconductor Microlithography IV, 22 (1979).

[37] F. H. Dill, A. R. Neureuther, J. A. Tuttle, and E. J. Walker, "Modelling Projection Printing of Positive Photoresists," *IEEE Trans. Electron Devices*, **ED-22**, 456 (1975)

[38] P. H. Berning, "Theory and Calculations of Optical Thin Films," in G. Hass, ed., *Physics of Thin Films*, Vol. 1, Academic, New York. 1963.

[39] D. F. Kyser and K. Murata, "Monte Carlo Simulation of Electron Beam Scattering and Energy Loss in Thin Films on Thick Substrates," in R. Bakish, *Proceedings of the 6th International Conference on Electron and Ion Beam Science and Technology*, The Electrochemical Society, Princeton, N.J., 1974.

[40] R. J. Hawryluk, A. M. Hawryluk, and H. I. Smith, "Energy Dissipation in a Thin Polymer Film by Electron Beam Scattering," *J. Appl. Phys.*, **45**, 2551 (1974).

[41] J. S. Greeneich, "Electron Beam Processes," in G. R. Brewer ed., *Electron-Beam Technology in Microelectronic Fabrication*, Academic, New York, 1980.

[42] D. F. Kyser and R. Pyle, "Computer Simulation of Electron-Beam Resist Profiles," *IBM J. Res. Develop.*, **24**, 426 (1980).

[43] D. F. Kyser, D. E. Schreiber, C. H. Ting, and R. Pyle, "Proximity Function Approximations for Electron-Beam Lithography From Resist Profile Simulation," in R. Bakish ed., *Proceedings of the 9th International Conference on Electron and Ion Beam Science and Technology*, The Electrochemical Society, Princeton, N.J., 1980.

[44] N. D. Wittels, "Fundamentals of Electron and X-Ray Lithography," in R. Newmann ed., *Fine-Line Lithography*, North-Holland, New York, 1980.

[45] M. Parikh, "Corrections to Proximity Effects in Electron-Beam Lithography. I: Theory. II: Implementation, III: Experiments," *J. Appl. Phys.*, **50**, 4371, 4378, 4383 (1979).

[46] T. H. P. Chang, "Proximity Effect in Electron-Beam Lithography," *J. Vac. Sci. Technol.*, **12**, 1271 (1975).

[47] M. Parikh and D. E. Schreiber, "Pattern Partitioning for Enhanced Proximity—Effect Corrections in Electron-Beam Lithography," *IBM J. Res. Develop.*, **24**, 530 (1980).

[48] L. Karapiperis, I. Adesida, C. A. Lee, and E. D. Wolf, "Ion Beam Exposure Profiles in PMMA-Computer Simulation," *J. Vac. Sci. Technol.*, **19**, 1259 (1981).

[49] R. E. Jewett, P. I. Hagouel, A. R. Neureuther, and T. van Duzer, "Line-Profile Resist Development Simulation Techniques," *Polym. Eng. Sci.*, **17**, 381 (1977).

[50] J. L. Reynolds, A. R. Neureuther, and W. G. Oldham, "Simulation of Dry Etched Line Edge Profiles," *J. Vac. Sci. Technol.*, **16**, 1772 (1979).

[51] A. R. Neureuther, C. Y. Liu, and C. H. Ting, "Modelling Ion Milling," *J. Vac. Sci. Technol.*, **16**, 1767 (1979).

[52] C. J. Mogab and W. R. Harshberger, "Plasma Processes Set to Etch Finer Lines with Less Undercutting," *Electronics*, **51**, 117 (1981).

[53] G. C. Schwartz, L. B. Rothman, and T. J. Schopen, "Competitive Mechanisms in Reactive Ion Etching in a CF_4 Plasma," *J. Electrochem. Soc.*, **126**, 464 (1979).

[54] W. G. Oldham, A. R. Neureuther, C. K. Sung, J. L. Reynolds, and S. N. Nandgaonkar, "A General Simulator for VLSI Lithography and Etching Processes: Part II—Application to Deposition and Etching," *IEEE Trans. Electron Devices*, **ED-27**, 1455 (1980).

[55] I. A. Blech, D. B. Fraser, and S. E. Haszko, "Optimization of Al Step Coverage through Computer Simulation and Scanning Electron Microscopy," *J. Vac. Sci. Technol.*, **15**, 13 (1978).

[56] C. H. Ting and A. R. Neureuther, "Applications of Profile Simulation for Thin Film Deposition and Etching Processes," *Solid State Technology*, **25** (2), 115 (1982).

[57] W. Fichtner and D. J. Rose, "On the Numerical Solution of Nonlinear Elliptic PDEs Arising from Semiconductor Device Modelling," in M. Schultz, ed., *Elliptic Problem Solvers*, Academic, New York, 1981.

[58] D. C. D'Avanzo, M. Vanzi, and R. W. Dutton, "One-Dimensional Semiconductor Device Analysis (SEDAN)," Stanford Electronics Laboratories Technical Report No. 6-201-5, October 1979.

[59] R. K. Watts, W. Fichtner, E. Fuls, L. R. Thibault, and R. L. Johnston, "Electron-Beam Lithography for Small MOSFETs," *IEEE Trans. Electron Devices*, **ED-28**, 1338 (1981).

[60] E. H. Nicollian and J. R. Brews, *MOS (Metal Oxide Semiconductor) Physics and Technology*, Wiley, New York, 1982.

PROBLEMS

1 Diffusion from a growing epitaxial layer into an undoped substrate.

Consider the case of an undoped silicon substrate. Suppose we deposit an epitaxial layer with concentration C_0 on top of this substrate. Since epitaxial deposition temperatures are usually high, impurities will diffuse out of the depositing layer into the substrate and vice versa. For a one-dimensional geometry, solve the diffusion equation.

2 Diffusion from a doped substrate into an undoped epitaxial layer.

This case concerns putting an undoped layer on a homogeneously doped semi-infinite substrate. The problem is more difficult than Prob. 1, because we have to incorporate the fact that some of the dopant that diffuses into the layer out of the substrate diffuses straight through and evaporates from the growth surface. A net loss of dopant occurs if the rate that dopant atoms leave the surface at $z = 0$ is greater than the rate at which they join it from the ambient atmosphere. For a one-dimensional geometry, solve the diffusion equation incorporating the evaporation case.

3 Solve the scattering integral

$$\theta = \pi - 2p \int_{r_0}^{\infty} \left(\frac{dr}{r^2 \sqrt{1 - \frac{V(r)}{E_c} - \frac{p^2}{r^2}}} \right) \tag{1}$$

for a repulsive Coulomb potential $V(r) = C_1/r$ $(C_1 > 0)$. Calculate the differential cross section

$$\sigma(\theta) = \frac{p}{\sin \theta} \frac{dp}{d\theta} \tag{2}$$

4 Write a FORTRAN program to calculate Pearson IV distributions.

5 Most frequently, the semiconductor substrate is subjected to a heat cycle after the implantation step, and the impurities are redistributed. Near the mask edge, this redistribution is obtained by solving the two-dimensional diffusion equation

$$\frac{\partial C}{\partial t} = D \left[\frac{\partial^2 C}{\partial x^2} + \frac{\partial^2 C}{\partial z^2} \right] \tag{1}$$

assuming intrinsic diffusion conditions. Calculate the analytical solution for the following initial condition (i.e., profile after implantation)

$$C(x, z, t = 0) = C_{max} \left[C_L(x, z, t = 0) + C_R(x, z, t, = 0) \right] \tag{2}$$

where C_{max} is the peak concentration and

$$C_L(x, t, 0) = \frac{1}{2} \exp \left[-\frac{(z - R_p)^2}{2 \Delta R_p^2} \right] \text{erfc} \left[\frac{x}{\sqrt{2} \Delta X} \right] \tag{3}$$

$$C_R(x, t, 0) = \frac{\Delta R_p}{2\sqrt{D_0}} \exp \left[\frac{-(\alpha_m x + z - R_p)^2}{2 D_0} \right]$$

$$\times \left\{ 1 + \text{erf} \left[\frac{x \Delta R_p^2 - \alpha_m(z - R_p) \Delta X^2}{\Delta X \Delta R_p \sqrt{2 D_0}} \right] \right\} \tag{4}$$

and

$$D_0 = \Delta R_p^2 + \alpha_m^2 \Delta X^2 \tag{5}$$

$\alpha_m = \tan \theta$ gives the edge slope of the mask, and R_p, ΔR_p, and ΔX are the usual Gaussian parameters.

6 Derive Eqs. 93 and 94 for the deposition rates of a cone source.

7 Solve the diffusion problem of redistributing donor and acceptor impurities between Si and SiO_2 during thermal oxidation at high temperature. Assume that the initial impurity distribution in the silicon is uniform.

8 Same as Prob. 7, but for a nonuniform initial impurity distribution. Use the method of finite differences to discretize the diffusion equation.

ELEVEN

VLSI PROCESS INTEGRATION

L. C. PARRILLO

11.1 INTRODUCTION

The integrated circuit (IC) was invented by Kilby[1] in 1958. The first ICs were phase-shift oscillators and flip-flops, fabricated in germanium substrates. The individual components in these circuits were isolated in mesa-shaped regions which had been etched in the substrate by using black wax (applied by hand) to mask the active regions. The individual devices were interconnected by wire bonding. These first working units were used for the first public announcement[1] of the "Solid Circuit" (integrated circuit) concept in March 1959. Other critical developments around the same time included the first modern diffused bipolar transistor by Hoerni.[1] This transistor was based on the planar diffused process, a cornerstone of modern IC fabrication, which uses silicon dioxide as a barrier to impurity diffusion. In 1958 a patent was filed on the first use of p-n junctions for device isolation and in 1959 a patent was filed for an IC that used evaporated aluminum metallization over an oxide layer to provide interconnections.[1]

From these early primitive forms, ICs have evolved into complex electronic devices containing hundreds of thousands of individual components on a single chip of silicon. The first ICs were based on contributions from many different fields including device physics, materials science, and chemistry. Interdisciplinary contributions continue to be sought today in the development of new IC technologies.

Since the most important part of the IC is the transistor, this chapter focuses on processing techniques which are used to optimize its characteristics. The major IC technologies discussed are standard bipolar (n-p-n), integrated injection logic (I^2L), n-channel MOS (NMOS), and complementary MOS (CMOS). Table 1 gives a generalized comparison of these various devices as integrated transistors.[2] We assume

Table 1 Characteristics of integrated transistors[2]

	NMOS	CMOS	n-p-n	I^2L
1. General				
Supply voltage range	+	++	−	+
Power	+	++	−	+
Speed	+	+	++	+
Transconductance	−	−	++	++
Circuit density	++	+	−	+
Drive capability	−	+	++	+
2. Digital				
Switching speed	++	++	++	+
Power	+	++	−	+
Noise margins	+	++	−	−
Logic swing	−	++	−	−
3. Analog				
Gain per stage	−	+	++	
Bandwidth	−	−	++	
Input impedance	++	++	−	
Power	++	++	−	
Output swing	−	+	++	
Linearity	−	+	++	
Analog switches	+	++	−	
Precision elements	++	++	+	

Note: The symbols represent moderate (−), good (+), and superior (++) behavior.

that the reader has a familiarity with the basic principles of operation of these devices. For reference, see texts on device physics written by Grove,[3] Muller and Kamins,[4] Streetman,[5] and Sze.[6]

11.2 BASIC CONSIDERATIONS FOR IC PROCESSING

11.2.1 Process Flow

Figure 1 illustrates the main steps in an n-channel, polysilicon-gate, metal-oxide-semiconductor (MOS) IC fabrication process.[7] The formation of the IC comprises many steps which have been discussed in previous chapters such as ion implantation, diffusion, oxidation, film depositions, lithography, and etching. These steps provide precisely controlled impurity layers in the silicon which form the individual circuit components (i.e., transistors, diodes, capacitors, resistors) as well as the dielectric and metallic layers used for interconnecting the individual components into an IC. The fabrication steps, which must proceed in a specific sequence, constitute an IC process flow (or process). When the process is properly executed, each wafer contains a number of individual ICs which will later be separated and packaged.

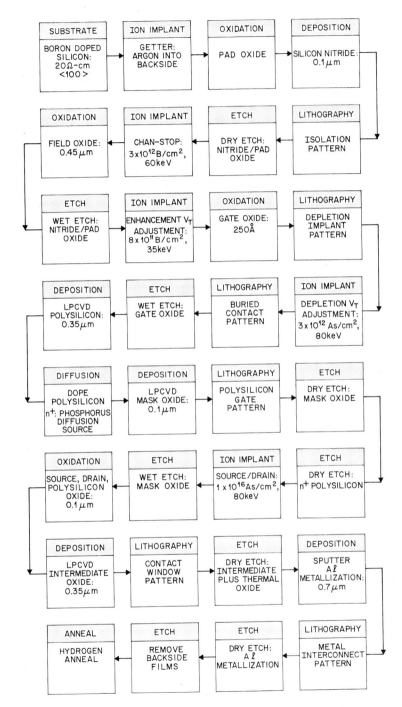

Fig. 1 Main steps in an n-channel, polysilicon-gate, MOS, IC process flow. *(After Siqusch et al., Ref. 7.)*

11.2.2 Interrelation of Process Steps

Virtually all the steps in a process are strongly interrelated—a few examples are cited here. Each of the thermal cycles in a process (e.g., oxidation, epitaxial layer growth, glass flow, gettering) add to affect the vertical and lateral diffusion of impurities. To obtain the desired impurity profiles, the total thermal cycle that the impurity undergoes must be taken into account. Because silicon dioxide is often grown at the same time on several different types of exposed silicon regions, its thickness can be different on the various regions—depending on how heavily doped the silicon is, the impurity type (n or p), whether it is single-crystal or polycrystalline silicon (polysilicon), and if an implanted area has been thermally annealed or not. The resulting variety of oxide thicknesses must be accounted for, if, for example, the oxide films are to be simultaneously removed, or if implants are to penetrate all oxide layers. Polysilicon's ability to mask a given implant is a function of the polysilicon's thickness and the size of its crystal grains, since ion channeling can take place through the grains (see Chapter 6). Grain size is a function of the polysilicon film's doping level and thermal history—hence the polysilicon's ability to act as an effective mask against implantation (e.g., source/drain) changes during the process.

Because the process steps are so interrelated, a fundamental rule in an established process is that *no process step is changed arbitrarily*. In developing a new process, especially one with new materials, identifying the interrelationships among the various steps can be very challenging.

11.2.3 Process Costs

Since the beginning of the IC era, the cost per electronic function has decreased by orders of magnitude because finer features and larger substrates have produced more complex IC chips per wafer. The key to the low cost is batch processing. Individual groups or "lots" of 20 to 50 wafers are processed together. Although a given process flow may have more than 100 individual steps, many individual ICs are fabricated simultaneously. For example, an individual wafer which costs about $200 to produce (labor, equipment, material, and overhead costs) may yield 100 good chips, for a chip cost of about $2. A goal of semiconductor processes continues to be to minimize device cost. This usually translates to simplifying the process flow, since the least complicated processes are the most reproducible and provide the highest yields.

11.3 BIPOLAR IC TECHNOLOGY

A major application of bipolar ICs is in the high-speed memory and logic needs of the computer industry. The bipolar device has recently taken a new form in integrated injection logic (I^2L) which is used extensively in low-power, high-density memory and logic circuits.[8] This section describes a basic fabrication sequence for bipolar ICs and additional considerations for key steps in the device formation. We introduce I^2L device fabrication as well as techniques to avoid emitter-collector leakage currents—a major yield limiting mechanism in bipolar ICs.

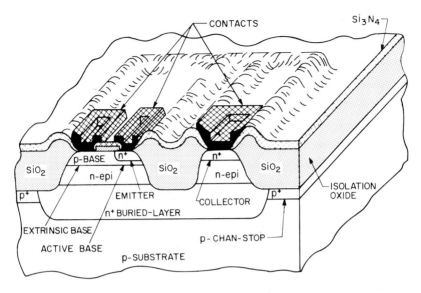

Fig. 2 Three-dimensional views of oxide-isolated bipolar transistor. *(After Labuda and Clemens, Ref. 9.)*

11.3.1 Illustrative Fabrication Process

Figure 2 shows an n-p-n bipolar transistor[9] that uses a thick silicon-dioxide layer to electrically isolate the transistor.[10] Earlier techniques for bipolar isolation relied on a reverse-biased p-n junction that surrounded the active device. Junction isolation consumed a larger area and introduced larger parasitic capacitances as compared to the oxide-isolation[10] techniques. Figure 3 shows a process sequence that can be used to fabricate the device of Fig. 2. Illustrations 3a through f show a top and side view of the device at several stages of the process.

The starting material is a lightly doped p-type substrate ($\sim 10^{15}$ atoms/cm^3), usually with $\langle 111 \rangle$ or $\langle 100 \rangle$ orientation. The substrate is oxidized and a window is opened in the oxide using the buried-layer mask. Arsenic or antimony is then implanted through this window, to serve as the heavily doped (n$^+$) portion of the collector (to reduce collector resistance). The implanted layer is driven into the substrate in an oxidizing ambient to form the "buried-layer." Because of the different rates of oxidation between the exposed buried-layer and the surrounding oxide-covered area, a step forms in the silicon surface at the periphery of the buried-layer (i.e., the buried-layer region is depressed). All oxide is then stripped and an n-epitaxial layer is grown on the substrate as shown in Fig. 3a. The step at the buried-layer periphery propagates up through the epitaxial layer and serves as an alignment mark for the next lithographic level.

After epitaxy growth, a pad layer of SiO$_2$ is grown (~ 500 Å) and a layer of Si$_3$N$_4$ (~ 1000 Å) is deposited on the wafer. The Si$_3$N$_4$ layer does not oxidize readily and thus prevents the oxidation of the underlying silicon. The thin oxide pad serves as a buffer layer to protect the silicon from stress-induced defects during subsequent high-temperature oxidation steps.[11] The isolation lithography is done next as

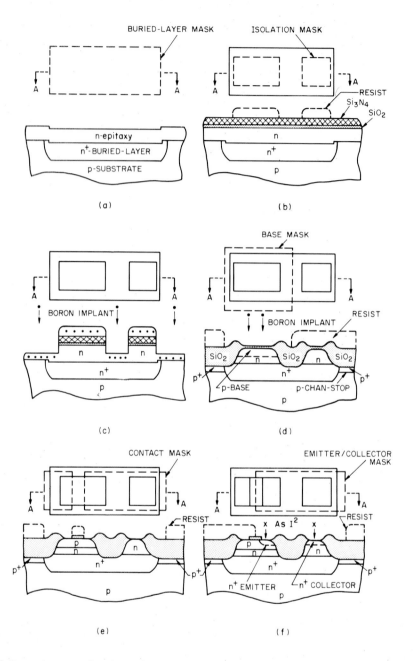

Fig. 3 Top and cross-section views of bipolar transistor fabrication. (a) Buried-layer mask and cross-section after n^+ buried-layer and n-epitaxy growth. (b) Isolation mask and cross section with isolation resist mask on nitride/pad oxide (depression in epitaxy surface is omitted for clarity). (c) Cross section after nitride/pad oxide/silicon etch and chan-stop implant. (d) Base mask and cross section after isolation-oxide growth and boron base implant. (e) Contact mask and cross section after base, emitter, and collector-contact opening. (f) Emitter/collector mask and cross section after arsenic emitter/collector ion implant (I^2).

shown in Fig. 3b. Figure 3c shows that the resist is used to mask the etching of the nitride/pad oxide layers as well as approximately half of the epitaxial layer. A boron channel stopper (chan-stop) implant can also be performed at this point (Fig. 3c). The purpose of the chan-stop implant is to raise the doping level of the p-type substrate directly under the isolation oxide. This will prevent surface inversion of the lightly doped p-type substrate which would electrically connect the buried layers. The resist is then removed and the wafers are oxidized so that all of the remaining epitaxy that is not protected by Si_3N_4 is converted to SiO_2 (Fig. 3d). The Si_3N_4 can then be stripped without disturbing the SiO_2. Up to this point, the thermal processes have involved high temperatures or long times. These long thermal cycles are performed before the active transistors are fabricated, to prevent the desired shallow junctions from being driven in too deeply.

After Si_3N_4 removal, the epitaxy can be oxidized and a base implant mask can be defined using resist (Fig. 3d). The base can be implanted through the oxide so that ion channeling of the base implant is attenuated and no subsequent oxidation of the base implant is necessary.[12] Contact holes to the intended base, emitter, and collector regions can then be made simultaneously with a single mask (Fig. 3e). In this fashion, the base-contact to emitter-contact separation is not influenced by an alignment step, but is set by the minimum spacing between metal contacts there.[13] This results in a reduced transistor area as well as a reduced base resistance. As shown in Fig. 3f, a resist mask can be used to protect the base contact area while the emitter and collector contact regions are implanted using a high arsenic dose at low energy. Note that the implanted emitter area is defined by the opening in the oxide over the exposed base region (Fig. 3f).

An option here is to implant an additional phosphorus dose selectively into the collector-contact region. The phosphorus can be diffused vertically into the buried-layer to minimize the vertical component of collector resistance. Also, a separate higher-dose extrinsic-base implant can be used (e.g., emitter and collector contact covered by a resist mask) to lower the base contact resistance and to decrease the lateral base resistance in the extrinsic-base region. Typically this portion of the extrinsic-base layer can be 50 to 200 $\Omega / \square$ in sheet resistance.

After the emitter is implanted, it is driven into the desired depth in a nearly inert ambient. A very thin oxide results on the emitter, base, and collector contacts and this can be washed off in a dilute HF solution. There is no reregistration of a contact window within an oxide-covered emitter area. The washed emitter process minimizes the emitter area at the risk of allowing the metallization to short out the emitter-base junction at the periphery. The periphery of the junction is only protected by a distance equal to the lateral diffusion of the emitter under the oxide.

After the contact areas are washed, a layer of Si_3N_4 can be deposited over the wafers. This layer protects the device from mobile-ion contaminants, such as sodium, that can diffuse through SiO_2 and result in junction leakage and surface inversion. A window can be cut in this nitride sealing layer (requiring another lithographic step), or the window can be formed in a self-aligned fashion. That is, the nitride can be electrochemically converted (anodized) to SiO_2 where it contacts the silicon, while the nitride on oxide layers remains intact.[14] The anodized oxide can then be stripped in dilute HF acid. The remaining nitride layer is undisturbed and has windows that are self-aligned to the base, emitter, and collector contacts (Fig. 2).

Finally, the metallization layer is deposited and defined as shown in Fig. 2. A variety of metallization systems can be used, including a single layer of aluminum or composite structures such as PtSi for contact followed by TiPtAu layers.[15] In addition to the contacts shown, Schottky-barrier diodes can also be formed in the device (e.g., PtSi, Pd_2 Si), which can be used to clamp the collector-base junction and also to lower the signal swing in Schottky I^2 L devices.

11.3.2 Key Steps in Device Formation

Buried-layer and epitaxial layer The buried-layer's sheet resistance should be as low as possible to reduce the collector resistance of the transistor; hence it is heavily doped. After the dopant is introduced into the substrate, it is driven in to spread out the doping profile and lower the impurity concentration from its initially high value. This results in lower sheet resistance of the buried-layer because of the inverse relationship between carrier mobility and dopant concentration. However, too heavily doped a buried-layer will cause excessive outdiffusion into the more lightly doped n-epitaxial collector, and can also cause defects in the epitaxial layer. Antimony or arsenic are commonly chosen as impurities in the buried-layer rather than phosphorus because of their smaller diffusion coefficients. Typical buried-layer sheet resistance values are about 15 to 50 $\Omega/\square$.

The n-epitaxial layer serves as the collector under the base and is doped in the 10^{15} to 10^{16} atoms/cm^3 range. The lighter the epitaxy doping, the less collector-base capacitance the device has. This is a key parasitic capacitance that limits the high-speed performance of bipolar transistors. A thick enough n-epitaxial layer must be grown so that the outdiffusion of the buried-layer impurity does not reach the base region and raise collector-base capacitance. However, too light a doping in the epitaxial layer is difficult to control in the epitaxy-growth process because of autodoping by the buried-layer impurity in the reactor. In addition, at high collector currents the conductivity of the lightly doped n-epitaxy collector becomes modulated by the collector current. This causes the base-collector junction to be pushed out into the lightly doped epitaxy, resulting in gain degradation as well as high-frequency performance degradation.[6] To avoid this base push-out effect, the doping level in the collector[16] should be greater than J/qv_s where J is the collector current density and v_s is the saturation velocity ($\sim 10^7$ cm/s).

Isolation Referring to Fig. 3c and d, the isolation oxide is typically grown to a thickness such that the top of the oxide and the silicon surface are in the same plane (approximately twice the epitaxy thickness) to minimize surface topography. However, at the periphery of the active silicon areas a "bird's beak" (lateral oxidation under the silicon nitride) and a "bird's head" (caused by the oxide growth at the corners of the etched silicon in Fig. 3c) form. These effects are not desirable because the bird's beak takes up lateral space and the bird's head produces surface topography. A variety of exploratory approaches to minimize these effects have been reported using atmospheric or high-pressure oxidation.[17, 18]

Base formation For a given emitter profile, the lower the total integrated charge is in the active base (Gummel number), the higher the current gain is.[6] However, if the base charge is too low it cannot support the reverse-bias voltage which is applied across the collector-base and/or emitter-base junctions. This results in unwanted "punchthrough" current. Also as the base charge is reduced, the output characteristics (i.e., collector current vs. collector-emitter voltage) exhibit steeper slopes (low output impedance), which may be undesirable in circuit applications. The narrower the base is, the shorter is the diffusion time of minority carriers across the base. As the base is made narrower for better performance, the doping level must therefore be raised to prevent punchthrough. In high-frequency bipolar transistors, the base doping profile is graded from a high to a low value from emitter to collector. This profile creates a built-in electric field in a direction that aids the transit of minority carriers across the base. Typical base Gummel numbers are in the 10^{12} to 10^{13} atoms/cm^2 range.

The extrinsic-base region (Fig. 2) must be adequately doped to provide a low-resistance path to the active-base region. If the extrinsic base is doped too heavily, however, excessive emitter-base capacitance will result along the emitter sidewall and a low emitter-base reverse-breakdown voltage may result. The active- and extrinsic-base regions can be formed by a single ion-implantation step, as shown schematically in Fig. 4a. Alternatively, a double-base implant can be employed as shown schemati-

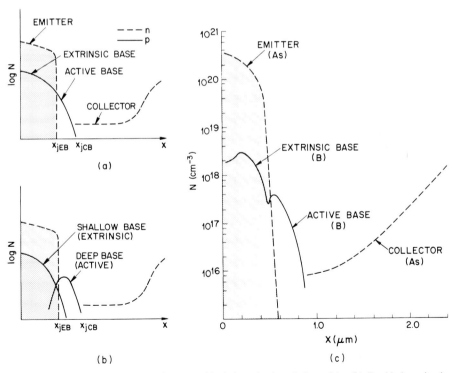

Fig. 4 n-p-n transistor doping profiles. (a) Single-base implant (schematic). (b) Double-base implant (schematic). (c) Single-base implant (actual).

cally in Fig. 4b. The higher-dose, low-energy (shallow-base) implant is used to provide the extrinsic-base properties while the higher-energy, lower-dose (deep-base) implant establishes the active-base properties. The common-emitter current gain varies inversely with deep-base implant dose.[19] The double-base implant technique allows more flexibility in designing the base structure and allows better control of its properties.[12] Figure 4c shows an actual doping profile of an n-p-n transistor that uses a single-base implant for both active and extrinsic bases. Because of strong cooperative diffusion effects between the arsenic emitter and the boron base, the base profile does not smoothly decrease into the silicon as is shown schematically in Fig. 4a. The active-base profile in Fig. 4c is both steeply graded and narrow in width for high-speed applications. Note that the collector doping profile increases with depth above the epitaxial doping level ($\sim 1 \times 10^{16}$ atoms/cm^3) because of outdiffusion of the buried-layer into the relatively thin epitaxial layer.

Emitter formation To improve current gain and to minimize emitter resistance, the emitter region is heavily doped. Consider a device with total charge in the emitter region of $Q_E \sim 10^{16}$ atoms/cm^2 (i.e., a doping level of 2×10^{20} atoms/cm^3 for 0.5 μm). A typical total active-base charge is of the order of $Q_B \sim 10^{12}$ atoms/cm^2. Hence the common-emitter current gain for this device can be of the order of 10^4. Such high values are generally not observed experimentally, however. To explain the observed current gain in actual transistors, both bandgap narrowing and Auger recombination must be included (see Chapter 5).

Very abrupt and shallow arsenic emitter profiles can be obtained because of arsenic's concentration-dependent diffusion. This makes it an attractive choice for an emitter impurity.[19] As emitters become shallower, the technique used to contact the emitter becomes increasingly important, since it can affect the current gain of the device. This is illustrated in Fig. 5, which shows the common-emitter current gain vs. collector current characteristics for three different shallow-emitter (0.2-μm) devices processed identically except for their emitter contacts.[16] The insert in Fig. 5 schematically illustrates the minority-carrier profiles in the emitters of each of these devices[20] at a given injection level (V_{BE}). With the aluminum contact, the hole concentration goes to zero near the original Al-Si interface, where carriers recombine with essentially an infinite recombination velocity. The gradient of holes in the emitter establishes the base current. The hole gradient is made steeper (more base current) when Pd_2 Si is used for the contact, since silicon is consumed during the silicide process. The base current is reduced when a thin layer (1000 Å) of arsenic-doped polysilicon is placed between the metal and the single-crystal emitter. The recombination velocity at the interface between the polysilicon and single-crystal silicon is no longer infinite, and the hole gradient in the emitter is reduced. Current gain can increase by three to seven times when polysilicon, rather than metal, is used to contact the emitter.[20]

Schottky clamps A metallization technique that can be used to keep bipolar transistors out of saturation is the application of Schottky-barrier-diode clamps to the collector region. Figure 6a illustrates a device without a Schottky clamp. When the collector-emitter potential is low enough, the collector-base junction becomes

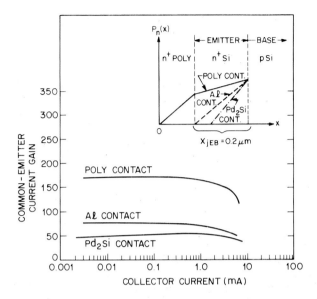

Fig. 5 Common-emitter current gain vs. collector current for shallow-emitter-junction (0.2-μm) n-p-n transistors. Contact to the emitter was made using n^+ polysilicon (poly), aluminum, and Pd_2 Si on separate devices. Insert shows a schematic representation of the minority-carrier profile in the emitter for the three contact schemes at fixed base-emitter potential. *(After Ning, Tang, and Solomon, Ref. 16; Ning and Isaac, Ref. 20.)*

forward-biased and minority carriers flood the base region as well as the n-epitaxial collector region. Since it takes time to remove these excess carriers from the base, a delay is introduced when trying to take the transistor out of saturation (i.e., to turn the transistor off) and circuit performance suffers. To prevent the collector-base from becoming forward biased, a Schottky-barrier diode to the collector can be formed as shown in Fig. 6b. A metal-silicide layer simultaneously contacts the n epitaxy and the p-type base. By choosing a metal-silicide having a high Schottky-barrier height to n material and, consequently, a low barrier height to p material (e.g., PtSi, Pd_2Si, etc.), the metal-silicide provides an ohmic contact to the base, and simultaneously produces a Schottky-barrier diode to the n collector (separate ohmic contacts are also made to the heavily doped n^+ emitter and collector by the metal-silicide). The Schottky diode conducts current at a smaller forward-bias potential (~0.3 V) than does the collector-base p-n junction diode (~0.7 V). Since the Schottky diode and collector-base diodes are in parallel, the collector-base junction is clamped to the lower potential and thus is prevented from becoming forward-biased enough to conduct significant current. As shown in Fig. 6b, at low V_{CE} potential, electrons injected from the emitter are essentially returned out of the forward-biased Schottky diode. The Schottky-barrier height, and hence the forward turn-on voltage of the Schottky diode, can also be modified by using ion-implantation techniques.[21]

A p^+ guard ring can also be used around the periphery of the Schottky diode to raise its reverse-breakdown voltage (the guard ring reduces the electric field at the

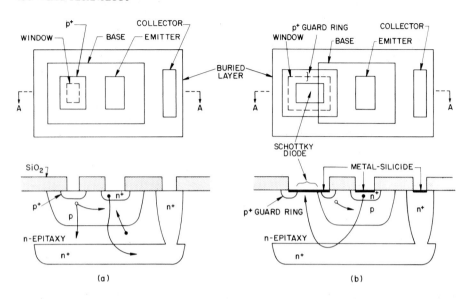

Fig. 6 Schottky-diode clamp technique. (a) Schematic top and side view of unclamped n-p-n transistor in saturation; open circles denote holes, closed circles denote electrons. (b) Schematic top and side view of Schottky-clamped transistor with p$^+$ guard ring to prevent saturation.

sharp corner of the unguarded diode). Incorporating Schottky clamps and guard rings causes the transistor area to increase. In addition to clamping the collector-base junction to prevent saturation, Schottky diodes are used in high-density Schottky transistor logic[22] and Schottky integrated injection logic circuits.[23]

11.3.3 Integrated Injection Logic

Integrated injection[24] logic (I^{2}L), also called merged transistor logic,[25] has become an increasingly important application for bipolar transistors. I^{2}L devices are used extensively[8] in high-density low-power memories, microprocessors, and custom-logic ICs. Figure 7a and b shows the electrical schematic diagram and cross section of an I^{2}L gate.[26] The basic logic cell is formed by integrating a lateral p-n-p transistor (Q_1) with a vertical n-p-n transistor (Q_2) having several collectors. The collector of the lateral p-n-p transistor also serves as the base of the vertical n-p-n transistor.

The n-p-n transistor operates in the inverse (upside-down) mode with the buried-layer serving as the emitter (and contact to the base of the p-n-p) and the top n$^+$ diffusions acting as multiple collectors. For logic implementation, the n-p-n device serves as an inverter and the collectors are wired to obtain specific logic functions. With a logic-high input signal ($V_{in} \sim 0.8$ V), current is injected from Q_1 into the base of Q_2, which is consequently driven into saturation. The outputs are then brought to a logic-low potential ($V_{CE} \sim 0.1$ V). Since resistors are replaced by p-n-p current sources and the inverse-mode n-p-n transistors have automatically isolated collectors and common emitters, the circuit packing density of I^{2}L devices can be very high.

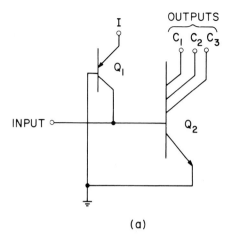

(a)

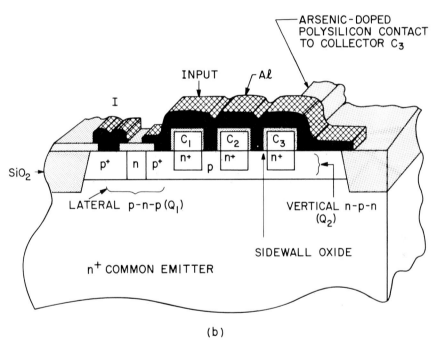

(b)

Fig. 7 Integrated injection logic. (a) Circuit diagram. (b) Three-dimensional view of structure with self-aligned base and collector contacts. *(After Tang et al., Ref. 26.)*

$I^2 L$ fabrication is compatible with conventional bipolar processing so that various bipolar circuit forms can be realized on the same chip.

The I^2L structure of Fig. 7b uses fully recessed oxide isolation. Arsenic-doped polysilicon is used to dope the n^+ collectors as well as to make contact to them. The base contacts are self-aligned to the polysilicon collector contacts so that the base area of the multicollector n-p-n transistor is minimized (the aluminum and polysilicon electrodes are isolated by the sidewall oxide). In addition, the extrinsic-base regions are connected by the low-resistance aluminum metal which reduces base resistance along the I^2L gate. These features improve the I^2L circuit performance.[26] Self-aligned techniques that employ polysilicon for contacting the emitter and/or base regions are being used more frequently to minimize transistor area and parasitics.[27] Propagation delays as low as 0.063 ns and speed-power products as low as 0.043 pJ/gate have been reported using these techniques.[28]

Special techniques for forming the active-base region are needed to improve upward current gain. One technique involves implanting or diffusing the boron directly into the buried-layer, prior to epitaxial growth. After later high-temperature steps, the boron up-diffuses into the n-epitaxial layer ahead of the slower-diffusing arsenic or antimony buried-layer and produces an active-base profile graded properly for upside-down operation.[29, 30]

The speed-power product of an I^2L gate is proportional[31] to CV_l^2, where C is the loading capacitance of a gate and V_l is the logic swing of the gate (logic-high minus logic-low). By reducing the logic swing, the speed-power product can be improved (the smaller logic swing, however, causes the I^2L gate to be more susceptible to being switched by spurious noise signals). The reduced logic swing can be achieved by substituting Schottky diodes for the heavy collector diffusions as shown[23] in Fig. 8a and b. The logic-low voltage is raised (V_l is reduced) because of the forward-bias

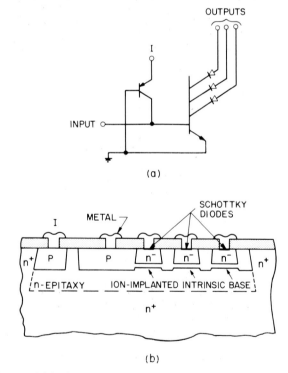

Fig. 8 Schottky integrated injection logic. (a) Circuit diagram. (b) Cross section of structure. *(After Hewlett, Ref. 23.)*

drop across the Schottky diode, which is in series with the lightly doped collector. To fabricate the device shown in Fig. 8b, the base is implanted below the surface of the epitaxial layer using a deep boron implant. The boron implant is deep enough so that a sufficient amount of lightly doped n epitaxy remains above it to form the Schottky-barrier diode with the metallization.

11.3.4 Emitter-Collector Leakage

One of the most severe yield-limiting mechanisms in the fabrication of bipolar integrated circuits is emitter-to-collector (E-C) leakage or shorts. Often a crystallographic defect which occurs in an emitter of a single transistor in a bipolar IC can cause the circuit to fail. This kind of failure mechanism does not generally occur in MOS ICs. This is a prime reason why MOS ICs can be fabricated with a higher yield than bipolar ICs.

Figure 9a illustrates an ideal device. The emitter and collector are isolated; with zero base current the collector current is negligible. Figure 9b shows a device that suffers from E-C leakage. The emitter has penetrated the base region in a localized area (the E-C pipe), which causes the emitter and collector to be effectively tied together. With zero base current, the collector current can be in the milliamp range

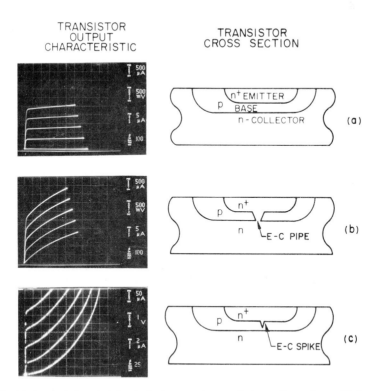

Fig. 9 Electrical output characteristics (collector current vs. collector-emitter voltage with four steps of base current) and schematic cross-sections of n-p-n bipolar transistors. (a) Ideal case. (b) Transistor with emitter-collector pipe. (c) Transistor with emitter-collector spike.

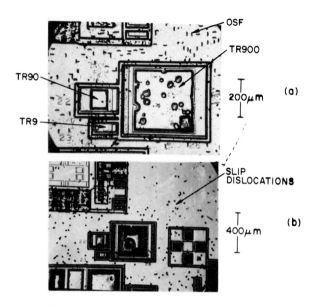

Fig. 10 Secco-etched wafers illustrating process-induced crystallographic defects in test-transistor array (TR9, TR90, TR900). (a) Defects are primarily oxidation-induced stacking faults (OSF). (b) Defects are primarily slip dislocations. *(After Parrillo et al., Ref. 33.)*

with several volts applied between collector and emitter. Figure 9c illustrates the effect of an E-C spike. The emitter impurity partially penetrates the base, and punchthrough current between collector and emitter occurs there at low values of V_{CE}.

E-C pipes are generally agreed to be formed by a locally enhanced diffusion of the emitter dopant through the base in the vicinity of material defects such as dislocations.[32] In narrow-base (easily penetrated) and shallow-emitter (sensitive to surface defects) structures, the problem becomes more severe. Figure 10a and b illustrates two types of material defects that can cause E-C leakage.[33] The defects appear after stripping all dielectrics and treating the silicon with an etch that delineates material defects. Oxidation-induced stacking faults (OSF) are crystallographic defects in the silicon formed during oxidation (Fig. 10a). If OSFs occur in an active emitter area, they can cause E-C shorts. In a particular study,[33] these were completely eliminated from the wafers by judicious choices of oxidation temperatures and material orientation. The yield with respect to E-C leakage on large test transistors was monitored[33] over time and showed no significant improvement after the elimination of the known fatal OSF defects (t_1 in Fig. 11). Slip dislocations are another type of crystallographic defect which can originate from thermal gradients in the wafers during the epitaxial-growth process (Fig. 10b). After the additional fatal defect of slip dislocations was identified and virtually eliminated, the yield jumped up dramatically (t_2 in Fig. 11). The lesson here is that when two types of fatal defects are present in about the same density ($\sim 10^4$ cm^{-2}) *both* have to be eliminated or reduced to see improvement in E-C leakage.

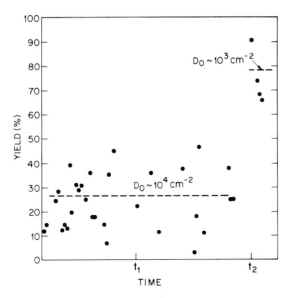

Fig. 11 Yield with respect to emitter-collector leakage of TR90 vs. time. OSFs were eliminated at t_1 and the density of slip dislocations was reduced at t_2. D_0 denotes the fatal defect density. *(After Parrillo et al., Ref. 33.)*

Another major source of dislocations comes about from the local oxidation process. In general, the thicker the pad oxide is and the thinner the silicon-nitride layer is, the less probable it is that dislocations will be generated in the silicon during the field-oxidation step.[34] Unfortunately, adjusting these dielectric thicknesses to minimize defects generally causes more lateral oxidation ("bird's beak"). A host of other process-induced material defects[33] can also cause E-C leakage, and these too must be eliminated to successfully fabricate bipolar VLSI circuits.

11.4 NMOS IC TECHNOLOGY

The metal-oxide-semiconductor field-effect transistor (MOSFET) is the dominant device used in VLSI circuits. The field-effect principle of operation was proposed in the early 1930s by Lilienfeld and Heil. The first working MOSFET, which used a thermally grown silicon-dioxide gate insulator, was demonstrated in 1960 by Kahng and Atalla.[6] ICs using MOSFETs were originally based on p-channel (PMOS) devices; however, n-channel MOS (NMOS) devices with their higher electron mobility with respect to hole mobility, have dominated the IC market since the early 1970s.

11.4.1 Illustrative NMOS Fabrication Process

Figure 12 shows a portion of an NMOS logic circuit with two enhancement-mode (normally off) devices (EMD_A and EMD_B) in series with a depletion-mode (normally on) device (DMD). A field oxide (FOX) surrounds the transistors, and the gate and

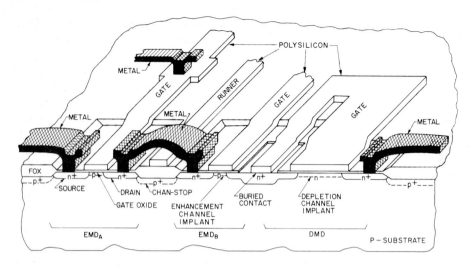

Fig. 12 Three-dimensional view of NMOS logic circuit containing two enhancement-mode devices (EMD) in series with a depletion-mode device (DMD). For clarity the intermediate dielectric is not shown.

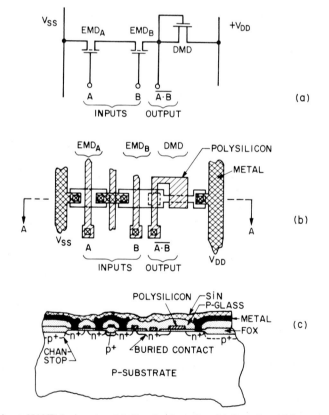

Fig. 13 Two-input NAND logic gate. (a) Circuit schematic. (b) Top view of layout. (c) Side view through cross section A-A. Plasma-deposited silicon nitride (SiN) covers the structure.

source of the DMD are connected together at the buried contact. An intermediate dielectric layer separates the overlying metal layer from the underlying layers. This structure can be used as a two-input NAND logic gate as shown in Fig. 13. Two EMDs are in series with a DMD and the three transistors are connected between the positive power supply V_{DD} and ground V_{SS} rails. The DMD is normally on ($V_{GS} = 0$) and acts as a current source for the two EMDs. Gates A and B of the two EMDs are inputs to the logic circuit, and the DMD's gate/source connection is the output electrode of the logic circuit. The output voltage of the two-input NAND circuit is low only when both EMDs are turned on (i.e., when inputs A and B are at their logic-high level).

Figure 14 shows a fabrication sequence for this circuit. The starting material is a lightly doped p-type substrate. The first lithographic step is isolation (Fig. 14a) where a composite silicon-nitride/pad oxide layer is defined using a resist mask and anisotropic dry etching. The silicon-nitride oxidation mask is retained over the active device area to prevent it from oxidizing later. A boron chan-stop layer is then implanted. After resist stripping and cleaning, the wafers are oxidized which causes a thick field oxide to grow outside of the active area and drives in the chan-stop implant (Fig. 14b). After stripping the nitride/pad oxide layers, the thin gate oxide layer (a few hundred angstroms) is then grown. This is a critical step; the integrity and cleanliness of the gate oxide is essential for proper device operation. A buried contact window is then patterned in the gate oxide (Fig. 14b). The polysilicon gate and the source of the DMD will be later connected via this window. A boron threshold-adjustment dose is then implanted through the gate oxide. This implant, together with the thickness of the gate oxide, sets the desired EMD threshold voltage. Many IC applications call for several enhancement-mode threshold voltages, which add to the complexity of the process. In Fig. 14c the EMDs are protected by resist and the depletion-mode threshold-adjustment dose is implanted using arsenic or phosphorus.

Next the polysilicon gate material is deposited and is doped n type. After the polysilicon is patterned (Fig. 14d), the source/drain regions are implanted with arsenic or phosphorus. The energy of this implant is high enough for the impurities to enter the silicon through the exposed gate oxide, but low enough to prevent their penetration through the polysilicon or field oxide. The sources and drains are thus self-aligned with respect to the gates. The self-alignment minimizes the overlap of the gate and the lateral source/drain diffusions, so that coupling capacitances are lowered there. Note that the n^+ polysilicon contacts the silicon at the buried contact. The phosphorus or arsenic used to dope the polysilicon can diffuse directly into the silicon below it, so that the polysilicon, the drain of the EMD, and the source of the DMD all become connected. The connection is made this way to avoid the additional space requirement that would be necessary if metal were used to strap the gate and source of the DMD together.

After implanting the source/drain, the wafers may be oxidized to provide a dielectric on the polysilicon (for isolation between two-level polysilicon structures) and the silicon substrate. A CVD oxide doped with phosphorus is then deposited at either low or atmospheric pressure. This P-glass intermediate dielectric serves several functions. The phosphorus in the glass protects the underlying devices from mobile-ion (Na^+) contamination, and also causes the glass to become viscous so that it can

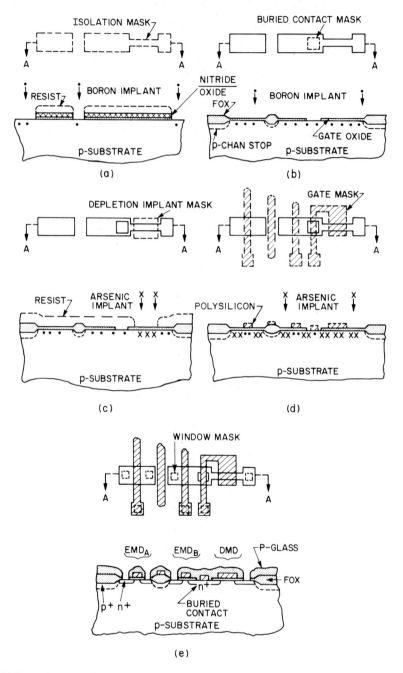

Fig. 14 Top and cross-section views of NMOS logic gate fabrication. (a) Isolation mask and cross section after nitride/oxide etch and boron chan-stop implant. (b) Buried contact mask and cross section after field oxidation (FOX), gate oxidation, buried contact window etch, and boron enhancement threshold-adjustment implant. (c) Depletion implant mask and cross section after resist-masked arsenic depletion implant. (d) Gate mask and cross section after polysilicon gate definition and arsenic source/drain implant. (e) Window mask and cross section after P-glass flow and window etch.

flow at an elevated temperature. This high-temperature process can also serve to activate and drive in the source/drain implants. The P-glass flow smoothes out the surface topography, which facilitates covering the steps with metal as well as aiding in metal patterning. The P-glass also isolates the metal from polysilicon runners. Contact windows are then etched in the P-glass, as shown in Fig. 14e. An additional high-temperature process after window etching (reflow) is often used to taper the steep sidewalls of the window in the P-glass, and this facilitates metal coverage of the window wall. Metal is then deposited and defined as depicted in Fig. 12. Contact to polysilicon is typically made outside of the active transistor area to avoid eroding the polysilicon in that area, and possibly causing damage in the underlying gate oxide in subsequent processing. Aluminum is nearly universally used as a metallization either alone or in combination with other metals. The contacts are later sintered at temperatures up to 500°C in a reducing ambient, to form good ohmic contacts to the silicon and also to anneal out radiation damage that may have been introduced during metal deposition and patterning. Finally, an overcoat layer, such as plasma-deposited silicon nitride[35] (SiN), is put down on the wafer to seal it from contaminants and to serve as a mechanical scratch protection. Windows are then etched in the top coating where external connections (wire bonds) will be made to the metallization layer.

11.4.2 Key Steps in Device Formation

Starting material Conventional bulk silicon consists of lightly doped ($\sim 10^{15}$ atoms/cm^3) p-type $\langle 100 \rangle$ substrates. The $\langle 100 \rangle$ orientation is preferred over $\langle 111 \rangle$ because it introduces about ten times less interface trap density.[6] The lighter the doping in the substrate is, the less sensitive the transistor threshold voltage will be to back-gate bias effects, and the lower the source/drain-to-substrate capacitance will be. If the substrate is too lightly doped, however, the depletion regions of the sources and drains of the same or adjacent transistors can punch through to each other.

In addition, lightly doped substrates have higher concentrations of minority carriers. These carriers diffuse long distances (hundreds of micrometers) to space-charge layers and are collected as reverse-bias leakage currents. This minority-carrier diffusion current can dominate over leakage current generated within the space-charge layers, especially at higher operating temperatures (≥ 40°C).[36] A technique that can be used to circumvent this problem is to form the lightly doped p layer ($\sim 10^{15}$ atoms/cm^3) as an epitaxial layer, grown on a heavily doped p$^+$ substrate ($\sim 10^{19}$ atoms/cm^3).[37, 38] The heavily doped substrate has few minority electrons, so minority carriers must originate primarily in the thin epitaxial layer. Hence diffusion currents in reverse-bias junctions are suppressed, even though the minority-carrier diffusion lengths are long. This is especially important in preserving holding times in dynamic nodes (e.g., dynamic random-access memories[37]). The added cost of growing well-controlled epitaxial films on heavily doped substrates must be weighed against improved device performance.

Isolation Figure 15a shows two adjacent n-channel transistors. The direction of active transistor conduction is perpendicular to the polysilicon gate. Under the polysilicon gate between the transistors is a parasitic transistor (see the cross sections

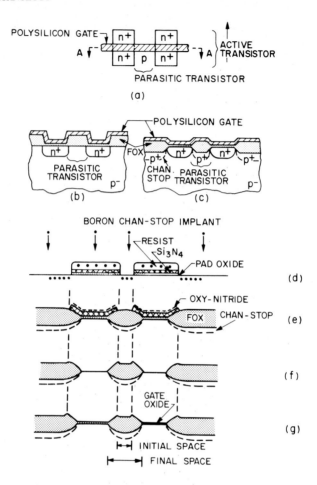

Fig. 15 MOSFET isolation. (a) Top view of adjacent NMOS transistors with common polysilicon gate, illustrating active- and parasitic-transistor conduction paths. Cross section through A-A for (b) etched field oxide isolation and (c) local oxidation isolation structures. Details of local oxidation process: (d) after nitride/pad oxide etch and chan-stop implant, (e) after field oxidation (FOX), which produces an oxy-nitride film on the nitride, (f) after oxy-nitride, nitride, and pad oxide removal, (g) after gate oxide growth.

of Fig. 15b and c). If the threshold voltage of the parasitic transistor is too low, an inversion layer can form between the induced n^+ regions of the individual transistors and tie them together. The parasitic threshold voltage should, in fact, be made high enough to avoid the onset of subthreshold conduction between the adjacent transistors. The parasitic threshold voltage can be made high by having a thick field oxide and/or by raising the substrate doping level between active transistors.

Two isolation techniques are illustrated in Fig. 15. Figure 15b shows a simple technique that consists of growing the thick field oxide everywhere, and then cutting windows in it where active transistors will be formed (the thin gate oxide is later grown in the windows). Figure 15c shows another technique—the local oxidation technique.[10] This technique has the advantage of recessing about half of the field oxide below the silicon surface, which makes the surface more planar (compare the topographies of Fig. 15b and c). Another advantage is that it allows chan-stop layers to be formed self-aligned to the active transistor area. Chan-stop doses are in the mid 10^{12} to 10^{13} atoms/cm^2 range and the depth of the implant is adjusted to allow sufficient boron to remain in the underlying silicon after oxidation. Too heavy a chan-stop doping increases the source/drain-to-substrate capacitance, reduces junction-breakdown voltage, and increases the sensitivity of the threshold voltage to narrow-width effects.

The local oxidation process is illustrated in detail in Fig. 15d to g. This process is similar to that used in bipolar isolation, except in MOSFET isolation the field oxide does not have to penetrate all the way through an epitaxial layer as in bipolar structures. After isolation definition and chan-stop implantation (Fig. 15d), the field oxide is selectively grown outside of the active area, typically to a thickness of several thousand angstroms. The thinner the field oxide is, the smaller is the bird's beak and the more planar is the surface. Too thin a field oxide, however, causes a low parasitic threshold voltage in the field region and increases the polysilicon-to-substrate capacitance. A disadvantage of the local oxidation process is that the silicon surface under the nitride can be damaged during field oxidation—that is, a thin layer of silicon nitride (or oxy-nitride) can form on the silicon surface due to the action of NH_3 generated from the masking silicon-nitride layer.[39] Where it occurs, the oxy-nitride film impedes the subsequent growth of the gate oxide and causes low-voltage breakdown of the gate oxide and other deleterious device effects. One method used to avoid the problem, is to grow a sacrificial oxide after stripping the masking nitride and then remove it before growing the final gate oxide.[40, 41]

The field oxidation and subsequent thermal cycles can cause significant lateral diffusion of the chan-stop layer. The diffusion raises the surface concentration of the substrate near the nitride periphery, and hence the threshold voltage of that portion of the device. The edges of the device will not conduct as much as the interior portion, and the transistor behaves as if it were narrower. For a given field oxide thickness, less lateral intrusion of the chan-stop layer occurs at lower field oxidation temperatures.[42] Another geometrical aspect can be seen by comparing the initial separation between nitride islands and the final space between active transistor areas (Fig. 15g). Because lateral oxidation occurs under the masking nitride, the space between transistors grows during processing. Since there is a limit to how small the initial space can be made (because of lithographic considerations), there is a limit on how small the final space between transistors will be. This limitation of local oxidation has been addressed by many researchers[10, 40, 43] and new approaches to forming steep-walled oxide-isolated islands have been investigated.[17, 18, 44]

Although the ideal steep-walled boxlike oxide-isolation region is attractive for many reasons, unwanted parasitic conduction can take place in the sharp corner region of the active transistor. The insert of Fig. 16 shows the sidewall of an oxide-isolated transistor in the direction from source to drain. Because of the electric field concentration at the sharp corner of the silicon boundary, the threshold voltage of the corner region is reduced and this part of the device turns on at a lower voltage than the

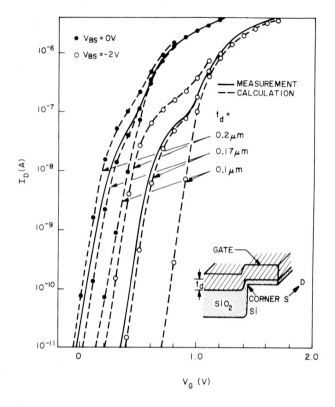

Fig. 16 Measured and simulated drain current vs. gate potential for devices having a downward step in the field oxide (t_d). The insert illustrates the geometry where conduction is from source (S) to drain (D). Parasitic conduction occurs at the corner. Results are shown for $t_d = 0.1$ to 0.2 μm and back-gate bias (V_{BS}) of 0 and −2 V. Devices have substrate doping of 1.4×10^{15} atoms/cm³, 5-μm channel length, 5-μm channel width, and applied drain voltage of 0.1 V. *(After Iizuka, Chiu, and Moll, Ref. 45.)*

interior portion away from the corner.[45] The situation becomes worse if there is a downward step in the field oxide (t_d in the insert of Fig. 16). The larger the step is, the lower the corner threshold becomes, and unwanted subthreshold conduction begins at progressively lower values of V_G. The calculated and measured subthreshold I-V curves for $t_d = 0.1$ μm to 0.2 μm are shown in Fig. 16 for a device with a field oxide 0.75 μm thick. The corner threshold can be made the same or higher than the threshold of the planar region by allowing a step-up in going from the active to the field oxide regions.[45] In addition to these effects, the geometry of the isolation oxide wall, as well as the doping levels in the active and parasitic regions of the underlying silicon affect the threshold sensitivity of the transistor to its physical width.[44]

Channel doping As channel lengths become shorter and gate oxides become thinner, a higher doping level under the gate is required to provide the desired threshold (and subthreshold) voltage characteristics.[46] Using a heavily doped substrate will provide the higher doping level; however, this increases the back-gate bias sensitivity of the threshold voltage and increases source/drain-to-substrate capacitances as well. A shallow ion implant is widely used to set the desired doping level in the channel region without raising the background substrate doping level. In this way the threshold sensitivity to back-gate bias can be minimized while still having the desired high surface concentration.[47]

A shallow implant may be sufficient to provide the desired transistor properties. Depending on the substrate doping, source/drain junction depth, and gate oxide thickness, however, a single shallow threshold-adjustment implant may not be sufficient to prevent punchthrough of the drain electric field to the source region. When the device is intended to be off ($V_G \ll V_T$), the path for punchthrough occurs *below* the silicon surface. This is because, owing to the influence of the gate fields, the source/drain depletion widths are reduced at the surface.[48] This is illustrated schematically in the insert in Fig. 17 which shows the source/drain depletion regions impinging on each other below the surface for $V_G \ll V_T$ and $V_D > 0$. To prevent the sub-

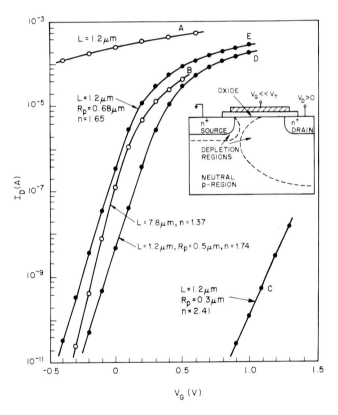

Fig. 17 Drain current vs. gate voltage for n-channel devices with a substrate doping of 1.9×10^{15} atoms/cm^3, source/drain junctions 0.47 μm deep, 575-Å gate oxide, drain voltage of 5 V, and back-gate bias of 0 V. Devices A and B have no channel implant, and devices C, D, and E have a boron channel implant of 8×10^{11} atoms/cm^2 at various energies. Insert schematically illustrates the source and drain depletion regions for $V_G \ll V_T$ and $V_D > 0$. (*After Nihira et al., Ref. 49; Bateman, Armstrong, and Magowan, Ref. 48.*)

surface punchthrough, a deep boron implant can be used to raise the substrate doping level at the appropriate depth. Figure 17 illustrates subthreshold I-V curves for a device with a light substrate doping.[49] A short-channel device (L = 1.2 μm) punches through badly (curve A). Curve B shows the desired behavior of a long-channel (L = 7.8 μm) transistor in the same substrate. A boron implant dose of 8×10^{11} atoms/cm^2 was implanted in the short-channel device at a projected range of 0.3 μm. This was too shallow, and the threshold voltage (curve C) rose well above the desired value. By increasing the implant energy (curves D to E), the deep boron implant became deep enough to allow the desired low threshold voltage (surface concentration undisturbed) and to prevent punchthrough (peak of implant at 0.68 μm below the surface) in the 1.2-μm-long device. Of course the back-gate bias sensitivity increases also. Note that the subthreshold swing factor $[n \equiv q/kT (dV_G/d \log I_D)]$ is large (n = 2.41) for the shallow implant (curve C) and becomes smaller (n = 1.65) as the implant is made deeper. This occurs because the high-doping region is pushed below the silicon surface. Two boron implants can be used to establish the surface doping for the desired threshold voltages of short-channel devices (shallow implant) and to tailor the subsurface profile to prevent punchthrough (deep implant).[50, 51]

A deleterious consequence of increasing the surface concentration is the accompanying reduction in carrier mobility at the surface. This reduction in surface mobility is caused by the increased vertical electric field experienced by the carriers in the channel—which in turn is a consequence of the more heavily doped substrate.[52, 53] Using ion implantation to form a shallow p-n junction at the surface, and choosing a gate material with an appropriate work function, can increase the electron[54] and hole[55] mobilities above the mobilities found in conventional structures.

The discussion, so far, has centered around surface-channel conduction. Buried-channel devices, however, are becoming increasingly common in ICs. They are generally of two types—normally on and normally off. An example of a normally on buried n-channel was discussed in connection with Fig. 13, where an arsenic surface layer provides the source-to-drain conduction at zero gate voltage. A normally off buried n-channel device can be fabricated in a similar manner (n-type surface layer in a p-type substrate). However, a gate material with a suitably large work-function is chosen (e.g., p$^+$ polysilicon[56] or MoSi$_2$[57]) to deplete the n-type surface layer of carriers, and to produce the normally off characteristics. When the device is turned on, much of the current is conducted below the surface (where bulk mobility is larger than surface mobility); thus the buried-channel device has significantly higher carrier mobility than conventional devices. However, since the conduction path is further removed from the gate than in conventional surface-channel devices, the transconductance $g_m = (\partial I_D/\partial V_G)|_{V_D}$ can suffer. These competing effects must be weighed against each other to ascertain the effect on transconductance.

Gate material Heavily doped n-type polysilicon has been widely used as a gate and as an interconnect because of its ability to withstand high-temperature processing. The resistance of the polysilicon ($>10 \ \Omega/\square$) may contribute significantly to the RC delay of signals that are routed along it. More recently, refractory metals and their silicides have been used in conjunction with polysilicon[58, 59] or alone[41, 57] to reduce the resistance. The technique of combining a refractory metal silicide on top of doped polysilicon (called polycide)[59] has the advantage of preserving the well-understood polysilicon-SiO$_2$ interface while lowering the overall sheet resistance of the polycide to about 1 to 3Ω/$\square$. The use of certain silicides directly on gate oxide results in larger work-functions than n$^+$ polysilicon, thus requiring corresponding adjustments in the channel-doping technique.[41, 57]

Source/drain formation For a shallow junction and minimal lateral diffusion, arsenic is used extensively as a source/drain impurity. Source/drain implants are typically in the high 10^{15} to 10^{16} atoms/cm^2 dose range to produce low-resistance source/drain regions. Figure 18 shows the details of a shallow arsenic source/drain

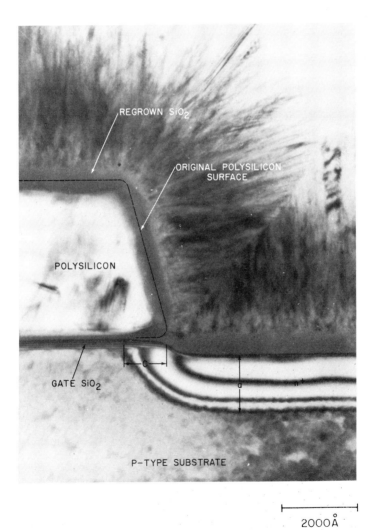

$\vdash\!\!\!-\!\!\!-\!\!\!-\!\!\!-\!\!\!-\!\!\!\dashv$
2000Å

Fig. 18 Transmission electron micrograph showing details of source/drain and gate regions. Parameter a is the junction depth and c is the lateral penetration of the junction from the original implant position before reoxidation. *(After Sheng and Marcus, Ref. 60.)*

diffusion at the edge of a polysilicon gate.[60] After the arsenic was implanted into the bare silicon source/drain regions, it was driven in in an oxidizing ambient (reoxidation). Reoxidation is sometimes used to provide a dielectric on the polysilicon for double polysilicon processes and also to help protect the source/drain regions from phosphorus penetration from the P-glass. The results of the reoxidation are shown by the bird's beak at the edge of the polysilicon gate and the depressed surface of the source/drain region (Fig. 18). Excessive reoxidation can enhance these effects and cause the shallow drain impurity profile in the silicon to be *electrically* deep because the silicon surface has been depressed.

The series source/drain resistance is becoming increasingly important as device dimensions shrink. As channel conductance increases with shorter channel length, the resistance of the shallow source/drain regions stays fixed or actually increases because of the need for shallower junctions. The result is that the resistance of the source/drain limits the current-delivering capability of short-channel devices and becomes an important parasitic resistance. Figure 19a shows a technique used to reduce the resistance of the source/drain and gate.[51] After forming the polysilicon gate and driving in the source/drain regions (to a depth of 0.23 μm), a CVD oxide is

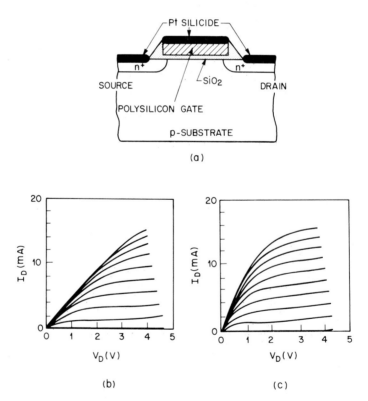

(a)

(b)　　　　　　　　　(c)

Fig. 19 Reduction of source, drain, and gate resistances. (a) Cross section of an n-channel transistor with platinum silicide on source, drain, and gate. (b) Output *I-V* characteristics without silicided source/drain. (c) Output *I-V* characteristics with silicided source/drain. *(After Shibata et al., Ref. 51.)*

deposited on the device and removed in the horizontal regions by reactive ion etching. The thicker oxide along the polysilicon's sidewall remains and a sidewall oxide spacer is formed. Platinum is deposited and then allowed to react with the exposed silicon in the source/drain and gate regions (Pt does not react with oxide). This lowers the sheet resistance of the source/drain regions[51] from 50 $\Omega/\square$ to 3 $\Omega/\square$. The resulting effect on the output I-V characteristics of a 0.5-μm electrical-channel-length transistor is shown in Fig. 19b and c. The increased current drive of the silicided device is most evident at lower drain voltages. Other techniques to reduce parasitic resistance include the use of a *selective* deposition of tungsten only on exposed silicon areas (source/drain and polysilicon gate) without the silicide reaction. Sheet resistances of 1 $\Omega/\square$ were achieved for these areas using 1500 Å of tungsten.[61]

11.4.3 Memory Technology

One of the most important VLSI products is the memory chip. This section considers some basic concepts in producing memory-related structures. Among memory chips, the random access memory (RAM) has the highest component density per chip. In a RAM any bit of information in a matrix of bits can be accessed independently. Individual rows of memory bits are accessed by a conductive word line which may be a diffusion, polysilicon, or metal line. Similarly, individual columns of bits in the matrix are accessed by a bit line. The acronym RAM is generally used to refer to randomly addressable memories into which data can be written and retrieved indefinitely. In contrast, the read only memory (ROM) has data permanently coded into it and new information cannot be entered. Static RAMs retain their data indefinitely, unless the power to the circuit is interrupted. Dynamic RAMs require that the charge (data) stored in each memory cell be "refreshed" periodically to retain the stored information.

Figure 20 gives an example of a single static RAM cell. Figure 20a shows a six-transistor (n-channel) cell, which uses a cross-coupled inverter pair (flip-flop) (T_1 to T_4) to store 1 bit of information.[62] A pair of access transistors (T_5 and T_6) transmit data into and out of the cell when the word and bit lines are simultaneously activated. The loads for the flip-flop are depletion-mode transistors (T_1 and T_2) with their sources and gates tied together as shown before in the NAND circuit of Fig. 13. The data (logic 1 or 0) is retained in the cell by the positive feedback existing in the flip-flop circuit. For example, with the gate of T_4 at a high potential, its drain is forced to a low potential ($\ll V_T$). This potential, in turn, is fed to the gate of T_3, and keeps T_3 off. The drain of T_3 is then tied to the high potential by T_1 (which is always on) and so is the gate of T_4. This arrangement ensures that the drain of T_3 is kept high in potential and the drain of T_4 is kept low in potential. This state of the cell defines a logic 1 or 0, which is retained unless new data is entered by T_5 and T_6.

Figure 20b shows a layout[62] for the circuit of Fig. 20a. The width-to-channel-length ratio of the depletion-mode load transistors (1/5) is adjusted to provide enough current drive to meet the speed requirements of the cell without causing excessive steady-state (quiescent) power dissipation (T_2 and T_4 are on simultaneously in our example and current flows between V_{CC} and V_{SS}). To minimize the cell area, buried contacts (diffusion to polysilicon contact) are required (see Fig. 12).

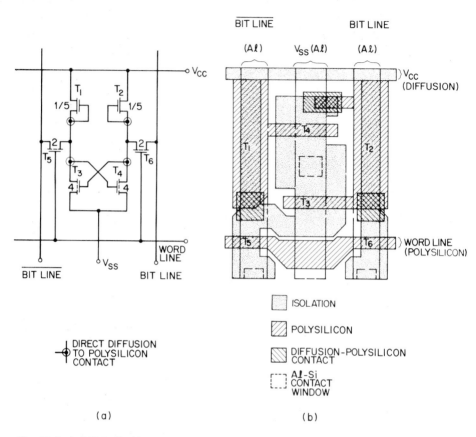

BIT LINE   BIT LINE

(Aℓ) V_{SS} (Aℓ)   (Aℓ)

V_{CC}
(DIFFUSION)

T_1 T_4 T_2

T_3

T_5 T_6 WORD LINE
(POLYSILICON)

ISOLATION

POLYSILICON

DIFFUSION–POLYSILICON
CONTACT

Aℓ-Si
CONTACT
WINDOW

V_{CC}

T_1 T_2

1/5   1/5

2   2

T_5 T_6

T_3 T_4

4   4

WORD
LINE

V_{SS}

BIT LINE   BIT LINE

DIRECT DIFFUSION
TO POLYSILICON
CONTACT

(a)   (b)

Fig. 20 Static RAM cell with transistor loads. (a) Circuit diagram of a six-transistor static RAM cell. V_{CC} and V_{SS} are the power supply and ground potentials, respectively. The numbers next to the transistors indicate the relative width-to-channel-length ratios. (b) Static RAM layout. *(After Hunt, Ref. 62.)*

The depletion-mode load transistors can be replaced by high-valued resistors,[63] as shown in the circuit schematic of Fig. 21a (resistor MOS or RMOS cell). A high value of resistance is desired to reduce the quiescent power dissipation in the cell. High-valued resistors can be made in a relatively small space by using polysilicon which has been ion-implanted to provide the proper resistance. Polysilicon is used because its sheet resistance can be modified by many orders of magnitude using ion implantation. Diffusions in the silicon would require too much area to produce the same high-resistance values ($>10^7 \ \Omega$). The polysilicon resistors can be made in the same single layer of polysilicon (gate and interconnect) by masking the polysilicon resistor regions from the high-impurity doping used in the gate and interconnect portions of the polysilicon level. Additional area can be saved by using a second level of polysilicon for the load resistors, and overlaying these resistors on the active area of the cell (Fig. 21b). Using this technique, static RAM cell areas can be reduced to half the cell area required in conventional transistor load cells.[63] To virtually eliminate

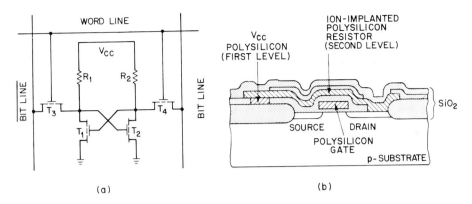

Fig. 21 Static RAM cell with resistor loads. (a) Circuit schematic of polysilicon resistor load (R_1 and R_2) static RAM cell. (b) Device cross section. First-level polysilicon is used for gate and routing power supply V_{CC}. Second-level polysilicon is used for resistor load directly over an active transistor. Connection to drain and V_{CC} is made directly from an implanted polysilicon resistor. *(After Ohzone et al., Ref. 63.)*

quiescent power consumption in static RAM cells, the depletion-mode load transistors of Fig. 20 can be replaced by p-channel transistors in a CMOS static RAM cell at the expense of area (see Section 11.5).

Because of the large number of devices needed in static RAM cells, large-capacity memories ($\geq$16 kilobit) require large chip areas. In addition static RAMs can dissipate a great deal of power. For these reasons large memory chips use dynamic memory (dynamic RAM) cells, which require only one transistor and one storage capacitor per bit of information. Additional circuitry is required to sense and refresh the data in the dynamic cells, but it is well worth the effort because of the much reduced chip area and power dissipation required for dynamic RAMs. Figure 22a shows the basic dynamic memory cell.[62] When a word and bit line are simultaneously addressed (brought to a high voltage), the access transistor is turned on and charge is transferred into the storage capacitor if it had no charge initially (stored "zero"), or little charge is transferred to the storage capacitor if it were fully charged initially (stored "one"). The amount of charge that the bit line must supply to the storage capacitor is measured by the sensing circuitry, and this information is used to interpret whether a "zero" or "one" had been stored in the cell. The sense circuitry then restores full charge in the capacitor if it had been there originally, or fully depletes the capacitor if little charge had existed originally. The information in the cell is thus "refreshed" after it is read.

An example[62] of a dynamic RAM cell layout is shown in Fig. 22b and a cross section of the cell through A-A is shown in Fig. 22c. A diffusion (source/drain) forms the bit line and also the source of the access transistor. The capacitance of the diffused bit line (i.e., the junction capacitance) and its resistance can be limiting factors in the performance of dynamic RAMs. Several approaches are used to minimize these parasitic effects, including the use of $MoSi_2$ for word lines and Al for bit lines in fabricating advanced memory chips such as the 256-kilobit dynamic RAMs.[57]

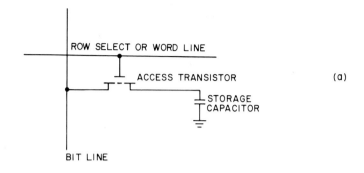

(a)

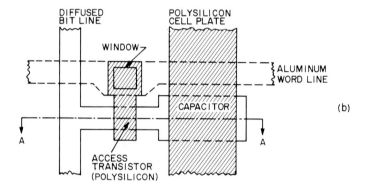

(b)

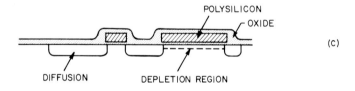

(c)

Fig. 22 Single-transistor dynamic RAM cell with storage capacitor. (a) Circuit schematic. (b) Cell layout. (c) Cross section through A-A. *(After Hunt, Ref. 62.)*

In scaling ICs to finer features, it is also desirable to shrink the area of the storage capacitors in dynamic RAMs. As the area of the storage capacitor decreases so too does its capacity to store charge. As less charge is stored in the cell, accurate interrogation of its contents becomes more difficult (small signals). In order to increase the charge-storage capacity, the use of thinner gate insulators with higher dielectric constants (e.g., $Si_3 N_4$ and $Ta_2 O_5$ with dielectric constants of ~8 and ~22, respectively) is being explored.[64, 65]

Other techniques include the use of the high-capacity (Hi-C) RAM cell[66] (Fig. 23). A shallow arsenic implant and deeper boron implant are used to increase the depletion-layer capacitance under the storage capacitor and to also increase its

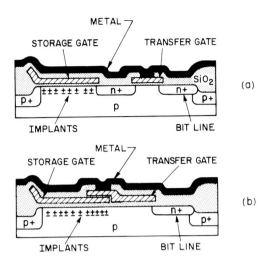

Fig. 23 High-capacity (Hi-C) dynamic RAM cell structure with shallow arsenic $(+)$ and deeper boron $(-)$ implants. (a) One-transistor cell with single-level polysilicon. (b) Double-level polysilicon cell. *(After Tasch, Jr., et al., Ref. 66.)*

charge-storage capacity. Simply increasing the substrate doping under the storage capacitor (boron implant) increases the depletion-layer capacitance there. This does not increase the cell's charge-storage capacity, however, because the surface potential difference $\Delta \phi_S$ between an empty and a full cell decreases. By also incorporating the shallow n-type implant (which acts like a positive oxide-fixed charge), the change in $\Delta \phi_S$ coupled with the increased depletion capacitance causes the charge-storage capacity of the cell to increase. Depending upon the back-gate bias applied to the cell, the Hi-C RAM cell can have up to twice the charge-storage capacity of conventional cells.[66]

The drain region of the access transistor acts as a conductive link between the inversion layers under the transfer and storage gates. This drain region can be eliminated by using the double-level polysilicon approach[66] shown in Fig. 23b. The second polysilicon electrode is separated from the first polysilicon electrode by a thin SiO_2 layer, thermally grown on the first-level polysilicon after it has been defined. The second-level polysilicon is then deposited and defined so that it closely overlaps the first polysilicon level. Charge from the bit line can therefore be transmitted directly to the area under the storage gate by the connection of inversion layers under the transfer and storage gates. The double-level polysilicon approach is widely used in dynamic RAMs because it reduces cell size; however, the complexity that having a second level of polysilicon adds to the process can be costly.

Thus far we have focused on techniques to fabricate individual memory cells. To successfully fabricate an IC having many thousands of cells requires that all components on the chip be free of defects. In memory ICs, many chips fail because of localized defects that cause failure in only single bits, or single rows or columns of bits in the memory array. The yield of large dynamic RAMs can be greatly increased by incorporating redundant (spare) rows and columns of bits which can be exchanged for the faulty ones.[67] Fusible links, that can be opened by laser programming or by electrical means, are used to disconnect the faulty rows or columns from the memory array. After the faulty row or column is disconnected, its previous identity in the memory array is transferred to the spare row or column by opening additional fusible links in the memory decoding circuitry. Redundancy techniques for large dynamic RAMs have very significantly lowered the manufacturing cost of these ICs.

The few examples of semiconductor memory structures mentioned here are all volatile—that is, data is lost when power is removed from the chip. An entire field of nonvolatile semiconductor memories[6] also exists, however. These devices semipermanently retain their data, which has been preprogrammed into them either electrically or by other means. The information in these devices can be electrically programmable (EPROM) and, more recently, electrically erasable and programmable (E^{2}PROM).[68]

11.5 COMPLEMENTARY MOS IC TECHNOLOGY

First introduced in 1963 by Wanlass and Sah,[69] complementary MOS (CMOS) technology provides both NMOS and PMOS transistors on the same chip. CMOS circuits consume low power when compared to NMOS circuits. By comparison, however, early CMOS processes were more complex and early circuit designs required larger chip areas (a PMOS was used for every NMOS). As NMOS circuits have grown in density, NMOS processes have grown in complexity to avoid excessive power consumption (e.g., additional masks are now used to produce a variety of threshold voltages). Modern CMOS processes have been simplified so that NMOS and CMOS technologies are now comparable in complexity. Current CMOS designs use more NMOS than PMOS transistors, which conserves chip area while still minimizing power consumption. CMOS technology has benefited from the advances of NMOS technology and has emerged as one of the most important VLSI technologies.

11.5.1 Special Considerations for CMOS

Figure 24 shows the operation of a CMOS inverter.[2] The p-channel transistor is formed in the n-type substrate. The n-channel transistor is formed in the p region, which in turn is formed in the n-type substrate. The p region acts as the n-channel transistor's substrate (back gate), and is commonly referred to as a tub or well. The gates of the n- and p-channel transistors are connected and serve as the input to the inverter. The common drains of each device are the output of the inverter. The threshold voltages of the n- and p-channel transistors are V_{Tn} and V_{Tp}, respectively ($V_{Tp} < 0$). Figure 24c shows the dependence of the output voltage V_O on the input voltage V_I of the CMOS inverter. For $V_I = 0$, the n-channel transistor is off ($V_I \ll V_{Tn}$), while the p-channel transistor is turned on heavily (the gate-to-source potential of the p channel is $-V_{DD}$, which is much more negative than V_{Tp}). Hence, $V_O = V_{DD}$. As V_I increases above zero, the n-channel transistor eventually turns on, while the p-channel transistor eventually turns off. When V_I is larger than ($V_{DD} - |V_{Tp}|$), then $V_O = V_{SS}$.

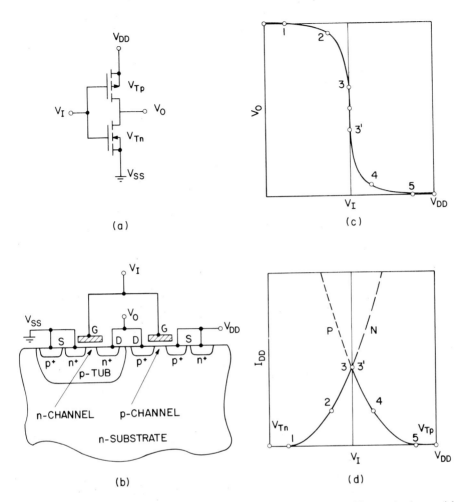

Fig. 24 CMOS inverter. (a) Circuit schematic. V_{DD} and V_{SS} are the highest and lowest circuit potentials, respectively. (b) Device cross section. (c) Output (V_O) vs. input (V_I) voltage of inverter. (d) Current through inverter as a function of input voltage (solid curve); *I-V* characteristics of n- and p-channel transistors (dashed curves). The numbers correspond to different points on the inverter transfer characteristic. *(After Hoefflinger and Zimmer, Ref. 2.)*

A key feature of this CMOS gate is that in either logic state ($V_O = V_{DD}$ or V_{SS}) one of the transistors is *off* and the current conducted between V_{DD} and V_{SS} is negligible. This feature is illustrated in Fig. 24d where current through the inverter (I_{DD}) is plotted as a function of V_I (solid curve). A significant current is conducted through this CMOS circuit only when both transistors are on at the same time (during switching). The low power consumption of CMOS is one of its most important attributes. Performance and ease of circuit design are other attractive features of CMOS circuits. CMOS provides the circuit designer with flexibility in designing circuits that are either static CMOS (a p-channel transistor for every n-channel transistor) or have more of one type of transistor than the other (dynamic).

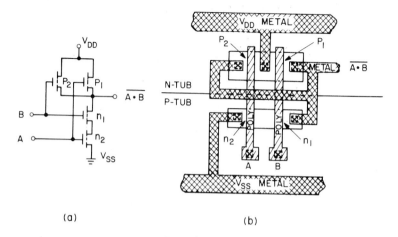

(a)

(b)

Fig. 25 CMOS two-input NAND circuit. (a) Circuit schematic. (b) Circuit layout.

The circuit schematic in Fig. 25a is an example of a static CMOS two-input NAND gate. As in the NMOS two-input NAND gate of Fig. 13a, the logic function is described by: the output is low ($\overline{}$) only when input A and ($\cdot$) input B are high ($\overline{A \cdot B}$). In the CMOS gate, when A and B are high, both n-channel transistors are on and both p-channel transistors are off. In the NMOS gate (Fig. 13a), however, when A and B are high, the two enhancement-mode devices are on and so is the depletion-mode load device. Hence the NMOS circuit dissipates power in this state while the CMOS circuit does not. CMOS is also desirable because the output voltage of the CMOS circuit makes a full excursion between V_{DD} and V_{SS} (a large excursion of output voltage is desirable for noise margins). This is not the case with the NMOS circuit.

A disadvantage of the static CMOS circuits is their additional input capacitance, which is due to the gate capacitance of the p-channel transistors in parallel with the n-channel gates. Also, static CMOS circuits require a significant amount of chip area as shown in Fig. 25b. A minimum separation is needed between n- and p-channel transistors to prevent leakage between them. Often this space can be used for wiring tracks, as shown by the metal output line of the circuit in Fig. 25b.

To avoid the area penalty of static CMOS circuits and to take advantage of CMOS's low power consumption, modern complex CMOS circuits are designed with many n-channel transistors (in a common p tub) and fewer p-channel transistors. Figure 26 shows an example of a dynamic logic circuit, called Domino CMOS.[70] When the clock signal is low (p_1 on, n_1 off), the signal to the output inverter is held high and the output of the circuit is low, regardless of any signal inputs on the many n-channel transistors. Negligible power is dissipated in this circuit when the clock is low. When the clock signal goes high, the circuit is activated (p_1 off, n_1 on). If a combination of input signals is applied to turn on a branch of the series n-channel transistors (A and B, for example), the signal to the output inverter is pulled down, causing the output of this circuit to go up. Since the overall IC consists of many dynamic circuits (like that in Fig. 26), which feed other similar circuits, the data cas-

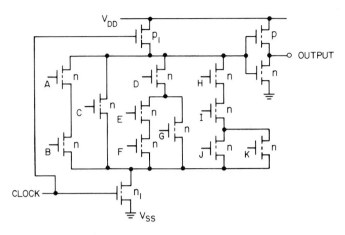

Fig. 26 Dynamic (Domino) CMOS circuit. Individual inputs, *A* through *K*, are labeled. *(After Krambeck, Lee, and Law, Ref. 70.)*

cades from one dynamic circuit to another, like a series of dominos. Very significant speed enhancement and savings in chip area can be obtained by using dynamic CMOS circuits.

A generic problem associated with CMOS structures has been their vulnerability to an undesirable conduction mechanism known as latchup. Latchup is a condition where high currents are conducted between V_{DD} and V_{SS}, which can cause the IC to cease functioning and even be destroyed. The CMOS inverter structure produces lateral p-n-p as well as vertical and lateral n-p-n bipolar transistors (Fig. 27a). The collectors of each of these bipolar transistors feed each others' bases and together make up a thyristor (p-n-p-n device) as shown by the insert in Fig. 27a. With the thyristor biased appropriately (or inappropriately in a CMOS circuit), the collector current of the p-n-p supplies base current to the n-p-n, and vice versa in a positive-feedback arrangement. A sustained current can then exist between the positive and negative terminals of the thyristor (i.e., the latchup). The latchup current is terminated when electric power to the thyristor is interrupted.

Figure 27b shows how a CMOS circuit can be induced to latchup.[71] If the output terminal is momentarily brought below the V_{SS} potential by about 0.7 V (by a spurious noise spike from electrostatic discharge, for example), then the n^+ drain (emitter of n-p-n) injects electrons into the p tub (base of n-p-n); the electrons reach the n substrate (collector of n-p-n), where they drift out of the positive V_{DD} terminal. If this electron current is high enough and if sufficient resistance exists between the V_{DD} contact and the p^+ source, an *IR* (current-resistance) drop develops, which lowers the potential of the substrate under the p^+ source by about 0.7 V. This drop in potential causes holes to be emitted from the p^+ source (emitter of p-n-p) into the n substrate (base of p-n-p); the holes reach the p tub (collector of p-n-p) and drift out of the V_{SS} terminal. If enough hole current exists in the p tub and if sufficient resistance exists between the V_{SS} contact and the n^+ source, an *IR* drop develops, which causes the n^+ source to inject electrons into the p tub. This electron current adds to the initial elec-

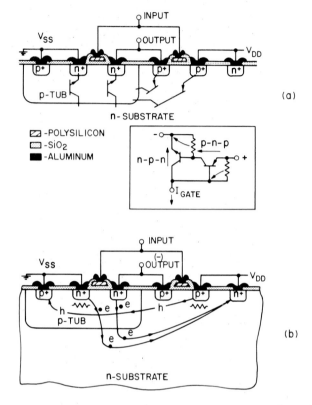

Fig. 27 CMOS inverter cross section. (a) Parasitic n-p-n and p-n-p bipolar transistors comprise a thyristor (shown in insert). (b) A latchup condition is induced by biasing the output below V_{SS}. *(After Grant, Ref. 71.)*

tron current and strengthens the positive feedback between the p-n-p and n-p-n transistors, which leads to the latchup condition. The initial disturbance can now be removed and the large latchup current will be self-sustained unless power to the CMOS circuit is interrupted (e.g., V_{DD} or V_{SS} is disconnected). In a similar fashion, latchup can be initiated by hole injection from a p^+ drain if the output is biased sufficiently above V_{DD}.

11.5.2 Illustrative Fabrication Process

An important consideration in fabricating CMOS structures is the technique for forming the substrates for the two types of MOSFETs. The early CMOS processes were developed to be compatible with the PMOS process; hence the n-channel transistor was formed in a p diffusion (tub) in the n substrate. Although some of the early processing constraints disappeared and NMOS circuits have dominated MOS ICs, the traditional p-tub approach has been the most widely used CMOS structure.

The p tub is implanted or diffused into the n substrate at a concentration that is high enough to overcompensate the n substrate and to give good control over the

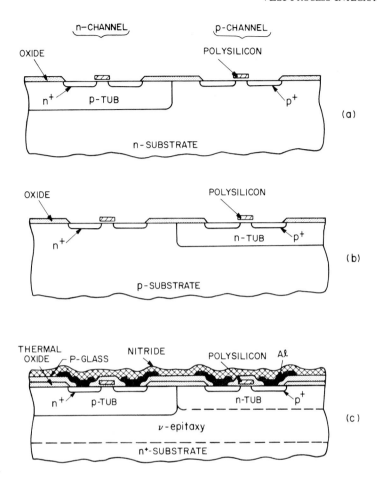

Fig. 28 Various CMOS structures. (a) p tub. (b) n tub. (c) twin tub. *(After Parrillo et al., Ref. 72.)*

desired net p-type doping (Fig. 28a). The doping level in the p tub is typically five to ten times higher than that in the n-type substrate to ensure this control. This excessive p-tub doping produces deleterious effects in the n-channel transistor, however, such as increased back-gate bias effects, and increased source/drain to p-tub capacitance.

An alternative approach is to use an n tub to form the p-channel transistors.[55] As illustrated in Fig. 28b, the n-channel device is formed in the p-type substrate and this n-tub approach is compatible with standard NMOS processing. In this case the n tub overcompensates the p substrate and the p-channel device suffers from excessive doping effects.

Figure 27c shows an approach that uses two separate tubs implanted into very lightly doped n-type silicon. This "twin-tub" CMOS approach[72] allows the doping profiles in each tub region to be tailored independently, so that neither type of device must necessarily suffer from excessive doping effects. This approach has been used on lightly doped n-type (ν-type)[72, 73] or p-type (π-type)[74] substrates.

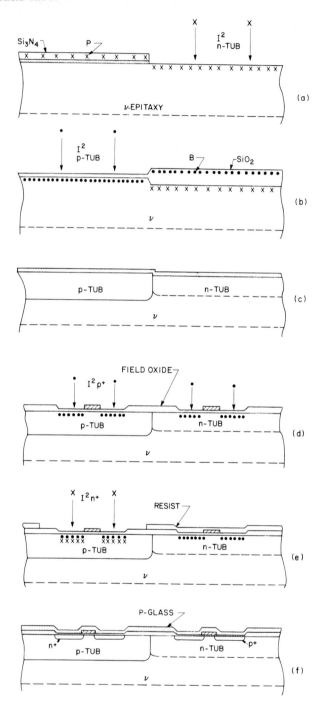

Fig. 29 Twin-tub CMOS structure at several stages of the process: (a) n-tub ion implant (I^2); (b) p-tub implant; (c) twin-tub drive-in; (d) nonselective p^+ source/drain implant; (e) selective n^+ source/drain implant using photoresist mask; (f) P-glass deposition. *(After Parrillo et al., Ref. 72.)*

The highlights of the twin-tub CMOS process[72] are illustrated in Fig. 29. The starting material is lightly doped n epitaxy over a heavily doped n^+ substrate. This structure, combined with proper layout techniques, produces CMOS circuits that are not prone to latchup.[75] Figure 29a to c shows how the self-aligned twin tubs are formed using one lithographic mask step. A composite layer of SiO_2 (pad) and Si_3N_4 is defined and silicon is exposed over the intended n-tub region. Phosphorus is implanted as the n-tub dopant at low energy, and enters the exposed silicon, but is masked from the adjacent region by the Si_3N_4 (Fig. 29a). The wafers are then selectively oxidized over the n-tub regions. The nitride is stripped and boron is implanted for the p tub (Fig. 29b). The boron enters the silicon through the thin pad oxide but is masked from the n tub by the thicker SiO_2 layer there. All oxides are then stripped and the two tubs are driven in (Fig. 28c).

After the tubs are driven in, the intra-tub transistor isolation is performed (a tub may contain tens of thousands of transistors of a given type within it) using the techniques described in Fig. 15. After field and gate oxides have been formed, threshold adjustment implants can be made into the channel regions of the devices.

Next, n^+ polysilicon is deposited and defined and the source/drain regions are implanted. To save another mask step, boron is first nonselectively implanted into *all* sources and drains (Fig. 29d). Following this, phosphorus is selectively implanted into the n-channel source/drain regions at a higher dose so that it overcompensates the existing boron (Fig. 29e). After processing, the boron profile in the n-channel source/drains is completedly covered vertically and laterally by the phosphorus. This technique has also been used with As and BF_2 for shallow junction n- and p-channel devices, respectively.[74] A phosphorus glass layer is later deposited (Fig. 29f) and flowed at high temperature. After windows are dry-etched in the P-glass, aluminum metallization is defined using dry etching. The final layer is a plasma-deposited silicon-nitride layer which seals the devices and provides mechanical scratch protection. Figure 28c shows the finished cross section.

11.5.3 Key Steps in Device Formation

Isolation The same principles as discussed previously (Section 11.4.2) apply in isolating the same types of MOSFETs from each other within a given tub region. However, in CMOS circuits there is the added concern of isolating the two different *types* of transistors. Figure 30a shows the top view of n- and p-channel transistors straddling the common tub border. Figure 30b shows a cross section of the structure beneath the polysilicon rail. A parasitic n-channel transistor exists between the n source (induced under the polysilicon gate) and the adjacent n tub. Similarly a parasitic p-channel transistor exists between the p source and the p tub. Single or twin tubs are typically driven in rather deeply to ensure that enough charge exists below the transistor to prevent punchthrough to the substrate, and to keep the h_{FE} of the vertical bipolar device from becoming too large (and hence susceptible to latchup). The long diffusion length associated with driving the tub in causes a reduction in surface concentration of each tub near the border, and hence a reduction in the parasitic transistor's threshold voltage. Figure 30c shows the threshold voltage of each type of

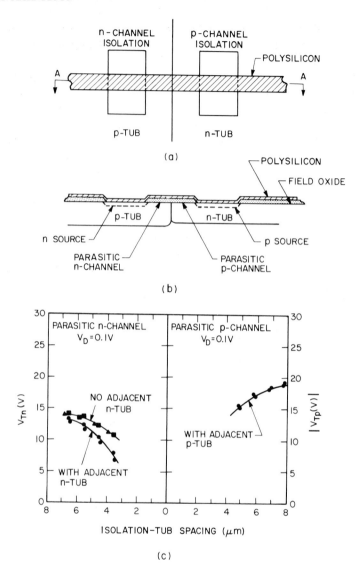

Fig. 30 Isolation of n- and p-channel transistors. (a) Top view of adjacent n- and p-channel transistors sharing a common polysilicon gate. (b) Cross section under the polysilicon rail. (c) Parasitic n- and p-channel threshold voltages vs. transistor-edge-to-tub spacing. *(After Parrillo et al., Ref. 72.)*

parasitic device as a function of the separation between the transistor edge and the tub border.[72] The upper curve on the left shows the parasitic n-channel threshold voltage reduction near the tub border, which occurs with no adjacent n tub and characterizes the effect of a long diffusion of a single p-tub-type process. The interdiffusion (i.e., the compensation) of the two types of tub impurities further reduces the net surface concentrations of each tub near the border, and the parasitic field thresholds are further reduced (Fig. 30c). The n- and p-channel transistors must be placed laterally far enough away from the tub border so that the field threshold voltages are adequately large in magnitude.

To avoid some of the problems associated with deep-tub drive-in cycles, very-high-energy (400- to 600-keV) p-tub implants can be used to place a sufficient charge below the n-channel transistors without the long thermal cycle.[76, 77] The deep boron implant, performed after local oxidation, also provides a high surface concentration under the field oxide, which serves as a chan-stop layer. Significant improvements in packing density and latchup susceptibility have been reported using this technique.[76, 77]

Threshold adjustment The threshold voltages for the two types of transistors often must be comparable and below 1 V in magnitude. This condition allows for both low-voltage operation of CMOS circuits ($V_{DD} > V_{Tn} + |V_{Tp}|$) and higher-current drive for the devices at higher values of V_{DD}. Meeting this condition requires some adjustment, however. If a given material (e.g., n^+ polysilicon) is used as the gate for each type of device, the work-function difference ϕ_{MS} will be different for the n- and p-channel transistors. This difference causes an asymmetry in the threshold voltages of the two types of transistors. Figure 31 shows the calculated threshold voltages of

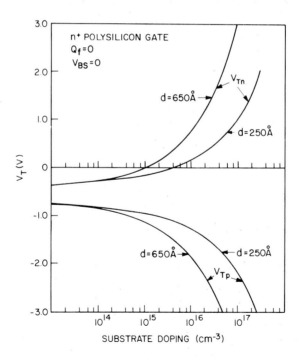

Fig. 31 Calculated threshold voltages of n-channel (V_{Tn}) and p-channel (V_{Tp}) transistors as a function of their substrate's doping, assuming an n^+-polysilicon gate, zero fixed charge Q_f, and zero back-gate bias V_{BS}. Curves for gate-oxide thicknesses (d) of 250 and 650 Å are shown.

n- and p-channel devices as a function of their substrate doping. Note that we cannot obtain $|V_{Tp}| \lesssim 0.7$ V by simply lowering the p channel's substrate doping, whereas we can obtain $V_{Tn} \lesssim 0.7$ V by adjusting the n channel's substrate doping.

To obtain the desired p-channel threshold voltage with n^+-polysilicon gates, a shallow boron layer is often implanted into the channel region of the p-channel device.[55, 72] The boron shifts the lower curves in Fig. 31 to more positive values. This boron threshold-adjustment dose can also be implanted into the n-channel device to raise the magnitude of V_{Tn}. With a judicious choice of n- and p-type background doping, a single, nonselective boron implant can be used to set the desired threshold voltage of each type of device. This technique is illustrated in Fig. 32 which shows a plot of V_{Tn} and V_{Tp} vs. the boron implant dose for devices having a 650-Å gate oxide and n^+-polysilicon gates.[55] This CMOS structure uses an n well implanted into a p substrate. V_{Tn} increases as the boron dose is increased, because the surface concentration of the p substrate is increased. The magnitude of V_{Tp} decreases primarily because of the negatively ionized charge (boron) in the silicon depletion layer. For lower n-well implant doses, $|V_{Tp}|$ decreases more quickly as the threshold-adjustment dose is increased.

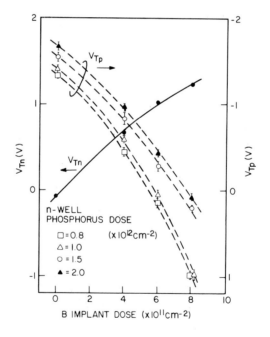

Fig. 32 Threshold voltages of n-channel (V_{Tn}) and p-channel (V_{Tp}) transistors as a function of boron threshold-adjustment dose. The CMOS structure uses an n well implanted into a p-type substrate whose doping level is 6×10^{14} atoms/cm^3. V_{Tp} results are shown for various implant doses of the n well. (*After Ohzone et al., Ref. 55.*)

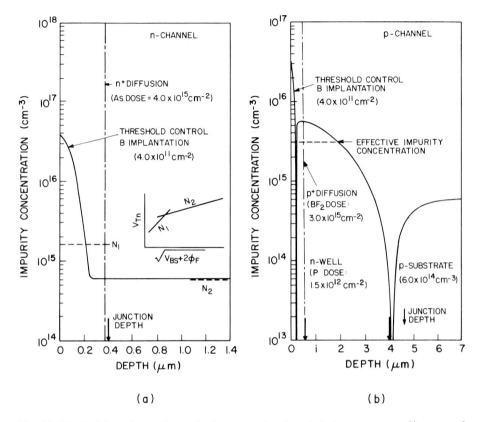

Fig. 33 Calculated impurity profiles under the gate. (a) n-channel device with 6×10^{14} atoms/cm^3 p-substrate doping, and 4×10^{11} atoms/cm^2 boron threshold-adjustment implant. The n$^+$ source/drain junction depth (0.4 μm) is indicated. The insert shows the threshold voltage sensitivity to back-gate bias due to the high-low doping profile. (b) p-channel device in an n well with the same boron threshold-adjustment implant. The p$^+$ source/drain junction depth (0.55 μm) is indicated. *(After Ohzone et al., Ref. 55.)*

The desired threshold voltages of $V_{Tn} = -V_{Tp} = 0.7$ V are obtained by using an n-well phosphorus dose of 1.5×10^{12} atoms/cm^2 and a threshold-adjustment boron dose of 4×10^{11} atoms/cm^2. For these conditions, the calculated impurity profiles under the gate for each type of device[55] are shown in Fig. 33a and b. The n MOSFET uses arsenic for the source/drain impurity. Its threshold sensitivity to back-gate bias is initially steep and then less sensitive at larger values of V_{BS} (see insert in Fig. 33a)— which is a consequence of the high-low doping profile under the gate. The p-channel device (Fig. 33b) uses BF$_2$ for the source/drain because it results in a shallower implant depth than B. Note that, as a result of the threshold-adjustment implant, a net p region exists at the surface of the p-channel device and connects the p$^+$-source/drain regions. This structure is analogous to a normally off, buried, n-channel MOSFET. That is, the work-function difference of the n$^+$-polysilicon gate in the device of Fig.

33b depletes the p region from the surface, while the underlying n well depletes the p region from below. Hence the boron threshold-adjustment layer is depleted of carriers (normally off). If a large enough threshold-adjustment dose is used (e.g., 8×10^{11} atoms/cm^2 in Fig. 32), the shallow p region will not be depleted for $V_G = 0$ and the device functions as a depletion-mode transistor (normally on).

The threshold-adjustment procedure discussed above is based on n$^+$-polysilicon gates. The use of different gate materials requires different threshold-adjustment techniques for the two types of devices. MoSi$_2$ gates have a work-function 0.8 V larger than that for n$^+$ polysilicon. CMOS devices using MoSi$_2$ have been made[78] using arsenic or phosphorus channel implants for the n channel (buried channel) and boron implants for the p channel to provide $V_{Tn} = -V_{Tp} = 0.8$ V.

Latchup prevention The critical device parameters in latchup can be described using the thyristor diagram in the insert of Fig. 27a. The current gains h_{FE} of the n-p-n and p-n-p bipolar transistors are key parameters. If the product of the current gains of the two devices exceeds unity, the device can latch. Several techniques have been used to lower the current gains of the two devices, including gold doping and neutron irradiation to reduce the minority-carrier lifetimes.[79] These techniques are difficult to control and cause other deleterious effects in device operation (excess leakage, for example). The vertical n-p-n gain can be reduced by the use of p$^+$ buried-layers under p wells,[79] or the use of high-dose, high-energy boron p well implants.[76, 77]

Another effective technique[79] is to reduce the resistances that shunt the emitter-base junctions of the two types of bipolar devices shown in Fig. 27a. If these shunt resistors are made small enough, a sufficient IR drop cannot be developed across them to forward-bias the emitter-base junctions, and the device will not latch. The shunt resistance of the lateral p-n-p emitter-base junction can be reduced by the use of an n epitaxy over an n$^+$ substrate.[75, 73] As can be seen from Fig. 27b, a more conductive substrate reduces the lateral resistance under the p$^+$ source. In addition, electrons injected from the n$^+$-source/drain regions into the p tub can be collected vertically out the back of the chip, which is solidly connected to V_{DD}. The additional processing expense to grow the epitaxial layer must be weighed against the benefit of this effective method of latchup control.

In addition to the n-epitaxy/n$^+$-substrate structure, proper circuit-layout techniques must be employed to prevent latchup in CMOS ICs.[75] Guard rings which surround n- and p-channel transistors in the input/output (I/O) circuitry can be used to divert minority carriers from creating lateral IR drops. Input protection is critical in guarding against external signals which induce latchup as well as overstress gate oxides. The I/O devices are generally large enough to provide high off-chip drive capability; in comparison, the additional area needed for guard rings is usually negligible for complex chips.

11.6 MINIATURIZING VLSI CIRCUITS

In this section, we discuss some basic guides for the miniaturization of individual devices and the ICs that they produce.

11.6.1 Basic Design Rules

The rules governing the dimensions of features that are permissible in designing and laying out an IC in a particular technology are referred to as design rules. They are generally a list of minimum feature sizes and separations between features (including overlaps), which are consistent with the patterning and device limitations of a particular technology. Many factors are considered in deriving a set of manufacturable design rules, and some of the important ones are described here.

Minimum features on individual levels For each individual level, the minimum linewidth of a feature and the minimum separation between these features on a fully processed wafer must be established. These dimensions are a function of (1) the minimum dimensions on a mask that can be *routinely* resolved in the lithographic process and (2) the change that the feature undergoes during the specific step in the process that defines the final feature. Item (1) is a function of the quality of the masks, the lithographic tool, the lithographic process, and the topography of the wafer at that particular level. Item (2) is a function of the pattern-transfer process. This process can take several forms. For example, the change in the size of an etched feature compared with the lithographic mask feature is small if anisotropic dry etching is used. Hence the mask feature can be made very close in size to the final feature on the wafer. The local oxidation process (Fig. 15), on the other hand, causes the space between active regions to grow, and hence the final separation between these features in the silicon is larger than that on the masks. The design rule governing the minimum separation between transistors must take this into account. The lateral diffusion of impurities and the lateral extent of the junction depletion layers (under maximum applied voltages) also govern the minimum separation between devices on a given level. These and many other considerations related to patterning and device physics must be taken into account to establish the minimum linewidths and spaces allowable on a given level.

Nesting of individual levels Equal in importance to minimum feature sizes are the distances required to *nest* the features on one level with respect to features on a previous level. The nesting of individual levels is a function of (1) the *overlay tolerance* in the lithographic process, (2) the *variation* in finished feature sizes, and (3) the *alignment sequence* of the individual levels. *Overlay tolerance* is again a function of the lithographic tool and its alignment accuracy (or operator ability). If Δ is the worst-case misregistration distance between two levels which are aligned to each other (e.g., two standard deviations, or sigmas, in a distribution of misregistration distances) and there are n alignment steps between levels A and B, the worst-case misregistration[80] distance M between levels A and B is $M_{A,B} = \sqrt{n_{A,B}}\ \Delta$, where Δ is assumed to be the same for all levels (although it is not necessarily so). According to this equation, as the intervening alignment steps become more numerous between two levels, the overlay precision decreases.

In addition to overlay error, the *variation* in the size of individual features— i.e., item (2)—places restrictions on how well levels can be nested. The reproducibility of the definition process (lithography, etching, diffusion, etc.) dictates the

Table 2 Matrix displaying the number of alignment steps between any two mask levels of Fig. 34

	Isolation	Buried contact	Depletion implant	Polysilicon gate	Window 1	Metal	Window 2
Isolation	⋯	1	1	1	2	3	4
Buried contact	1	⋯	2	2	3	4	5
Depletion implant	1	2	⋯	2	3	4	5
Polysilicon gate	1	2	2	⋯	1	2	3
Window 1	2	3	3	1	⋯	1	2
Metal	3	4	4	2	1	⋯	1
Window 2	4	5	5	3	2	1	⋯

FABRICATION SEQUENCE   ALIGNMENT SEQUENCE

1. ISOLATION
2. BURIED CONTACT
3. DEPLETION IMPLANT
4. POLYSILICON GATE
5. WINDOW 1
6. METAL
7. WINDOW 2

Fig. 34 Fabrication and mask-level alignment sequences for the process of Fig. 14. The arrows indicate the levels to which the subsequent levels are aligned.

feature size variation. If $\pm\delta l_A$ is the worst-case variation of an *edge* of a feature on level A (a two-sigma limit) and $\pm\delta l_B$ is the worst-case variation of an *edge* of a feature on level B (a two-sigma limit), then the minimum separation S which must be allowed between them to ensure that the edges do not coincide is $S_{A,B} = \sqrt{n_{A,B}\, \Delta^2 + \delta l_A^2 + \delta l_B^2}$. Therefore, to minimize the separation between critical levels, the following must be minimized: overlay error, variation in feature size, and number of alignment steps between them.

The *fabrication* process must proceed in a given sequence; however, the optimum *alignment sequence*, item (3), can be quite different. For example, Fig. 34 lists the fabrication sequence for the seven-mask process of Fig. 14 which must be adhered to. The figure also gives an example of an alignment sequence for this process. Notice that consecutive levels in the fabrication sequence are not necessarily aligned to each other (because buried contact and depletion-implant levels leave very vague, if any, marks on the wafer). By knowing the alignment sequence, we can deduce the number of alignments between any two levels. The number of alignments between any two levels of Fig. 34 are displayed in the matrix of Table 2. If the overlay tolerance and feature size variations are known for each level, the worst-case separation between edges of features on any two levels can also be conveniently displayed in the matrix.

Some of the separations between features in Table 2 are critical and others are not. Figure 35 illustrates how devices can fail if insufficient space is allowed between certain features. Two layouts are shown with the same minimum dimensions of individual features; however, the area that they consume differs considerably because of nesting considerations. Figure 35a is an idealized "aggressive" layout of isolation, window, and gate levels for an MOS transistor. Figure 35b illustrates how a worst-case misalignment and change in feature size might affect the device. For illustrative purposes, the windows on the left side of the figure are drawn undersized and those on the right side are drawn oversized. The undersized windows (which have become rounded in processing) provide minimal contact to the source/drain region, since a significant part of their area overlaps the thick field oxide. Excessive contact resistance or even open contacts can result. The gate overlap over the field oxide at the top of Fig. 35a is inadequate; hence the source-drain regions are separated by a very small distance, or they can touch, causing a short between them (Fig. 35b). Because of the misalignment between window and gate and the growth of the window (and/or gate) dimension, the source/drain shorts to the gate (Fig. 35b) after metallization. A

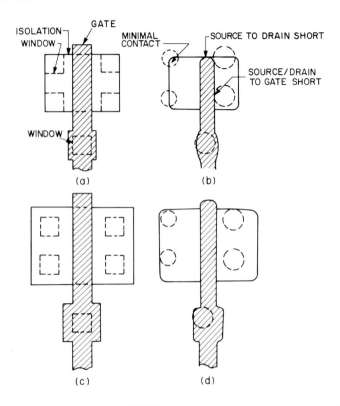

Fig. 35 Effects of nesting tolerances on MOSFET layout. (a) Ideal alignment and feature definition in an aggressive layout of an MOS transistor. (b) After worst-case alignment and feature definition in aggressive layout. (c) Ideal conservative layout. (d) After worst-case alignment and feature definition in conservative layout. (For illustrative purposes only, the windows on the left are shown small while the windows on the right of b and d are shown large.)

conservative layout suffers less from these process variations (Fig. 35c and d). Packing density (chip area) and ease of processing (defect density) therefore must be measured against one another to determine a manufacturable set of design rules for a given technology.

11.6.2 Scaling Principles

The goal of reducing device dimensions (scaling) for improved IC packing density and performance can be achieved in a variety of ways. The basic objective in scaling is to preserve the original device characteristics in miniaturized geometry.

Scaling prinicples for bipolar devices have been based on optimizing the power-delay product of particular bipolar circuits. Some of the major considerations in achieving this goal are described here.[16] As the lateral dimensions of the bipolar IC are reduced by a given factor, the vertical device dimensions (i.e., epitaxy thickness, base width, emitter depth) are reduced by nearly the same factor, and the current density increases by about the square of that factor. The collector doping (in the epitaxy) is kept proportional to the emitter current density to avoid base push-out. As the base width is reduced, the base doping level is raised as 1 over the square of the base width, to prevent punchthrough. The emitter depth is reduced to maintain control over the base width and the base charge. Heavy doping effects, such as bandgap narrowing, become significant for the base as well as the emitter in the scaled devices. The operating voltage is kept constant in bipolar device scaling, because the turn-on voltage of p-n junctions is relatively insensitive to device geometries and doping levels.[16]

The original guidelines for MOSFET device miniaturization,[46] summarized here, are based on scaling the physical dimensions of a given reference device by a factor K, where $K > 1$. In order to keep the internal electric fields of the scaled device constant, the operating voltages are also reduced by K. The physical dimensions including oxide thickness, gate length, transistor width, and junction depths are reduced by K. The threshold voltage is reduced by approximately K as are the junction depletion widths. Using these uniform scaling principles, the current of the scaled device is reduced by K; hence the "resistance" of the scaled device V/I is unchanged. Because of the reduction in all dimensions by K (interconnect as well as active devices), the number of devices per unit area will increase by K^2. To first order, the capacitances of circuit elements are reduced by K (area reduced by K^2, but dielectric thickness reduced by K); hence circuit delay times (VC/I) are reduced by K. The power dissipation (VI) is reduced by K^2 and the power-delay product ($VI \cdot VC/I$) is reduced by K^3.

Since the original MOSFET scaling principles were reported,[46] several workers have examined them in greater detail. As the devices become smaller, second-order effects become more important. For example, the interconnection capacitance in scaled devices becomes dominated by fringing effects so that it decreases logarithmically rather than linearly with scaling.[81] The built-in junction potentials do not scale, so in order to proportionately shrink depletion regions, the doping levels must increase more rapidly than linearly with the scaling factor.[95]

A different scaling approach focuses on preserving long-channel behavior in a given MOSFET.[82] A criterion for long-channel behavior is that the subthreshold drain current of the MOSFET at $V_G = V_T$ will increase by less than 10% as V_D is increased (the subthreshold current for a long-channel device is independent of V_D for $V_D > 3\ kT/q$). The minimum channel length L_{min} of a MOSFET where long-channel behavior is still observed is given by the empirical relation $L_{min} = 0.4[r_j d (W_S + W_D)^2]^{1/3}$, where r_j is the source/drain junction depth in micrometers, d is the oxide thickness in angstroms, and $W_S + W_D$ is the sum of the source and drain depletion widths in micrometers using a one-dimensional abrupt junction analysis. For a uniform doping density N_A in the substrate

$$W_D = \sqrt{\frac{2\epsilon_S}{qN_A}} (V_D + V_{bi} + V_{BS})$$

where V_{BS} is the back-gate bias, and V_{bi} is the built-in potential of the junction ($W_S = W_D$ for $V_D = 0$). Using this model, the various device parameters can be adjusted independently to be consistent with the desired L_{min} in a more flexible way than just scaling them by a common factor.[6] However, devices in VLSI circuits can and do exhibit short-channel effects. As long as the circuit design takes these effects into account, the ICs can be successfully manufactured.

As MOSFETs are scaled to finer dimensions, PMOS devices begin to perform as well as NMOS devices—because the saturation velocity of holes becomes comparable to that of electrons.[81, 83] In addition, holes have a much lower impact-ionization coefficient than electrons, so substrate and gate currents caused by the high electric fields at the drain are several orders of magnitude lower for PMOS devices than for comparable NMOS devices.[81, 83]

In general, the power supply voltages have not been scaled because, in a system, ICs have to interface with a variety of other IC families that are designed to operate at a standard power supply voltage (e.g., 5 V). The performance of miniaturized devices has been studied by reducing the supply voltage by a factor K (constant electric field CE), keeping the voltage constant (CV), and a quasiconstant voltage scaling (QCV) where the supply voltage is reduced by $\sqrt{K}$. Based on the specific assumptions in the particular scaling model, different conclusions can be reached as to the optimum voltage scaling scenario.[81, 95] Hot-carrier injection into gate oxides is becoming a severe problem in fine-line devices because it causes gate-oxide charging and threshold-voltage instabilities. For these reasons alone, it is expected that operating voltages will probably be reduced to 3 V and below, since 3.1 V is approximately the energy required for an electron to surmount the Si-SiO$_2$ energy barrier.[6]

11.6.3 Multilevel Metallization

As chip functionality has grown, intra-chip wiring complexity has also grown so that most of the area of complex random-logic ICs (e.g., microprocessors) is devoted to interconnect wiring. Two independent levels of interconnect naturally arise in MOSFET ICs because of the refractory-gate and metallization layers. An additional metal-interconnect level has also been used on complex MOSFET chips.[84, 85] The two levels of interconnect in bipolar technology generally consist of metal layers[22, 30]; however, in a manner similar to MOSFET technology, silicided polysilicon can also be used as one of the interconnect levels.[27]

The advantages of multilevel metal interconnections are reduced chip size and improved performance. The disadvantages are the technological difficulties and yield loss associated with providing the additional layers. However, as chips become more complex, more than three levels of interconnect will be required to avoid inordinately large chip areas and wiring capacitances. A four-level metal approach can shrink chip areas to half the size achieved with a three-level approach.[86] With four levels of interconnect, two levels can be used to internally wire portions of the IC. The IC is thus laid out as modules in the silicon. The modules can then be interconnected globally on a chip by using the second pair of interconnect levels. Future VLSI technologies will focus increasing attention on multilayered interconnect structures.

11.7 MODERN IC FABRICATION

Throughout this chapter, attention was focused on various techniques to produce specific components of fine-line semiconductor structures. To successfully manufacture VLSI circuits, these process steps must be carried out in an environment that is meticulously controlled with respect to cleanliness, temperature, humidity, and orderliness. Process monitoring and VLSI testing are other important areas of IC fabrication, and these are briefly considered in this section.

11.7.1 IC Fabrication Facilities

Pure water system IC fabrication uses large quantities of water (e.g., frequent overflow rinses for cleaning). The use of ultrapure, electronic-grade water is thus a fundamental requirement in fabricating ICs. Particulates in the water, as well as contaminants such as sodium, copper, and iron, are deposited on the wafers and lead to device degradation. The water system must be engineered to deliver a continuous and large supply of ultraclean water with a low ionic content, as measured by monitoring the water's resistivity. Water with a specific resistivity of 18 MΩ-cm is considered to have a low ionic content.

A pure water system consists of several sections.[87] Inflowing water is passed through charcoal filters and into electrodialysis units that filter and demineralize the water. It then passes through resin tanks, followed by mixed-bed ion-exchange resin units—to remove more minerals—and nuclear-grade resin cannisters to minimize sodium ion content. Undissolved solids, bacteria, and other organic matter are removed by a series of filters ranging in pore size from 10 μm down to 0.01 μm. Bacterial content (which can foul the system) is minimized by continuously circulating the water through an ultraviolet sterilization unit. The resistivity of the water is monitored continuously by electrical means and particulates in the water are monitored continuously with an automatic light-blockage type of liquid-borne particle counter.[87]

Clean room In general, all process steps are implemented in a clean room. Here the air is maintained at a well-controlled temperature and humidity and is continuously filtered and recirculated. Particulates must be avoided on wafers since they cause improperly defined features, undesirable surface topography, leakage through insulating layers, and other deleterious effects. Air in the clean room is monitored and classified with respect to particulates. A "class 100" environment has a maximum of 100 particles per cubic foot ($\sim$3500 particles per cubic meter) with particle size larger than 0.5 μm, and a maximum of 10 particles per cubic foot with particle size larger than 5.0 μm.[88] In the highly critical lithography area, particle densities are typically maintained below 10 particles per cubic foot. The filtered air flows from ceiling to floor at more than 85 linear feet per minute ($\sim$26 linear meters per minute), which is approximately the threshold for laminar flow (i.e., for uniform velocity of air following parallel flow lines without turbulence). Particulates can emanate from process equipment (e.g., in CVD and etch reactors) as well as from humans (from street clothes, skin flakes, etc.); hence personnel must wear proper clothing to protect the wafers.

11.7.2 Process Monitoring

The electrical, mechanical, and visual tests used to evaluate and characterize the various process steps are an integral part of IC process development and manufacture. Nondevice wafers are used to monitor the calibration of equipment and to ensure that uniform results are obtained across wafers and lots of wafers. For example, sheet resistance measurements on monitor wafers are used to check ion-implantation machines, the thermal cycles of furnaces, and film-deposition equipment. Oxide thicknesses are monitored using MOS and ellipsometric measurements. Capacitance-voltage measurements at elevated temperature are routinely used to monitor mobile-ion contamination in oxides which may emanate from furnaces, film-deposition equipment, and other sources.

Visual inspections of device wafers at key steps in the lithographic and etching processes are also an integral part of IC fabrication. Such inspections help to identify yield-limiting sources such as residues from incomplete cleaning, particulates from specific apparatus, and variations in the deposition and definition of layers.

In addition to these in-process monitoring techniques, many measurements are performed on test chips which are located at several sites on completed IC wafers. Test chips are designed to measure the electrical characteristics of the various structures in the particular technology such as MOS properties, capacitances of the periphery as well as the area of p-n junctions, sheet resistances, and transistor properties. Parasitic as well as active transistor properties are routinely monitored using automated computer-controlled data-acquisition systems. The device information is also used to generate model parameters, which play back the device characteristics using computer-aided design tools. These model parameters are used to design and diagnose circuits fabricated in a given technology. In addition to device parameters, electrical test patterns can be designed to characterize linewidth, linewidth uniformity, lithographic alignment quality, wafer distortion, and random fault densities, and the results can be displayed as wafer maps.[89]

11.7.3 VLSI Circuit Testing

As the functional capability of ICs has become more complex, the techniques and equipment required to test the circuits have also grown in complexity. Modern test systems consist of sophisticated computers that communicate with the test device. Testing digital ICs involves inputting sequences of logic 1-0 patterns (driving voltages), which cause the device being tested to respond with a sequence of 1-0 signals on the outputs. The output patterns are then compared with the expected results, which may be generated using circuit simulations. The input logic 1-0 patterns (test patterns) are designed to identify where possible faults may exist in the chip as well as to exercise the chip in its intended fashion. Test patterns for memory chips repetitively store ones and zeroes in the memory cells. Test patterns for logic chips are much more complex, since the logic circuit may contain thousands of logic gates with only a few tens of inputs.

Individual gates on logic circuits are not physically accessible, so they cannot be tested directly.[90] Logic networks are either combinational or sequential. Combinational logic networks contain no memory and the output at any time is a function only of the logic input at that time (i.e., it can be derived from truth tables). To test combinational logic networks, test patterns can be generated that use algorithms to achieve virtually 100% fault coverage.[90]

Sequential logic contains memory elements (e.g., latches) and the output is a function of previous as well as current input patterns. Most logic devices are sequential (e.g., microprocessors) and general algorithms for generating test patterns for sequential networks have not been established.[90] However, "design for testability" methods have been developed, where additional circuitry, used only for self-testing, is implemented on the chip. Using additional shift-register latch circuitry, sequential logic circuits can be transformed into combinational logic circuits so that more than 98% of the circuitry can be tested with truth tables.[91] Of course, such a transformation requires additional silicon area, which may reduce chip yield and increase the chip cost.

A significant part of the overall IC cost is incurred in separating and packaging chips from wafers. Fully processed wafers are tested and only the good chips are selected for packaging (see Chapter 13). The chips are retested in packaged form, the circuit boards containing individual packaged chips are tested, and finally the machine or system containing several circuit boards is tested. The cost of detecting, isolating, and repairing a fault escalates greatly from the IC level to the system level (e.g., the cost can be one dollar at the IC level, compared with thousands of dollars at the system level). Hence extensive testing for fault coverage at the IC level (which may increase IC cost by 10 to 20%) minimizes the number of faults that will be carried to higher-level systems, and therefore reduces the overall system cost.[90]

11.8 SUMMARY AND FUTURE TRENDS

In this chapter we have considered some of the basic techniques currently used for bipolar, NMOS, and CMOS IC fabrication. Over the last few years we have seen originally exploratory techniques such as ion implantation, local oxidation, and anisotropic dry etching, become standard IC fabrication techniques.

Current exploratory techniques which may also be adopted include very-high-energy (MeV) implants that place impurity layers deep in the silicon and eliminate the need for epitaxial layers in certain applications.[92] In the future, more emphasis will be placed on producing high-quality silicon on insulating substrates to reduce parasitic junction capacitances and to provide high packing density of devices.[93] Refractory silicided MOSFET gates will probably be replaced by higher-conductivity metal gates. Both low-temperature, long-duration and high-temperature, short-duration processing will be used to provide shallow impurity-profiles. New techniques are expected for forming active devices in recrystallized layers of polysilicon,[94] so that complex vertically stacked circuits can be fabricated. In addition, new lithographic techniques will remove the necessity of having planar topography. The trend toward multiple levels of high-conductivity interconnects will continue to improve packing density and circuit performance. Fabrication equipment will become interconnected and fully automated, so that wafers will be processed with minimal human handling to avoid contamination and operator errors. With new developments occurring at an ever-increasing rate, future IC fabrication promises to be even more productive and challenging than in the past.

REFERENCES

[1] J. S. Kilby "The Invention of the Integrated Circuit," *IEEE Trans. Electron Devices*, **ED-23**, 648 (1976).

[2] B. Hoefflinger and G. Zimmer, "New CMOS Technologies," in J. Carrol, Ed., *Solid-State Devices*, Ser. 57, Institute of Physics Conf., 1980.

[3] A. S. Grove, *Physics and Technology of Semiconductor Devices*, Wiley, New York, 1967.

[4] R. S. Muller and T. I. Kamins *Device Electronics for Integrated Circuits*, Wiley, New York, 1977.

[5] B. G. Streetman, *Solid-State Electronics*, Prentice-Hall, Englewood Cliffs, N.J., 1980, 2d ed.

[6] S. M. Sze, *Physics of Semiconductor Devices*, Wiley, New York, 1981, 2d ed.

[7] R. Siqusch, K. H. Horninger, W. A. Müeller, D. Widman, and W. G. Oldham, "A 1-μm Process: Linewidth Control Using 10:1 Projection Lithography," IEEE Int. Electron Device Meet., Wash., D.C., 1980, p. 429.

[8] J. L. Stone, "I²L: A Comprehensive Review of Techniques and Technology," *Solid State Technol.* p. 42 (June 1977).

[9] E. F. Labuda and J. T. Clemens, "Integrated Circuit Technology," in R. E. Kirk and D. F. Othmer, Eds., *Encyclopedia of Chemical Technology*, Wiley, New York, 1980.

[10] J. A. Appels, E. Kooi, M. M. Paffen, J. J. H. Schlorje, and W. H. C. G. Verkuylen, "Local Oxidation of Silicon and its Application in Semiconductor Technology," *Philips Res. Rep.*, **25**, 118 (1970).

[11] W. A. Westdorp and G. H. Schwuttke, *Thin Film Dielectrics*, in F. Vratny, Ed., Electrochem. Soc., 1969, p. 546

[12] L. C. Parrillo, G. W. Reutlinger, R. S. Payne, A. R. Tretola, and R. T. Kraetsch, "The Sensitivity of Transistor Gain to Processing Variations in an All Implanted High-Speed Bipolar Technology," IEEE Int. Electron Device Meet., Wash., D.C., 1977, p. 265A.

[13] J. Lohstroh, "Devices and Circuits for Bipolar (V)LSI," *IEEE Proc.*, **69**, 812 (1981).

[14] W. D. Ryden and E. F. Labuda, "A Metallization Providing Two Levels of Interconnect for Beam-Leaded Silicon Integrated Circuits," *IEEE J. Solid State Circuits*, **SC-12**, 376 (1977).

[15] M. P. Lepselter, "Beam Lead Technology," *Bell Syst. Tech. J.*, **45**, 233 (1966).

[16] T. H. Ning, D. D. Tang, and P. M. Solomon, "Scaling Properties of Bipolar Devices," IEEE Int. Electron Device Meet., Wash., D.C., 1980, p. 61.

[17] T. Takemoto, T. Fujita, K. Kawakita, H. Sakai, and T. Komeda, "A Vertically Isolated Self-aligned Transistor," IEEE Int. Electron Device Meet., Wash., D.C., 1981, p. 708.

[18] K. Y. Chiu, J. L. Moll, and J. Manoliu, "A Bird's Beak Free Local Oxidation Technology Feasible for VLSI Circuits Fabrication," *IEEE Trans. Electron Devices*, **ED-29**, 536 (1982).

[19] R. S. Payne, R. J. Scavuzzo, K. H. Olson, J. M. Nacci, and R. A. Moline, "Fully Ion-Implanted Bipolar Transistors," *IEEE Trans. Electron Devices*, **ED-21**, 273 (1974).

[20] T. H. Ning and R. D. Isaac, "Effect of Emitter Contact on Current Gain of Silicon Bipolar Transistors," IEEE Int. Electron Device Meet., Wash., D.C., 1979, p. 473.

[21] J. B. Bindell, W. M. Moller, and E. F. Labuda, "Ion-Implanted Low-Barrier PtSi Schottky-Barrier Diodes," *IEEE Trans. Electron Devices*, **ED-27**, 420 (1980).

[22] S. A. Evans, S. A. Morris, L. A. Arledge, J. O. Englade, and C. R. Fuller, "A 1-μm Bipolar VLSI Technology," *IEEE Trans. Electron Devices*, **ED-27**, 73 (1980).

[23] F. W. Hewlett, Jr., "Schottky I²L," *IEEE J. Solid State Circuits*, **SC-10**, 343 (1975).

[24] K. Hart and A. Slob, "Integrated Injection Logic—A New Approach to LSI," IEEE Int. Solid State Circuits Conf., Philadelphia, Pa., 1972, p. 92; *IEEE J. Solid State Circuits*, **SC-7**, 346 (1972).

[25] H. H. Berger and S. K. Wiedmann, "Merged Transistor Logic—A Low Cost Bipolar Logic Concept," IEEE Int. Solid State Circuits Conf., Philadelphia, Pa., 1972, p. 90; *IEEE J. Solid State Circuits*, **SC-7**, 340 (1972).

[26] D. D. Tang, T. H. Ning, S. K. Wiedmann, R. D. Isaac, G. C. Feth, and H. N. Yu, "Sub-Nanosecond Self-Aligned I²L/MTL Cirucits," IEEE Int. Electron Device Meet., Wash., D.C., 1979, p. 201.

[27] H. Nakashiba, I. Ishida, K. Aomura, and T. Nakamura, "An Advanced PSA Process for High-Speed Bipolar VLSI," *IEEE Trans. Electron Devices*," **ED-27**, 1390 (1980).

[28] T. Sakai, Y. Kobayashi, Y. Yamamoto, and H. Yamaguchi, "High-Speed Bipolar LSI Technology SST," in J. Nishizawa, Ed., *Semiconductor Technology*, Ohmsha, 1982.

[29] R. D. Davies and J. D. Meindl, "Poly I²L- A High-Speed Linear Compatible Structure," *IEEE J Solid State Circuits*, **SC-12**, 367 (1977).

[30] J. Agraz-Güereña, P. T. Panousis, and B. L. Morris, "OXIL—A Versatile Bipolar VLSI Technology," *IEEE Trans. Electron Devices*, **ED-27**, 1397 (1980).

[31] A. B. Glaser and G. E. Subak-Sharpe, *Integrated Circuit Engineering*, Addison-Wesley, Reading, Mass., 1977.

[32] F. Barson, "Emitter-Collector Shorts in Bipolar Devices," *IEEE J. Solid State Circuits*, **SC-11**, 505 (1976).

[33] L. C. Parrillo, R. S. Payne, T. E. Seidel, M. Robinson, G. W. Reutlinger, D. E. Post, and R. L. Field, "The Reduction of Emitter-Collector Shorts in a High Speed All Implanted Bipolar Technology," *IEEE Trans. Electron Devices*, **ED-28**, 1508 (1981).

[34] K. Shinada, S. Shinozaki, K. Kurosawa, and K. Taniguchi, "Nature and Mechanism of Emitter-Collector Shorts in Oxide Isolated Bipolar Integrated Circuits," IEEE Int. Electron Device Meet., Wash., D.C., 1979, p. 344.

[35] A. K. Sinha, H. J. Levinstein, T. E. Smith, G. Quintana, and S. E. Haszko, "Reactive Plasma Deposited Si-N Films for MOS-LSI Passivation," *J. Electrochem. Soc.*, **125**, 601 (1978).

[36] R. C. Sun and J. T. Clemens, "Characterization of Reverse Bias Leakage Currents and Their Effect on Holding Time Characteristics of MOS Dynamic RAM Circuits," IEEE Int. Electron Device Meet., Wash., D.C., 1977, p. 254.

[37] J. T. Clemens, D. A. Mehta, J. T. Nelson, C. W. Pearce, and R. C. Sun, *MOS Dynamic Memory in a Diffusion Current Limited Semiconductor Structure*, U.S. Patent No. 4,216,489 (Aug. 5, 1980).

[38] J. W. Slotboom, H. A. Harwig, and M. J. M. Pelgrom, "Leakage Current in High Density CCD Memory Structures," IEEE Int. Electron Device Meet., Wash., D.C., 1981, p. 667.

[39] E. Kooi, J. G. van Lierop, and J. A. Appels, "Formation of Silicon Nitride at the Si-SiO₂ Interface During Local Oxidation of Silicon and During Heat-Treatment of Oxidized Silicon in NH₃ Gas," *J. Electrochem. Soc.*, **123**, 1729 (1976).

[40] T. A. Shankoff, T. T. Sheng, S. E. Haszko, R. B. Marcus, and T. E. Smith, "Bird's Beak Configuration and Elimination of Gate Oxide Thinning Produced During Selective Oxidation," *J. Electrochem. Soc.*, **127**, 216 (1980).

[41] S. Nakajima, K. Kiuchi, K. Minegishi, T. Araki, K. Ikuta, and M. Oda, "1 μm 256K RAM Process Technology Using Molybdenum-Polysilicon Gate," IEEE Int. Electron Device Meet., Wash., D.C., 1981, p. 663.

[42] A. N. Lin. R. W. Dutton, and D. A. Antoniadis, "The Lateral Effect of Oxidation on Boron Diffusion in ⟨100⟩ Silicon," *Appl. Phys. Lett.*, **35**, 799 (1979).

[43] E. Bassous, H. N. Yu, and V. Maniscalco, "Topology of Silicon Structures with Recessed SiO₂," *J. Electrochem. Soc.*, **123**, 1729 (1976).

[44] K. Kurosawa, T. Shibata, and H. Iizuka, "A New Bird's-Beak Free Field Isolation Technology for VLSI Devices," IEEE Int. Electron Device Meet., Wash., D.C., 1981, p. 384.

[45] T. Iizuka, K. Y. Chiu, and J. L. Moll, "Double Threshold MOSFETs in Bird's-Beak Free Structures," IEEE Int. Electron Device Meet., Wash., D.C., 1981, p. 380.

[46] R. H. Dennard, F. H. Gaensslen, H. Yu, V. L. Rideout, E. Bassous, and A. R. LeBlanc, "Design of Ion Implanted MOSFET's with Very Small Physical Dimensions," *IEEE J. Solid State Circuits*, **SC-9**, 256 (1974).

[47] V. L. Rideout, F. H. Gaensslen, and A. LeBlanc, "Device Design Consideration for Ion-Implanted n-Channel MOSFETs," *IBM J. Res. Dev.*, p. 50 (Jan. 1975).

[48] I. M. Bateman, G. A. Armstrong, and J. A. Magowan, "Drain Voltage Limitations of MOS Transistors," *Solid State Electron.*, **17**, 539 (1974).

[49] H. Nihira, M. Konaka, H. Iwai, and Y. Nishi, "Anomalous Drain Current in n-MOSFET's and its Suppression by Deep Ion Implantation," IEEE Int. Electron Device Meet., Wash., D.C., 1978, p. 487.

[50] P. P. Wang, "Double Boron Implant Short-Channel MOSFET," *IEEE Trans. Electron Devices*, **ED-24**, 196 (1977).

[51] T. Shibata, K. Hieda, M. Sato, M. Konaka, R. L. M. Dang, and H. Iizuka, "An Optimally Designed Process for Submicron MOSFETs," IEEE Int. Electron Device Meet. Wash., D.C., 1981, p. 647.

[52] A. G. Sabnis and J. T. Clemens, "Characterization of the Electron Mobility in the Inverted ⟨100⟩ Si Surface," IEEE Int. Electron Device Meet., Wash., D.C., 1979, p. 18.

[53] S. C. Sun and J. D. Plummer, "Electron Mobility in Inversion and Accumulation Layers on Thermally Oxidized Silicon Surfaces," *IEEE Trans. Electron Devices*, **ED-27**, 1497 (1980).

[54] E. Sun, B. Hoefflinger, J. Moll, C. Sodini, and G. Zimmer, "The Junction MOS (JMOS) Transistor—A High-Speed Transistor for VLSI," IEEE Int. Electron Device Meet., Wash., D.C., 1980, p. 791.

[55] T. Ohzone, H. Shimura, K. Tsuji, and T. Hirao, "Silicon-Gate n-Well CMOS Process by Full Ion-Implantation Technology," *IEEE Trans. Electron Devices*, **ED-27**, 1789 (1980).

[56] H. Oka, K. Nishiuchi, T. Nakamura, and H. Ishikawa, "Two Dimensional Numerical Analysis of Normally Off Buried Channel MOSFETs," IEEE Int. Electron Device Meet. Wash., D.C., 1979, p. 31; *IEEE Trans. Electron Devices*, **ED-27**, 1514 (1980).

[57] T. Tonaka, H. Ishiuchi, Y. Takeuchi, M. Ishikawa, T. Mochiuzuki, and O. Ozawa, "Characterization of MOSi$_2$-Gate Buried Channel MOSFET for a 256K DRAM," IEEE Int. Electron Device Meet., Wash., D.C., 1981, p. 659.

[58] S. P. Murarka, D. B. Fraser, A. K. Sinha, and H. J. Levinstein, "Refractory Silicides of Titanium and Tantalum for Low-Resistivity Gates and Interconnects," *IEEE Trans. Electron Devices*, **ED-27**, 1409 (1980).

[59] H. J. Geipel, Jr., N. Hsieh, M. H. Ishaq, C. W. Koburger, and F. R. White, "Composite Silicide Gate Electrodes — Interconnections for VLSI Device Technologies," *IEEE Trans. Electron Devices*, **ED-27**, 1417 (1980).

[60] T. T. Sheng and R. B. Marcus, "Delineation of Shallow Junctions in Silicon by Transmission Electron Microscopy," *J. Electrochem. Soc.*, **128**, 881 (1981).

[61] P. A. Gargini and I. Bienglass, "WOS: Low Resistance Self-Aligned Source, Drain and Gate Transistors," IEEE Int. Electron Device Meet., Wash., D.C., 1981, p. 54.

[62] R. W. Hunt, "Memory Design and Technology," in M. J. Howes and D. V. Morgan, Eds., *Large Scale Integration*," Wiley, New York, 1981.

[63] T. Ohzone, T. Hirao, K. Tsuji, S. Horiuchi, and S. Takayanagi, "A 2K × 8 Static RAM," IEEE Int. Electron Device Meet., Wash., D.C., 1978, p. 360.

[64] M. Taguchi, T. Ito, T. Fukano, T. Nakamura, and H. Ishikawa, "Thermal Nitride Capacitors for High Density RAMs," IEEE Int. Electron Device Meet., Wash., D.C., 1981, p. 400.

[65] K. Ohta, "VLSI Dynamic Memory Cell Using Stacked Ta$_2$O$_5$ Capacitor," in J. Nishizawa, Ed., *Semiconductor Technology,* Ohmsha, 1982.

[66] A. F. Tasch, Jr., P. K. Chatterjee, H. S Fu, and T. C. Holloway, "The Hi-C RAM Cell Concept," *IEEE Trans. Electron Devices*, **ED-25**, 33 (1978).

[67] R. P. Cenker, D. G. Clemons, W. R. Huber, J. B. Petrizzi, F. J. Procyk, and G. M. Trout," A Fault-Tolerant 64K Dynamic Random-Access Memory," *IEEE Trans. Electron Devices*, **ED-26**, 853 (1979).

[68] D. Frohman-Bentchkowsky, "Non-Volatile Semiconductor Memories," IEEE Int. Electron Device Meet., Wash., D.C., 1981, p. 14.

[69] F. M. Wanlass and C. T. Sah, "Nanowatt Logic Using Field-Effect Metal-Oxide Semiconductor Triodes", IEEE Solid State Circuits Conf., Philadelphia, Pa., 1963, p. 32.
[70] R. H. Krambeck, C. M. Lee, and H. F. S. Law, "High-Speed Compact Circuits with CMOS," *IEEE J. Solid State Circuits*, **SC-17**, 614 (1982).
[71] W. N. Grant, private communication.
[72] L. C. Parrillo, R. S. Payne, R. E. Davis, G. W. Reutlinger, and R. L. Field, "Twin-Tub CMOS—A Technology for VLSI Circuits," IEEE Int. Electron Device Meet., Wash., D.C., 1980, p. 752.
[73] Y. Sakai, T. Hayashida, N. Hashimoto, O. Mimato, T. Masuhara, K. Nagasawa, T. Yasui, and N. Tanimura, "Advanced Hi-CMOS Device Technology," IEEE Int. Electron Device Meet., Wash., D.C., 1981, p. 534.
[74] D. B. Scott, Y. C. See, C. K. Lau, and R. D. Davies, "Considerations for Scaled CMOS Source/Drains," IEEE Int. Electron Device Meet., Wash., D.C., 1981, p. 539.
[75] R. S. Payne, W. N. Grant, and W. J. Bertram, "The Elimination of Latchup in Bulk CMOS," IEEE Int. Electron Device Meet., Wash., D.C., 1980, p. 248.
[76] R. D. Rung, C. J. Dell'Oca, and L. G. Walker, "A Retrograde P-Well for Higher Density CMOS," *IEEE Trans. Electron Devices*, **ED-28**, 1115 (1981).
[77] S. R. Combs, "Scaleable Retrograde P-Well CMOS Technology," IEEE Int. Electron Device Meet., Wash., D.C., 1981, p. 346.
[78] Y. Mizutani, S. Taguchi, M. Nakahara, and H. Tango, "Hot Carrier Instability in Submicron $MoSi_2$ Gate MOS/SOS Devices," IEEE Int. Electron Device Meet., Wash., D.C., 1981, p. 551.
[79] A. Ochoa, W. Dawes, and D. Estreich, "Latchup Control in CMOS Integrated Circuits," *IEEE Trans. Nucl. Sci.*, **NS-26** (6), 5065 (1979).
[80] R. J. Kopp and D. J. Stevens, "Overlay Considerations for the Selection of Integrated Circuit Pattern-Level Sequence," *Solid State Technol.*, p. 79 (July 1980).
[81] P. K. Chatterjee, W. R. Hunter, T. C. Holloway, and Y. T. Lin, "The Impact of Scaling Laws on the Choice of n-channel or p-channel for MOS VLSI," *IEEE Electron Device Lett.*, **EDL-1**, 220 (1980).
[82] J. R. Brews, W. Fichtner, E. H. Nicollian, and S. M. Sze, "Generalized Guide for MOSFET Miniaturization," *IEEE Electron Devices Lett.*, **EDL-1**, 2 (1980).
[83] W. Fichtner, R. M. Levin, K. K. Ng, and G. W. Taylor, "Experimental and Theoretical Results on Fine-Line P-channel MOSFETs," IEEE Int. Electron Device Meet., Wash., D.C., 1981, p. 546.
[84] R. A. Larsen, "A Silicon and Aluminum Dyamic Memory Technology," *IBM J. Res. Dev.*, p. 268 (May 1980).
[85] J. M. Mikkelson, L. A. Hall, A. K. Malhotora, S. Dana Seccombe, and M. S. Wilson, "An NMOS VLSI Process for Fabrication of a 32b CPU Chip," IEEE Int. Solid State Circuits Conf., New York, N.Y., 1981, p. 106.
[86] E. Berndlmaier, "Multi-Level Metal Leverage," IEEE Int. Conf. on Circuits and Computers, 1980, p. 1112.
[87] G. E. Helmke, "Anatomy of a Pure Water System," *Semicond. Int.*, p. 119 (Aug. 1981).
[88] E. C. Douglas, "Advanced Process Technology for VLSI Circuits," *Solid State Technol.*, p. 65, (May, 1981).
[89] G. P. Carver, L. W. Linholm, and T. J. Russell, "Use of Microelectronic Test Structures to Characterize IC Materials, Processes and Processing Equipment," *Solid State Technol.*, p. 133 (Aug. 1980).
[90] E. I. Muehldorf, "High Speed Integrated Circuit Characterization and Test Strategy," *Solid State Technol.*, p. 93 (Sept. 1980).
[91] J. D. Hutcheson, "Semiconductor Testing Requirements in the 1980s," *Solid State Technol.*, p. 133 (Aug. 1980).
[92] M. Doken, T. Unagami, K. Sakuma, and K. Kajiyama, "Novel Bipolar Process Utilizing MeV Energy Ion Implantation," IEEE Int. Electron Device Meet., Wash., D.C., 1981, p. 586.
[93] E. W. Maby, M. W. Geis, Y. I. Le Coz, D. J. Silversmith, E. W. Mountain, and D. A. Antoniadis, "Stress Enhanced Mobility in MOSFETs Fabricated in Zone-Melting-Recrystallized Poly-Silicon Films," *IEEE Electron Device Lett.*, **EDL-2**, 249 (1981).
[94] J. F. Gibbons and K. F. Lee, "One-Gate-Wide CMOS Inverter on Laser-Recrystallized Polysilicon," *IEEE Electron Device Lett.*, **EDL-1**, 117 (1980)
[95] H. Shichijo, "A Re-examination of Practical Scaling Limits of N Channel and P Channel MOS Devices for VLSI," IEEE Int. Electron Device Meet., Wash., D.C., 1981, p. 219.

PROBLEMS

1 The n-p-n bipolar transistor of Fig. 2 has the following lateral and vertical dimensions (in μm): base area $= 6 \times 4$, emitter area $= 2 \times 4$, collector contact area $= 2 \times 4$, emitter-base junction depth $= 0.4$, collector-base junction depth $= 0.7$, n-epitaxy thickness $= 2.0$ (including out-diffusion of the buried-layer).

(a) Calculate the doping level in the n epitaxy needed to prevent base push-out for a collector current (I_C) of 2 mA.

(b) Calculate the contribution to collector resistance of the n epitaxy in the collector contact region, and calculate the voltage drop there at $I_C = 2$ mA.

(c) Comment on the usefulness of a deep-collector contact diffusion.

2 Consider an n-p-n bipolar transistor with a 0.2-μm-wide uniformly doped active base (excluding emitter- and collector-base space-charge layers). Under maximum applied collector to emitter voltage, 5 V will be applied across the active base (excluding built-in potentials and the voltage dropped in the collector).

(a) Calculate the minimum doping level in the base to prevent collector-emitter punchthrough.

(b) Estimate the base charging time.

(c) Calculate the base Gummel number.

(d) Assuming that the emitter is degenerately doped and the collector doping is $1 \times 10^{16}/cm^3$, calculate the total base charge (including emitter-base and collector-base space-charge layers) required to produce the desired Gummel number.

3 Plot the change in threshold voltage (ΔV_T) vs. back-gate bias (V_{BS}) for MOSFETs with a 350 Å gate oxide and uniformly doped substrates of

(a) $2 \times 10^{16}/cm^3$ and

(b) $2 \times 10^{15}/cm^3$.

(c) Comment on the advantages of using a channel implant in a lightly doped substrate to provide the desired threshold voltage.

(d) Comment on how back-gate bias effects can influence the NAND gate of Fig. 13.

4 The electron inversion layer mobility μ_n can be estimated[95] by the empirically derived expression:

$$\mu_n = \frac{850 \; cm^2/V\text{-s}}{\left[1 + \dfrac{\mathcal{E}_x}{4.2 \times 10^5 \; V/cm}\right]\left[1 + \left(\dfrac{\mathcal{E}_y}{8.7 \times 10^3 \; V/cm}\right)^{2.9}\right]^{1/2.9}}$$

where $\mathcal{E}_x = (1/\epsilon_s)(Q_B + Q_n/2)$ is the transverse electric field and $\mathcal{E}_y \approx V_D/L$ is the longitudinal electric field. For an n-MOSFET with n$^+$ polysilicon gate, $V_D = 0.1$ V, and $L = 1 \; \mu$m.

(a) Plot μ_n vs. $\mathcal{E}_x$ for $\mathcal{E}_x = 0$ to 6×10^5 V/cm.

(b) For a device with uniform substrate doping of $N_A = 2 \times 10^{16}/cm^3$ and gate oxide thickness $d = 350$ Å, calculate μ_n for an applied gate voltage V_G near the threshold voltage V_T.

(c) Calculate μ_n for the device of (b) at $V_G = 5V$.

(d) Repeat (b) for a device with $N_A = 6 \times 10^{16}/cm^3$ and $d = 200$ Å.

(e) Repeat (c) for a device with $N_A = 6 \times 10^{16}/cm^3$ and $d = 200$ Å.

(f) Compare the ratio of channel conductances for the two devices of (c) and (e) assuming that they have identical widths. Comment on the trade-offs between scaling oxide thickness and surface doping with respect to channel conductance.

5 The two-input NAND gate of Fig. 13 has a capacitive load of 0.2 pF, $V_{DD} = 5$ V, and $V_{SS} = 0$ V.

(a) Calculate the currents which must be supplied by the DMD to charge the load to 2.5 V in 0.5, 1.0, 10, and 100 ns as inputs A and B go from a high to a low state.

(b) Assuming that an IC chip has 10,000 gates like that in Fig. 13, where inputs A and B are simultaneously high, calculate the chip power dissipation for the DMD currents in (a).

(c) Assuming that the packaged chip temperature rises 20°C/W above the ambient, calculate the rise in chip temperature for the power dissipation results in (b).

(d) Discuss the advantages of the CMOS NAND gate of Fig. 25 over the NMOS gate of Fig. 13.

6 A particular dynamic RAM must operate at 80°C with a minimum refresh time of 4 ms. The storage capacitor in each cell is 10 μm square, has a capacitance of 50 fF (50×10^{-15} F), and is fully charged at 5 V.

(a) Calculate the number of electrons that are stored in each cell.

(b) Estimate the worst-case leakage current that the dynamic capacitor node can tolerate (e.g., 50% loss in stored voltage).

(c) Assuming that the leakage is uniformly generated in the storage node's depletion region, estimate the maximum leakage current density that is tolerable at 25°C.

(d) Comment on the leakage requirements of the access transistor.

7 (a) Describe the logic function of the Domino CMOS circuit of Fig. 26.

(b) The clock signal C goes from low to high at time $t = 0$, and down to low at $t = t_f$. Input D goes from high to low at $t = -t_1$ and low to high at $t = +t_1$. Input G goes from high to low at $t = -(t_1 + \Delta)$ and low to high at $t = t_1 + \Delta$. Sketch the clock, D, G, and output signals in time from $t = -(t_1 + \Delta)$ to $t > t_f$ (all other inputs stay low).

(c) The output of the gate of Fig. 26 feeds input M of gate 2 which is described by $L + M + N$. Gate 2 feeds input Q of gate 3, which is described by $P + Q$. Sketch the outputs of gates 2 and 3 in response to the signals of part (b).

8 The p-channel MOSFET of Fig. 33(b) has an n^+ polysilicon gate and a 650 Å thick gate oxide.

(a) Calculate the threshold voltage of this device without the threshold-adjustment implant.

(b) Figure 32 shows $V_{Tp} \simeq -1.54$ V (no threshold-adjustment) and $V_{Tp} \simeq -0.82$ V (threshold adjustment $= 4 \times 10^{11}$ atoms/cm²). Calculate the charge density at the Si-SiO₂ interface that would produce this threshold voltage shift, and compare it with the implanted dose.

9 (a) Assuming $\Delta = \pm 1.0$ μm and $\delta l = \pm 0.4$ μm for all levels, construct a matrix of worst-case separations between edges of features on all levels in Table 2 (round up to the nearest 0.05 μm).

(b) Repeat part (a) assuming $\Delta = \pm 0.5$ μm and $\delta l = \pm 0.2$ μm.

(c) Assuming minimum dimensions of 2.0 μm for window and polysilicon, lay out separate minimum-size MOSFETs, including isolation, polysilicon, and window, using the assumptions of part (a) and then the assumptions of part (b).

(d) Compare the total areas of each of the two MOSFET layouts in part (c) and comment on the importance of nesting tolerances in IC layout.

TWELVE

DIAGNOSTIC TECHNIQUES

R. B. MARCUS

12.1 INTRODUCTION

This chapter summarizes instrumental methods found to be useful for solving problems that arise in VLSI technology development efforts, and explains their application to the problems. Some of the instrumental methods described in this chapter are most applicable to the analysis of circuit structures; others are more applicable to problems generated during experiments on the preparation of new materials for VLSI processing programs. Of the large class of diagnostic techniques available, only those are discussed which are presently useful or which show high potential for becoming useful in the immediate future.

Four areas of application of instrumental methods to VLSI problems can be described: the determination of morphology, chemical analysis, the determination of crystallographic structure and mechanical properties, and electrical mapping of sites of device leakage and breakdown. The application of instrumental methods to these problems is summarized in Table 1. Crosses indicate cases where the instrumental methods are primary sources of information for the indicated areas of application; the cross marks in parenthesis ($\times$) are cases where special accessory equipment is needed for the indicated application. The four areas of application constitute four main sections of this chapter, and instrumental methods constitute subsections.

In a number of the analytical procedures described in Table 1 the sample is bombarded with a beam of x-rays or electrons, and analysis requires a measure of the resulting radiation. The procedures based on these interactions and typical energy ranges of incident and secondary radiations are given in Table 2.

Table 1 Application of instrumental methods to VLSI problems

Instrumental method	Acronym(s)	Morphology determination	Chemical analysis	Crystallographic structure and mechanical properties	Electrical mapping
Auger electron spectroscopy	AES		×		
Electron beam induced current microscopy	EBIC				×
Laser reflectance	LR			×	
Neutron activation analysis	NAA		×		
Normarski interference contrast optical microscopy	. . .	×			
Rutherford backscattering spectroscopy	RBS		×	(×)	
Scanning electron microscopy	SEM	×	(×)		(×)
Secondary ion mass spectroscopy	SIMS		×		
Transmission electron diffraction	TED			×	
Transmission electron microscopy	TEM	×	(×)	×	(×)
Voltage contrast microscopy	VC				×
X-ray diffraction	XRD			×	
X-ray emission spectroscopy	XES		×		
X-ray fluorescence	XRF		×		
X-ray photoelectron spectroscopy	XPS, ESCA		×		

12.2 MORPHOLOGY DETERMINATION

One of the first steps in most diagnostic efforts is examination of the shapes of relevant features: edge acuity of patterned lines, proximity between features, misalignment, and so on. These features are examined by optical microscopy, scanning electron microscopy, and transmission electron microscopy. The maximum useful magnification of these three methods is approximately $1,000\times$, $50,000\times$, and $500,000\times$, respectively. Since the magnification ranges overlap, few questions on morphology of device features can escape scrutiny.

12.2.1 Nomarski Interference Contrast Optical Microscopy

Nomarski interference contrast microscopy[1, 2] is the most generally useful form of optical microscopy for solving VLSI processing problems. With this method, surface

Table 2 Types of radiation resulting from electron or x-ray bombardment of a sample surface and instrumental methods of analysis based on these interactions.

Incident beam			Secondary radiation				
			Electron			X-ray	
Radiation	Energy E_0 (keV)	Type	Energy (eV)	Analytical Procedure		Energy	Analytical Procedure
Electron	2–10	Auger	20–2000	AES			
	2–40	Secondary	<10	SEM(VC)			
	2–40	Back-scattered	$<E_0$	SEM(BS)			
	20–200					$<E_0$	XES
X-Ray	<2	Primary ionized	20–2000	XPS			
	<50					$<E_0$	XRF

features of different elevations appear as different colors or as different shades of gray. This contrast is achieved by splitting the illuminating beam into two beams displaced by a short distance on the sample surface, followed by reflection and reconstitution of the reflected beams. The optical path length changes because of the presence of a step or a change in the index of refraction (caused by a phase boundary). These changes in optical path length produce a contrast change in the reconstituted beam, which appears in the microscope image.

The resolution of an optical microscope is governed by the wavelength of illuminating light λ and the numerical aperture (NA) of the objective lens:

$$r = \frac{0.61\lambda}{NA} \tag{1}$$

The resolution limit of the optical microscope is approximately 0.25 μm. If visual resolution with the unaided eye is assumed to be 0.1 mm, then loss of image sharpness can begin to be detected above 400× magnification, and the upper limit to useful magnification (when images become too blurred to be "acceptable") is between 1000× and 2000×. For many problems requiring an analysis of the morphology of VLSI features, a lateral resolution of 0.25 μm or even 1.0 μm is quite acceptable. A level of vertical (depth) resolution is also desirable, however, which permits the clear identification of features that are about 200 Å thick (for some MOSFET gate oxides) or thicker. This depth resolution is obtained by using Nomarski interference contrast optics.

Figure 1 shows the essential features of a Nomarski interference contrast microscope operating in the reflectance mode. Light passes through a polarizer and is reflected downward toward birefringent crystals that together make up a Wollaston

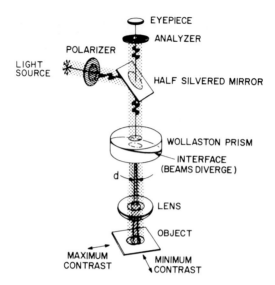

Fig. 1 Features of a Nomarski interference contrast microscope operating in the reflectance mode. The divergence of the two beams that illuminate the sample is given by d.

prism, that is, a prism in which the light is split into two mutually perpendicular polarized components that move at different velocities with an angular divergence d. After emerging from the prism and reflecting off the sample, the two beams recombine by passing once again through the Wollaston prism in the opposite direction. The reconstituted beam then passes through an analyzer, where its intensity changes are observed.

The microscope image contains contrast effects that depend on differences in optical path length caused by changes in the geometrical profile of the surface, and differences in phase changes resulting from variations in index of refraction, as may occur across a phase boundary. Figure 2 illustrates intensity variations seen in the microscope when monochromatic light illuminates a substrate. Figure 2a shows a cross section of the sample. The sample consists of two phases with different refractive indices; B represents their phase boundary. Figure 2b shows the wave fronts of the reflected beams after they pass through the Wollaston prism. The optical path differences between the two reflected light beams give rise to the intensity variation shown in Fig. 2c. Both the polarizer and analyzer settings and the position of the prism can be varied. The resulting shapes of the emerging wave fronts and intensity profiles reflect these settings as well as the optical properties of the phases. Figure 2b and c, therefore, describes the situation for one specific setting of the three adjustments for a given sample orientation.

Interference contrast is maximized in a direction parallel to the maximum displacement of the two beams and is essentially zero in the orthogonal direction. Figure 3 shows two micrographs of a sample consisting of oxide features. The micrographs are taken under identical polarizer, analyzer, and Wollaston prism settings, but the sample in Fig. 3b has been rotated 90° compared with Fig. 3a. The oxide pattern contains a slight offset, which gives rise to an approximately 100-Å step along C − − C

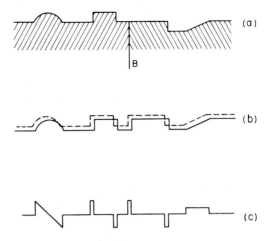

Fig. 2 (a) Representation of a cross section of a sample at a surface. (b) The wave fronts of the reflected beams after emerging from the prism, and (c) an intensity distribution in the image plane in a Nomarski interference contrast microscope. The sample is assumed to consist of two phases (with different refractive indices) that meet at the boundary B.

(Fig. 3a). The interference contrast image in Fig. 3a clearly shows the offset, since the polarizer, analyzer, and prism are adjusted for maximum contrast for the sample position used for that photograph; the contrast disappears when the sample is rotated 90° and the offset becomes extremely hard to see (Fig. 3b). The contrast features in Fig. 3b are essentially the same as those that would be seen in a microscope with ordinary optics. Figure 3 excellently illustrates the use of Nomarski interference microscopy in revealing subtle changes in morphology.

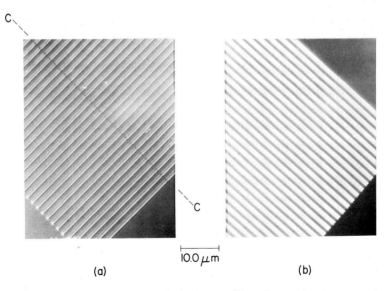

(a) (b)

Fig. 3 (a) Nomarski interference contrast reflection images of 1-μm features showing contrast due to slight pattern error at C−−C. (b) This contrast feature is absent when the sample is rotated 90°.

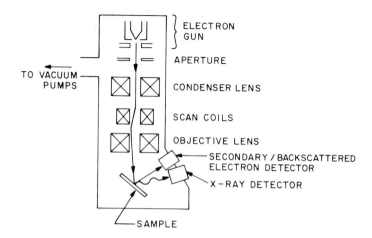

ELECTRON GUN

APERTURE

TO VACUUM PUMPS

CONDENSER LENS

SCAN COILS

OBJECTIVE LENS

SECONDARY / BACKSCATTERED ELECTRON DETECTOR

X–RAY DETECTOR

SAMPLE

Fig. 4 Schematic drawing of a scanning electron microscope. The incident electron beam of energy 2 to 40 keV produces low-energy secondary and higher-energy backscattered electrons as well as x-radiation (see Table 1), all of which can be analyzed to provide useful information.

12.2.2 Scanning Electron Microscopy (SEM)

Scanning electron microscopy is a standard analytical method in VLSI laboratories, mainly because it provides increased spatial resolution and depth of field compared with optical microscopy, and because chemical information can be obtained from the x-ray spectra generated by electron bombardment (see Section 12.3.7). Resolution better than about 100 Å can be achieved under optimum conditions, and typical depths of field are 2 to 4 μm at 10,000$\times$ magnification and 0.2 to 0.4 mm at 100$\times$ magnification.

Figure 4 gives a schematic drawing of a scanning electron microscope. An electron gun, usually consisting of a tungsten or LaB$_6$ filament, generates electrons, which are accelerated to an energy of 2 to 40 keV. A combination of magnetic lenses and scan coils produces a small-diameter beam which is rastered across the sample surface. The electron bombardment produces three useful types of radiation: x-rays, secondary electrons, and backscattered electrons (Table 2).

Figure 5 shows the energy spectrum of electrons emitted from a sample that has been bombarded with an electron beam. A large fraction of the spectrum consists of electrons with energy less than 50 eV peaking at less than 5 eV. These electrons are referred to as secondary electrons. The other electrons peak at an energy close to E_0 and are referred to as backscattered electrons.

The secondary or backscattered electron current is used to modulate the intensity of an electron beam in a cathode ray tube (CRT). Since the CRT electron beam moves in synchronism with the rastering incident beam of the scanning electron microscope, the CRT beam produces an image of the sample surface whose contrast is determined by variations in the secondary or backscattered electron flux. The x-ray signal (Table 2) is useful for chemical analysis, and a discussion of this diagnostic procedure appears in Section 12.3.7.

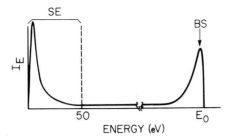

Fig. 5 Energy distribution of electrons emitted from sample bombarded by electron beam of energy E_0. A peak of low-energy secondary electrons (SE) occurs near 5 eV, and the backscattered electron (BS) peak occurs near E_0. Auger electron peaks (not shown) occur between these two peaks.

The incident electron beam undergoes multiple collisions as it penetrates a sample; incident electrons that are not backscattered finally come to rest after traversing a range R that can be calculated or measured. Figure 6 gives the electron ranges calculated from one set of expressions[3] for silicon, aluminum, and gold. The electron range increases with decreasing atomic number and with increasing incident beam energy E_0. The electron trajectory changes with each collision, causing a narrow incident beam to spread as it penetrates into a sample. Figure 7 shows the maximum penetration depth of incident electrons at four different beam energies. The penetration is described by a series of pear-shaped envelopes, which increase in depth and width with increasing energy.

For normal incidence on a bulk sample, the surface area contributing backscattered electrons is a disc with a diameter approximately equivalent to the electron

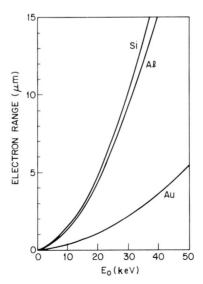

Fig. 6 The electron range in silicon, aluminum, and gold as a function of incident beam energy. *(After Everhart and Hoff, Ref. 3.)*

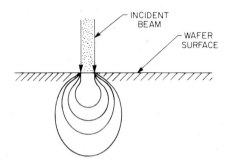

Fig. 7 Illustration of volume spread of incident electrons within sample after sufficient multiple scattering to reduce electron energy to nearly zero. Envelopes of increasing size represent maximum electron penetration at increasing beam energy.

range. The resolution of a backscattered electron image improves as the sample thickness decreases; thus, the resolution of a backscattered electron image from a thin metal film on silicon or SiO_2 is better than the resolution from the same metal in bulk form. Resolution of images formed from secondary electrons is partly determined by the lateral size of material within which most secondaries are generated at a depth less than the escape depth. The escape depth[4] of electrons in metals reaches a minimum of 4 Å at 70 eV, and increases with decreasing energy to a value of 25 Å at 10 eV. The escape depth is more than 50 Å for insulators. The lateral resolution of secondary electron images from a flat surface is therefore given by the diameter of the incident beam plus a lateral increment that is due to the electron mean free path.[5] Note that secondary electrons generated by emerging backscattered electrons within an escape depth of the surface come from a larger area and they therefore contribute to resolution degradation.

The contrast of both backscattered and secondary electron images depends on variations in the flux of electrons arriving at the detector. The yield of backscattered electrons increases with increasing atomic number Z and is 10 times higher for gold than for carbon (Ref. 6, pp. 75 to 77). Because the yield depends on the atomic number, contrast is produced in the backscattered electron image between regions with different atomic numbers. The contrast between adjacent pairs of elements in the periodic table decreases with increasing Z, and for aluminum and silicon the contrast is 6.7% (Ref. 6, p. 159). Thus, backscattered electron detection can be used to distinguish aluminum particles from a silicon background.

The secondary electron yield is a weaker function of Z than is the yield of backscattered electrons; the secondary electron yield only increases by about a factor of 2 from carbon to gold. Yield is a stronger function of the work function of the material,[7] and is significantly higher for oxides and other wide-band gap materials than for silicon.[8] This source of contrast makes the use of secondary electron SEM imaging in VLSI studies a major advantage, since metallization, oxide, and silicon regions are easily distinguishable from each other. A second source of contrast in secondary electron images is due to the dependence of secondary electron yield on surface curvature. The secondary electron flux from a surface of changing slope varies with the secant of the tilt angle. Therefore, surfaces that differ significantly in slope can be clearly distinguished. The secondary electron flux that is detected is also a strong function of the orientation of the emitting surface with respect to the detector; surface regions that face the detector appear significantly brighter than other surface regions.

Spatial resolution depends on the size of the sample surface contributing to secondary or backscattered electrons, and on local changes in phase, composition, and sample orientation which influence the secondary or backscattered electron flux as discussed above. Resolution also depends on the condition of the scanning electron microscope. These factors interact intimately and cannot be separated. For example, the diameter of the electron probe decreases with decreasing beam current and increasing beam energy, and is smaller for a LaB_6 source than for a tungsten filament. A 30-keV electron beam with a current of 10^{-11} A has probe diameters of about 40 and 90 Å for a LaB_6 and tungsten filament, respectively; these values change to 60 and 130 Å, respectively, at 10 keV. The actual resolution achieved in SEM study, however, can be considerably poorer than 40 to 130 Å. The minimum beam current I_{min} needed to produce a detectable contrast C between adjacent regions is given by (Ref. 6, p. 174)

$$I_{min} = \frac{4 \times 10^{-12}}{\epsilon C^2 t_f} \qquad (2)$$

where ϵ is the efficiency of signal collection and t_f is the time needed to scan an SEM frame. "Difficult" samples have low contrast between adjacent regions, say 1 to 5%. For a thermionic tungsten filament the minimum beam diameter needed to provide sufficient current (I_{min}) to detect this brightness difference is 2300 Å for $C = 1\%$ and 460 Å for $C = 5\%$ (Ref. 6, p. 177). Thus, a major limitation on spatial resolution is the need to have sufficient beam current for the experimenter to be able to distinguish regions that have similar brightness.

Three common difficulties which occur during SEM study of VLSI circuits are sample contamination, electron beam induced damage, and surface charging during examination. The major type of sample contamination is hydrocarbon polymerization, which occurs as the electron beam strikes the surface. Although modern microscopes are well pumped and maintain a vacuum (at the specimen chamber) of about 10^{-6} torr, contamination cannot be avoided. A related problem is damage to oxides by electron beams and the effect of this damage on device performance. Electron irradiation produces positive oxide charges and interface traps which can be avoided by maintaining a sufficiently low beam energy to prevent penetration to an active area (such as a gate oxide). Once they have formed, defects can be annealed out[9] at temperatures between 400 and 550°C.

The third problem frequently encountered in SEM studies of insulating surfaces is surface charging. This problem sometimes occurs when beam energies are above the second secondary electron yield crossover point (the incident beam energy above which the secondary electron yield is less than 1). The surface becomes negatively charged, disturbing the trajectory of the incident electron beam and degrading the image. A method for avoiding this problem uses a low-energy incident beam. A field emission source is the only electron source that seems capable of producing high-resolution images at low-beam energies,[10] and some commercially available SEM units have these sources. Applying grounded metallic coatings approximately 100 Å thick to sample surfaces is another method for avoiding charge buildup during study. The metallic coatings provide a conducting path to ground. Since the escape depth of secondary electrons is much smaller in metals (~5 Å) than in insulators, a thin metal surface coating also greatly improves spatial resolution in secondary electron imaging. Unfortunately, such coatings make the samples unsuitable for further processing.

12.2.3 Transmission Electron Microscopy (TEM)

Transmission electron microscopy (TEM) is a useful tool for solving problems in VLSI technology that require high spatial resolution. TEM offers a resolution of about 2 Å. In a transmission electron microscope an electron beam passes through the thin-film sample, and forms an image that displays morphological and crystallo-graphic features of the film components. Commercial TEM instruments use electron beams with energy between 60 and 350 keV. Higher beam energy permits greater sample penetration; the maximum thickness of silicon which permits TEM image formation is 1.5 μm for a 200-keV beam, but only 0.5 μm for an 80-keV beam.[11]

VLSI specialists are usually concerned with the morphology of features whose phase boundaries extend to both surfaces of the TEM sample. These boundaries limit further the maximum sample thickness. For example, a well-oriented thin-film TEM sample which contains a 1-μm-wide polysilicon runner on oxide produces an image of two lines, one corresponding to each interface (Fig. 8a). In Fig. 8 it is assumed

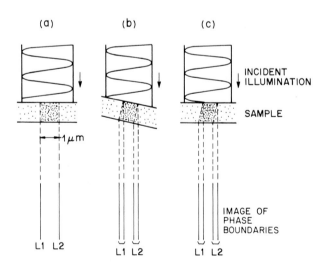

Fig. 8 Illustration of problems of angular misorientation of a sample in TEM study of VLSI circuits. (a) TEM study of a well-oriented vertical cross section of a two-phase region (such as 1-μm-wide polysilicon runner over oxide) produces a clean image of two lines, corresponding to the two interfaces. (b) A misoriented sample or (c) a misoriented feature in an oriented sample produces line doubling.

that the sample is a vertical cross section through a chip (e.g., the plane of the page is perpendicular to the original surface of the chip). If the sample is tilted slightly during study, the image of the interfaces L1 and L2 doubles (Fig. 8b). Such doubling becomes harder to avoid as the sample becomes thicker, and places a severe practical limit on permissible sample thickness: Although a $0.5°$ misorientation of the sample causes a doublet spacing of 10 Å from a 1000-Å-thick film, the same misorientation produces a doublet spacing of 50 Å from a 5000-Å-thick film.[12] Figure 8c illustrates another related problem caused either by misorientation of the sample initially cut from the wafer (or chip), or by texture at an interface. Texture often results at the polysilicon-oxide interface (after the polysilicon has thermally oxidized) and at the edges of patterned features.

Contrast in a TEM image can be described for two situations. One is TEM study of crystalline materials (such as silicon, aluminum, polysilicon, and various silicides), and the other is the study of amorphous materials.

In crystalline materials, the incident electron beam is diffracted by the material, and local variations in diffraction intensity produce contrast in an image from the undiffracted beam (bright field image) or from one or more diffracted beams (dark field image). The intensity of the emergent beam is periodic with sample thickness. The sample thickness corresponding to one period in silicon is 602 Å for a (111) reflection and 757 Å for a (220) reflection.[13] Thus, a thicker region of the sample does not necessarily look lighter in the negative; a wedge-shaped crystalline sample produces a TEM image that has alternate light and dark bands called thickness extinction contours. Dark bands are also caused by a bent sample and are then called bend extinction contours. Abrupt changes in thickness, phase structure, or crystallographic orientation cause corresponding abrupt changes in contrast, and these crystallographic features can be easily imaged at high resolution.

In the case of nearly amorphous material, contrast is determined by local changes in electron scattering which result from differences in sample thickness or from differences in chemical or phase composition. A sample region whose thickness varies continuously produces a corresponding continuous variation in image intensity, unlike the case of diffraction contrast. TEM images obtained from oxides, nitrides, and other amorphous materials are therefore somewhat easier to interpret intuitively than images obtained from crystalline samples.

Sample preparation Difficulties in sample preparation have been the main factor limiting the application of TEM methods to VLSI programs. One difficulty is preparing a sample that is sufficiently thin for TEM study. Another difficulty is that, after a thin-film sample is prepared, the morphological feature of interest to the diagnostician must be present in the thinned region. The solutions of these problems are briefly discussed in the following paragraphs.

Thin-film sections most useful to VLSI diagnosticians are those made orthogonal to the wafer surface (vertical cross section). TEM study of such samples provides information on the relationship among multiple layers as well as information on the shapes of steps created by edges and contacts. Figure 9 describes the main features of vertical cross section sample preparation. Samples can be prepared in less than 2

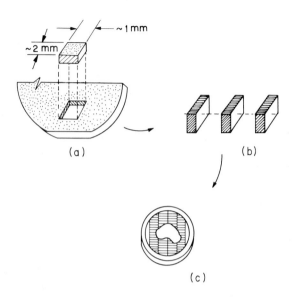

Fig. 9 Preparation of a vertical cross section for TEM study. (a) A small piece (~1 mm × ~2 mm) is cleaved from the wafer. (b) A number of such pieces are epoxy bonded face to back. (c) After lapping and ion milling a hole is produced. A region sufficiently thin to permit TEM study is usually produced within a 50- to 100-μm area around the hole perimeter.

days, particularly when ion milling machines are used with hole detection capability which obviates the need for constant attention. More details on this procedure as well as methods for the preparation of horizontal sections (parallel to the wafer surface) are described in Refs. 12 and 14.

The second problem, assuring that the morphological feature(s) of interest is (are) in the thin part of the TEM preparation, has been solved by incorporating special TEM test chips on each processed wafer; the test chip includes every morphological feature relevant to the particular VLSI technology (see Fig. 10). Each feature appears within 1 to 3 μm in one dimension, and extends about 2 mm in the orthogonal direction; in the figure all the features are contained within a 23-μm distance. This "repeat unit" is replicated a number of times over a distance of 2 mm. Using the method of sample preparation described in Fig. 9, at least one complete set of morphological features in one test-chip repeat unit can be produced within a region of sample that is sufficiently thin for TEM study.

Figure 11 shows an application of TEM to a VLSI problem. The photograph is of a vertical section through a charge-coupled device test structure and shows the presence of very thin oxide at the tips of "horns" at the polysilicon corners. These thin oxide regions are sites of potential failure; after a second layer of metallization is deposited over the oxide, the oxide breakdown voltage is significantly lowered. This failure mode would have been extremely difficult to detect with any other diagnostic technique.

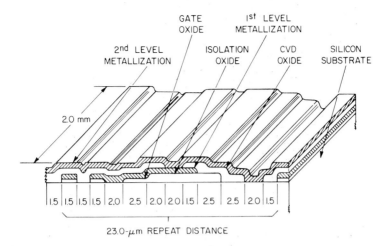

Fig. 10 Schematic drawing of a TEM test pattern. All morphological features essential to the technology appear within a short (1.5- to 2.5-μm) distance, comprising a "repeat" unit of 23.0 μm for the case shown. This unit is repeated 87 times to cover a distance of about 2 mm; each feature is extended 2 mm in the orthogonal direction.

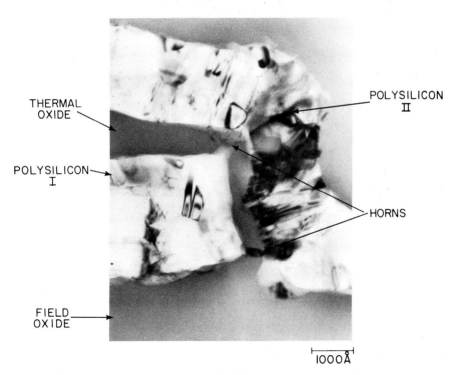

Fig. 11 TEM photograph of cross section through a CCD test structure showing "horns" which constitute a failure mode. This defect would have been extremely difficult to detect by any other technique.

12.3 CHEMICAL ANALYSIS

Many different methods are required for the chemical analysis of materials used in VLSI technology. Spatial resolution requirements vary from atomic dimensions, as in the depth profiling of intermetallics or dopants, to essentially macroscopic dimensions, as in the bulk analysis of large area films or substrates. Vertical (depth) and lateral spatial resolution requirements for these studies are often quite different. Sensitivity requirements range from 10^{11} atoms/cm^3 to 10^{21} atoms/cm^3. The chemicals usually sought in these studies are silicon dopants (arsenic, phosphorus, boron), oxygen, carbon, resist residue, various components of metallizations, and metallic impurities. Thus, the chemicals run the gamut from light elements to heavy elements, such as gold, platinum, and tungsten.

12.3.1 Auger Electron Spectroscopy (AES)

Auger electron spectroscopy (AES) analyzes a certain class of electrons, called Auger electrons, that are generated when either an electron beam or a photon beam strikes the surface of the sample; in its usual mode, an incident electron beam is used (Table 2). Auger electrons are ejected from an atom in response to core-level ionization.[15, 16] As shown in Fig. 12, an incident electron or photon of sufficient energy can eject an

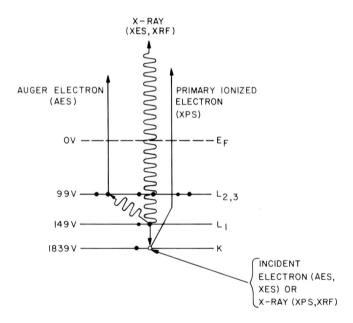

Fig. 12 Partial energy level diagram of the singly ionized silicon atom. An incident electron or x-ray photon is assumed to eject an electron from the K level into the vacuum. An electron transition from the L$_1$ level to the empty position in the K level releases energy, which may be detected directly as x-radiation, or may eject a second electron into the vacuum; this latter electron is called an Auger electron. This diagram illustrates the types of interactions detected and analyzed by four types of spectroscopy: AES, XES, XRF, and XPS. The transitions shown are only one set of a number of possible transitions.

Table 3 Escape depth[†] of Auger electrons from a silicon matrix

Element	Transition energy (eV)	Escape depth (Å)
Phosphorus	120	5
	1859	32
Boron	179	6
Oxygen	507	12
Arsenic	1228	23
Aluminum	1396	26
Silicon	92	4
	1619	29

[†]Escape depths are interpolated from data presented in Table 2 of Ref. 17.

electron in the K shell of a target atom (the material assumed in Fig. 12 is silicon). An electron transition from the L_1 to the empty position in the K level releases energy, which ejects an Auger electron from the $L_{2,3}$ level.

The incident electron beam usually falls between 2 and 10 keV and penetrates only a short distance into the sample (Fig. 6). Most Auger electron energies fall within 20 to 2000 eV, and appear in the spectral range between the low-energy (secondary electron) and high-energy (primary backscattered electron) peaks shown in Fig. 5. The escape depth of Auger electrons is generally less than 50 Å and is considerably less for the lower-energy transitions (see Table 3).[17] Thus, a chemical analysis of surface regions can be made from AES data.

Many diagnostic problems require depth analysis beyond the escape depth of Auger electrons, and the sample must be ion milled in order to continuously create a new surface which essentially moves through the sample during AES analysis. Data are obtained by interrupting the milling process at regular intervals and obtaining Auger spectra, or by continuously recording spectra during ion milling. The Auger peak heights can then be plotted as a function of milling time or milling depth, and a depth profile can be obtained as shown in Fig. 13.

For quantitative analysis, the concentration C_i of element i in the host material (matrix) is calculated from the expression

$$C_i = \frac{\alpha_i I_i}{\sum_j \alpha_j I_j} \tag{3}$$

where I_i is the intensity of the Auger peak of element i and I_j is the Auger peak intensity from an element in the matrix. The proportionality constants α are most easily determined from known standards. Methods for quantitative analysis are discussed in Refs. 4, 15, and 16. Table 4 gives the approximate sensitivity of AES detection of

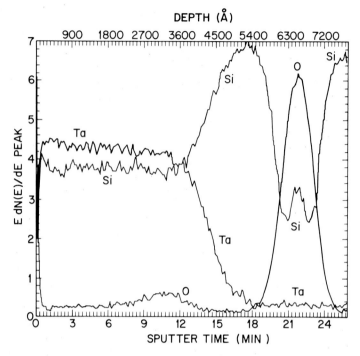

Fig. 13 AES depth profile from a tantalum silicide film on polysilicon on an oxidized silicon wafer showing the changing distribution with depth. *(Courtesy C. C. Chang, Bell Laboratories.)*

some species important to VLSI technology.[18] Sensitivity is given as the detection limit for an element homogeneously distributed within a silicon matrix.

Lateral spatial resolution is governed mostly by the size of the incident beam which, in turn, is related to the minimum current necessary to generate useful data (about 5 nA). In modern commercial instruments the resolution can be higher than 0.1 µm. Depth resolution in nonprofiling AES study is limited by the Auger electron

Table 4 Detectability limits for AES analysis of various species in silicon[18]

Element	C_{min} (atoms/cm^3)
Phosphorus	1×10^{19}
Arsenic	5×10^{18}
Oxygen	5×10^{17}
Carbon	5×10^{17}

escape depth and is essentially one monolayer for lower-energy Auger transitions. The greater escape depth of higher-energy transitions permits the thickness of very thin surface films to be measured. The relative heights of a 1619-eV Si Auger peak from elemental silicon in the substrate and the 1607-eV chemically shifted silicon peak from a surface oxide film can be used to measure the oxide thickness.[19] The method is useful for measuring films with oxide thickness less than about four times the escape depth of the Auger electrons.

12.3.2 Neutron Activation Analysis (NAA)

Neutron activation analysis (NAA) is the most sensitive analytical method for many elements. This method uses nuclear irradiation of a sample to produce radioisotopes that can then be analyzed. NAA is most useful for solving problems involving gettering, or problems involving surface or bulk appearance of trace impurities on whole wafer samples. It is also useful in measuring contamination introduced by processing furnaces. Only light elements (such as boron, oxygen, nitrogen, and carbon) do not produce radioactive isotopes suitable for analysis.

Irradiation is performed with a thermal neutron flux of 10^{13} to $10^{14}/cm^2$-s over a period of 0.5 to 12 h. During irradiation of silicon wafers a number of radioactive species are produced, including ^{31}Si. The half-life of ^{31}Si is 2.6 h. After irradiation the sample is allowed to "cool" for 24 to 48 h to permit the radiation level from the silicon to fall below the levels from the other elements.

The species most commonly monitored during these experiments is γ radiation, with energies between 0.1 and 2.5 MeV. This radiation is detected by a lithium-drifted germanium detector and analyzed by a multichannel analyzer.

Both the energy of γ emission and the measured half-life are used to identify the isotope giving rise to the radiation. To measure the amount of the element present, several factors must be known: the radiation count over a time interval, the detector and emitter efficiencies for a particular peak, the thermal neutron flux, the time lapse since irradiation, the irradiation time, and other factors listed in Table 5. The number of atoms of the particular element in the counted sample can be determined from known formulas,[20] and the corresponding volume concentration can be computed from the original sample shape.

The minimum detectable limits for a number of species are listed[21] in the last column of Table 5. Although the sensitivity varies with sample size, irradiation time, flux, and other factors, the last column indicates the extremely high sensitivity of this analytical method.

By counting before and after removal of layers of controlled thickness, the depth distribution of elements within a sample can be determined. Thus, by repeatedly etching and counting, it can be shown that phosphorus gettering is successful in reducing bulk gold impurity atoms by a factor of 1/50 from an initial concentration of 3×10^{14} atoms/cm^3, and that gettering confines gold atoms to within 2 μm of the back surface.[22]

Table 5 Radioisotope parameters and sensitivity limits

Element	Atomic mass	$\tau_{1/2}$	f	σ(b)	γ Energy (MeV)	C_{min} (atoms/cm^3)
Arsenic	75	26.4 h	100	4.3	0.560	7.1×10^{11}
Copper	63	12.75 h	69.17	4.5	0.511	2.3×10^{12}
Gold	197	2.69 d	100	98.8	1.34 0.411	1.1×10^9
Sodium	23	15.0 h	100	0.53	1.37	6.3×10^{12}
Tantalum	181	115 d	100	21.0	1.121 1.221	1.1×10^{12}
Tungsten	186	23.9 h	28.41	40.0	0.686	4.9×10^{11}

Notes:

$\tau_{1/2}$ = the half-life.

f = the percent abundance of the radioisotope.

σ = the thermal neutron cross section in barns.

C_{min} = the minimum detectable concentration assuming analysis was per-formed on an entire 7.3-cm-diameter silicon wafer 510 μm thick. Detection limit values also assume a 10-h irradiation with a neutron flux of 1×10^{13}/cm^2-s and a 40-h delay before counting.

12.3.3 Rutherford Backscattering Spectroscopy (RBS)

RBS uses an incident 1- to 3-MeV ion beam (usually He$^+$) directed on the surface of a sample. The beam diameter is typically 10 μm to 1 mm. Elastic collisions between the incident ion and a target atom cause the ion to lose energy. The kinematic factor K relates the incident ion energy E_0 to the energy of the ion after backscattering E_0' as:

$$E_0' = KE_0 \qquad (4)$$

The scattered ions are detected by an energy-dispersive silicon barrier detector, and the signal is fed into a multichannel analyzer (see Fig. 14). Since values of K are known for each element,[23] the chemical composition at the surface of a sample can be determined by measuring the energies of the backscattered ions.

Incident ions suffer energy loss as they penetrate a sample and scatter. Ions scattered at a depth ΔZ have to return through the same path before they can escape and be detected, and additional energy is lost on this return trip. The total energy difference between ions scattered at the surface and ions emerging from the sample after scattering at a depth ΔZ is

$$\Delta E = KE_0 - E_1 = [\epsilon]N\Delta Z \qquad (5)$$

where $[\epsilon]$ is the stopping cross-section factor, and N is the atomic density of the sample. A depth profile is obtained by monitoring the number of backscattered ions as a function of backscattered ion energy E_1, as illustrated by the RBS spectrum[24] in Fig.

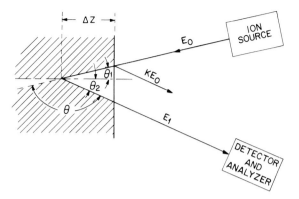

Fig. 14 Features of a Rutherford backscattering spectrometer. The incident beam energy E_0 becomes KE_0 after scattering; the emergent beam from scattering at a depth ΔZ has an energy E_1 because of the loss of energy of the incident beam during penetration to ΔZ, and loss of energy of the scattered beam during the return trip. The scattering angle is θ.

15. Note that the abscissa is also a depth axis (for a particular phase) by the relationship given in Eq. 5.

The sample used to generate the RBS spectrum of Fig. 15 is shown in the upper part of the figure. An aluminum film was used as the top layer of the sample, and the energy position of the leading (high-energy) edge of the Al peak is at $K_{Al}E_0 = 1.1$ MeV, while the energy position of the trailing edge is less by ΔE and appears at 0.950 MeV. The leading edge of the Ti peak in TiN is not given by $K_{Ti}E_0$, since the incident beam energy at that depth has been attenuated while passing through the aluminum layer, and the scattered beam undergoes further attenuation as it passes back through the aluminum. The leading edges of the titanium, silicon, and platinum peaks are all lower than their positions would have been in the absence of surface films; these positions are indicated by the vertical arrows in the figure.

RBS is one of the few chemical analysis methods that can provide quantitative information without the use of standards. The total number of counts detected in an RBS spectrum is a product of the differential scattering cross section of the scattered species $d\sigma/d\Omega$, the number of scattering centers per cm^2 ($N\Delta Z$), the acceptance angle of the detector ($\Delta\Omega$), and the beam current Q:

$$H = \frac{d\sigma}{d\Omega} N\Delta ZQ\Delta\Omega \tag{6}$$

Values for $d\sigma/d\Omega$ have been tabulated[23] for all elements as a function of scattering angle θ for ^{4}He$^+$ and ^{1}H$^+$ incident ions. For the special but frequently occurring case of the analysis of a homogeneous film that contains only one compound of unknown composition A_mB_n, the ratio of the peak heights H_A and H_B can be found from Eq. 6:

$$\frac{H_A}{H_B} = \frac{\dfrac{d\sigma_A}{d\Omega}}{\dfrac{d\sigma_B}{d\Omega}} \frac{m}{n} \frac{[\epsilon]_B}{[\epsilon]_A} \tag{7}$$

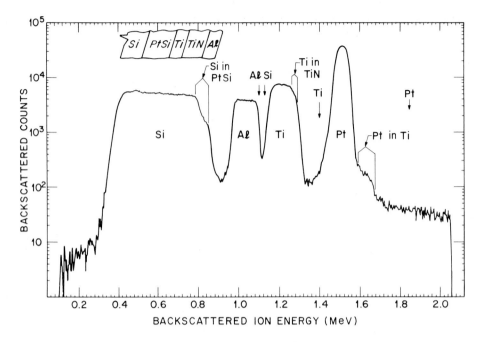

Fig. 15 RBS data from the structure diagramed at the top of the figure. The positions of the leading edges of each peak are shown by arrows for the case of no absorption. Only the leading edge of the aluminum peak coincides with its arrow; the other edges are shifted to lower energies because of energy loss by absorption by surface layer(s). *(Courtesy R. Schutz, Bell Laboratories.)*

Since $d\sigma_i / d\Omega = kZ_i^2$ where Z_i is the atomic number of element i, Eq. 7 can be written as

$$\frac{m}{n} = \frac{H_A}{H_B} \left[\frac{Z_B}{Z_A} \right]^2 \frac{[\epsilon]_A}{[\epsilon]_B} \tag{8}$$

By measuring H_A / H_B and finding the appropriate values for $[\epsilon]$ in a table, we can determine the quantity m/n.

The energy resolution of modern detectors is about 15 keV, which translates to a depth resolution of ~300 Å for silicon, and a depth resolution of about 100 Å for heavier metals such as those present in silicides. Unfortunately, the large incident ion-beam diameter (typically 10 μm to 1 mm) precludes the use of RBS for the analysis of most VLSI features. The sensitivity of RBS is determined by ion straggling, mass separation of peaks, and beam current. Table 6 gives estimates of the sensitivity limits for the detection of elements relevant to VLSI technology. The sensitivity for phosphorus is poor because of the close proximity of the phosphorus and silicon peaks.

Table 6 Detectability limits for RBS analysis of various species in silicon

Species	C_{min} (atoms/cm^3)	Reference
Arsenic	9×10^{18}	Calculated
Oxygen	5×10^{21}	25
Antimony	4×10^{18}	Calculated
Phosphorus	Poor	

12.3.4 Scanning Electron Microscopy (SEM)

The scanning electron microscope provides information on chemical composition by use of x-ray spectrometer attachments; an SEM-like instrument specifically designed for quantitative chemical analysis has been called an electron probe. A description of x-ray emission spectroscopy is found in Section 12.3.7.

12.3.5 Secondary Ion Mass Spectroscopy (SIMS)

In the SIMS method, an ion beam sputters material off the surface of a sample, and the ionic component is mass analyzed and detected (Fig. 16). Sputtered ions are extracted and mass analyzed with a magnetic prism or quadrupole analyzer. In a system that uses a magnetic prism, a two-dimensional image of the distribution of an ionic species across the surface can be obtained by directing the secondary ion beam onto a channel plate. In quadrupole instruments the image is formed by recording the changing secondary ion-beam current as the primary beam is rastered across the sample surface. The intensity of the detected signal is related to the mass concentration.

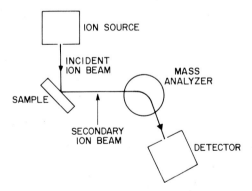

Fig. 16 Schematic diagram of a secondary ion mass spectrometer. An ion source creates a beam which rasters across a sample surface and sputters material off that surface. The ion fraction of sputtered material is mass analyzed, and displayed as a current intensity for a particular mass, or as a two-dimensional image of the distribution of that mass species.

The detected signal can be displayed as a mass spectrum during ion milling, giving a depth profile of the chemical species.

Both positive and negative incident ions are used with a beam energy typically between 5 and 15 keV. Since only the ionic fraction of sputtered material produces a SIMS signal, ion beams are chosen that produce the highest ion yields of the species under study. Positive cesium ion beams are generally useful for producing high negative ion yields of electronegative species from a target,[26] and O_2^+ ion beams are usually used for generating high positive ion yields from electropositive species.[27]

The incident beam is rastered across a small area of the surface to create a crater with a nearly flat bottom. Mass analysis is performed on the ionic fraction of sputtered material only from a central portion of the crater. When very low primary ion currents are used, sputtering rates are lowered to the point where data can be collected from a few monolayers, and surface analysis can be performed. Depth profiles are obtained by using higher primary ion currents. Lateral spatial resolution depends on the type of ion optics used in the instrument. Increased resolution must be traded off with sensitivity. SIMS achieves lateral resolutions of about 0.5 μm which is useful for the analysis of some problems with patterned VLSI chips. Vertical (depth) resolution is controlled by many factors, such as texture at the bottom of a crater, the contribution of signals from the crater wall, and impurity redistribution during ion milling.

Sensitivity limits are set by the factors described above and by problems of mass interference. The mass resolution of an experiment is defined by $M/\Delta M$, where ΔM is the minimum mass difference that can be detected at a mass level M. Typical values of $M/\Delta M$ are 250 to 5000. An example of the influence of mass resolution on sensitivity is the difficulty in detecting phosphorus in silicon with background water vapor present in the system. The $^{31}P^+$ peak is very close to the peak from $^{31}SiH^+$ (formed from the interaction of water with silicon). At the mass resolution of >3500 required to distinguish the two masses, the sensitivity limit becomes 1×10^{19} atoms/cm^3. Table 7 lists sensitivity for the detection of a number of species, plus it identifies the most useful primary ion(s) and the detected ion species. Note that because of the complexity of the SIMS process, standards are required to apply SIMS to a problem requiring quantitative analysis.

Table 7 SIMS parameters for detection of some elements relevant to VLSI problems[28]

Element[†]	Primary beam	Detected species	C_{min} (atoms/cm^3)
Arsenic	Cs^+	$^{75}As^-$	5×10^{14}
Phosphorus	Cs^+	$^{31}P^\pm$	5×10^{15}
Boron	O_2^+, O^-	$^{11}B^+$	1×10^{13}
Oxygen	Cs^+	$^{16}O^-$	1×10^{17}
Hydrogen	Cs^+	$^1H^-$	5×10^{18}

[†]In all cases a silicon matrix is assumed.

12.3.6 Transmission Electron Microscopy (TEM)

The transmission electron microscope provides information on chemical composition by use of an x-ray spectrometer attachment. Information on this instrumental method is found in the following section.

12.3.7 X-Ray Emission Spectroscopy (XES)

Both SEM and TEM instruments can generate x-rays by bombarding a sample with electrons. Electron bombardment of a sample produces an x-ray continuum, as well as x-ray peaks which are characteristic of the material. The intensity of background continuum at a particular wavelength λ is given by[29]

$$I_{\lambda,c} \sim \frac{iZ(E_0-E_\lambda)}{E_\lambda} \tag{9}$$

where i is the electron beam current, Z is the atomic number of the material under bombardment, E_0 is the beam energy, and E_λ is the x-ray energy at wavelength λ. For many elements the intensity of an x-ray peak occurring at an energy E_λ is[30]

$$I_{\lambda,p} \sim i \left[\frac{E_0-E_\lambda}{E_\lambda}\right]^n \tag{10}$$

where n is a constant, typically 1.7. The peak-to-background ratio becomes

$$\frac{I_{\lambda,p}}{I_{\lambda,c}} \sim Z \left[\frac{E_0-E_\lambda}{E_\lambda}\right]^{0.7} \tag{11}$$

Thus, a higher beam energy produces stronger signals. A high beam energy is also used to stimulate x-ray emission of higher-energy peaks. Offsetting these arguments for higher beam energy, however, is the need to maintain a sufficiently low primary beam energy in studies on multilayer structures, where the stimulation of x-ray signals from lower layers is to be avoided. Typical beam energies fall between 15 and 40 keV.

Spatial resolution depends on the sample volume contributing x-ray spectra. X-rays are generated during scattering of the incident beam and backscattered electrons; they originate from a sample volume determined by the electron range (Fig. 7). This volume is quite large for bombardment of silicon, as can be determined from the electron range values shown in Fig. 6.

The loss in lateral spatial resolution caused by the size of the irradiated volume is not a major problem for qualitative analysis of patterned features or small particles on wafers, as long as the feature's chemical composition and "background" (i.e., the region beneath or adjacent to the feature) are known to be different. For example, identifying a gold particle on a boron nitride x-ray mask is easy, but identifying a particle in a 1.0-μm-wide space between two aluminum lines may be difficult if the particle contains aluminum.

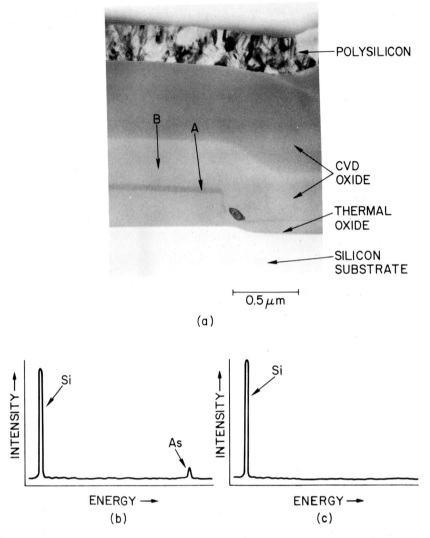

Fig. 17 (a) TEM cross section through an NMOS structure. (b) XES analysis of the 300-Å-wide dark band [region A in (a)] shows the presence of arsenic. (c) Analysis of background region B shows the absence of arsenic.

For thin (~ 0.1-μm) films studied by TEM, lateral spatial resolution is determined by the diameter of the electron beam; small-diameter (<100-Å) probes are available in modern TEM and STEM instruments. Figure 17a shows a TEM image of a cross section through part of an NMOS TEM test pattern structure made with arsenic source-drain implantation through a thin thermal oxide.[31] The top edge of the dark band correlates with the top edge of the thermal oxide. XES analysis of the band, region A, using an electron probe with diameter about 100 Å, and analysis of the background region B, produced the spectra shown in Fig. 17b and c, respectively. The figure shows that the band is a region of oxide which contains arsenic.

The sensitivity and accuracy of quantitative analyses are determined by the efficiency of x-ray generation, interference from other peaks and from background, and by detector and other instrumental parameters. When all conditions are optimal, the sensitivity of XES is often better than 10^{-15} g on bulk samples (Ref. 30, p. 355) which corresponds to a gold particle with diameter about 100 Å or a silicon particle with diameter about 950 Å. On a thin-film TEM sample the sensitivity is described by the expression[32]

$$M_{min} = \frac{1}{P_A \tau J} \tag{12}$$

where M_{min} is the minimum detectable mass, P_A is a constant for a given element and detector geometry, τ is the counting time, and J is the current density of the electron beam. With a thermionic electron source operating at 100 keV, a beam current density of 20 A/cm^2, and a counting time of 100 s, M_{min} for silicon is calculated to be 2×10^{-20} g, which corresponds to a silicon particle with diameter 20 Å.

Quantitative XES analysis generally requires the use of standards. The main problem in such analysis is the determination of the proportionality constant k, which relates the peak height (or integrated counts) H to the concentration C of element i in matrix j:

$$C_{ij} = k_{ij} H \tag{13}$$

Refer to Refs. 6 and 30 for an excellent discussion of these topics for the case of XES during SEM analysis, and to Ref. 33 for the case of XES during TEM and STEM analysis.

Both energy-dispersive and wavelength-dispersive detectors are used for XES studies. Energy-dispersive detectors are lithium-drifted silicon diodes, which convert an x-ray photon to a voltage pulse. The detector is kept at the temperature of liquid nitrogen to reduce noise, and must be kept in a vacuum. A thin window of beryllium or mylar (or other low-Z material) is usually used to provide vacuum isolation from the SEM. The energy distribution of the voltage pulses is proportional to the incident photon energy, and is displayed on a screen or read into a computer for storage and analysis. The main advantages of an energy-dispersive detector are its ability to detect x-ray peaks over a broad spectral range simultaneously and its ease of operation. A major limitation is the absorption of low-energy x-rays by the thin window, which effectively prevents x-ray peaks from low-Z elements (fluorine and below) from reaching the detector. A second disadvantage of energy-dispersive detectors is their limited ability to discriminate between two adjacent x-ray peaks. Typical detector linewidths (full-width at half-maximum) are 150 to 170 eV. Although this energy resolution is acceptable in many analyses some cases require greater resolution. For example, using an energy-dispersive detector to determine the stoichiometry of a tantalum silicide film is difficult, because the silicon K_α peak is at 1.74 keV and the strong tantalum M_α peak is at 1.71 keV; the separation is only 31 eV.

Wavelength-dispersive detection is achieved by using a crystal analyzer which intercepts the emitted x-radiation. The analyzer is adjusted until maximum Bragg reflection occurs. The resulting signal is then detected by a gas proportional counter. Such an analyzer can only be used to detect one peak at a time and is, therefore, more tedious to operate than an energy-dispersive analyzer. However, this type of detector is attractive because of its high-energy resolution ($\sim$5 to 10 eV) and ability to detect low-Z elements (the analyzing crystal is directly exposed to the x-radiation and need not be isolated by a vacuum).

12.3.8 X-Ray Fluorescence (XRF)

A primary x-ray beam of sufficient energy can stimulate the emission of x-radiation from a bombarded sample (Fig. 12), and both qualitative and quantitative analysis may be obtained from the spectra. Methods for detection of x-ray spectra are similar to the case of x-rays generated by electron bombardment. Refer to Section 12.3.7 for details.

Three features of XRF and XES, however, are quite different. Insulating samples, such as oxides and polymers used in packaging, are often difficult to examine by XES, because they charge (negatively) or decompose during electron bombardment. These types of samples may be easily examined by XRF. The large width of the primary x-ray beam makes XRF useless for analyzing small features on a VLSI chip, but quite useful for large-area analyses. Finally, in XRF, the characteristic x-radiation originates from a sample volume deeper than the volume which emits x-rays during electron stimulation (XES), because of the greater attenuation of electrons. This deeper penetration must be considered when applying XRF to the study of multilayered structures.

12.3.9 X-Ray Photoelectron Spectroscopy (XPS, ESCA)

X-ray bombardment of a sample can stimulate the emission of core-level electrons if the incident x-ray energy is sufficiently high (Fig. 12). Letting E_0 be the energy of the incident x-radiation, E_b the binding energy of the emitted electron, $\Delta \phi$ the work function difference between the sample and spectrometer surfaces, and E_{xp} the energy of the emitted electron (photoelectron kinetic energy),

$$E_{xp} = E_0 - E_b - \Delta \phi \qquad (14)$$

with constant $\Delta \phi$ and E_0 in a given experiment, electrons of different binding energies give rise to separate peaks in the photoelectron spectrum. The application of this phenomenon to chemical analysis of the bombarded surface is called x-ray photoelectron spectroscopy (XPS); this procedure is also called electron spectroscopy for chemical analysis (ESCA).

The incident x-ray beam is usually generated by low-energy electron bombardment of an aluminum or magnesium anode. K_α radiation from magnesium has an energy of 1253.9 eV and a linewidth of 0.7 eV, and K_α radiation from aluminum has an energy of 1487.0 eV and a linewidth of 0.85 eV. Since $\Delta \phi$ is about 1 eV, photoelectron energies from aluminum or magnesium sources are sufficiently low so that escape depths are less than 50 Å. The two methods, XPS and AES, are therefore similar in providing chemical information from a region within a few monolayers of the surface.

The electron detection and analysis instrumentation used in XPS is similar and, in some cases, nearly identical to that used for AES. As in AES, ion milling is used to obtain chemical depth profiles, and the depth resolutions of XPS and AES are similar. Lateral spatial resolution is quite poor in XPS, however, because x-ray beams used have cross sections 1 to 2 mm in diameter.

XPS is often used as a complement or back-up to AES, because of the following three advantages. First, radiation sensitive material can be nondestructively studied, because the scattering cross sections for x-ray induced desorption and dissociation are significantly lower than the corresponding cross sections for electron bombardment. Second, insulators can be studied with less surface charging, since a neutral incident beam is used, and third, information on chemical bonding can be obtained from XPS data. The energy levels of core electrons are affected by the valence state and type of chemical bonding. The energy resolution of XPS peaks is typically about 0.5 eV, and since different chemical bonds often produce shifts in the binding energy by larger amounts, these shifts can be detected and the bond identified.[34]

12.4 CRYSTALLOGRAPHIC STRUCTURE AND MECHANICAL PROPERTIES

An important aspect of device materials and process development programs is the analysis of crystallographic and mechanical properties of films and substrates. These analyses include the determination of substrate orientations and the determination of the degree of preferred orientation and crystallite size in grown and deposited films, the identification of phases and determination of unit cell parameters, the identification of amorphous regions and characterization of crystallographic defects, and the measurement of film stress. Five types of analyses are considered: laser-reflectance measurements of wafer curvature used to compute film stress, Rutherford backscattering spectroscopy (channeling), x-ray diffraction (camera and diffractometer methods), transmission electron diffraction, and transmission electron microscopy. A summary of the degree of success of each method in solving problems relating to structure is given in Table 8.

12.4.1 Laser Reflectance (LR)

The stress of a film on a thick substrate, assuming uniform film thickness and otherwise isotropic film stress, is given as[35]

$$\sigma = \frac{E}{6(1-v)} \frac{D^2}{Rt} \tag{15}$$

where E and v are Young's modulus and Poisson's ratio, respectively, for the substrate, D and t are the substrate and film thicknesses, respectively, and R is the radius of curvature of the composite. By convention R is negative for a convex wafer surface (compressive film stress) and positive for a concave surface (tensile film stress). The quantity $E/(1-v)$ has the value 1.8×10^{12} dyn/cm^2 for {100} silicon.[36]

Table 8 Applicability of various analytical methods to solution of structure related problems

	XRD					
Type of problem	Camera	Diffracto-meter	TED	RBS (channeling)	LR	TEM
Phase identification	Good	Good	Good	N.A.	N.A.	N.A.
Preferred orientation	Good	Poor	Good	N.A.	N.A.	N.A.
Unit cell parameter	Fair	Good	Fair	N.A.	N.A.	N.A.
Presence of amorphous regions	Good	Good	Good	Good	N.A.	Good
Lattice location of impurity atoms	N.A.	N.A.	N.A.	Good	N.A.	N.A.
Substrate orientation	Good	Fair	N.A.	N.A.	N.A.	Poor
Crystallographic defect analysis	N.A.	N.A.	N.A.	N.A.	N.A.	Good
Film stress	Poor	Good	Poor	N.A.	Good	Poor

Note: N.A. means either that the method would be a very poor choice or that it cannot be used.

The radius R can be determined by measuring the deflection of a light beam reflected off the wafer surface as the wafer is moved a fixed distance. This method uses a collimated laser beam which reflects from a wafer surface and projects on a screen over 10 m away. The wafer is moved a known and fixed distance while being illuminated by the incident laser beam. Wafer curvature causes the image of the beam to appear shifted on the screen; the shift is magnified by the optical leverage of the system. A measure of the wafer translation x, the corresponding translation of the position of the reflected beam d, and the reflected beam path length L gives the radius of curvature R

$$R = 2L\frac{x}{d} \qquad (16)$$

The minimum detectable curvature for the case $L = 10$ m, $x = 7$ cm, and $d = 0.1$ cm is 1400 m, assuming simple wafer curvature. Unprocessed wafer surfaces are sometimes not sections of spheres but saddle shapes and other complex shapes that make detection of such large radii of curvature difficult. Because of this problem, film stress is best determined by measuring wafer curvature before and after film deposition (or removal), using approximately the same endpoints for wafer translation. The curvature used for film stress calculation is computed from these measurements by

$$\frac{1}{R_f} = \frac{1}{R_T} - \frac{1}{R_S} \qquad (17)$$

where R_f, R_T, and R_S are curvatures due to the film alone, composite, and substrate alone, respectively.

Coefficients of thermal expansion can be obtained by curvature measurements of the wafer at different temperatures. Film stress σ is usually resolvable into two components, σ_i and σ_{th}, which are the intrinsic stress and the stress introduced by differential thermal expansion of the film and substrate:

$$\sigma = \sigma_i + \sigma_{th} \tag{18}$$

In Eq. 18,

$$\sigma_{th} = \frac{E_f}{1 - \nu_f} \int_{T_1}^{T_2} (\alpha_s - \alpha_f)\, dT \tag{19}$$

E_f and ν_f are Young's modulus and Poisson's ratio, respectively, for the film; α_s and α_f are the coefficients of thermal expansion of the substrate and the film, respectively. Assuming both coefficients of thermal expansion are constant over the temperature range,

$$\frac{d\sigma}{dT} = \frac{E_f}{1 - \nu_f} \Delta\alpha \tag{20}$$

where $\Delta\alpha = \alpha_s - \alpha_f$. When α_s is known, α_f can be determined. This method has been used for determining values of α for various silicides.[37]

12.4.2 Rutherford Backscattering Spectroscopy (Channeling)

Rutherford scattering intensity is reduced by a factor of about 100 when the incident ion beam is coincident with a low index crystallographic axis or with a plane of high symmetry. In such a configuration the beam is "steered" down crystallographic channels by low-angle repulsive interaction with the cores of lattice atoms, and penetrates to a depth about ten times greater than depths achieved with off-axis (random) orientations. If the incident beam angle is changed slightly from the value for optimum channeling, the backscattered yield increases sharply as shown in Fig. 18a. Yield versus energy is plotted in Fig. 18b, and shows the strong attenuation of the yield when channeling conditions are exactly met. The ratio H_A/H, measured just to the left of the leading edge of the substrate signal in a spectrum such as shown in Fig. 15, is known as the minimum yield. Disturbances to crystalline perfection, such as those caused by interstitial atoms or linear defects, cause an increase in the yield for an aligned orientation. In the extreme case of an amorphous layer, the yield is identical to that from a random orientation of a perfect lattice.

The distribution of dopant atoms between interstitial and substitutional sites can be determined by taking random and aligned RBS spectra and comparing the minimum yield H_A/H for dopant and silicon signals. Dopant atoms positioned substitutionally are shielded by silicon atoms from the channeled beam, and H_A/H is the same for both silicon and for the dopant. As the dopant atoms begin to occupy interstitial positions, H_A/H rises proportionally, and the fractional interstitial component can be computed (see Ref. 23, p. 269). RBS (channeling) has been used to compute the distribution of arsenic implanted into silicon after various anneals.[38, 39]

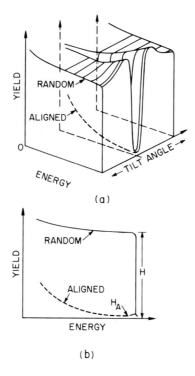

Fig. 18 (a) Three-dimensional representation of RBS yield as a function of tilt angle. (b) Typical yield plots for channeled (aligned) and random orientation. *(After Chu, Mayer, and Nicolet, Ref. 23.)*

12.4.3 X-Ray Diffraction (XRD)

An analysis of the angular position and intensity of x-ray beams diffracted by crystalline material gives information on the crystal structure (phase) of the material. Four types of x-ray diffraction methods have been found to be useful: the back-reflection Laue technique for determining wafer orientation, the Read camera technique, the Huber-Seemann-Bohlin camera technique, and diffractometer methods. All x-ray methods depend on establishing conditions that satisfy the Bragg requirement for diffraction of x-rays by a crystal lattice:

$$n\lambda = 2d \sin\theta \qquad (21)$$

where λ is the x-ray wavelength, d is the interplanar spacing, θ is the Bragg diffraction angle, and n is an integer giving the order of the diffraction. Diffraction occurs only when Eq. 21 is satisfied.

Substrate orientation can be quickly determined from a back-reflection Laue photograph. An incident beam of "white" radiation (the continuous unfiltered output of an x-ray tube) is collimated and strikes a wafer placed parallel to a photographic film. The incident beam passes through a hole in the film, and the interaction with the wafer produces a large number of diffracted beams. Diffraction occurs simultaneously from a large number of crystallographic planes and leaves an image of spots on the film. The spots lie on hyperbolae, and the distribution and symmetry of spots directly relate to the crystallographic orientation of the substrate. Back-reflection Laue patterns from {100}, {111}, and {110} silicon wafer surfaces are clearly distinguishable. A pattern from a wafer of unknown orientation can be quickly compared with standard patterns to determine the orientation.

A version of the classical x-ray powder diffraction camera that is particularly suited for analysis of thin films is the Read camera, illustrated in Fig. 19. The camera is a cylinder with a 5-cm radius and 13 cm high and has an entrance hole for the x-ray beam. Photographic film is laid against the inside wall of the cylinder. During sample irradiation by a monochromatic x-ray beam, a cone of diffraction is created for each set of diffracting planes in a polycrystalline film sample. The intersection of each cone with the film produces a curved line of intensity.[40] The cone axis is coincident with the incident beam direction, and the angle at the cone apex is four times the Bragg angle. A template that matches the camera dimensions can be laid over the developed photographic film to identify the lines. A line has an associated "d-value," which is the interplanar spacing which produced the diffraction. These "d-values" plus a semiquantitative measure of their intensity are usually sufficient to either identify the material by referring to standard x-ray crystallographic tables, or establish the spacings of a newly discovered phase. Interplanar spacings measured by this method are usually accurate to 0.01 Å.

The two features of the Read camera that make it particularly useful for thin-film analysis are the shallow angle of incidence, and the large photographic area. The shallow angle of incidence permits greater beam penetration into the sample film material than would be the case with normal incidence. The large photographic film

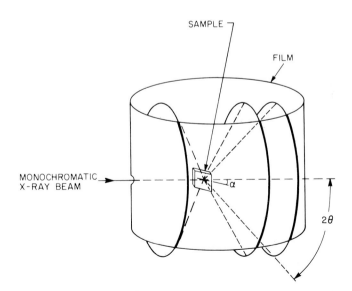

Fig. 19 Read x-ray diffraction camera method. A monochromatic x-ray beam passes through a hole in the camera and strikes a sample surface at an angle α, typically 15°. Diffraction cones which intersect the film produce lines of intensity (dark lines); the angle 2θ is twice the Bragg angle.

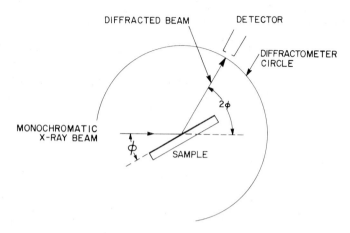

Fig. 20 X-ray diffractometer configuration. As the sample rotates over an angle φ the detector moves along a circumference an angular distance 2φ; a diffraction maximum occurs when φ is equivalent to a Bragg angle.

width permits information to be recorded on the degree of preferred orientation of the sample film. Preferred orientation has the effect of producing nonuniform cone intensities. Instead of curved lines of approximately constant intensity from a polycrystalline sample, preferred orientation introduces elongated spots and arcs. Most of the information on preferred orientation would be lost with a Huber-Seemann-Bohlin camera or with diffractometer methods.

The Huber camera with Seemann-Bohlin geometry is more sensitive to very thin films than is the Read camera. This improvement in sensitivity is due to decreasing the angle of incidence to a few degrees which permits more of the material to diffract. The sample is located on the wall of a cylinder, and a convergent monochromatic x-ray beam becomes incident over an elongated path on the sample surface. The camera is designed so that beams diffracted from the same *d*-spacing at the near and far edge of the sample both converge at the same point on the film.

More precise measurements of lattice parameters result from using a diffractometer geometry, shown in Fig. 20. A monochromatic x-ray beam strikes the film surface at an angle φ. The sample is slowly rotated and a detector simultaneously moves, at twice the rate, along the circumference of a circle with the same center as the sample. Diffraction maxima appear wherever φ coincides with a Bragg angle. A measure of 2φ is used to identify the interplanar spacing (*d*-values) of the crystalline material giving rise to those peaks. Data are usually fed into a strip chart recorder where the peak positions, intensities, and half-widths can be easily read. The identification of a phase on the basis of calculated lattice parameters and the determination of lattice parameters of a new phase both depend on the precision of the measurement of θ. The highest precision results from diffractometer measurements near 2φ = 180° (see Fig. 20). Most diffractometer measurements can give precision better than 0.01 Å.

12.4.4 Transmission Electron Diffraction (TED)

Equation 21 establishes the basic condition for obtaining the diffraction of electrons as well as of x-rays. A typical transmission electron diffraction (TED) pattern is a set of concentric rings from polycrystalline material (see Fig. 28) or spots from a single-crystal sample (Fig. 21a). The rings are formed by diffraction cones intersecting the screen, analogous to the formation of diffraction arcs in a Read camera as previously described. TED patterns can be generated with a transmission electron microscope. In electron microscopes that permit diffraction patterns to be obtained from regions no smaller than about 1 μm in diameter, diffraction rings are continuous for crystalline sizes less than 100 Å, and become increasingly spotty as grain size increases. In the limiting case where only one crystallite contributes to the diffraction pattern, a single set of diffraction spots appear. Other TEM instruments, which include scanning transmission electron microscope (STEM) capability, can probe regions of a film smaller than 100 Å diameter; the shape of the diffraction pattern in this case is partly determined by the size, number, and shape of the crystallites within this small region, and by beam spot size. Microdiffraction studies of small areas are particularly useful in the study of individual layers within TEM samples of VLSI circuits.

The most common structural problem that occurs during TEM study of metallization layers or crystallographic inclusions is identifying unknown phases. These phases are identified by measuring the diameters of a number of diffraction rings of a pattern, and computing the corresponding interplanar spacings by using a pattern of a known standard taken under identical conditions. The phase is identified with techniques similar to those used to identify unknown phases by x-ray analysis.

Figure 21 shows how TED can be used to determine the orientation of a small silicon crystallite. One of the features produced during thermal oxidation of polysilicon is inclusions of silicon within the polysilicon oxide. Figure 21b shows the TED pattern from the inclusion shown in the micrograph of Fig. 21c. The TED pattern from the silicon substrate (Fig. 21a) is compared with the pattern from the inclusion. The comparison shows that the inclusion is oriented with a $\langle 110 \rangle$ axis normal to the wafer surface. This information was used to develop a model of inclusion formation.[41]

12.4.5 Transmission Electron Microscopy (TEM)

Information on structure is obtained from a transmission electron microscope by an analysis of electron diffraction patterns (see Section 12.4.4) and by an analysis of the microscope image (see Section 12.2.3).

12.5 ELECTRICAL MAPPING

In electrical mapping, an electron beam is used to locate regions in a device structure that differ in electrical activity, and in some cases, to measure the difference. Two mapping methods are used. In one, the energy distribution of secondary electrons, produced when an electron beam strikes a device sample, is influenced by the local

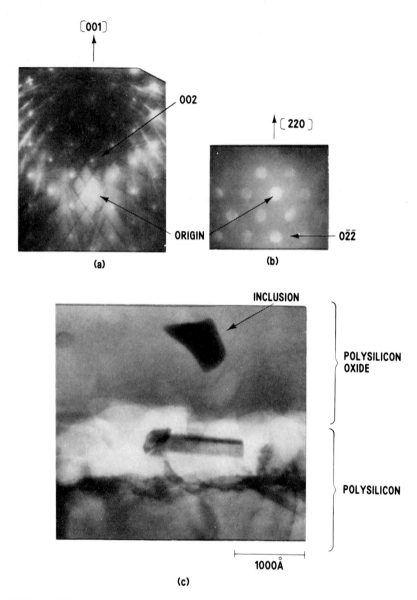

Fig. 21 Electron diffraction patterns (a) from silicon substrate and (b) from silicon inclusion within polysilicon oxide. (c) A TEM photo of the polysilicon, polysilicon oxide, and inclusion. The orientation of the inclusion was determined from the diffraction patterns.

surface potential, and the secondary electron flux reaching the detector reflects this potential. This phenomenon, essential to voltage contrast imaging, can be used to determine the potential of an element on a device surface. The other involves generation of charges in the device by the electron beam, and the collection of the charges via a capacitor or junction. Local changes in morphology, material properties, and junction electric field cause corresponding local modulations of the collected current.

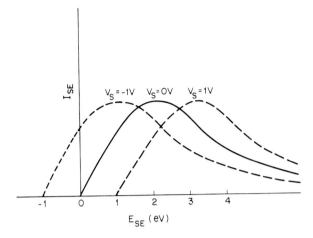

Fig. 22 Secondary electron current (I_{SE}) as a function of secondary electron energy (E_{SE}) for cases where the surface potential V_s is -1, 0, and $+1$ V.

This procedure is called EBIC (electron beam induced current) or charge-collection microscopy. Both types of electrical mapping can be performed with a scanning electron microscope; a transmission electron microscope can be used for EBIC studies.

12.5.1 Voltage Contrast Microscopy

The energy distribution of secondary electrons, produced when an electron beam strikes a surface, shows a peak at an energy near 5 eV (Fig. 5). This energy is sufficiently low that changes in surface potential on the order of 1 V have a significant effect on the energy distribution as illustrated in Fig. 22. In Fig. 22, a secondary electron distribution from a surface at zero potential is assumed to peak at 2.1 eV. A change of surface potential V_s of ± 1 V shifts the distribution right or left by an approximately proportional amount. Such shifts can be qualitatively observed for voltage contrast analyses of images, or quantitatively measured for voltage contrast analyses of surface potentials.

The secondary electron image in most SEM instruments is created by an extraction field of a few hundred volts that brings the electrons to a detector where a strong field attracts them to a scintillator surface. A light pipe transfers the signal to a photomultiplier tube. Not all electrons reach the scintillator. Local fields at the sample surface affect the trajectories of the secondary electrons and permit only those above a threshold energy, about 1 eV, to reach the detector. The total flux of electrons reaching the detector and, therefore, the intensity of the signal depend on the number of secondary electrons that exceed the threshold energy. This number is strongly affected by the surface potential as shown in Fig. 22. The minimum intensity difference that can be noted on a screen corresponds to a surface potential difference of about 1 V, which can be considered a practical limit.

Because of this sensitivity of the secondary electron image intensity to surface potential, an SEM can be applied to detect electrical discontinuities in conducting

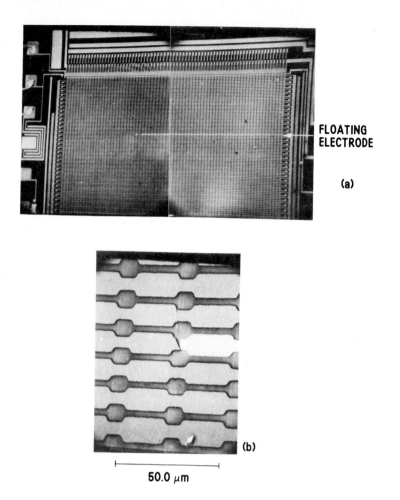

FLOATING
ELECTRODE

(a)

50.0 μm

(b)

Fig. 23 (a) Voltage contrast study of a failure in CCD metallization using a 25-V potential to locate the failure. (b) Higher-magnification study shows the failure to be due to a loss in metallization caused by a mask error.

regions of VLSI circuits. Electrical discontinuity can be easily seen by applying a dc bias to a metallization, as shown in Fig. 23. The bias can be applied by using a movable mechanical probe inside the SEM chamber or by fixed contact to input pads in a packaged device.

A bias may also be applied as a pulse,[42] i.e., the device can be operated dynamically inside the SEM and the voltage contrast image obtained by stroboscopically pulsing the incident electron beam in synchronism with the circuit.[43] Time resolutions of 0.2 ps have been achieved in the stroboscopic voltage contrast mode.[44]

Circuit testing, analysis, and design verification often require the characterization of waveforms at selected internal nodes of the circuit. Such measurements are usually made by placing a fine mechanical probe at these modes or at pads connected to them.

As metallization linewidths approach 1 µm, placing a probe at a selected site and avoiding metallization damage during probing becomes increasingly difficult. A second problem with mechanical probes is that probe capacitance, generally larger than 0.1 pF, can interfere with the measurement by placing a parasitic capacitance on the node under test. These problems can be solved by using an electron beam as a small-diameter (~0.1-µm) probe; the beam does not capacitively load the circuit and can be easily positioned to small circuit elements.[43]

12.5.2 Electron Beam Induced Current (EBIC) Microscopy

An electron beam penetrating into a device can generate carriers which, in turn, can be collected as current to modulate the intensity of a CRT image. This mode of electron beam induced current (EBIC) or charge-collection scanning electron microscopy is useful in spatially locating device failure sites within a junction or capacitor region. Figure 24 shows the arrangement for EBIC imaging using a Schottky barrier, a pn junction, and a capacitor. A mechanical probe inside the SEM specimen chamber applies the bias. An electron beam current of 10 nA is usually sufficient. The choice

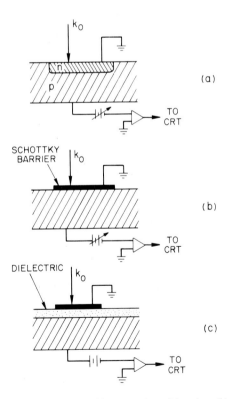

Fig. 24 Arrangement for EBIC measurement (a) on a np (or pn) junction, (b) on a Schottky-barrier junction, and (c) on a capacitor. The incident electron beam is k_0.

of beam energy is a compromise between the need to penetrate to a depth that maximizes EBIC signal and the loss of spatial resolution with increasing sample penetration. Excessive beam energy can damage a device although this damage is annealable (see Section 12.3.1); such a consideration may be an additional factor in limiting beam energy. For pn-junction or Schottky-barrier analysis, the beam energy is usually set high enough so that the electron range exceeds the thickness of the top electrode (or layers). This technique generates the highest electron-hole pair within or below the space-charge region. Capacitor analysis is somewhat more complicated. Useful EBIC information can be obtained by generating carriers in the field plate, oxide layer, or substrate, depending on the specimen geometry, applied bias, and beam energy. These points are discussed in the following paragraphs.

Useful information on junction leakage and breakdown sites is obtained by EBIC studies of Schottky barriers and pn junctions under zero or reverse bias conditions. The incident beam generates charges within silicon that diffuse to the space-charge layer where they are swept across by the field to contribute a current pulse. The current is given by

$$I_{EBIC} = \frac{I_0(E_0 - E_{BS})}{E_{eh}} \eta \tag{22}$$

where I_0 and E_0 are the incident beam current and energy, respectively, E_{BS} is the energy lost due to backscattering, E_{eh} is the energy needed to create an electron-hole pair in silicon (3.6 eV), and η is the charge collection efficiency.[45] At an incident beam energy of 10 keV and with $E_0 - E_{BS}$ approximately $0.9E_0$,

$$\frac{I_{EBIC}}{I_0} \sim 2500\eta \tag{23}$$

Since η is essentially unity in the space-charge charge region, EBIC currents can be large multiples of the incident beam current.

The efficiency of charge collection falls off with increasing electron beam penetration, due partly to recombinations that limit the lifetimes of minority carriers, and partly to loss of carriers that diffuse away from the junction. Recombination centers in the path of carriers diffusing to the space-charge layer decrease the current collected locally by the junction. These recombination centers are seen as regions of decreased EBIC signal. The magnitude of recombination efficiency of defects is characterized by the "defect strength"; the EBIC contrast, created by the defect, is a function of the defect strength, as well as the depth of the defect.[46] Although the defect strength of a dislocation is a function of the orientation of the dislocation,[47] defect strength appears to be dominated by the amount of impurity decoration at the crystalline defect.[48, 49]

Figure 25a and c shows Nomarski interference contrast micrographs of epitaxial stacking faults on silicon substrates of two different orientations, and Fig. 25b and d shows the corresponding EBIC images; gold Schottky barriers were used to collect the EBIC current. The most obvious contrast features in these EBIC images are at the dislocations at the junction of stacking faults. The defect strengths of two opposite dislocations in the stacking fault tetrahedron on the {100} surface (Fig. 25b) are

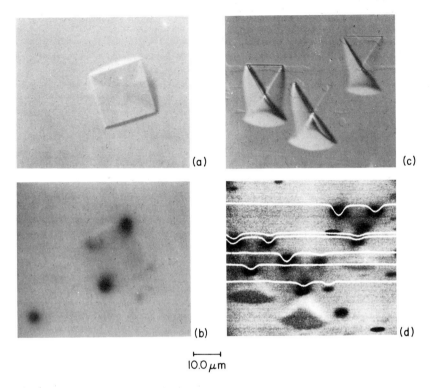

10.0 μm

Fig. 25 Nomarski interference contrast micrographs of epitaxial stacking faults (a) on {100} silicon, and (c) on {111} silicon. EBIC photographs of the same regions made with gold Schottky barriers and a beam energy of (b) 15 keV, and (d) 12 keV. The line scans in (d) give a quantitative measure of the EBIC "defect strength" of the partial dislocations. The EBIC images were made at 0 V applied bias.

clearly larger than the defect strengths of the remaining two dislocations. The difference in defect strengths of the dislocations in the pyramidal stacking faults on the {111} surface can be clearly seen with the aid of the EBIC line scans (Fig. 25d).

EBIC studies of Schottky barriers and pn junctions have been used to reveal silicon defects, especially stacking faults, dislocations, and inhomogeneities segregated during crystal growth. Except for the case of analysis of defects resulting from pn-junction processing, Schottky-barrier junctions are preferred for EBIC defect analysis because of the ease in forming such junctions. Thin films of gold and aluminum are generally employed for forming Schottky barriers on p-type and n-type silicon, respectively.

Defects responsible for capacitor leakage and breakdown can also be spatially located with EBIC methods.[50] EBIC current in this case is controlled by four mechanisms: secondary electron emission from the field plate or substrate across an oxide interfacial barrier, current generation within the oxide, electron tunneling across an interfacial barrier, and displacement currents induced in a space-charge layer. Fowler-Nordheim tunneling begins to be important as a current generation mechanism at higher fields and as oxide films become very thin. EBIC microscopy can be used

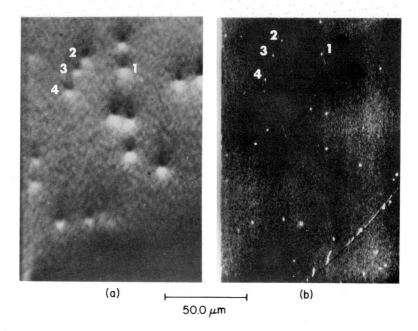

(a)

50.0 μm

(b)

Fig. 26 (a) EBIC image of a capacitor, and (b) EBIC image of a Schottky barrier at the same site showing EBIC defect sites. The appearance of EBIC signals at the same locations (see, for example, sites 1−4) shows that the defect is within the substrate.

to reveal defects in thin oxide capacitors because of locally enhanced tunneling due to higher fields created at these sites.[51] In all cases defects create local disturbances of current, which create contrast in the image.

The displacement current mechanism is operative only for capacitors under reverse bias or very weak forward bias conditions which maintain a space-charge layer. Figure 26a shows an EBIC image taken with a beam energy of 8 keV of a capacitor which was biased at nearly 0 V. The capacitor had an oxide thickness of 250 Å and a 1000-Å phosphorus doped polysilicon field plate. The white-on-black features are EBIC signals due to uncanceled displacement currents at the defects.[52] After removing the field plate and oxide and depositing a thin aluminum film over this region, Schottky-barrier EBIC imaging shows the same sites, indicating that the defects are in the substrate. Subsequent TEM analysis showed the defects to be heavily decorated stacking faults.[52]

12.6 SUMMARY AND FUTURE TRENDS

The diagnostic procedures described are useful in solving problems occurring in a VLSI technology development program. Future developments will undoubtedly improve the capabilities of these methods as well as produce new analytical methods.

SEM stages are now available that can handle wafers 125 mm and larger in diameter; full translation coverage of large-diameter wafers including eucentric tilts with short working distances (for improved resolution) are expected to become available. Scanning electron microscopes will thereby become more generally useful for a variety of VLSI diagnostic studies.

As methods for preparing TEM samples and the use of TEM test chips become better known, it is expected that TEM methods will become an integral part of VLSI programs, and XES studies using TEM methods will be more widely used. EBIC studies will also gain in importance since an electrically active defect spatially located by EBIC is (often) most easily analyzed by TEM.

An increased ability to perform quantitative measurements through the use of more and better standards will be a major improvement in AES, XPS, and SIMS. The application of these methods to VLSI problems is still in its early stages.

A major growth in the field of electrical mapping is expected to be in the increased use of electron beams for waveform analysis and voltage contrast imaging. Once the barrier to accepting this qualitatively different method of circuit diagnostics and testing is overcome, the advantages of having an automated, nondestructive, high-speed, low-capacitance probe method for circuit analysis should result in further growth.

REFERENCES

[1] G. Nomarski and A. R. Weill, "Application á la Métallographie des Méthodes interférentielles á Deaux Ondes Polarisées, *Rev. Metall.*, **52**, 121 (1955).

[2] D. C. Miller and G. A. Rozgonyi, "Defect Characterization by Etching, Optical Microscopy and X-Ray Topography," in S. P. Keller, Ed., *Handbook on Semiconductors*, North-Holland, New York, 1980, p. 217.

[3] T. E. Everhart and P. H. Hoff, "Determination of Kilovolt Electron Energy Dissipation vs. Penetration Distance in Solid Materials," *J. Appl. Phys.*, **42**, 5837 (1971).

[4] A. Joshi, I. E. Davis, and P. W. Palmberg, "Auger Electron Spectroscopy," in A. W. Czanderna, Ed., *Methods of Surface Analysis*, Elsevier, New York, 1975, p. 164.

[5] T. E. Everhart, O. C. Wells, and C. W. Oatley, "Factors Affecting Contrast and Resolution in the Scanning Electron Microscope," *J. Electron. Control*, **7**, 97 (1959).

[6] J. I. Goldstein, D. E. Newbury, P. Echlin, D. C. Joy, C. Fiori, and P. Lifshin, *Scanning Electron Microscopy and X-ray Microanalysis*, Plenum, New York, 1981.

[7] K. G. McKay, "Secondary Electron Emission," *Adv. Electron.*, **1**, 65 (1948).

[8] P. R. Thornton, *Scanning Electron Microscopy*, Chapman and Hall, London, 1968, p. 105.

[9] J. M. Aitken, "1 μm MOSFET VLSI Technology: Part VIII. Radiation Effects," *IEEE Trans. Electron Devices*, **ED-26**, 372 (1979).

[10] L. M. Welter and V. J. Coates, "High Resolution Scanning Electron Microscopy at Low Accelerating Voltages," in O. Johari, Ed., *Scanning Electron Microscopy/1974*, IITRI, Chicago, 1974, p. 59.

[11] Gareth Thomas, "Electron Microscopy at High Voltages," *Philos. Mag.*, **17**, 1097 (1968).

[12] R. B. Marcus and T. T. Sheng, *Transmission Electron Microscopy of Silicon VLSI Devices and Structures*, Wiley, New York, 1983

[13] J. M. Oblak and B. H. Kear, "Analysis of Microstructures in Nickel-Base Alloys: Implications for Strength and Alloy Design," in G. Thomas, R. M. Fulrath, and R. M. Fisher, Eds., *Electron Microscopy and Structure of Materials*, University of California Press, Berkeley, 1972, p. 566.

[14] T. T. Sheng and R. B. Marcus, "Advances in Transmission Electron Microscope Techniques Applied to Device Failure Analysis," *J. Electrochem. Soc.*, **127**, 737 (1980).

[15] C. C. Chang, "Analytical Auger Electron Spectroscopy," in P. F. Kane and G. B. Larrabee, Eds., *Characterization of Solid Surfaces*, Plenum, New York, 1974, Chap. 20.

[16] J. M. Morabito, "A First-Order Approximation to Quantitative Auger Analysis in the Range 100−1000 eV Using the CMA Analyzer," *Surf. Sci.*, **49**, 318 (1975).

[17] D. R. Penn, "Electron Attenuation Lengths for Free-Electron-Like Metals," *J. Vac. Sci. Technol.*, **13**, 221 (1976).

[18] C. C. Chang, Bell Laboratories, unpublished data.

[19] C. C. Chang and D. M. Boulin, "Oxide Thickness Measurements up to 120 Å on Silicon and Aluminum Using the Chemical Shifted Auger Spectra," *Surf. Sci.*, **69**, 385 (1977).

[20] Paul Kruger, *Principles of Activation Analysis*, Wiley, New York, 1971, pp. 44−47.

[21] S. P. Murarka, Bell Laboratories, unpublished data.

[22] S. P. Murarka, "A Study of the Phosphorus Gettering of Gold in Silicon by Use of Neutron Activation Analysis," *J. Electrochem. Soc.*, **123**, 765 (1976).

[23] Wei-Kan Chu, James W. Mayer, and Marc-A. Nicolet, *Backscattering Spectroscopy*, Academic, New York, 1978.

[24] R. Schutz, Bell Laboratories, unpublished data.

[25] G. Mezey, E. Kotai, T. Nagy, L. Lohner, A. Manuba, J. Gyulai, V. R. Deline, C. A. Evans, Jr., and R. J. Blattner, "A Comparison of Techniques for Depth Profiling Oxygen in Silicon," *Nucl. Instrum. Methods* **167**, 279 (1979).

[26] Peter Williams, R. K. Lewis, Charles A. Evans, Jr., and P. R. Hanley, "Evaluation of a Cesium Primary Ion Source on an Ion Microprobe Mass Spectrometer," *Anal. Chem.*, **49**, 1399 (1977).

[27] J. A. McHugh, "Secondary Ion Mass Spectrometry," in A. W. Czanderna, Ed., *Methods of Surface Analysis*, Elsevier, New York, 1975, p. 223.

[28] Vaughn Deline, Chas. Evans. Associates, San Mateo, California, private communication.

[29] H. A. Kramers, "On the Theory of X-ray Absorption and of the Continuous X-Ray Spectrum," *Philos. Mag.*, **46**, 836 (1923).

[30] E. P. Bertin, *Principles and Practice of X-ray Spectrometric Analysis*, Plenum, New York, 1970, p. 27.

[31] E. K. Kinsbron, W. Fichtner, T. T. Sheng, and R. B. Marcus, Bell Laboratories, to be published.

[32] D. C. Joy and D. M. Maher, "Sensitivity Limits for Thin Specimen X-ray Analysis," in O. Johari, Ed., *Scanning Electron Microscopy/1977*, IITRI, Chicago, 1977, Vol. 1, p. 325.

[33] J. I. Goldstein, "Principles of Thin Film X-ray Microanalysis," in J. J. Hren, J. I. Goldstein, and D. C. Joy, Eds., *Introduction to Analytical Electron Microscopy*, Plenum, New York, 1979, Chap. 3.

[34] C. R. Brundle and A. D. Baker, *Electron Spectroscopy*, Academic, New York, 1977, Vols. 1−4.

[35] R. W. Hoffman, "The Mechanical Properties of Thin Condensed Films," in G. Haas and R. E. Thun, Eds., *Physics of Thin Films*, Academic, New York, 1966, Vol. 3, p. 211.

[36] W. A. Brantley, "Calculated Elastic Constants for Stress Problems Associated with Semiconductor Devices," *J. Appl. Phys.*, **44**, 534 (1973).

[37] T. F. Retajczyk and A. K. Sinha, "Elastic Stiffness and Thermal Expansion Coefficients of Various Refractory Silicides and Silicon Nitride Films," *Thin Solid Films*, **70**, 241 (1980).

[38] C. E. Christodoulides, R. A. Baragiola, D. Chivers, W. A. Grant, and J. S. Williams, "The Recrystallization of Ion Implanted Silicon Layers. II. Implant Species Effect," *Radiat. Eff.*, **36**, 73 (1978).

[39] D. G. Beanland and J. S. Williams, "The Damage Dependence of the Epitaxial Regrowth Rate During the Annealing of Amorphous Silicon Formed by Ion Implantation," *Radiat. Eff.*, **36**, 15 (1978).

[40] B. D. Cullity, *Elements of X-Ray Diffraction*, Addison-Wesley, Reading, Mass., 1978, pp. 96−99.

[41] R. B. Marcus, T. T. Sheng, and P. Lin, "Polysilicon/SiO_2 Interface Microtexture and Dielectric Breakdown," *J. Electrochem. Soc.*, **129**, 1282 (1982).

[42] J. K. Rodman and J. T. Boyd, "Examination of CCD, Using Voltage Contrast with a Scanning Electron Microscope," *Solid State Electron.*, **23**, 1029 (1980).

[43] Peter Fazekas, Hans-Peter Feuerbaum, and Eckhard Wolfgang, "Scanning Electrom Beam Probes VLSI Chips," *Electron.*, **54**, 14, July 14, 1981, p. 105.

[44] T. Hoskowa, H. Fujioka, and K. Ura, "Generation and Measurement of Subpicosecond Electron Beam Pulses," *Rev. Sci. Instrum.*, **49**, 624, (1978).

[45] H. J. Leamy, L. C. Kimmerling, and S. D. Ferris, "Silicon Single Crystal Characterization by SEM," in O. Johari, E., *Scanning Electron Microscopy/1976*, IITRI, Chicago, 1976, Vol. 2, p. 529.

[46] C. Donolato, "Contrast Formation in SEM Charge-Collection Images of Semiconductor Defects," in O. Johari, Ed., *Scanning Electron Microscopy/1979*, IITRI, Chicago, 1979, Vol. 2, 257.

[47] M. Kittler and W. Seifert, "On the Characterization of Individual Defects in Silicon by EBIC," *Crystal Res. Technol.*, **16**, 157 (1981).

[48] R. B. Marcus, M. Robinson, T. T. Sheng, S. E. Haszko, and S. P. Murarka, "Electrical Activity of Epitaxial Stacking Faults," *J. Electrochem. Soc.*, **124**, 425 (1977).

[49] H. Blumtritt, R. Gleichmann, J. Heydenreich, and H. Johansen, "Combined Scanning (EBIC) and Transmission Electron Microscopic Investigations of Dislocations in Semiconductors," *Phys. Status Solidi, A*, **55**, 611 (1979).

[50] W. R. Bottoms, P. Roitman, and D. C. Guterman, "Electron Beam Imaging of the Semiconductor-Insulator Interface," *CRC Critical Reviews in Solid State Sciences*, **5**, 297 (1975).

[51] P. Lin and H. Leamy, *Appl. Phys. Lett.*, to be published.

[52] P. S. D. Lin, R. B. Marcus, and T. T. Sheng, "Gate Oxide Leakage and Breakdown Caused by Oxidation Induced Decorated Stacking Faults," *J. Electrochem. Soc.*, to be published.

PROBLEMS

1 Determine the stoichiometry of tantalum silicide from the AES data given in Fig. 13. Use the following values for the proportionality constant α: $\alpha_{Ta} = 1.00$ and $\alpha_{Si} = 0.37$.

2 A device chip is examined by SEM at $0°$ tilt; that is, the wafer surface is normal to the incident electron beam. Assume that the secondary electron current that arrives at the detector from the sloped wall of a 300-Å-thick polysilicon runner is 75% of the secondary electron current from the runner surface, (i.e., $I_w = 0.75 I_s$). Defining contrast C as $C = ((I_s - I_w)/I_s)$ and using the data in Table 4.5 of Reference 6 (p. 177), determine the minimum wall slope (ϕ, Fig. 27) that can be detected; a vertical wall is $90°$.

3 A 2400-Å-thick film of tantalum has been deposited over an oxidized {100} silicon wafer. Assume that loss of adhesion of the metal film to the substrate occurs within a week if the film stress exceeds 8×10^9 dyn/cm^2. The laser reflectance technique is used to measure film stress, and the laser beam is found to

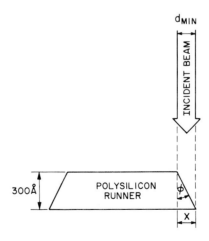

Fig. 27

move 1.30 cm when the wafer is translated 4.0 cm; before the tantalum was deposited a similar measurement showed no motion of the reflected laser beam during wafer translation. The wafer thickness is 510 μm and L (Eq. 16) is 10 m. Is loss of adhesion likely to occur?

4 Calculate the thickness of the aluminum layer from the RBS data given in Fig. 15. The energy limits for ΔE should be chosen at points (on the leading and trailing edges of the aluminum peak) where the backscattered ion counts decrease to half the maximum value; note that the ordinate has a logarithmic scale. Use the following information: $[\epsilon]_{Al} = 7.68 \times 10^{-14}$ eV cm²; $N_{Al} = 6.02 \times 10^{22}$ atoms/cm³.

5 Determine the concentration of platinum in the titanium layer from the RBS data given in Fig. 15. The background ion count is significant and must be subtracted from the platinum ion count. Use the following information: $[\epsilon]_{Pt} = 230.7 \times 10^{-15}$ eV-cm²; $[\epsilon]_{Ti} = 144.6 \times 10^{-15}$ eV-cm².

6 Oxide of 300-Å thickness is formed on a silicon wafer and doped (conducting) polysilicon dots with diameter of 500 μm are deposited over the oxide. The polysilicon is 2000 Å thick. One of these polysilicon-oxide-silicon capacitors is found to pass a high leakage current, and a SEM EBIC study has located the leakage in a 1-μm-diameter region near the center of the capacitor. The SEM study shows no discernible textural features at the EBIC site or anywhere else on the capacitor. A thin-film horizontal section (parallel to the wafer surface) is to be made of this capacitor for TEM study, in order to identify the defect which is the source of the leakage. The defect is thought to be less than 500 Å. Devise a method for marking the position of the EBIC site so that it can be located and its center found during TEM study.

7 The transmission electron diffraction pattern shown in Fig. 28 was taken from a thin foil containing a mixture of gold and an unknown phase. The gold consisted of large grains and gave rise to the ''spotty'' or textured rings; the unknown phase gave rise to the remaining set. Assuming that the first three gold rings (starting with the smallest-diameter ring) came from the {111}, {200}, and {220} lattice planes, respectively, determine the camera constant and identify the unknown phase. The camera constant RL is given by

$$RL = rd$$

where r is the radius of a diffraction ring and d is the interplanar spacing corresponding to that ring. The unknown phase is suspected to be silicon, aluminum, or palladium. Use the table of interplanar spacings given in the *Powder Diffraction File* (published by the JCPDS International Center for Diffraction Data, Swarthmore, Pennsylvania, 1981 and earlier years) or any other compilation of crystallographic data.

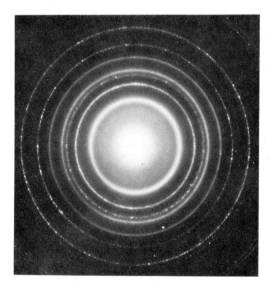

Fig. 28

THIRTEEN

ASSEMBLY TECHNIQUES AND PACKAGING

C. A. STEIDEL

13.1 INTRODUCTION

Integrated circuit packaging is a broad field that includes assembly operations on the silicon die (also called a chip), such as die bonding and wire bonding, fabrication technologies for the packages that house the die, and the package's performance characteristics. Cost or performance considerations of the packaged device usually dictates the assembly and packaging details. In many cases, packaging can significantly affect, if not dominate, the overall cost, performance, and reliability of the packaged IC. Appropriately, packaging is now receiving more attention by both packaged-device vendors and system builders.

The rapid increase in the number of devices per IC chip and the performance of these devices create the major challenges that face packaging designers. Silicon's increased capability naturally leads to chips with higher input/output (I/O) count, and, therefore, more connections to the silicon and more package pins. Chips with higher I/O count require larger packages which often dissipate more heat. Both package size and thermal dissipation must be considered to preserve a balance between component cost and system cost. More demands will be placed on packaging,[1] because system interconnection costs increase at each higher level of interconnection (Fig. 1). Although the processing costs for silicon are high on a unit area basis, the interconnection density is so superior to other interconnection media that silicon actually has the lowest cost on a per interconnection basis. An optimum overall packaging strategy integrates as much of the interconnection as possible in the silicon and then in the package (e.g., hybrid).

This chapter discusses all of the above mentioned packaging areas. It focuses on single-chip packaging techniques and thermal, electrical, and interconnection issues associated with the use of the single-chip package in a system's environment, particularly at the printed wiring board level.

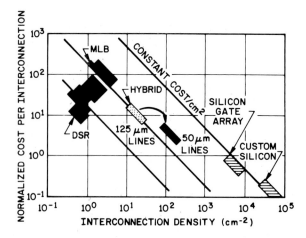

Fig. 1 The relative cost of interconnection for DSR (double-sided rigid printed wiring board), MLB (multilayer printed wiring board), ceramic hybrids, gate array, and custom silicon. The curves clearly show the advantage of silicon integration in decreasing interconnection costs. *(After Goddard, Ref. 1.)*

13.2 WAFER SEPARATION AND SORTING

Silicon dice may be separated by the use of many different technologies. Techniques are required to scribe the silicon surface, to break the wafer into separate dice if the scribe does not penetrate the entire wafer thickness, and to sort the electrically good, separated dice into a form useful for the next assembly step.

Partial scribing may be accomplished with a pulsed laser beam, a diamond-tipped scribing tool, or a diamond-impregnated saw blade. Diamond sawing is preferred[2] for partial scribing and for full wafer separation, because it produces a much straighter edge with significantly less chipping and cracking.

The sorting process depends on whether the die is to be removed from its separated wafer array (sorted array) before interconnection[3] or during interconnection to the silicon bond pads (bonding and sorting simultaneously out of the array).[4]

13.3 DIE INTERCONNECTION

Die interconnection consists of two steps. In the first step, the back of the die is mechanically attached to an appropriate mount media. This attachment sometimes enables electrical connections to be made to the back of the die. The two common die-bonding methods are hard solder or eutectic, and polymers. In the second step the bond pads on the circuit side of the chip are electrically interconnected to the package. The three common interconnection schemes to the chip bond pads are wire bonding, tape automated bonding, and flip-chip solder bonding.

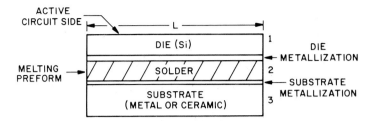

Fig. 2 The basic structure of a silicon device die bonded with a metal preform.

13.3.1 Eutectic Die Bonding

Figure 2 illustrates the fundamental aspects of a die bond.[6] Eutectic die bonding metallurgically attaches the die to a substrate material[6] (typically a metal leadframe made of a Cu or Fe-Ni alloy or to a ceramic substrate usually 90 to 99.5% Al_2O_3). Metallization is often required on the back of the die to make it wettable by the die-bonding preform, which is a thin sheet (usually <0.05 mm) of the appropriate solder-bonding alloy.[7] The substrate material is usually metallized with plated Ag (leadframes) or Au (leadframes or ceramic). Table 1 lists the compositions and melting points for solder-preform materials.[5]

Solder die bonding to refractory ceramic packages, which are to be hermetically sealed, or to Ni-Fe leadframes is usually performed with a Au or Au-2% Si preform. In both cases, the expansion coefficient difference between silicon and the substrate is relatively small (Table 2).[6-10] Also in both cases, in the presence of some mechanical scrubbing and temperatures above 370°C (the eutectic temperature), the preform reacts to dissolve the silicon. The Au-3.6% Si eutectic composition is reached and then exceeded. As the composition of the composite structure becomes more silicon rich, it freezes and the die bond is completed. Many applications require lower temperature or more ductile die-bonding solders to be compatible with other process steps.[6]

Table 1 Compositions and melting points for die attach preforms[5]

	Temperature (°C)	
Composition	Liquidus	Solidus
80% Au 20% Sn	280	280
92.5% Pb 2.5% Ag 5% In	300	
97.5% Pb 1.5% Ag 1% Sn	309	309
95% Pb 5% Sn	314	310
88% Au 12% Ge	356	356
98% Au 2% Si	800	370
100% Au	1063	1063

Table 2 Linear expansion coefficients, elastic moduli and thermal conductivities of materials used in packaging[6-10]

Material	Expansion coefficient α [(cm/cm/°C $\times$ 10^6)]	Elastic modulus E(GPa)	Thermal conductivity κ(W/cm-°C)	Application
90-99% Al_2O_3	6.5	262	0.17	Substrate
Beryllia (BeO)	8.5	345	2.18	Substrate
Common Cu alloys	16.3−18.3	119	2.64	Leadframe
Ni-Fe Alloys (42 alloy)	4.14	147	0.15	Leadframe
Au-20% Sn	15.93	59.2	0.57	Die bond Adhesive and lid Sealant
Au-3% Si	12.33	83.0	0.27	Die bond Adhesive
Pb-5% Sn	29	7.4	0.63	Flip-chip
Silicon	2.6	13.03	1.47	Electronic Circuit
Au	14.3	78	3.45	Wire metallurgy
Ag-loaded epoxy	53	3.5	0.008	Die bond Adhesive
Epoxy (Fused silica filler)	22	13.8	0.007	Molding compound

13.3.2 Polymer Adhesive Die Bonding

The chemical and structural properties of polymer adhesives for both lid sealing and die bonding have been reviewed.[9, 11, 12] Silver-filled epoxy adhesives are of major interest today as die bond materials, although polyimide adhesives are of increasing interest because of their ability to withstand higher processing temperatures than epoxies. The silver filling makes these materials both electrically conductive for low resistance between the die and the substrate, and thermally conductive to allow a good thermal path between the die and the rest of the package.

Epoxy die bond adhesives are used, because they are less expensive than the high-gold-content hard solders, they are flexible and soft, and they lend themselves to automation. For example, in the manufacture of plastic dual-in-line packages (DIPs), epoxy can be screened or stenciled onto die paddles at very high rates, and the die can be placed with high-speed pick and place tooling without the need for scrubbing. Metallization of the die and the substrate are still required. Because the epoxies are thermosetting polymers (crosslink when heated), they must be cured at elevated temperatures to complete the die bond. Typical cure temperatures range from 125 to 175°C. Table 2 lists several properties of a typical silver-filled epoxy.

In general, epoxy die bonds are as good or better than their metal counterparts, except in the most demanding applications, that is, applications that require high tem-

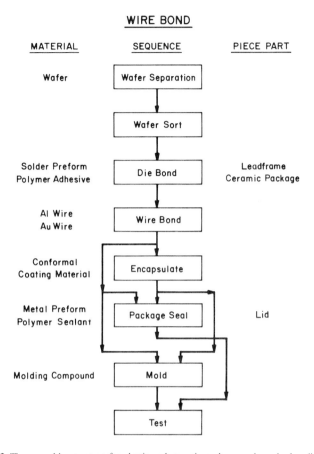

Fig. 3 The assembly sequence for plastic and ceramic packages using wire bonding.

peratures, high current through the die bond, critical thermal performance requirements, or with very surface sensitive devices bonded in a hermetic package.

13.3.3 Wire Bonding

Wire bonding is the most important interconnection method to silicon ICs. Figure 3 shows the assembly sequence for packages using wire. Wire bonding is always performed on ICs after they have been sorted from their separated array and die bonded. Wire bonding can be performed with gold wire by thermocompression, ultrasonic or thermosonic techniques, or with aluminum wire by the ultrasonic technique.

Typically, gold wire is ball-wedge bonded,[3] that is, ball bonding on the chip and wedge bonding on the package substrate as shown in Fig. 4, while aluminum wire is wedge-wedge bonded. One of the advantages of ball-wedge bonding is that a wedge bond can be performed on an arc around a ball bond (Fig. 4f). The physical shape of a wedge bond suggests that wedge-wedge bonds must be in line, that is, the orienta-

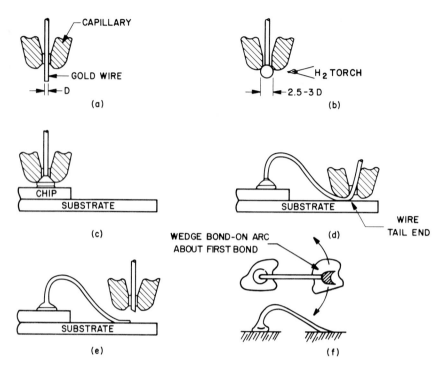

Fig. 4 Thermosonic ball-wedge bonding of a gold wire. (a) Gold wire in a capillary. (b) Ball formation accomplished by passing a H_2 torch over the end of the gold wire or by capacitance discharge. (c) Bonding accomplished by simultaneously applying a vertical load on the ball while ultrasonically exciting the wire (the chip and substrate would be heated separately to about 150°C). (d) A wire looped and a wedge bonded under load and ultrasonic excitation. (e) A wire broken at the wedge bond ready to go to step (a). (f) The geometry of the ball-wedge bond that allows high-speed bonding. Because the wedge can be on an arc from the ball, the bond head or package table does not have to rotate to form the wedge bond. *(After Stafford, Ref. 3.)*

tion of the wedge bond at the chip determines the direction of the wire that terminates in the package wedge bond. The versatility of ball-wedge bonding leads to greater automation potential.

Because a pure metal interconnection system is advantageous,[13] aluminum bonding to aluminum is better than gold wire bonding to aluminum. Intermetallic compound formation occurs in the gold to aluminum system. This compound formation is particularly important if the bond will go through high-temperature processing after wire bonding, such as in glass lid sealing.

In the following discussion we will review the theory of the ultrasonic bonding process,[14, 15] some metallurgical problems associated with gold-aluminum interactions, and design rules for wire bonding. Although thermocompression bonding of gold wire is widely used, it is being replaced by thermosonic bonding, and therefore will not be discussed further. Section 13.3.4 covers the thermocompression bonding process itself and also tape automated bonding.

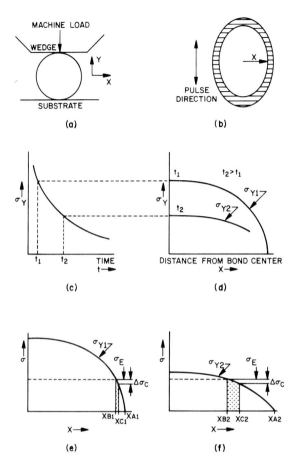

Fig. 5 The ultrasonic bonding process. (a) A wire cross section during wedge bonding. (b) Pulse direction during bonding. (c) The normal stress at the wire center as a function of bond time. (d) The vertical stress distribution for two different times as a function of distance from the bond center. (e) The interaction of the normal stress σ_{y1} at time t_1 and the pulse-induced horizontal stress σ_E which together define the bond zone XB_1 to XC_1. (f) The same as part (e) at a later time showing the growth in the bonded zone. *(After Mitchell and Berg, Ref. 15.)*

Ultrasonic bonding process Figure 5a shows a wire cross section during wedge bonding.[15] While the wire is under load, mechanical motion or pulsing of the bonding tool causes a wave pattern on the aluminum bond pad (Fig. 5b). Wave propagation through the wire causes the wavy pattern by creating a cutting action in the aluminum bond pad perpendicular to the pulsing direction. Before or during this wave motion, ultrasonic energy absorption by the wire softens it and under load it flows, breaking up surface oxides and exposing a fresh surface of both the wire and bond pad. These freshly exposed surfaces of metal cold-weld readily.

The bond is formed in a toroidal-shaped area around the center of the wire-pad contact inside the wavy pattern area. The center portion of the contact area is unbonded and shows no wavy pattern. Figure 5c to f shows a model illustrating the bond mechanics. As the wire softens and deforms, the vertical stress decreases with time. (The vertical stress is also a decreasing function from the center of contact outward). At the same time, the pulsing action of the bonding tool imposes a horizontal stress. At a high vertical stress, no bonding can occur because lateral motion is constrained. At a low vertical stress, the horizontal motion of the wire disrupts bonding. At intermediate vertical stresses, bonding does occur, and the bond area grows with time as the vertical stress decreases (Fig. 5f).

Metallurgical interaction of gold and aluminum The interaction of gold and aluminum during gold wire bonding has been studied extensively, especially for thermocompression bonding where significant intermetallic growth can occur. A purple intermetallic phase, Au_5Al_{12} (called purple plague), is commonly found in bonding, and it was thought that this phase was brittle and could lead to brittle failure of the bonded wire. Although there are still conflicting ideas about the rate of growth of the several possible intermetallics and the role of accelerators to that growth other than temperature, vacancy coalescence into voids (called Kirkendall voids) along the bond line[13] is now considered to be the cause of the brittleness of the bond. The vacancies are generated by the differential rate of diffusion of aluminum into gold and vice versa.

Design rules As the I/O count grows and the active device size shrinks, the space required for interconnection could represent a major fraction of the chip area. To avoid this problem, the effective bonding pad sizes and spacings must be reduced. Figure 6 shows the consequences of increasing I/O within the constraints of minimal extra use of silicon. For small chips and large I/O, the required effective bond pad centers will be considerably lower than today's 150 to 200 μm. Not only must the centers decrease, but the tolerances to produce quality bonds must also decrease.[16, 17]

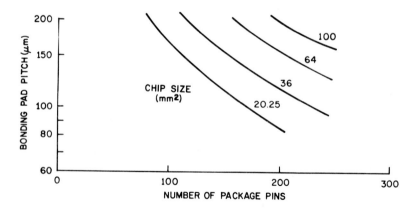

Fig. 6 Bonding-pad pitch plotted against chip lead count for several chip sizes. For each number of package pins the pitch is the minimum that can be accommodated on each chip with minimum increase of silicon area. *(After Otsuka and Usami, Ref. 16.)*

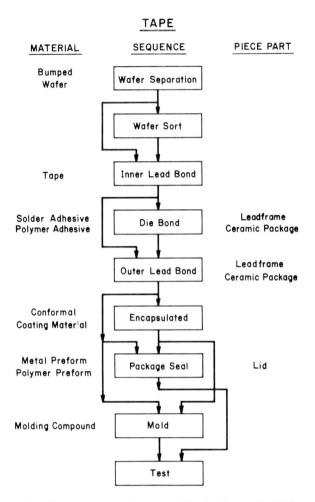

Fig. 7 Process sequence for the assembly of packages using TAB.

13.3.4 Tape Automated Bonding

Figure 7 shows the sequence for tape bonding the chip and assembling the package. Tape automated bonding (TAB) is a process in which chemically etched, prefabricated copper fingers, in the form of a continuously etched tape of repetitive sites, are simultaneously bonded by thermocompression or gold-tin eutectic methods, to every bump on a bumped silicon device. A bumped device is one in which gold bumps are deposited on the aluminum bond pads to provide a good surface for thermocompression bonding. The bumping process is described below. Figure 8 outlines the mechanics of making this bond.[4] Note that the sorting process is simultaneous with the bonding process (Fig. 8d and e) in this illustration. In some cases, it may be advantageous to bond the chips into tape from a previously sorted array.

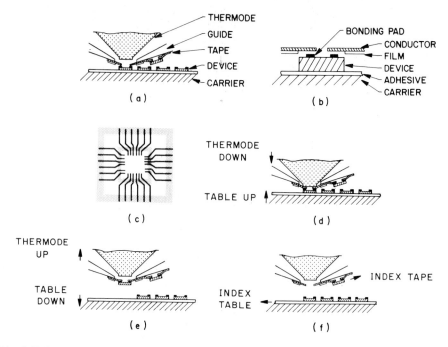

Fig. 8 TAB inner lead bonding process out of a wafer array. The tape shown in this figure is type b, cast polyimide on copper, as described in Fig. 10. (a) Tape and bumped die in alignment. (b) Magnified side view of the tape fingers and device after alignment. (c) Top view of the tape site showing polymer-supported tape fingers. (d) Heated thermode forcing tape fingers against bumps causing a thermocompression bond. (e) Heat from the bond releasing die from array. (f) Tape and wafer indexing to the next bond site. *(After Keizer and Brown, Ref. 4.)*

In one variant of this process the bumps are etched onto the tape rather than deposited on the wafer. Several aspects of tape technology require description including: the thermocompression bonding process; eutectic bonding; tape formation, type, and trade-offs; wafer processing; and application of the tape-bonded device to a package structure.

Thermocompression bonding process Figure 9 shows the thermocompression bonding process.[18] The lead to be bonded is modeled as a slab of width W_0 and height H_0 pressed between two rough, rigid, parallel plates with force, temperature, and time as variables (Fig. 9a). This model uses the case of a perfectly plastic material, that is, one in which the derivative of stress with respect to strain is zero, but which has a finite yield point. Good bonds are always accompanied by surface extension, that is, plastic shear strain.

Slip-line plasticity theory describes this process.[19] Figure 9b shows the slip-line field developed in the compressed slab. The slip-line field technique is a method of integrating the partial differential equations that describe flow of a rigid, perfectly plastic material. The resulting slip-lines represent the two orthogonal families of

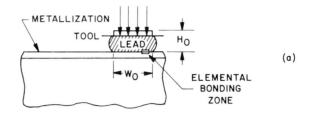

(a)

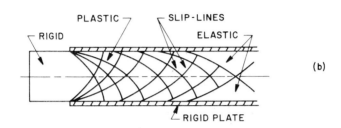

(b)

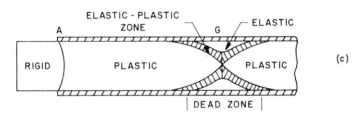

(c)

Fig. 9 Thermocompression bonding process. (a) Lead of initial width W_0 and height H_0 undergoing thermocompression bonding. (b) Slip-line formation as a result of deformation. (c) Plastically and elastically deformed zones of lead. No bonding occurs in the elastic zone (dead zone of the interface). *(After Ahmed and Svitak, Ref. 18.)*

curves whose directions at every point coincide with those of maximum shear strain rate. Since the directions of maximum shear stress and shear strain rate coincide for an isotropic material, the tangent at any point along the curve must be a maximum shear stress direction. The elastic-plastic boundary coincides with one of the contours of maximum shear stress in the elastic region, namely, along the contour on which the maximum shear stress is equal to the yield value (Fig. 9c).

The region of elastic strains is known as the dead zone. No bonding occurs in this zone, because it has no interfacial plastic shear strain. This situation is analogous to that noted in wire bonding in which the center portion of the lead is not bonded. For any specified vertical deformation, slip-line field theory predicts that the larger the ratio of W_0/H_0, the smaller the percentage of bond in the dead zone.

Bonding gold-plated copper tape ($W_0/H_0 \approx 2.3$) to a gold bump requires significant vertical deformation to achieve a small dead zone and to break through atmospheric contaminants, such as carbon, which are always present on real surfaces. Typical bonding parameters are 450 to 550°C thermode temperature, 0.5- to 2-s bond time, and 275- to 480-MPa bond pressure.[4]

Eutectic bonding Another approach to bonding TAB leads to gold bumps employs tin-coated copper tape (0.5 μm of tin) that can be eutectically soldered to the bump.[20] Because the Au-Sn eutectic forms at 280°C, this bonding technique requires relatively high temperatures. Low pressures (deformation) can be used, however, since the bond is formed by metallurgically reacting the tin and gold of the bump through a liquid intermediary.

Tape formation and types Figure 10 schematically shows the five common types of tape. One-layer tape (Fig. 10a) is made by single-sided patterning and etching. The metal is typically rolled, annealed copper that is 33 μm thick. It is usually plated with a thin layer of gold to improve its bondability.[21] This tape is ideally suited for low-cost DIP assembly, because its sprocket holes are etched simultaneously with its bonding fingers, making it very precise for high-speed inner lead bonding. Note that the wafer must be bumped and the device cannot be tested after bonding, because the tape is all metal. The bonding fingers can also be very fragile if they are long.

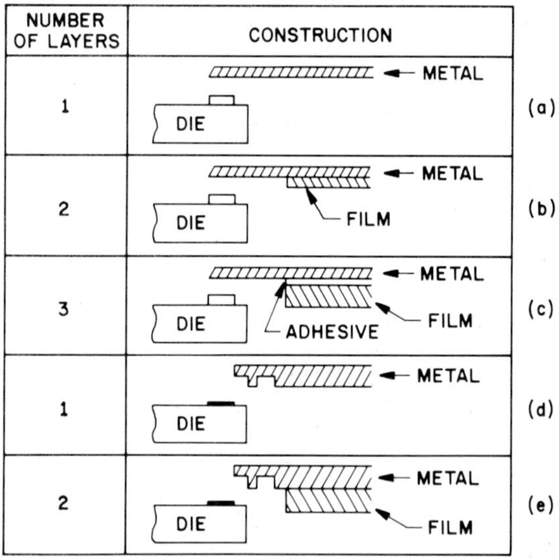

NUMBER OF LAYERS	CONSTRUCTION	
1	METAL / DIE	(a)
2	METAL / FILM / DIE	(b)
3	METAL / FILM / ADHESIVE / DIE	(c)
1	METAL / DIE	(d)
2	METAL / FILM / DIE	(e)

Fig. 10 Various tape types and silicon die configurations. (a) One-layer tape made by single-sided patterning and etching. This type is suitable for low-cost DIP assembly. (b) Two-layer tape made by plating copper onto a polyimide film, by casting a polyimide film onto copper, or by screening a polyimide film on copper. The polyimide increases rigidity. (c) Three-layer tape made from prepunched 75- to 125-μm-thick polyimide film bonded with epoxy to the copper. This tape is testable and is, thus, likely to be the most widely used for VLSI applications. (d) One-layer bumped tape made by photopatterning and etching 66-μm copper from both sides. (e) Two-layer film supported bumped tape made by the screened film technique. This tape is testable whereas the one-layer bumped tape is not. *(After Cain, Ref. 21.)*

Two-layer tape (Fig. 10b) can be made by plating copper onto a polyimide film, by casting a polyimide film onto copper, or by screening a polyimide film on copper. The first two processes use photopatterning and etching of the polyimide. The polyimide in the two-layer tape makes it more rigid than the one-layer tape. The two-layer tape provides better localized support for long tape fingers (Fig. 8c).

Three-layer tape (Fig. 10c) is made from prepunched 75- to 125-μm-thick polyimide film adhesively bonded with epoxy to copper. Polymers other than polyimide can be used. After bonding, devices made with this tape construction can be tested. Since the dielectric carrier is relatively stiff, the tape fingers can be extended to probe points on the dielectric to provide testing access.

Figure 10d shows one-layer bumped tape. This tape is usually made by photopatterning and etching 66-μm copper from both sides. Advantages include increased rigidity of the tape compared to 33-μm-thick one-layer tape and the potential to eliminate extra wafer metallization steps. This potential has not been fully realized as will be discussed later.

Two-layer film-supported bumped tape (Fig. 10e) is made by the screened film technique. This tape, unlike the one-layer bumped tape, forms devices that can be tested.

Both the two-layer and three-layer tape have the greatest potential for VLSI applications. Because these tapes can be made with 50-μm-wide bond fingers on 100-μm centers, very high bonding density to the IC can be achieved. In addition, the three-layer tape offers testability of the chip and the potential burn in of the chip prior to being committed to a package. Figure 11 shows a 144-lead testable, three-layer tape.

Wafer processing Two basic methods are used for preparing chips for tape bonding. The method chosen depends on the tape-die assembly approach.

The first and simplest method involves no additional preparation of the aluminum bond pads beyond opening up the passivation in the bond pad areas in the usual way. This type of wafer preparation method would be used with one- and two-layer bumped tapes. A bumped tape has a low value of W_0/H_0 (Section 13.3.4) for bonding; thus large deformations are required to achieve good bonds. Large deformations imply large vertical stresses that can fracture the underlying dielectric layers or the silicon. The input of ultrasonic energy into the bond would be of great assistance, but this technique is not yet available. The major advantage of bumped tape is that no additional wafer processing is required beyond that typically used for wire bonding.[22]

The most successful TAB processing method uses plated-gold or gold-capped copper bumps over the aluminum bond pads with one-, two-, or three-layer tape formats (Fig. 10a, b, or c).[23] Figure 12 shows the process sequence to make gold bumps on aluminum metallurgy devices after the passivation over the bond pads is opened. The selective etching of the evaporated or sputtered layers (Fig. 12f) is a critical step in preserving yield. If chemical strippers are used for the sputtered or evaporated layers, they can penetrate defects in the passivation and dissolve the easily corroded aluminum-interconnect metallurgy. Several processes, including plasma stripping, will avoid this yield loss. All of the extra wafer processing costs for TAB are loaded

Fig. 11 A 144-lead, testable TAB site. The site was made with three-layer tape. The leads are tin plated for eutectic bonding to the die.

onto the good die in a wafer. This cost, plus losses due to aluminum etching, could make wafer bumping very unattractive in a one-to-one comparison with wire bonding.

Application to package structures Tape automated bonding is preferred in two major areas: For small, high-yield die used in DIP assembly[4] where the cost per device to bump is low and the yield loss due to metal stripping is also low (because the dice are small), and for VLSI where there are overriding systems advantages. An example would be a multichip module for a computer where the TAB devices are assembled in testable tape. Each device can be thoroughly exercised electrically or burned-in before assembly[24] which is not possible with a bare aluminum-wire bond device.

13.3.5 Flip-Chip Technology

Flip-chip technology has been well described in a series of papers.[25, 26, 27] Figure 13 shows the essence of the flip-chip technology on which the flip-chip interconnect,

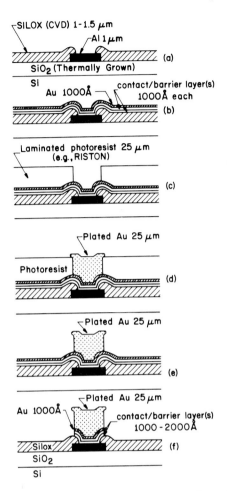

Fig. 12 Process sequence for making gold bumps on aluminum metallurgy devices. (a) A wafer (as-received) that is cleaned and sputter etched. (b) A contact barrier layer (which also serves as a conductive film for electroplating) sputter-deposited with a layer of gold for oxidation protection. (c) A thick-film photoresist (25 μm) laminated and developed. (d) Gold electroplated to a height of approximately 25 μm to form the bumps. (e) The resist being stripped. (f) The sputter-deposited conductive thin films are removed chemically or by back-sputtering. *(After Liu, Rodrigues de Miranda, and Zipperlin, Ref. 23.)*

made with solder balls, is built. Figure 14 outlines the assembly sequence for a flip-chip package. The process begins with sequential evaporation of Cr, Cu, and Au through a metal mask onto all of the aluminum bonding pads on a wafer. The pads may be located at any point on the active die surface subject to some constraints which will be discussed later. The gold protects the thin-film structure from oxidation until a second evaporation of Pb-Sn is made onto the Cr-Cu-Au cap. The Pb-Sn evaporation is deposited over a larger area than the capped bond pad. The area and height of this deposition determines the ultimate ball size. The structure after evaporation is shown by the dotted line in Fig. 13. After evaporation, the structure is reflowed in a reducing atmosphere where the Pb-Sn recedes from oxide surface due to surface

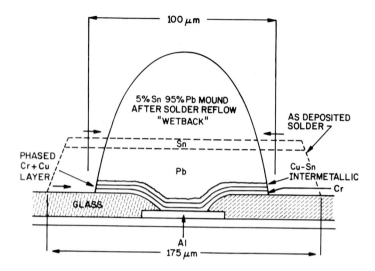

Fig. 13 The cross section through a solder ball as deposited and after reflowing. *(After Totta and Sopher, Ref. 25.)*

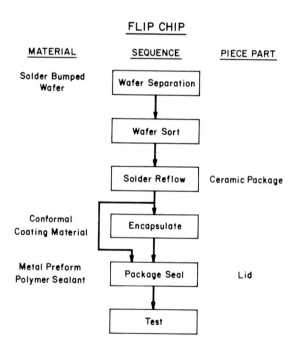

Fig. 14 Process sequence for the assembly of a flip-chip device to a ceramic substrate.

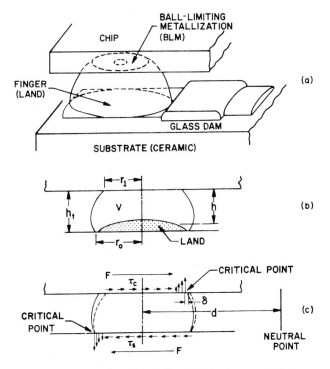

Fig. 15 A flip-chip technology procedure. (a) A flip-chip joint is connected to a ceramic substrate. (b) A cross section through the joint showing the three critical design parameters h, r_0, and r_1. (c) The stresses τ_c at the chip-bump interface and τ_s at the chip-substrate interface, and displacements δ that arise during a thermal excursion. *(After Goldmann, Ref. 26.)*

tension forces and forms a mound or solder ball with its base determined by the Cr-Cu-Au cap (also called the ball-limiting metallization, BLM).

Figure 15a shows schematically a solder-bumped flip-chip attached to a ceramic substrate. This section discusses this structure with respect to geometric optimization of the solder bump structure.[26] The objective in optimization is to minimize shear strain in the bulk of the solder during a temperature cycle and to achieve the highest possible strength at the interfaces between the chip and the substrate. Only the main results are presented here.

The fracture strength of the interfaces is optimized by selecting a joint-substrate area πr_0^2, such that in a temperature cycle test or a torque test, the two interfaces fracture in equal proportions (Fig. 15b). This condition is achieved by achieving equality of the interfacial stresses, τ_c and τ_s (Fig. 15c). The value of r_1 is usually fixed by many considerations, including the minimization of the silicon area allocated to interconnection. The optimum ratio of r_0/r_1 is defined as K. The size of the square bond pad (land) is then

$$2r_0^* = \sqrt{K}\ r_1 \tag{1}$$

where $2r_0^*$ is the edge length of the bonding land.

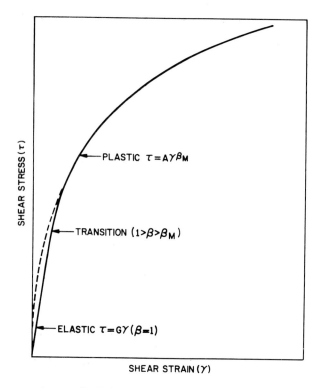

Fig. 16 Shear stress-strain curve for Pb-Sn solder. In the plastic region the work-hardening exponent is β_M, where M is the power law coefficient in the Coffin-Manson fatigue equation. *(After Totta and Sopher, Ref 25.)*

The next step is to determine the value of the strain-hardening exponent β of the solder material. Figure 16 shows a stress-strain curve for solder in shear. We assume that the Coffin-Manson[28] equation applies for solder in fatigue in the plastic zone, and that the strains generated in any fatigue cycle of interest are mainly plastic strains. The Coffin-Manson fatigue relationship may be written as

$$N_F \propto \left[\frac{1}{\gamma_p} \right]^m \qquad (2)$$

where N_F is the number of cycles to failure in a cycling experiment, γ_p is the constant plastic strain in each cycle, and m is an observed parameter equal to about 2. Because we are dealing with the plastic strain region, the strain-hardening exponent is denoted by β_m to indicate the strain region where Eq. 2 is applicable.

The number of cycles to failure for the solder ball case is given by[26]

$$N_F = \left[\frac{1}{\gamma} \right]^m C_T C_S \qquad (3)$$

where γ is the local shear strain; m is 1.9 for Pb-5% Sn solder[27]; C_T is a function of temperature cycles, strain rate, and dwell times; and C_S is a function of material properties. In a mechanical destructive test the number of cycles to failure is $\frac{1}{2}$ cycle and the shear strain is γ_μ, the ultimate strain. Thus

$$\frac{1}{2} = \left[\frac{1}{\gamma_\mu} \right]^m C_T' C_s \tag{4}$$

where C_T' is a function only of the test cycle. Dividing Eq. 3 by Eq. 4 gives

$$N_F = K_T \left[\frac{\gamma_\mu}{\gamma} \right]^m \tag{5}$$

where K_T is a function only of the destructive test method used to determine γ_μ.

In a cycling test, failure begins at that cross section where γ_μ / γ is a minimum. Fracture is usually observed to be near the interfaces. To determine γ_μ, observe from Fig. 16 that

$$\tau = A \gamma^\beta \tag{6}$$

where A and β are constant. We use β here rather than β_m to simplify the algebraic expressions. Equation 5 becomes

$$N_F = K_T \left[\frac{\tau_u}{\tau} \right]_f^{m/\beta} \tag{7}$$

where the f denotes the critical interface where fracture begins, and τ_μ is the shear stress associated with γ_μ through Eq. 6.

The average shear stress is assumed to be the shear force F divided by the cross-sectional area V/h (Fig. 15b), and the average shear strain is δ/h, where δ is the shear deformation of the solder pillar (Fig. 15c). Thus, substituting these values into Eq. 6 gives

$$\frac{Fh}{V} = A \left[\frac{\delta}{h} \right]^\beta \tag{8}$$

or

$$F = \frac{AV \delta^\beta}{h^{1+\beta}} \tag{9}$$

The average shear stress τ at an interface of cross-sectional area πr^2 is then

$$\tau = A \delta^\beta \frac{V}{\pi r^2 h^{1+\beta}} \tag{10}$$

Substituting into Eq. 7, we have

$$N_F \;=\; K_T \left[(\tau_\mu \pi r_f^2) \left[\frac{h^{1+\beta}}{AV} \right] \right]^{m/\beta} \left[\frac{1}{\delta} \right]^m \tag{11}$$

where r_f is the radius of the critical fracture interface.

In this expression $\tau_\mu \pi r_f^2$ is a measure of the ultimate shear strength of the joint. This term should be as large as possible. The term $h^{1+\beta}/AV$ is related to the joint compliance which should be maximized. The value of the shear deformation is easily shown to be

$$\delta \;=\; (\alpha_{cer} - \alpha_{Si}) \Delta Td \tag{12}$$

where α_{cer} is the expansion coefficient of the substrate, α_{Si} is the expansion coefficient of the silicon, and d is the distance from the neutral point of the chip (usually one-half the chip diagonal).

A simple approximation of the solder bump shape may be obtained by assuming that it is a right circular cylinder with different top and bottom areas. The volume is given by

$$V \;=\; \frac{\pi h}{6} [h^2 + 3(r_0^2 + r_1^2)] \tag{13}$$

This equation shows the interdependence of V and h for a given r_0 and r_1, and allows an optimization of the compliance term $h^{1+\beta}/V$.

In VLSI chips several complications arise because not all of the solder bumps are located at equal distances from the neutral point (Fig. 15c). In addition, as d increases and no change is made in joint dimensions, fatigue life decreases. If fatigue life is to be maintained, r_1 can be increased, along with other dimensions, to maintain optimization. Larger r_1 means a larger clearance area for solder deposition and thus an increased distance between bond pads. Increasing the distance requires more silicon area. The choice of solder ball size and shape is then a complex trade-off of silicon cost, reliability, circuit speed, and many other variables.

Two major benefits of flip-chip technology are the ability to have area-array bond pads (compared to bond pads just around the edge) and very short lengths of interconnect that minimize their inductance. Two major shortcomings of this technology are the inferior thermal performance (compared to a die-bonded chip) and the difficult routing scheme required to access the area array of bonding pads in the package.

13.4 PACKAGE TYPES AND FABRICATION TECHNOLOGIES

Single-chip packages are generally based either on refractory ceramic technology or on metal leadframes and molded plastics. Ceramic packages are usually used for state-of-the-art devices where maximum reliability is required. For more mature products where low cost is critical but which still require a hermetic seal, Cerdip technology is employed. This technology uses a combination of leadframe, dry-pressed refractory ceramic parts, and glass sealing. Plastic packaging is usually better for more mature products where cost is paramount and hermeticity is not required. Until recently, plastic had implied lower reliability; now, however, plastic technology has improved and is highly reliable with proper process control.

This section discusses processing technologies and the types of packages. The processing technologies are generally applicable to a variety of package types. The end of this section briefly examines high-performance ceramic packages.

13.4.1 Ceramic Package Technology

Figure 17 illustrates multilayer ceramic technology.[29, 30] A dispersion or slurry of ceramic powder and a liquid vehicle (solvent and plasticized resin binder) is first prepared, then cast into thin sheets by passing a leveling or doctor blade over the slurry. After drying, the sheets are cut to size, via holes (holes through the dielectric layers through which interconnects are made) and cavities are mechanically punched into the sheet, custom wiring paths are screened onto the surface (usually a slurry of tungsten powder), and the via holes are filled with metal. Several of these sheets are press-laminated together in a precisely aligned fixture, and the entire structure is fired at 1600°C to form a monolithic sintered body. The refractory ceramic technology is a complex process requiring careful process control throughout.

After the laminate is sintered, it is ready for the finishing operations of lead attachment and metallization plating.[8] Nickel is plated over the tungsten in preparation for lead brazing. The lead material is a Fe-Ni-Co alloy called Kovar; the brazing material is a silver-copper eutectic alloy. All exposed metal surfaces are electroplated

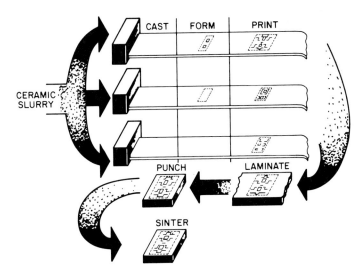

Fig. 17 Process sequence to create a laminated, refractory ceramic product starting from a ceramic slurry. *(After Gardner and Nufer, Ref. 29.)*

or electrolessly plated (usually gold over nickel) for bondability and environmental protection. Multilayer ceramic packages can be made up to 100 by 100 mm in lateral dimensions to a tolerance of ±0.5% and with up to 30 tape layers in the most advanced processes.

Ceramic package technology is very effective for constructing complex packages with many signal, ground, power, bonding, and sealing layers. It does have, however, three technical areas of weakness: hard to control tolerances caused by the high shrinkage in processing, high dielectric constant of ($\epsilon_i/\epsilon_o = 9.5$), and modest thermal conductivity of Al_2O_3 (Table 2). The tolerance problem makes it difficult to use edges as accurate references, and the high dielectric constant affects signal-line capacitive loading. The use of beryllia (BeO) modules, instead of Al_2O_3, would produce greatly superior thermal performance and a significantly lower dielectric constant. Although not in common use today, BeO packages will be used more in the future because of their potential for superior performance.

13.4.2 Refractory Package Technology

Figure 18a to f illustrates a variety of potential refractory package types. Packages can have brazed pins or leads and edge or array pinouts. Figure 19a, b, and c shows three packages: DIPs, chip carriers, and pin-arrays, respectively. These packages have certain standardized features. DIPs have leads along their edges on 2.54-mm (100-mil) centers; pin-grid arrays typically have pins on a 2.54-mm area-array grid. If the die cavity is on the side with the pins, a portion of the area-array would be miss-

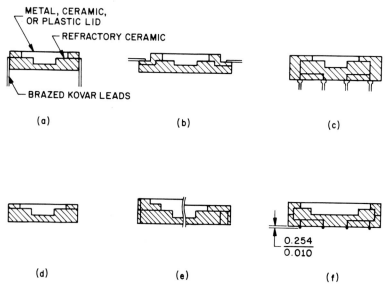

Fig. 18 Cross-sectional sketches of several package types. (a) Side brazed. (b) Top brazed. (c) Pin-array. (d) Leadless with edge metallization. (e) Leadless with via holes. (f) Leadless-array package with standoffs. *(After Kyocera, Ref. 8.)*

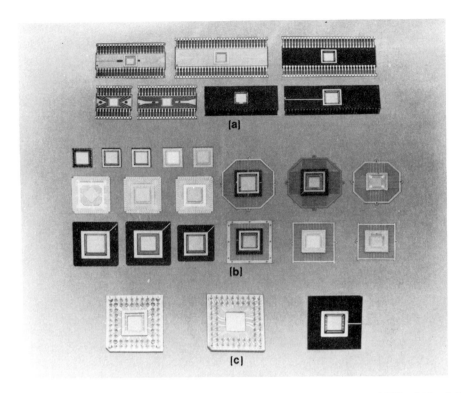

Fig. 19 Several refractory ceramic package types. (a) Both top-brazed and side-brazed DIPs. (b) Leaded and leadless chip carriers. (c) Pin-arrays with cavity up or down. *(After Kyocera, Ref. 8.)*

ing as shown for the pin-grids in Fig. 19c. Standard chip carriers are four-sided devices with leads or contacts on 1.02-mm (40-mil) or 1.27-mm (50-mil) centers. These packages have been standardized by the Joint Electron Device Engineering Council (JEDEC).[31] At present, packages have been standardized up to about 156 leads. Chip carriers with 0.51 and 0.635-mm (20- and 25-mil, respectively) pitches will also be standardized. Figure 20 shows various standard leaded and leadless carriers.

13.4.3 Glass-Sealed Refractory Technology and Packages

Figure 21 illustrates a lower-cost ceramic technology for single-layer DIPs. This technology relies on glass sealing a leadframe between two pressed ceramic units using low temperature glass. The glasses used for glass sealing are basically PbO-ZnO-B_2O_3 types. Both crystallizing and noncrystallizing glasses are used. Sealing is usually performed above 400°C in an oxidizing ambient. Since gold-aluminum intermetallic growth would be severe at these temperatures, aluminum wire is used. The leadframes usually have a stripe of sputtered aluminum deposited on their lead tips so that an all-aluminum bonding system is possible. The glass-sealed package, which is hermetic, has the potential of being highly automated, competing with plastic technology for low-cost packaging.

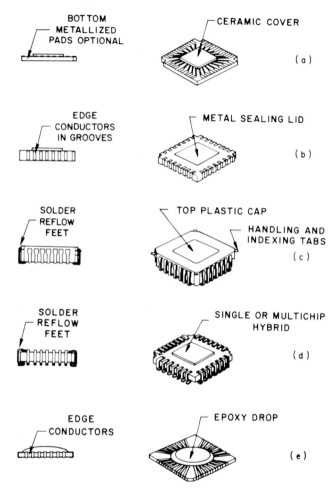

BOTTOM
METALLIZED
PADS OPTIONAL

CERAMIC COVER

(a)

EDGE
CONDUCTORS
IN GROOVES

METAL SEALING LID

(b)

SOLDER
REFLOW
FEET

TOP PLASTIC CAP

HANDLING AND
INDEXING TABS

(c)

SOLDER
REFLOW
FEET

SINGLE OR MULTICHIP
HYBRID

(d)

EDGE
CONDUCTORS

EPOXY DROP

(e)

Fig. 20 Standard JEDEC chip carriers. (a) Leadless type A carrier. The particular leaded type A package shown here is premolded. (b) Leadless type B carrier. Both the leaded A and B types show the leads bent under the package per the JEDEC standard. (c) Leaded type A carrier. (d) Leaded type B carrier. (e) Mini-Pak (an epoxy glass printed circuit with die-bonded and wire-bonded chip, epoxy encapsulated).

13.4.4 Plastic Molding Technology

Figure 22 shows a cross section of a plastic DIP. A few of the critical steps in successfully producing a molded product are selection of a molding material that meets the reliability objectives of the molded device and control of the molding process itself. Overall integrated automation of the assembly process, in particular, the automation of the die bonding, wire bonding, and molding steps, is also important.

Two basic types of molded devices are postmolded[32] as illustrated for a DIP in Fig. 22, and premolded[33] as shown for a leaded chip carrier in Fig. 23. The postmolded device uses thermosetting (cross linking) silicone, epoxy silicone, or epoxy resins and is molded around the leadframe-chip assembly after the chip is bonded to

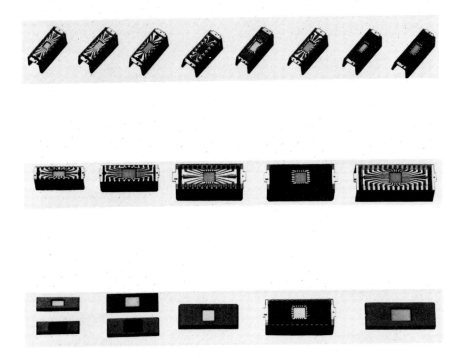

Fig. 21 Pieceparts and partially assembled glass-sealed Cerdip packages. The ceramic bodies are made by dry pressing alumina. The leadframe material is Kovar, an Fe-Ni alloy.

the leadframe. The postmolding process is relatively harsh. To avoid the exposure of the die and its wire or tape bonds to viscous molding material, the premolded package concept was developed.[33] Here the package is molded first, and then the chip and interconnects are added. The molding may be performed with the thermosets mentioned above or with thermoplastic (melting) polymers, such as polyphenylene sulfide. The premolded package is the plastic equivalent of the refractory ceramic cavity package and has great future potential for VLSI devices. Only the postmolded device and its material, and construction will be discussed here in detail.

The polymers most commonly used for IC packaging are epoxies and silicones.[34] The original epoxy resin, used to mold ICs, was made by condensing epichlorohydrin with bisphenol-A to produce a material called Epoxy-A. An excess of epichlorohydrin is used to leave epoxy groups on each end of the low-molecular-weight polymer.

Today, novolac epoxies are generally preferred because their higher functionality gives them heat resistance, derived from having each repeating group contain an

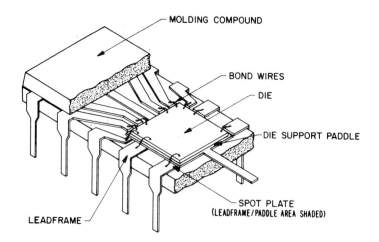

MOLDING COMPOUND

BOND WIRES

DIE

DIE SUPPORT PADDLE

SPOT PLATE
(LEADFRAME/PADDLE AREA SHADED)

LEADFRAME

Fig. 22 Ball- and wedge-bonded silicon die in a plastic (DIP). The die support paddle may be connected to one of the external leads. For most commercial products, only the die paddle and the wedge bond pads are selectively plated. The external leads are solder plated or dipped after package molding. *(After Howell, Ref. 32.)*

Fig. 23 Two different premolded package sizes. (a) A 24-lead carrier as molded in its carrier frame. (b) 44-lead carrier with leads trimmed and formed. (c) 24-lead chip carrier socket. The footprint of this socket is identical to the footprint of the carrier shown in (d). (d) 24-lead chip carrier in completed form with lid applied. (e) 24-lead carriers in its socket. *(After Levinthal. Ref. 33.)*

epoxy group. These resins, called Epoxy-B, are made by reacting epichlorohydrin with novolac phenolic resin and a base. The synthesis of the resins produces sodium chloride as a by-product. Both sodium and chloride ions are deleterious to device reliability, and, therefore, these by-products must be carefully washed from the resins before they are compounded into useful molding compounds.

Silicones are basically polysiloxanes which can be represented by the formula

$$\left[\begin{array}{c} R_1 \\ | \\ - \; Si \; - \; O \\ | \\ R_2 \end{array} \right]_n$$

where R_1 and R_2 are simple organic groups. Silicone resins are made by cohydrolysis of mixtures of mono-, di-, and tri-methylchlorosilanes or mono-, di-, and tri-phenylsilanes (or alkoxysilanes) in solvents. Usually the silicone cure has no by-products, and silicones are relatively free of sodium and chloride ions.

Fillers (usually amorphous or crystalline SiO_2, glass fibers, and sometimes Al_2O_3) are added to the resin so that the resultant mixture is 65 to 73% by weight filler. Amorphous SiO_2 is used when a minimum expansion coefficient is desired at some sacrifice of thermal conductivity, while crystalline SiO_2 improves thermal conductivity at the expense of the expansion coefficient. Al_2O_3 has high thermal conductivity (as a filler material) but is very abrasive to molds. Fillers greatly improve the mechanical strength of the resin, reduce its expansion coefficient, and therefore reduce its shrinkage after molding. Small amounts of pigments, mold release agents, antioxidants, water getters, plasticizers, and flame retarders must also be added to complete and optimize the molding resin.

The rheological, chemical, and thermophysical properties of molding compounds are important to understand both for the molding process and for their interrelation to the reliability of the finished device. The two most important characteristics that must be evaluated for a molding compound are its moldability and expansion coefficient. The molding process is not cost effective unless it has high yield, and the most common reliability problems can be related to the mechanical quality of the molded part and its ability to withstand thermal stresses. As will be discussed, the expansion coefficient is the critical variable in the thermal stress problem.

Molding process Thermoset molding materials are usually transfer molded (Fig. 24). After entering the pot, the preheated molding compound, under pressure and heat, melts and flows to fill the mold cavities containing leadframe strips with their attached ICs. The IC leadframe often has long, fragile lead fingers, and the die is interconnected to these thin, fragile leads with 25- to 30-μm-diameter gold wire. To avoid damaging this fragile structure, the viscosity and velocity of the molding compound must fall within certain ranges. Commercial molding compounds are designed to meet these requirements when molded at approximately 175°C and at pressures of about 6 MPa. The mold cycle requires from 1 to 5 minutes.

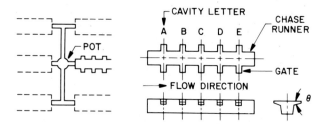

Fig. 24 Schematic of a typical multicavity mold with uniform runners and gates. *(After Kaneda et al., Ref. 36.)*

To control the velocity of the molten molding compound, each device cavity has a gate or restriction to slow the material flow. The fluid mechanics of molding are relatively complex, because the materials are non-Newtonian.[35] In addition, partial cross linking can occur during the molding process affecting the material viscosity. Elaborate mold designs have been proposed to compensate for these variables.[36]

Thermomechanical stresses in molded parts Mismatch between the expansion coefficient of the plastic molding compound and of the silicon die or gold wire are directly responsible for stresses induced after molding and also during temperature cycling. The stresses are induced in the bulk of the die and its surface passivation and in the interconnection to the die, typically gold wire.

Figure 25a and b shows a DIP cross section and an enlarged view of the wire bond.[35] For modeling purposes,[37] chip and leadframe may be combined as a structural member that resists the shrinkage of the plastic (Fig. 25c).

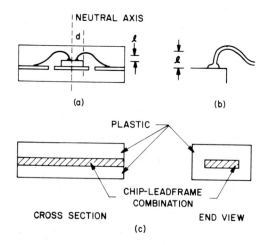

Fig. 25 Cross sections of a device in plastic. (a) The actual DIP cross section. (b) Close-up of a wire. (c) Cross section and end view of the device for purposes of modeling the die bond. The chip and leadframe are assumed to act as one member as described in the text. *(After Usell and Smiley, Ref. 37.)*

The strain in the chip-leadframe combination, γ_c, at temperatures below the mold and cure temperature is given by

$$\gamma_c = \frac{1}{E_c} \left[\frac{F_c}{A_c} \right] + \alpha_c \, \Delta T \tag{14}$$

where F_c is the force exerted by the plastic, E_c is the effective modulus, A_c is the effective cross-sectional area, α_c is the thermal expansion coefficient, and ΔT is $T_{actual} - T_{mold}$.

The expansion coefficients of the leadframe and chip are assumed to be equal in this analysis; thus,

$$A_c \, E_c = A_s \, E_s + A_L E_L \tag{15}$$

where E_L is the modulus of the frame, E_s is the modulus of the chip, A_s is the area of the chip, and A_L is the area of the leadframe.

The assumption of equal expansion coefficient of silicon and leadframe is reasonable for silicon rigidly attached to 42 alloy. However, for silicon attached to copper with epoxy, $A_c \, E_c \simeq A_s \, E_s$, because the silicon and copper would essentially be decoupled due to plastic deformation in the epoxy die bond. Continuing with the above analysis, the strain in the plastic is given by

$$\gamma_p = \frac{1}{E_p} \left[\frac{F_p}{A_p} \right] + \alpha_p \, \Delta T \tag{16}$$

where the subscripts refer to plastic. Since the plastic and the chip are joined,

$$\gamma_p = \gamma_c \tag{17}$$

and equilibrium requires that

$$F_p = - F_c \tag{18}$$

Combining Eqs. 15, 16, 17, and 18 gives

$$\gamma_c = \frac{(\alpha_p - \alpha_c) \, \Delta T}{1 + \dfrac{E_c \, A_c}{E_p \, A_p}} + \alpha_c \, \Delta T \tag{19}$$

The second term, $\alpha_c \, \Delta T$, can usually be neglected. Because α_p and E_p are both functions of temperature, Eq. 19 should be rewritten as

$$\gamma_c = \int_{T_{mold}}^{T} \frac{[\alpha_p(T) - \alpha_c] \, dT}{1 + \dfrac{E_c \, A_c}{A_p \, E_p(T)}} \tag{20}$$

Curve a in Fig. 26 shows the strain as a function of temperature for a molding compound in which both $\alpha_p(T)$ and $E_p(T)$ were measured. Curve b in Fig. 26 shows

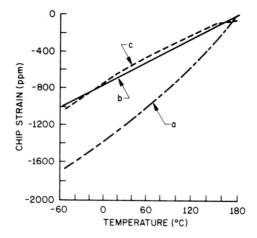

Fig. 26 Measured and calculated values of strain in a chip, molded in epoxy, as a function of temperature. Curve a was calculated analytically with the simple model described in the text, curve b was calculated from finite element modeling, and curve c shows the measured results. *(After Usell and Smiley, Ref. 37.)*

the results of a more sophisticated analysis using finite elements in which temperature dependencies of α_p and E_p are ignored. Curve c in the same figure shows the strain calculated from actual measurement. Measured strains were found to be within 10 to 15% of the finite element results and therefore much lower than predicted by a simple one-dimensional model. The model is useful, however, in estimating the magnitude of the effect and clearly shows the advantage of a plastic with low α_p. The actual strains are large enough to possibly cause fatigue of surface metallurgy on the die or parametric shifts in device parameters.

Referring again to Fig. 25a and b, it is clear that the relative motion of plastic and wire caused by temperature cycling induces strain in the wire. For vertical motion of the wire, we have[10]

$$\gamma_{gold} = (T - T_0)(\alpha_p - \alpha_{gold}) \qquad (21)$$

where T is the temperature and T_0 is the ambient temperature. This analysis ignores any compressive stresses in the wire that result from cooling down from the molding temperature. The wire and the plastic are assumed to be tightly adherent. For a $T - T_0$ of 100°C and an $\alpha_p - \alpha_{gold}$ of 10×10^{-6} cm/cm-°C, the strain in the wire would be approximately 1000 ppm. Considering that 2000 ppm strain is usually defined as the yield point of a metal, this value puts the wire close to the plastic zone. The wire should be amenable to cyclic fatigue analysis with the classical Coffin-Manson approach,[28] although the exponent m in the fatigue equation may be much larger than the value of 2 we observed for solder.

For large die or large values of d (Fig. 25a), lateral strains in the wire are also important. If the plastic is not adherent to the die, the lateral displacement of the plastic versus the silicon causes large strains in the wire as

$$\gamma_{Au} = d(\alpha_p - \alpha_{gold})\Delta T / l \qquad (22)$$

where l is the distance from the die surface as shown in Fig. 24b. This equation works only for l values above the ball. When the molding compound is adherent to the silicon, the situation is much more complex, but the strains are much lower. These analyses clearly show that selecting a plastic with as low an expansion coefficient as possible helps in minimizing thermal-mechanical reliability problems in molded plastic devices.

13.4.5 High-Performance Packages

A high-performance package is designed to meet one or several difficult design characteristics such as size, pin count, operating frequency, or thermal capability. The assembly techniques already described, and the applications considerations (Section 13.6) suggest the ways in which such design objectives may be met. One specific example of a high-performance package is outlined below.

Figure 27 shows a package proposal that is aimed at meeting high-performance objectives.[38, 39] This package uses tape technology to contact an area array of bond pads. The figure shows what is basically a die-bonded flip-chip in a pin-grid array in which the die and the package are interconnected by a metallized polymer. This type of package has several important features. The pin-grid array is the densest package type that has contacts on a 2.54-mm spacing. The die bonding of the flip-chip device solves some of its inherent thermal dissipation problems, yet the area-array contacts maintain short lead lengths. The polymer interconnect has a lower dielectric constant than ceramic and can yield lower capacitive loading. The polymer interconnect should have ground and power planes for optimum electrical performance.

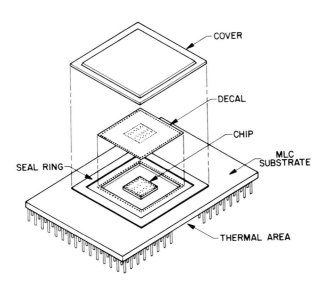

Fig. 27 Pin-grid package with an array flip-chip like device and a polymer interconnect between the die and the package. *(After Lyman, Ref. 38, and Robbins, Ref. 39.)*

13.5 SPECIAL PACKAGE CONSIDERATIONS

This section looks at four aspects of packaging that must be addressed to have a complete package assembly methodology. They are cleaning, package or lid sealing, encapsulation, and α-particle protection.

13.5.1 Cleaning

The most critical cleaning step the die undergoes is performed before bonding, encapsulation, or final lid seal. The cleaning process must be chemically compatible with the die metallurgy. Aluminum has a very narrow range of pH values in which its oxide protects it from corrosion in aqueous solution.[40] Corrosion reactions, with metal dissolution, can occur in both basic and acidic solutions. Two objectives are usually sought in cleaning. One is the removal of organic species that can affect bondability; an organic solvent is usually required for this. The second is the removal of ionic species that can cause corrosion during the life of the device, or in an unusual instance, contribute to surface charge accumulation. Water is a very good solvent for ionic species. Because environmental considerations have curtailed the use of powerful organic solvents such as trichloroethylene, the Freons[†] are often used for organic and ion removal. Studies have been made on the efficacy of several combinations of water, Freon TA (contains 11% acetone), and oxygen plasma, cleaning both with respect to initial corrosion of aluminum and to subsequent reliability of aluminum test devices in a humid environment.[40]

Although a preference for oxygen plasma and water rinse over a single Freon cleaning can be stated, a Freon cleaning alone is probably adequate for surfaces that are to be sealed into a low moisture environment, such as a hermetic package.

13.5.2 Package Lid Sealing

The major objective of package sealing is to protect the device from external contaminants during its design lifetime. Further, any contaminants present before sealing must be removed to an acceptable level before or during sealing. Packages may be hermetically sealed with glass or metals or lidded with polymers. The definition of hermeticity used here includes not only the ability to pass a fine vacuum leak test, but also to exclude environmental contaminants for long time periods.

Figure 28 shows the relative capabilities of several materials to exclude moisture over long time periods.[41] Clearly organic sealants are not good candidates for hermetic packages. However, in some cases, organic sealants, properly integrated into the package design, do meet the operational definition of a hermetic seal given above.[42] For almost all high-reliability applications the hermetic seal is made with glass or metal. Glass sealing was mentioned previously for Cerdip, and the process is essentially the same for lid sealing. Many of the metals used for die bonding are

---

[†] Freon is the registered trademark for Dupont's fluorinated solvents.

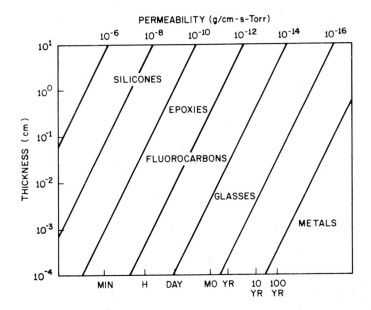

Fig. 28 The time for moisture to permeate various sealant materials (achieving 50% of the exterior humidity) in one defined geometry has been calculated to illustrate how much time is involved for various materials. Organics are orders of magnitude more permeable than materials typically considered for hermetic seals. *(After Traeger, Ref. 41.)*

suitable for lid sealing. A leak-tight metal seal that excludes the external environment can be made without difficulty. The real difficulty has been freeing the package of contaminants, especially water, before sealing.[43]

Particular problems can arise in a device at low operating temperatures where water can condense inside the package.[43] The following reactions can then lead to corrosion in the presence of small amounts of halide contamination:

$$6HCl + 2Al \rightarrow 2AlCl_3 + 3H_2 \uparrow \qquad (23)$$

$$AlCl_3 + 3HOH \rightarrow Al(OH)_3 + 3HCl \qquad (24)$$

$$2Al(OH)_3 + Aging \rightarrow Al_2O_3 + 3H_2O \qquad (25)$$

Note that HCl is regenerated in these reactions until the aluminum is completely consumed. A relative humidity of 65% is apparently necessary to sustain these reactions. The level of moisture depends on the particular case, but no package should be sealed with greater than 6000 ppm water by volume (dew point = 0°C).[43]

The technology to measure such moisture levels has been thoroughly studied.[44] Yet today the military specification limit is still 5000 ppm by volume mainly because of the technical difficulties in measuring moisture in small packages. Although 5000 ppm is the specification limit. the package assembler wants to achieve as low a moisture level as possible. A particularly effective method to achieve low moisture levels has been suggested.[45] This method makes use of the ability of atomic silicon to react with water to form SiO_2. The atomic silicon is formed during the lid seal through the melting of the die bond material or an added gettering preform. The proposed reaction is

$$Si + H_2O \rightarrow SiO_2 + 2H_2 \tag{26}$$

13.5.3 Encapsulation

Often the cost or difficulty of making a hermetic seal in a particular package type is prohibitive. In these cases, in addition to polymer lid sealing, surface die coatings can be used to protect aluminum metallurgy from atmospheric contaminants such as water. Silicones are very effective for this purpose.[46, 47] A thin layer of silicone (0.25 mm) is not a diffusion barrier to water based on the data in Fig. 28. Qualitatively, the polymer must prevent moisture condensation at the surface and thus inhibit leakage between adjacent metal lines. One proposed mechanism suggests that partially oriented siloxane dipoles at the substrate surface, working in concert with polar groups in the substrate, would sufficiently increase the work of adhesion of water molecules to prevent water from entering the interfacial region.[48]

Silicone encapsulants are used by several electronic systems companies with excellent success for both aluminum and gold metallurgy devices. Not yet established is a lower limit on line spacing below which these materials are no longer effective.

13.5.4 α-Particle Protection

Soft errors in memory circuits due to α-particles emanating from packaging materials were first reported[49] in 1978; there have been many papers on the subject since then. The α-particles are emitted by the decay of uranium and thorium atoms contained as impurities in the packaging materials.

Decreasing design rules suggest that future devices will be more sensitive to this problem.[50] Packaging materials will probably not be pushed below the 0.001 to 0.01 α-particles/cm^2-h level. Because α-particles have low penetrating power in solids, low α materials have been suggested as α-absorbing coatings on silicon dice. A 0.001 α-particle/cm^2-h emission rate, the lowest level anticipated, has been reported using silicone coatings.[51] This rate will eventually be achieved in packages of ceramics, glass, and metals. Although great progress can be made by reducing α-particle emission in current packaging materials, error correction is required to control soft errors caused by α-particles.

13.6 PACKAGE APPLICATION CONSIDERATIONS

This section considers thermal and electrical design and package interconnection to the next higher level. As devices have greater pinout, power dissipation, and speed, these issues are increasingly important in the choice of package type and the design details. These considerations are already well known in the design of high-performance computers.[52] No package design has yet resolved these issues of thermal and electrical design and interconnection completely.

13.6.1 Thermal Considerations

The objective of thermal design is to keep the operating junction temperature of a silicon die low enough to prevent the failure rate due to temperature-activated failure mechanisms from exceeding the acceptable limit for a particular application of that silicon die. Only in the simplest possible applications can this objective be met considering the packaged silicon device alone. Usually the packaged device environment must be established for most of the following variables: printed wiring board (PWB) temperature, total power dissipation on the board, local neighbor power dissipation, degree of forced-air cooling, space usable by the package both laterally and vertically (between boards), conductivity of the PWB, and ideal performance of the isolated package.

Conventionally, either junction to ambient or junction to case (the body of the packaged device) thermal resistance (in °C/W) has been used. These resistances are usually derived by suspending the device in still air or mounting the device in a particular socket arrangement. For noncritical thermal design, these thermal resistance values are adequate and allow reasonable calculations of junction temperature rise based on expected ambient temperatures and device power dissipation. We will now look at the heat flow paths in both a ceramic and a plastic package.

To begin the analysis in a ceramic package, we start at the junction on a chip and work out through the silicon into the package body and ultimately into the environment. Figure 29 illustrates a hybrid module.[7] This module contains both die-bonded

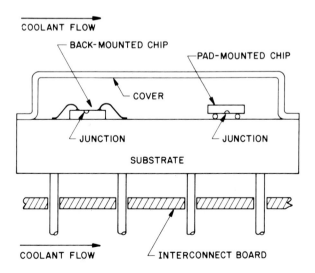

Fig. 29 A hypothetical ceramic module containing a flip-chip and a die-bonded chip. Coolant can flow over or under the package depending on the system configuration. *(After Buchanan and Reeber, Ref. 7.)*

and flip-chip devices. The module as drawn would be best suited to an application where the external heat removal system is under the package, for example, convection cooling of the board on which the package is mounted. This arrangement is typical for military applications where the convective air flow is kept away from the packages. An inverted package like the pin array in Fig. 19c is better suited to the more conventional system where the air flows over the packages.

Two thermal problems are associated with the module in Fig. 29: (1) the transfer of energy from the heat-producing junctions to the package external surfaces, defined in terms of an internal thermal resistance, and (2) the removal of heat from the package surfaces by convection exchange with the ambient (free or forced air, or liquid) and by conduction through a printed wiring board, typically pins into the package.

The general expression for three-dimensional heat flow is given by the Fourier equation as[7]

$$\partial Q / \partial t = -\kappa A \Delta T \qquad (27)$$

where Q is the total heat flow, κ is the thermal conductivity of the solid (W/cm-°C), ΔT is the temperature difference (°C) along a thermal path ΔX (cm), and A is the area normal to the heat flow in the solid (cm^2). The negative sign indicates a positive flow of heat toward colder temperatures. For steady-state conduction in one dimension, Eq. 27 reduces to

$$Q' = \kappa A \frac{\Delta T}{\Delta X} \qquad (28)$$

For steady-state convection from a surface, the analogous expression is given by

$$Q' = (\kappa'/\delta)A'\Delta T = hA'\Delta T \qquad (29)$$

where Q' is the heat flow per unit time (W), κ' is the thermal conductivity of the fluid environment at distance δ from a surface of effective area A' (cm^2), and h equals κ'/δ which is the convection heat transfer coefficient (W/cm^2-°C).

Analogous to an electrical resistance, thermal resistance R can be defined for conduction as

$$R = \frac{\Delta T}{Q'} = \frac{\Delta X}{\kappa A} \quad °C/W \qquad (30)$$

and for convection

$$R = \frac{\Delta T}{Q'} = \frac{1}{hA} \quad °C/W \qquad (31)$$

The rise of junction temperature of a chip can be written as

$$T_j = T_0 + \Delta T_M + R_{S-M} P_S + R_{C-S} P_C + R_{J-C} P_J \qquad (32)$$

where T_j the junction temperature of the device, R_{J-C}, R_{C-S}, and R_{S-M} are thermal resistance from junction to chip, chip to substrate and substrate cooling media, and substrate to cooling medium, respectively; P_j, P_C, and P_S are the powers dissipated by the junction, chip, and substrate, respectively; ΔT_M is the temperature rise of the cooling medium, and T_0 is the inlet temperature of the cooling medium.

Equation 32 incorporates three thermal design considerations which are linearly related: the removal of heat to the package surface (internal), the removal of heat from the package surface (external), and the specification of the operating environment (T_0). The desired T_j can be achieved through variations in any of these three considerations.

Thermal resistance to the external environment has been discussed in detail for heat-finned packages in a convective environment.[54] Both computer analysis using Green's function techniques and experimental studies were used to define package thermal resistance as a function of most package assembly variables and the velocity of the convective cooling air stream. The convective cooling issue will not be discussed further, but note that in many cases the thermal resistance of a package can be reduced by as much as five times going from an unfinned part in natural air convection to a finned part in a forced air velocity of 10 m/s. Assuming that all package cooling is due to convection off a fin, and that the chip is in a cavity down package constructed of alumina (Fig. 19c), then junction to air thermal resistances of 5 to 10°C / W are possible.

To achieve these low thermal resistance values, a low internal thermal resistance must be achieved. The internal thermal resistance is the sum of resistances that serially make up the total resistance between the junction and the package surface.

For heat flow in the die, the thermal resistance is composed of two parts

$$R_C = R_j + R_C' \tag{33}$$

where R_j is the thermal resistance between the junction and some reference point on the die, and R_C' is the thermal resistance of the die.

Figure 30a, a diagram of a die, shows a power-dissipating junction. The emitter-base junction is considered to be a hemisphere in the semi-infinite die media which is dissipating heat with reference to another concentric-reference point at r_2 to which heat is transported. For a rectangular junction of size ab and with $R_2 = \infty$, R_j is given by

$$R_j = \frac{1}{\pi a \kappa} \ln 4\frac{a}{b} \tag{34}$$

where κ is the thermal conductivity of the junction area. Thermal resistances can reach several hundred °C / W for a junction; however, each individual junction rarely dissipates more than a few milliwatts.

The thermal resistance of the die, considered to be composed of many dissipating junctions spread uniformly over its surface, is given by

$$R_C' = \frac{\Delta X}{\kappa A} \tag{35}$$

where ΔX is path length for heat flow, A is the area normal to the heat flow path, and κ is the intrinsic thermal conductivity of the chip. Both ΔX and A are significantly different depending on chip mounting (Fig. 30b and c).

For a die-bonded chip, ΔX is the chip thickness, and A is the die area, whereas for a flip-chip, these parameters are not so clearly defined. For a flip-chip the heat must spread downward and outward through the die in the case of peripheral bond

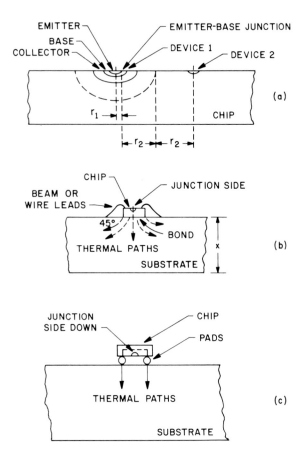

Fig. 30 Thermal heat flow paths in a chip. (a) Heat spreading from a junction. (b) Heat spreading from a die-bonded device. The effective spreading zone for the die-bonded device is taken as a truncated cone beginning at the base of the device and spreading at a 45° angle through the substrate thickness X. (c) Heat spreading from a flip-chip. *(After Buchanan and Reeber, Ref. 7.)*

pads because the heat must ultimately flow through the solder-bonding pillars. For area-array solder pillars, the thermal flow through the die is similar to the die-bonded part.

The bond thermal resistance for both cases may be examined using Eq. 35. In the die bond case the thermal path is through a thin layer of Au-Si eutectic, solder, or epoxy. Figure 31 shows the die bond thermal resistance as a function of die-bond-material thermal conductivity. Die size is shown as a parameter. Similarly, the bonds of an area-array flip-chip can be assumed to be an array of cylinders in parallel.

Heat transported through the die and bond must be spread into the substrate. For a die-bonded device on an infinite media, the spreading resistance is given by

$$R_S = \frac{1}{\pi a \, \kappa_s} \ln 4\frac{a}{b} \tag{36}$$

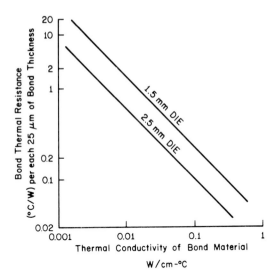

Fig. 31 Bond thermal resistance as a function of the thermal conductivity of the bond material for two die sizes. The thermal resistance is expressed per 25 μm of bond thickness. *(After Buchanan and Reeber, Ref. 7.)*

Here a and b are the chip dimensions, and κ_s is the effective thermal conductivity of the mount media. Usually the mount substrate is finite, and Eq. 35 is more useful to calculate the thermal resistance to the external surface. The area is taken as the median area between the source and sink (the effective area defined in Fig. 30b). For a large, die-bonded device on a ceramic substrate, the thermal spreading resistance through the alumina to the external surface is a large contributor to the total junction to external surface thermal resistance. The analysis of ceramic packages is continued in Section 13.6.3 where the trade-off in package density and thermal capability is reviewed.

Figure 32 shows the heat flow paths in a plastic DIP.[55] Path (a) depends mainly on choosing a plastic with good thermal conductivity; the trade-offs here were mentioned previously. A large die paddle is best to increase the effective area for heat spreading. Paths (b) and (d) depend on having a high-conductivity leadframe and on having a highly conductive plastic. For example, the use of copper versus 42 alloy decreases by approximately half the thermal resistance for a 16-pin DIP (R junction to ambient = 80 and 160°C/W, respectively). Path (c) depends on the leadframe material.

13.6.2 Electrical Performance

Electrical performance at the IC package level has only recently been of general interest for silicon devices.[56] However, with the increasing speed of today's circuits and their potentially reduced noise margins, package design must be considered more carefully. Several electrical performance criteria are of interest, namely: low ground

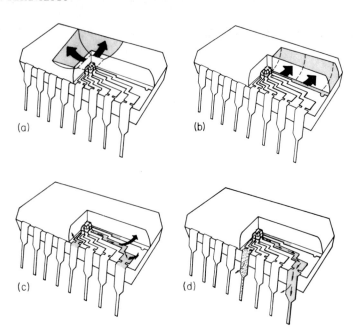

Fig. 32 Schematic representation of four major paths of heat flow in a 16-pin plastic DIP. Sketches expose one-eighth of package with paths shaded or clarified by arrows. (a) Heat flows from the chip to the die paddle to the plastic above and below the paddle. (b) Heat flows from the chip to the die paddle to the tie bar to the plastic above and below the tie bar. (c) Heat flows from the chip to the die paddle and the tie bar to the plastic to the leads to the plastic above and below the leads. (d) Heat flows from the chip to the die paddle and tie bar to the plastic to the leads to the socket. *(After Mitchell and Berg, Ref. 54.)*

resistance (minimum power-supply voltage drop), short signal leads (minimum self-inductance), minimum power-supply spiking due to signal lines simultaneously switching, short-paralleled signal runs (minimum mutual inductance and cross talk), short-length signal runs near a ground plane (minimum capacitive loading), and the maximum use of matched impedances to avoid signal reflection. These criteria are, of course, not all mutually independent. They may be related through simple geometric variables, such as conductor cross section and length, dielectric thickness, and dielectric constant of the packaging body. These problems are usually handled with transmission line theory in printed circuit boards, where the lengths of circuit paths are more obviously a cause for concern. A wealth of papers and books cover this topic, and all of the techniques described are generally applicable at the package level.[57-60]

Table 3 gives a basic summary of package characteristics for various implementations of 64- to 68-pin packages. All of these packages are assumed to contain wire-bonded silicon devices where the wire is 25.4 μm in diameter and 1.25 mm in length. This wire contributes 1.25 nH to the inductance.[56] The ceramic packages are constructed of alumina with standard refractory technology. A trade-off between inductance and capacitance can be made here because inductance is minimized near ground planes while capacitance is increased.

Table 3 Summary of electrical package characteristics (64 to 68 pins)[8, 53, 61]

Parameter	Package type			Ceramic chip carrier, leadless	Plastic chip carrier, leaded
	Ceramic DIP	Plastic DIP	Pin grid		
Max. lead resistance (R in Ω)	1.1	0.1	0.2	0.2	0.1
Max. lead capacitance* (C in pf)	7	4	2	2	2
Max. lead inductance* (L in nH)	22	36	7	7	7
Thermal resistance— junction to pin at ambient temperature (R_{j-p} in °C/W)	32	35	20	13	28
Printed wiring board area Consumption (cm²)	18.7	18.7	6.45	6.45	6.45

*For the ceramic packages, the capacitance and inductance are for a line running in dielectric over a ground plane at a separation distance of 0.5 mm. The capacitance for the plastic packages is line to line.

The most important practical electrical design problem in IC packages is noise reduction. Basically, when a line switches, the voltage induced in the ground line is given by

$$V_i = L_g \frac{di}{dt} \qquad (37)$$

where V_i is the induced voltage, L_g is the inductance of the ground lead, and di/dt is the derivative of current with respect to time. If j lines are switching, then V is given by

$$V_i = L_g \sum_j \frac{di_j}{dt} \qquad (38)$$

To reduce V_i, multiple grounds must be used to lower L_g. If m ground leads are used, the total inductance is approximately L_g/m. The practical result is that up to 25% of the leads may have to be grounded to control noise. This high percentage has a large impact on packaging density and creates an incentive to reduce L_g. Inductance can be reduced significantly through the use of large area power and ground planes within the package.

13.6.3 Interconnection

Most of the packaged devices that we have discussed are designed to be plugged into a PWB or attached at its surface. Some of the packages are also suited to hybrid

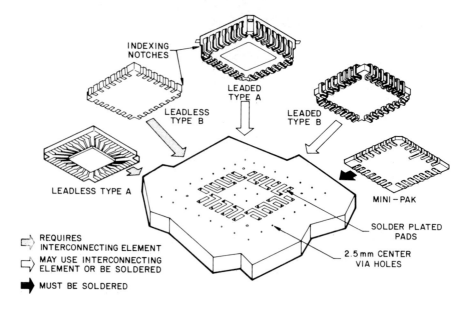

INDEXING NOTCHES

LEADED TYPE A

LEADLESS TYPE B

LEADED TYPE B

LEADLESS TYPE A

MINI – PAK

REQUIRES INTERCONNECTING ELEMENT

MAY USE INTERCONNECTING ELEMENT OR BE SOLDERED

MUST BE SOLDERED

SOLDER PLATED PADS

2.5 mm CENTER VIA HOLES

Fig. 33 Interfacing of JEDEC, standard chip carrier types to a common printed circuit board land configuration (2.54-mm centers). Note the staggered fanout on the printed circuit board to mate with the 1.27-mm center chip carriers.

assembly, for example, the leadless chip carrier. Which packages provide the best compromise in size, routability, and thermal performance is not an easy question to answer; in fact, opinions vary. We outline some of the technical arguments in the following discussions.

The four major packaging contenders are DIPs, leadless chip carriers, leaded chip carriers, and pin-grids. DIPs are really not contenders for high-density packaging, because they are large. The centers for lead spacings on the other package types may be considered to be variables, although 2.54-mm (100-mil) centers are typical today for pluggable packages (DIP and pin-grids), while 1.02-mm (40-mil) and 1.27-mm (50-mil) centers are typical for surface mountable devices.

Figure 33 shows the standard footprint for chip carriers. The leads on 1.27-mm (50-mil) centers may be fanned out conveniently to the grid of holes on 2.54-mm (100-mil) centers which would be used for pluggable devices. The holes themselves are not required for surface mountable devices, only vias to underlying interconnect levels. Thus, the space used for holes for pluggable packages is much greater than the space used for vias for chip carriers.

Interconnection density can be viewed several ways. Figure 34 shows density on the basis of terminals per unit area.[62] This density measurement does not address other issues such as interconnectibility, or power handling, or cost. Figure 34 shows that on the basis of terminal density alone, a pin-grid would be preferred over a 1.02-mm (40-mil) pitch-chip carrier at 100 to 120 pins and over a 0.635-mm (25-mil) pitch-chip carrier above about 200 pins. Comparison is made here to the less dense

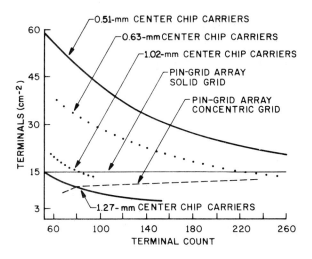

Fig. 34 Package terminal density as a function of package pin count for a number of potential VLSI packages. *(After Amey, Ref. 61.)*

concentric grid package, because it has better power-handling capability (cavity down).

Interconnection density may be viewed on the basis of gates also.[63] It has been shown empirically in many studies that for random logic, the gates on a chip and the package signal leads are related by an expression of the form

$$N = \alpha G^\beta \tag{39}$$

where N is the number of signal terminals, G is the number of gates, α equals 2 to 4.4, and β equals 0.5 to 0.7.

Using this relationship and the density of pins on the various packages, the curves for perimeter- and area-array packages with a pin spacing of 2.54 mm (100 mils) have been derived (Fig. 35). Perimeter packages show decreasing efficiency of gate packing with increased chip gate count, while area packages show an increasing efficiency for the observed range of β values.

Figure 36 shows that power-handling capability and gate density are also related. The power dissipation of the chip (5 W) has been chosen somewhat arbitrarily as representative of high-performance devices. The power densities (0.077 W/cm^2 for air cooling and 1.08 W/cm^2 for liquid cooling) are supposed to apply over large PWBs. Isolated packages with much higher power densities than 0.077 W/cm^2 can be used in a forced-air environment.

For a given chip integration level (e.g., 1500 gates/chip), the optimum package is the one that allows the greatest gate count per unit area,[62] thus producing the greatest gate count at the module or PWB level. In Fig. 36 we see that an air-cooled system has about 12 gates/cm^2 for a 2.54-mm (100-mil) perimeter package and is limited by cooling to about 36 gates/cm^2 for a 1.27-mm (50-mil) perimeter package or any other dense package. If air cooling is to be preserved, larger gate counts per chip are

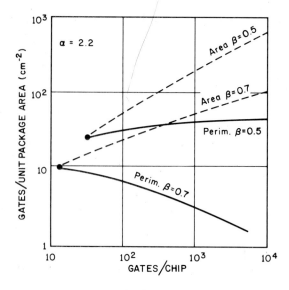

Fig. 35 Packing efficiencies for 2.54-mm center area-array and peripheral packages for $\beta = 0.5$ and 0.7 (see Eq. 39) and $\alpha = 2.2$. For β values above 0.5, gate density goes down as integration level increases for perimeter packages. For area-array packages the number of terminals omitted for the chip is taken as a constant equal to 4. *(After Steele, Ref. 62.)*

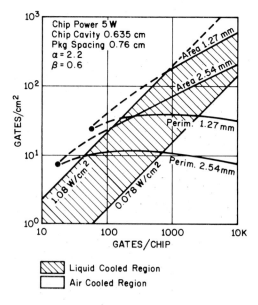

Liquid Cooled Region
Air Cooled Region

Fig. 36 Gate density versus gate count in a package for several potential area- and perimeter-VLSI packages. The limitations of air and liquid cooling are overlaid for silicon devices dissipating 5 W of power. *(After Steele, Ref. 62.)*

required, which, in turn, requires a lower power dissipation per gate to maintain the 5-W chip dissipation.

13.7 SUMMARY AND FUTURE TRENDS

Package pin counts will undoubtedly continue to increase with increasing silicon capability. Alumina ceramic packages will dominate the high-performance packaging technology until factors such as its high dielectric constant or modest thermal conductivity force a change to beryllia packages. If chip integration keeps pace with decreasing feature size (keeping the relative proportions of chip size and interconnection area), today's chip interconnection technologies, especially wire bonding, will be adequate without major change. Wire bonding will require improvements in speed and accuracy, particularly in the bonding of the package leads. For devices where decreased feature size is used to decrease silicon size, serious interconnection problems could arise—the area of a device devoted to interconnect could far exceed the active area.

TAB and flip-chip will continue to be specialty assembly approaches. Both technologies lend themselves to hybrids, especially TAB with its pretestability features. Flip-chip has the advantage of a very low inductance connection to the package and the best capability in terms of the density of interconnection (using an area-array) to the device compared to wire or TAB. Combinations of area flip-chip and TAB as described in Section 13.4.5 have the potential for achieving the best thermal and electrical characteristics and best density of any packaging scheme thus far proposed. Specialized metallurgical treatment of the wafers and tapes will make this a fairly-high-technology area.

The design of systems will depend more and more on a systematic optimization of the entire interconnection scheme to achieve the potential benefits of improved silicon capability. This optimization may lead to completely new requirements for assembly and packaging but more than likely will lead to a sorting out of the various existing assembly technologies and packages.

REFERENCES

[1] Charles T. Goddard, "The Role of Hybrids in LSI Systems," *IEEE Trans. Components, Hybrids, Manuf. Technol.*, **2**, 367 (1979).

[2] Robert C. Cook, Sandy Madden, and Harvey Williamson, "New Approaches to Sawing Microelectronic Materials," *Semicond. Int.*, p. 65, Dec. (1980).

[3] J. W. Stafford, "The Implications of Destructive Wire Bond Pull and Ball Bond Shear Testing on Gold Ball-Wedge Wire Bond Reliability," Semicond. Int. p. 82, May (1982).

[4] Dr. Alan Keizer and Don Brown, "Bonding Systems for Microinterconnect Tape Technology," *Solid State Technol.*, p. 59, Mar. (1978).

[5] C.E.T. White and J. Slatery, "An Update on Preforms," *Circuits Manufact.*, p. 78, Mar. (1978).

[6] Dennis R. Olsen and Howard M. Berg, "Properties of Die Bond Alloys Relating to Thermal Fatigue," *Proc. 27th Electronic Components Conf.*, p. 193 (1977).

[7] R. C. Buchanan and M. O. Reeber, "Thermal Considerations in the Design of Hybrid Microelectronic Packages," *Solid State Technol.*, p. 39, Feb. (1973).

[8] Kyocera, *Design Guidelines for Multilayer Ceramics*, Publ. Number A-125E.

[9] Peter J. Planting, "An Approach for Evaluating Epoxy Adhesives for Use in Hybrid Microelectronic Assembly," *IEEE Trans. Parts, Hybrids, Packag.*, **11**, 305 (1975).

[10] Clark N. Adams, "A Bonding-Wire Failure Mode in Plastic Encapsulated Integrated Circuits," *Proc. Int. Rel. Phys. Symp.*, p. 41 (1973).

[11] James J. Licari, Kenneth L. Perkins, and Salvadore V. Caruso, "Evaluation of Electrically Insulative Adhesives for Use in Hybrid Microcircuit Fabrication," *IEEE Trans. Parts, Hybrids, Packag.*, **9**, 199 (1973).

[12] Curtis Mitchell and Howad Berg, "Use of Conductive Epoxies for Die Attach," *Int. Microelectron. Symp.*, p. 52 (1976).

[13] Elliott Philofsky, "Purple Plague Revisited," *Solid State Electron.*, **13**, 1391 (1970).

[14] George G. Harmon and John Albers, "The Ultrasonic Welding Mechanism as Applied to Aluminum- and Gold-Wire Bonding in Microelectronics," *IEEE Trans. Parts, Hybrids, Packag.*, **13**, 406 (1977).

[15] Vern H Mitchell, II and Howard M. Berg, "Enhancing Ultrasonic Bond Development," *IEEE Trans. Components, Hybrids, Manuf. Technol.*, **1**, 211 (1978).

[16] Kanji Otsuka and Tamotsu Usami, "Ultrasonic Wire Bonding Technology for Custom LSI's with Large Number of Pins," *Proc. 31st Electronic Components Conf.*, p. 350 (1981).

[17] Mike B. McShane, "Device Design and Lead Frame Layout Techniques for Automatic Wire Bonding," *Semicond. Int.*, p. 85, June (1980).

[18] N. Ahmed and J. J. Svitak, "Characterization of Gold-Gold Thermocompression Bonding," *Solid State Technol.*, p. 25, Nov. (1975).

[19] R. Hill, E. H. Lee, and S. J. Tupper, "A Method of Numerical Analysis of Plastic Flow in Plane Strain and its Application to the Compression of a Ductile Material Between Rough Plates," *J. Appl. Mech.*, **18**, p. 46 (1951).

[20] Dr. Tien-Shih Liu and Dr. Henry S. Fraenkel, "Metallurgical Considerations in Tin-Gold Inner Lead Bonding Technology," *Int. J. Hybrid Microelectron.*, **1**, 69 (1978).

[21] Roger L. Cain, "Beam Tape Carriers—A Design Guide," *Solid State Technol.*, p. 53, Mar. (1978).

[22] Robert F. Unger and John W. Kanz, "BTAB's Future—An Optimistic Prognosis," *Solid State Technol.*, p. 77, Mar. (1980).

[23] Dr. T. S. Liu, W. R. Rodgrigues de Miranda, and P. R. Zipperlin, "A Review of Wafer Bumping for Tape Automated Bonding," *Solid State Technol.*, p. 71, Mar. (1980).

[24] John L. Kowalski, "Multichip Hybrid Assembly Using Tape Automated Bonding," *Electron. Packag. Prod.*, p. 107, Jan. (1979).

[25] P. A. Totta and R. P. Sopher, "SLT Device Metallurgy and its Monolithic Extension," *IBM J. Res. Dev.*, **13**, 226 (1969).

[26] L. S. Goldmann, "Geometric Optimization of Controlled Collapse Interconnections," *IBM J. Res. Dev.*, **13**, 251 (1969).

[27] K. C. Norris and A. H. Landzberg, "Reliability of Controlled Collapse Interconnections," *IBM. J. Res. Dev.*, **13**, 266 (1969).

[28] S. S. Manson, *Thermal Stress* and *Low Fatigue*, McGraw-Hill, New York, 1966.

[29] R. A. Gardner and R. W. Nufer, "Properties of Multilayer Ceramic Green Sheets," *Solid State Technol.*, p. 38, May (1974).

[30] Kanji Otsuka, Tamatsu Usami, and Mosao Sekihata, "Interfacial Bond Strength in Alumina Ceramics Metallized and Cofired with Tungsten," *Ceram. Bull.*, **60**, 540 (1981).

[31] Daniel I. Amey, "The JEDEC Chip Carrier and LSI Standard: A Summary," *Semicond. Int.*, p. 103, June (1981).

[32] J. R. Howell, "Reliability Study of Plastic Encapsulated Copper Lead Frame Epoxy Die Attach Packaging System," *Proc. Int. Rel. Phys. Symp.*, p. 104 (1981).

[33] Donald S. Levinthal, "Semiconductor Packaging Trends," *Semicond. Int.*, p. 33, Apr. (1979).

[34] C. M. Melliar-Smith, S. Matsuoko, and P. Hubbauer, "Plastic Encapsulation of Integrated Circuits," *Plast. Rubber Mater. Appl.*, **5**, 49 (1980).

[35] Preston J. Heinle, "Flow Patterns in Thermoset Mold Compounds," SPE ANTEC, p. 426 (1979).

[36] Anoki Kaneda et al., "Modifications of Flow Channels in Multi-Cavity Mold Die for Resin Molding of IC Devices," *Proc. Int. Conf. Polymer Processing*, p. 349 (1979).

[37] R. J. Usell, Jr. and S. A. Smiley, "Experimental and Mathematical Determination of Mechanical Strains within Plastic IC Packages and Their Effect on Devices During Environmental Tests," *Proc. Int. Rel. Phys. Symp.*, p. 65 (1981).

[38] Jerry Lyman, "Packaging VLSI," *Electronics*, p. 66, Dec. (1981).

[39] M. A. Robbins, "VHSIC Packaging," *Proc. Int. Electron. Packag. Soc.*, p. 483 (1981).

[40] Melanie Iannuzzi, "Development and Evaluation of a Preencapsulation Cleaning Process to Improve Reliability of HIC's with Aluminum Metallized Chips," *IEEE Trans. Components., Hybrids, Manuf. Technol.*, **4**, 429 (1981).

[41] R. K. Traeger, "Hermeticity of Polymeric Lid Sealants," *Proc. 25th Electronics Components Conf.*, p. 361 (1976).

[42] Irving Memis, "Quasi-Hermetic Seal for IC Modules," *Proc. 30th Electronics Components Conf.*, p. 121 (1980).

[43] Robert W. Thomas, "Moisture, Myths, and Microcircuits," *IEEE Trans. Parts, Hybrids, Packag.*, **12**, 167 (1976).

[44] *Moisture Measurement Technology for Hermetic Semiconductor Devices*, March 22–23, 1978, NBS (National Bureau of Standards) Publ. 400-69 (1981).

[45] M. L. White, K. M. Striny and R. E. Sammons, "Attaining Low Moisture Levels in Hermetic Packages," *Proc. Int. Rel. Phys. Symp.*, p. 253 (1982).

[46] Ching-Ping Wong, "Room Temperature Vulcanized (RTV) Silicone as Integrated Circuit (IC) Coating," *Proc. ISHM*, p. 315 (1981).

[47] R. G. Mancke, "A Moisture Protection Screening Test for Hybrid Circuit Encapsulants," *IEEE Trans. Components, Hybrids, Manuf. Technol.*, **4**, 492 (1981).

[48] George Cvijanovich, "Active Protection of IC Surfaces," *Semicond. Int.*, p. 57, May (1979).

[49] T. C. May and M. H. Woods, "Alpha-Particle-Induced Soft Errors in Dynamic Memories," *IEEE Trans. Electron Devices*, **26**, 2 (1979).

[50] Timothy C. May, "Soft Errors in VLSI: Present and Future," *IEEE Trans. Components, Hybrids, Manuf. Technol.*, **2**, 377 (1979).

[51] Malcolm S. White, Joseph W. Serpiello, Kurt M. Striny, and W. Rosensweig, "The Use of Silicone RTV Rubber for Alpha Particle Protection on Silicon Integrated Circuits," *Proc. Int. Rel. Phys. Symp.*, p. 43 (1981).

[52] Hisao Kanai, "Low Energy LSI and Packaging for System Requirements," *IEEE Trans. Components, Hybrids, Manuf. Technol.*, **4**, 173 (1981).

[53] Capt. Roger E. Settle, Jr., "A New Family of Microelectronic Packages for Avionics," *Solid State Technol.*, p. 54, June (1978). Gordon N. Ellison, "Thermal Design of an LSI Single-Chip Package," *IEEE Trans. Parts, Hybrids, Packag.*, **12**, 371 (1976).

[54] Curtis Mitchell and Howard M. Berg, "Thermal Studies of Plastic Dual-In-Line Packages," *IEEE Trans. Components, Hybrids, Manuf. Technol.*, **2**, 500 (1979).

[55] Leonard W. Schaper, "The Impact of Inductance on Semiconductor Packaging," *Proc. Int. Rel. Phys. Symp.*, p. 38 (1981).

[56] A. J. Rainal, "Transmission Properties of Various Styles of Printed Wiring Boards," *Bell Syst. Tech. J.*, **58**, 995 (1979).

[57] Frederick W. Grover, *Inductance Calculations : Working Formulas and Tables*, Van Nostrand, New York, 1949.

[58] Y. M. Hill, N. O. Reckford, and D. R. Winner, "A General Method for Obtaining Impedance and Coupling Characteristics of Practical Microstrip and Triplate Transmission Line Configurations," *IBM J. Res. Dev.*, **13**, 314 (1969).

[59] John A. DeFalco, "Reflections and Crosstalk in Logic Circuit Interconnections," *IEEE Spectrum*, p. 44, July (1970).

[60] Timothy G. O'Neill, "VLSI Packaging Requirements and Trends," *Semicond. Int.*, p. 43, Mar. (1981).

[61] Daniel J. Amey, "Integrated Circuit Package Selection: Pin Grid Array vs. Chip Carriers," *Proc. Int. Electron. Packag. Soc.*, p. 1 (1981).

[62] Thomas S. Steele, "Terminal and Cooling Requirements for LSI Packages," *IEEE Trans. Components, Hybrids, Manuf. Technol.*, **4**, 187 (1981).

PROBLEMS

1 Derive Eq. 21.

2 Based on the discussion of ultrasonic bonding, how would you expect the bonded zone to vary if you increased the vertical stress? How would you expect it to vary if you increased the pulse power? Show your answers in the graphical format of Fig. 5.

3 How many grams of silicon are required to react with 5000 ppm by volume of water in a 0.1-cm^3 die cavity? Assume room temperature and atmospheric pressure in the cavity.

4 Find the spreading resistance of a die-bonded silicon die on 0.6-mm beryllia and alumina substrates for the two die sizes 1 mm and 7 mm.

5 Suppose you have a 5-V, 1-ns output rise-time silicon chip with 68 signal leads and you want to package it such that inductive noise in the ground line is kept to 0.2 V. You want to switch 18 lines simultaneously and you want each package lead to have approximately 7-nH inductance. The output buffers draw 25 mA. How many package leads do you need for this device excluding power?

6 Find the optimum value N_F for a flip-chip system, assuming that all terms have been optimized except volume and height. Assume that r_0 and r_1 are predetermined.

7 What minimum bonding pad pitch could be achieved with 25-μm wire assuming 30% deformation of a ball (three times original wire diameter)? The deformed ball must be on the metal portion of the bond pad. Assume 10 μm of overlap of surface passivation on the bond pad. Assume a ball placement accuracy of ±12.5 μm and spacing between pads (metal to metal) of 25 μm. If the chip function can fit into a 4.5 $\times$ 4.5-mm active area and 150 leads are required, what percentage additional non-active area is required using the design rules obtained above?

8 Figure 36 overestimates the number of gates implementable in a particular package type because it neglects power and ground leads. According to Fig. 36, how many gates can be achieved in an air-cooled 1.27-mm perimeter device? If 20% of the leads of a device are power and ground, how many gates/cm^2 could be achieved in practice in this device?

FOURTEEN

YIELD AND RELIABILITY

W. J. BERTRAM

14.1 INTRODUCTION

Previous chapters discussed the science and technology of VLSI circuit fabrication. This chapter considers two conditions that must be satisfied in order for VLSI to be a useful, growing technology. First, the fabricated circuits must be capable of being produced in large quantities at costs that are competitive with alternative methods of achieving the circuit and systems function. Second, the circuits must be capable of performing their function throughout their intended useful life. To produce circuits that meet these two requirements we must understand the mechanisms that lead to high cost and unreliable devices. Once these mechanisms are understood and quantified, we can develop the technology required to meet our goals.

The optimum size of an IC, with respect to the number of circuit functions, is a compromise between several competing considerations: partitioning of the system (or subsystem), the expected yield of good circuits, packaging and system assembly cost, and the overall reliability of the complete system. If the system is divided into a large number of small ICs, then the yield of these circuits may be very high; however, the total cost of packaging a large number of these chips and assembling them on a circuit board may be greater than the cost of manufacturing a smaller number of larger circuits, at a reduced yield, and packaging and assembling these circuits on a circuit board. To divide the complete system into an optimum number of ICs, we must be able to predict the yield and cost of an IC as a function of circuit size. This optimization is further complicated, because the number of required circuit functions may increase as the system is divided into smaller and smaller ICs. Overall system performance may also suffer, because the time to propagate signals between different parts of the system is greater if the different parts are on separate ICs than if all circuit functions are contained on one IC. Similarly, to make an optimum division of the system, we must be able to predict the total system reliability as a function of the number of ICs of varying size.

In this chapter we will discuss the various methods which have been developed to quantify the measurements of VLSI circuit yield and reliability. The various mechanisms which have been identified as seriously limiting the yield and reliability of VLSI circuits will be reviewed. Finally we will discuss the techniques which are used to assure that new device technologies are not limited in their usefulness by yield or reliability reducing mechanisms.

14.2 MECHANISMS OF YIELD LOSS IN VLSI

Ideally, in a properly fabricated wafer of VLSI circuits, we would expect all of the circuits on the wafer to be good, functional circuits. In practice the number of good circuits may range from close to 100% to only one or fewer good circuit per wafer. Usually, the causes for less than perfect yield fall into three basic categories: parametric processing problems, circuit design problems, or random point defects in the circuit. Each of these categories is discussed below.

14.2.1 Processing Effects

If one looks at a map or photograph of a tested wafer, one of the most obvious features is that the wafer may be divided into regions with a very high proportion of good chips and other regions where the yield of good chips is very low or even zero. A photograph of such a wafer is shown in Fig. 1. This section discusses processing effects that may lead to the existence of low yield regions. These effects include: variations in the thickness of oxide or polysilicon layers, in the resistance of implanted layers, in the width of lithographically defined features, and in the registration of a photomask with respect to previous masking operations. Many of these variations depend upon one another. For example, regions where the polysilicon layer is thinner than average become overetched when the wafer is etched for the length of time needed to clear the polysilicon in regions where the layer is thicker than average. Polysilicon gates are shorter in the thinner polysilicon regions than in the thicker polysilicon regions. This effect may result in regions in which the channel lengths are too short and the transistors can not be turned off when the appropriate gate voltage is applied. Consequently the circuits may not function or may have excessive leakage currents.

Variations in the doping of implanted layers can lead to variations in the contact resistance to the implanted layers. Variations in the thickness of the deposited dielectric can lead to variations in contact window size. Both of these effects can lead to nonoperative circuits if the circuit contains paths whose performance is dependent upon having a low value of contact resistance.

During the processing of a wafer various operations are carried out which result

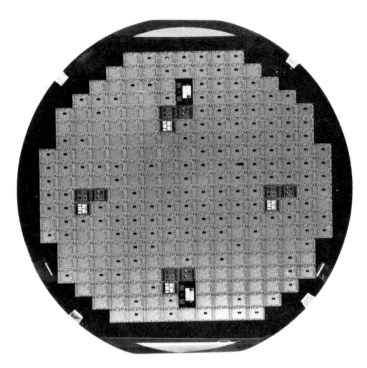

Fig. 1 Photograph of an IC wafer showing regions of high and low yield. The chips with a black ink dab are bad. The chips in the four groups are test chips.

in small but critical changes in the size of the wafer. For example, when a wafer is oxidized, the SiO_2 formed has approximately twice the volume of the silicon consumed in the oxidation process. The resultant composite wafer consists of an interior silicon layer which is in tension, and oxide layers on the two surfaces which are in compression. The resultant wafer is larger in diameter than the original silicon wafer. If the degree of tension exceeds the elastic limit of the material, deformation occurs. If the oxide is then removed from one surface, the wafer will have a convex bow on the remaining oxide surface.

During processing, it is not uncommon for wafer size to vary in excess of 20 ppm. Thus, a 125-mm wafer changes size by 2.5 μm, which is far greater than the desired realignment tolerance. If such variations in wafer size are not compensated for, areas of the wafer may have inoperative circuits because of misalignment. In addition we may find areas of the wafer where improper cleaning has left chemical residue which can lead to excessive formation of oxidation-induced stacking faults. These stacking faults may result in excessive leakage currents and subsequent circuit failure.

As advanced processes and processing techniques are developed many of these yield-limiting effects will be reduced or eliminated; however, new ones may be introduced!

14.2.2 Circuit Sensitivities

In addition to areas of a wafer where the yield is low because processing difficulties have led to device parameters being outside of the specified range, certain areas of a wafer may have low device yield because the design of the circuit has failed to take into account the expected variation in device parameters and the correlation between variations in different parameters.

Threshold voltage (V_T) and channel length (L) of the MOS transistors are two of the most important parameters in MOS circuit design. Variations in substrate doping, ion implantation dose, and gate oxide thickness will cause variations in the threshold voltage. Variations in gate length and source and drain junction depth cause the channel length to vary. The threshold voltage and channel length variations are generally not correlated with each other. However, the speed of a circuit generally increases as the threshold voltage and channel length decrease. Circuit performance is often simulated under high speed (small V_T and L) and low speed (large V_T and L) conditions. It is important that circuit performance should also be simulated for small V_T and large L, and large V_T and small L. Circuit design must also consider variations in other device parameters, such as the resistance of implanted regions, capacitance of conductors to substrate, contact resistance, and leakage currents.

Often two circuits with nominally the same size and complexity, processed in the same technology, have vastly different yields. The low yield in the one case is due to a lack of understanding of the sensitivity of the circuit to device parameters. Here, higher yield requires a cooperative effort between the circuit designer, who identifies the specific device parameters to which the circuit is sensitive, and the process engineer, who optimizes both the value and range of variation of those parameters. Once the circuit sensitivities to specific process parameters have been determined, redesign of the circuit to reduce these sensitivities produces high yield and low cost with minimal attention from the process engineer.

14.2.3 Point Defects

We typically find that the yield of circuits is not 100%, even after we have identified those wafer areas in which all processing parameters are not only within the expected range but are also within the range where the circuit performs satisfactorily. The cause of this less-than-perfect yield is generally attributed to "point defects." A point defect is a region of the wafer where the processing is imperfect, and the size of this region is small compared to the size of the chip. For example, consider a chip 2000 μm square with features whose size is 2 μm. A 3-μm-diameter dust particle on the wafer can cause a break in a metal conductor. Similarly a 200-μm dust particle could cause a large area of metal to be missing from the chip. Both these particles could be considered to be point defects. A region of the wafer that is 1 cm in diameter, covering 20 or 30 chips, where the metal is missing from the chip, is not considered a point defect.

Many types of processing defects are considered to be point defects. One of the most common is dust or other particles in the environment. These particles may fall on the wafers while they are being transported around the processing facility, or they may be generated in a film-deposition operation and be incorporated in the film. They may also be present in photoresist solutions and deposited on the wafer during the photoresist operation, or they may be silicon particles, knocked out of the wafer during handling, which adhere to the wafer surface. Isolated oxidation-induced stacking faults that cause excessive junction leakage and circuit failure can also be considered point defects, as can isolated spikes in an epitaxial film or pinholes in a dielectric film.

Point defects can occur on lithographic masks as well as on silicon wafers. Dust or particles on the mask substrate during the mask-generation process can cause permanent defects in the mask. Particles which adhere to the mask during use cause a gradual increase in the density of point defects that are reproduced on the wafers. Periodic cleaning of the mask will remove these defects.

Successful IC fabrication requires continued monitoring of the density of point defects. Monitoring may be carried out by visual or SEM observation of circuit quality during all steps of fabrication. When any variation in the defect density of a particular operation is observed, the appropriate corrective action can be taken. Special operations may also be performed to monitor defects. Wafers can be etched to remove all films, and the silicon substrate treated to reveal stacking faults, whose density is then monitored. Other wafers may be patterned by using special masks which allow electrical measurement of the incidence of dielectric breakdown defects. As new types of defects are identified as the cause of circuit failure, methods of monitoring the density of these defects must be instituted.

The circuit yield and the cause of circuit failure must also be monitored. The use of a circuit, such as a memory chip, in which a failure can be related back to a specific area of the chip (for example, one of the six transistors of a static memory cell) has been most useful. The successful operation of an IC facility requires that the density of point defects be continuously monitored, controlled, and reduced.

14.3 MODELING OF YIELD LOSS MECHANISMS

Modeling IC yield in terms of fundamental parameters that are independent of the particular IC and characteristic of the process and processing line is important for several reasons. By accurately modeling the yield we can predict the cost and availability of future circuits, provided they are related in technology and design to the circuits used to develop the yield modeling parameters. Once we know the yield modeling parameters, we can compare the processing quality of different process lines and thus indicate where improvements in process facilities are required. Yield modeling parameters, for a given process and process line, that show significant variation for a particular circuit may indicate a sensitivity to device parameters in that circuit. Further study of the circuit design may reveal a possible change in design or process that will result in a significant improvement in device yield.

Generally, IC yield is expressed in the form

$$Y = Y_0 Y_1(D_0, A, \alpha_i) \tag{1}$$

where Y is the ratio of good chips per wafer to the number of chip sites per wafer, $(1 - Y_0)$ is the fraction of chip sites that yield bad chips either because of process-related effects or because of circuit sensitivities, Y_1 is the yield of good chips on the remaining chip sites, D_0 is the density of point defects per unit area, A is the area of the chip, and α_i are parameters unique to different models of the yield.

All of the models, which will be discussed in Sections 14.3.1 to 14.3.4, predict that the yield decreases monotonically as the area of the chip increases. The models can be quite useful in predicting IC yield, provided we stay within the range of parameters developed and verified in the models. IC yield modeling is notorious for underestimating the yield of future ICs; IC processing is an ever-developing art. Yield modeling can identify those processes and mechanisms that limit the yield of present ICs. As the yield-limiting features are identified the processes are improved or eliminated as needed. For example, projection optics has replaced contact printing, resulting in lower defect density; dry etching has replaced wet chemistry, resulting in better feature size control; and ion implantation has replaced diffusion, resulting in better control of resistance and junction depth.

Yield modeling is not a predictor of the future; rather, it is a tool for improving the present generation of processes and designs. The following sections discuss the features of some of the more common methods for modeling the effect of point defects on IC yield.

14.3.1 Uniform Density of Point Defects

In the area of the wafer where the yield has not been degraded by either processing effects or circuit sensitivities, the remaining cause of loss of good chips is randomly distributed point defects. Figure 2 shows a grid of 24 chip sites with 10 defects randomly distributed over the area. In this example, 16 of the 24 sites have zero defects, that is, they are good chip sites. Of the remaining eight sites, six have one defect,

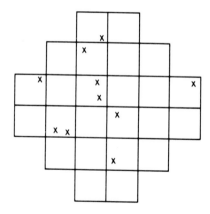

Fig. 2 An array of 24 chip sites with 10 defects (indicated by an X randomly distributed on the sites.)

and two have two defects; no sites have more than two defects. The problem of determining the yield of good chips is identical to the classical statistics problem of placing n balls in N cells, and then calculating the probability that a given cell contains k balls.[1] If n defects are distributed randomly among N chips, the probability that a given chip contains k defects is given by the binomial distribution

$$P_k = \frac{n!}{k!(n-k)!} \frac{1}{N^n} (N-1)^{n-k} \tag{2}$$

In the limit when N and n are both large, and the ratio $n/N = m$ remains finite, the binomial distribution can be approximated by the more tractable Poisson distribution

$$P_k \cong e^{-m} \frac{m^k}{k!} \tag{3}$$

The probability that a chip contains no defects, which is the yield, is given by

$$Y_1 = P_0 = e^{-m} \tag{4}$$

and the probability that a chip contains one defect is given by

$$P_1 = me^{-m} \tag{5}$$

If the area of a chip is A, the total area of the chips in the good part of the wafer is NA, and the density of defects is $n/NA = D_0$. The average number of defects per chip m is

$$m = \frac{n}{N} = D_0 NA/N = D_0 A \tag{6}$$

and

$$Y_1 = P_0 = e^{-D_0 A} \tag{7}$$

The Poisson estimate of the yield of good chips was used to predict the yield during the early manufacture of ICs. However, the actual yield obtained on larger circuits was considerably greater than that estimated using the D_0 values measured by fitting the yield of smaller circuits to the Poisson formula. In fact, the low yields estimated by the Poisson formula no doubt delayed early work in the development of ICs.

14.3.2 Simple Nonuniform Distributions of D

The discrepancy between the measured and predicted yields of early ICs led researchers to investigate the effect of nonuniform distributions of D_0 across a wafer on the yield of that wafer. The yield of chips on a wafer, where D_0 is nonuniform across the wafer, can be expressed as[3]

$$Y = \int_0^\infty e^{-DA} f(D) \, dD \tag{8}$$

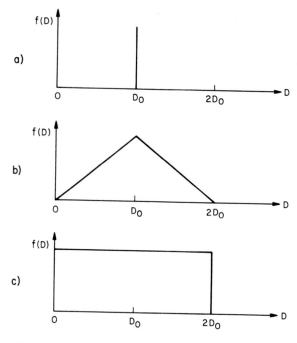

Fig. 3 Distribution of defect density. (a) Delta function. (b) Triangular. (c) Rectangular. *(After Murphy, Ref. 3.)*

where

$$\int_0^\infty f(D)\, dD \equiv 1$$

Three distributions of D_0 are evaluated, the delta function, a triangular distribution, and a rectangular distribution (Fig. 3). The results are as follows:

Delta function: $\quad Y_1 = e^{-D_0 A}$ (9)

Triangular: $\quad Y_2 = \left[\dfrac{1 - e^{-D_0 A}}{D_0 A}\right]^2$ (10)

Rectangular: $\quad Y_3 = \dfrac{1 - e^{-2 D_0 A}}{2 D_0 A}$ (11)

If we look at the form of these expressions for values of $D_0 A \gg 1$ we find that

$$Y_1 \cong e^{-D_0 A}$$ (12)

$$Y_2 \cong \frac{1}{(D_0 A)^2}$$ (13)

$$Y_3 \cong \frac{1}{2 D_0 A}$$ (14)

Equations 13 and 14 give much less pessimistic estimates of the yield than the original Poisson estimate. These expressions found considerable use, with Y_3 being the expression that most closely fit the observed yield of large ICs. The nonphysical characteristics of the above distributions led workers to investigate other, more physical distributions that could also predict the observed high yield of large ICs.

14.3.3 The Gamma Distribution of D

A more physical distribution for calculating IC yield is the gamma distribution.[4-6] The probability density function (pdf) of this distribution is

$$f(D) = \frac{1}{\Gamma(\alpha)\beta^\alpha} D^{\alpha-1} e^{-D/\beta} \qquad (15)$$

where α and β are the two distribution parameters, $\Gamma(\alpha)$ is the gamma function, and D, α, and β are all greater than zero. In this distribution the average density of defects is given by $D_0 = \alpha\beta$, the variance of D is given by var $(D) = \alpha\beta^2$, and the coefficient of variation is given by $\sqrt{\text{var}\ (D)}\ /D_0 = 1/\sqrt{\alpha}$.

The probability of a chip having k defects is given by

$$P_k = \int_0^\infty e^{-m} \frac{m^k}{k!} f(D)\ dD$$

$$= \frac{\Gamma(k + \alpha)}{k!\ \Gamma(\alpha)} \frac{(A\beta)^k}{(A\beta + 1)^{k+\alpha}} \qquad (16)$$

where $\int_0^\infty f(D)\ dD = 1$.

The probability of having no defects on a chip is

$$Y_4 = P_0 = \frac{1}{(A\beta+1)^\alpha}$$

$$= \frac{1}{(1+SAD_0)^{1/S}} \qquad (17)$$

where

$$S = \frac{\text{var}\ (D)}{D_0^2} = \frac{1}{\alpha}$$

In Eq. 17 the yield is a function of the average defect density D_0, the chip area A, and the square of the coefficient of variance S of the distribution of D. In general, the gamma distribution of D is a skewed distribution stretching from zero to infinity. Figure 4 shows the shape of the distribution function for several values of the shape parameter S. In the limiting case of $S \to 0$, the gamma distribution reduces to a delta function

$$f(D) = \delta(D - D_0) \qquad (18)$$

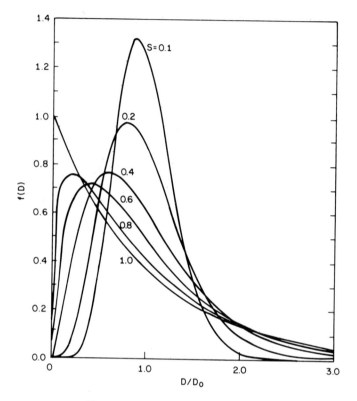

Fig. 4 Gamma distribution of defect density.

where $S \cong 0$, $\beta = SD_0$, and the yield Y_4 is given by the expected Poisson yield

$$Y_4 = e^{-D_0 A} \qquad S \to 0 \qquad (19)$$

Figure 5 is a plot of the gamma yield function of Eq. 17. The yield functions Y_2 and Y_3 are plotted for comparison. The yield function for a uniform density of D_0, Y_1, is identical to the gamma yield function with $S = 0$. The gamma yield function is clearly a good approximation to the functions Y_2 and Y_3 over a wide range of $D_0 A$ values. Furthermore, the gamma yield function has the potential to represent quite a large variation in the shape of an experimental yield versus area curve. For these reasons the gamma yield function is the most common function for representing IC yield. Values of the shape parameter S vary considerably between the different types of product manufactured in different processing facilities. Also, many different types of defects affect the yield, and the parameters D_0 and S vary considerably for different types of defects.

14.3.4 Yield of Chips with Redundant Circuitry

Many large MOS memory chips are designed with redundant circuitry,[2] which can be switched in to replace defective circuit elements. This replacement is particularly

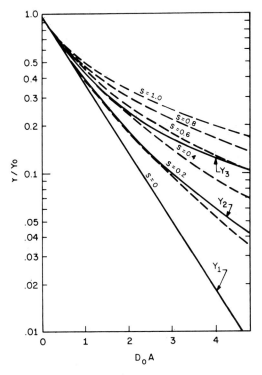

Fig. 5 Yield of good chips as a function of D_0A for the gamma function (broken curves) and the delta-function, triangular, and rectangular distributions (solid curves).

simple in the case of the MOS memory, which in large part is composed of reiterative circuitry. The replacement of a defective circuit is usually accomplished using fusible links. These links are then fused as needed using laser or other techniques.

If a memory chip were designed with one redundant column, for example, any chip with a defect in a column could be repaired by electrically substituting the redundant column for the defective column. The yield of this chip would then be given by

$$Y_1 = P_0 + \eta P_1 \tag{20}$$

where P_0 is the probability of a chip containing no defects, P_1 is the probability of a chip containing one defect, and η is the probability that a chip containing one defect can be repaired by using the single redundant column. This yield model can be extended to chips containing more complex redundant circuitry.

14.3.5 Multiple Types of Defects

IC yield is affected by many types of defects. Each type affects different circuits to a different extent. For example, defects in the gate oxide occur only in the gate region of the transistors. Junction leakage caused by oxidation-induced stacking faults can occur in both the gate region and the source and drain regions. Shorts between metal runners are important only in those regions of the chip with a high density of closely packed metal runners.

Since yield loss so strongly depends on the different defect mechanisms, each type should be considered independently. Each defect mechanism should be characterized by its mean defect density D_{n0}, the shape factor of the distribution of defects S_n, and the portion of the total chip area A_n that is susceptible to that particular type of defect. Making use of the gamma yield function, the yield for each type of defect is

$$Y_n = \frac{1}{(1 + S_n A_n D_{n0})^{1/S_n}} \tag{21}$$

The overall yield is then the product of the yield for each known type of defect, that is,

$$Y = \prod_{n=1}^{N} Y_n = \prod_{n=1}^{N} (1 + S_n A_n D_{n0})^{-1/S_n} \tag{22}$$

When chip yield is determined by several major types of defects, the parameters S_n, A_n, and D_{n0} must be determined individually, and the yield of the overall process estimated[7-9] using Eq. 22.

For a mature product processed in a well-controlled, high yield fabrication line, all of the major yield limiting defects have probably been controlled or eliminated. Thus the yield (represented by Eq. 22) is the product of many terms, all close to unity, that is, $S_n A_n D_{n0} \ll 1$. The yield is

$$Y = \prod_{n=1}^{N} (1 + S_n A_n D_{n0})^{-1/S_n}$$

$$\ln Y = \sum_{n=1}^{N} -\frac{1}{S_n} \ln (1 + S_n A_n D_{n0}) \tag{23}$$

Since $S_n A_n D_{n0} \ll 1$,

$$\ln (1 + S_n A_n D_{n0}) \approx S_n A_n D_{n0}$$

and

$$\ln Y = \sum_{n=1}^{N} -A_n D_{n0} \tag{24}$$

or

$$Y = \exp \left[- \sum_{n=1}^{N} A_n D_{n0} \right]$$

$$= \exp (- A \overline{D}) \tag{25}$$

where

$$\overline{D} = \frac{1}{A} \sum_{n=1}^{N} A_n D_{n0}$$

Thus for a mature, high yield chip, the yield is again represented by an exponential, independent of the shape parameters for each type of defect. The composite defect density $\overline{D}$, however, is a function of the type of circuit, and circuit layout, through the fractional areas A_n. The $\overline{D}$ obtained from one chip design is not applicable to another chip design. The yield expressed in Eq. 25 does, however, apply to the type of experiment that studies the yield of multiples of a single chip.[7, 10, 11] In that case we would expect the yield of M-multiple chip sites, for a mature, high yield chip, to be

$$Y_M = e^{-MA\overline{D}} \tag{26}$$

as is observed in many experiments.

14.3.6 Radial Distribtuion of Defects

In the previous sections we assumed that the density of defects is not uniform, but varies across a wafer and from wafer to wafer. Several researchers have performed detailed studies on the origin of the defect density variation for their own particular process.[8, 9, 12, 13] They find that certain types of defects show a strong radial dependence in the frequency of their occurrence. This relationship holds especially for cosmetic defects, such as handling damage, misalignment errors, photoresist residue, and so forth.

The radial variation in defect density can be represented by the expression

$$D(r) = D_0 + D_R \, e^{(r-R)/L} \tag{27}$$

where D_0 is the defect density associated with the center of the wafer, D_R is the increase in defect density at the edge of the wafer, r is the radial coordinate, R is the radius of the wafer, and L is the characteristic length associated with the edge-related defects.

The yield of a wafer with a radial distribution of defects may be obtained by integrating the Poisson yield function over the area of the wafer:

$$Y_R = \frac{2}{R^2} \int_0^R e^{-D(r)A} \, r \, dr \tag{28}$$

Equation 28 can be evaluated exactly, but not in a simple, closed form. Computer calculations of Y_R for specific cases have been performed.[9]

The radial distribution of defect density in Eq. 27 can also be transformed to a distribution function of D, $f(D)$. The yield can then be calculated by using Eq. 9, or $f(D)$ can be approximated by a gamma distribution, and the results of Section 14.3.3 are then applicable.

14.3.7 Summary

The previous sections have examined some causes which lead to loss of yield in the manufacture of ICs. One characteristic of the yield loss mechanisms is the dependency upon the technology; the type of circuit; the particular circuit design; and the maturity of the processing technology, facility, and circuit design. Yield loss mechanisms change with time. As the technology matures, the major causes of yield loss are identified and eliminated. Similarly as the sensitivities of a circuit to process variations are identified, the process control is improved, and the circuit will be redesigned to reduce the sensitivities. Finally, when both process and design have been optimized, the remaining yield loss mechanism is a low level of random defects. Identification and elimination of yield loss mechanisms will continue with the development of new technologies.

14.4 RELIABILITY REQUIREMENTS FOR VLSI

Before we begin a detailed study of reliability, it will be instructive to consider several examples of the effects of device failure on system performance. These examples will give the reader perspective as to the acceptable values of device failure rates and point out the disastrous effects of excessive device failures on the system's economic viability.

A series of studies on semiconductor-device failure rates and mechanisms began shortly after the development of the transistor, when serious consideration was given to the construction of large solid-state computer systems and electronic telephone switching systems (ESS). One of these early systems could contain on the order of 100,000 discrete transistors, plus other components. If we take one device failure per month as a goal, then the devices must have a failure rate of

$$\lambda < \frac{1 \text{ failure}}{10^5 \text{ devices} \times 720 \text{ hours}}$$

$$= 14 \times 10^{-9} \text{ failure/device hour} \qquad (29)$$

We will define the unit of failure rate to be 1 <u>F</u>ailure un<u>IT</u> $\equiv$ 1 FIT $\equiv$ 1 failure/10^9 device hour. Thus we see that our goal for the device failure rate of our hypothetical system is

$$\lambda < 14 \text{ FIT}$$

(Note that because the system will be designed with redundant elements, a single device failure most likely will not result in a system failure.) Over the expected 10-year life of the typical system 120 devices ($\approx 0.1\%$ of the devices in the system) will fail. A total of 120 service calls will have to be made. Table 1 summarizes the effect of different device failure rates on the number of failures per month and the total percentage of devices that will fail in the system's 10-year life. It can be seen that a device

Table 1 Effect of transistor failure on system performance

Failure rate (FIT)	Failure/month*	Total % of device failing in 10 years
10	0.7	0.1
100	7	1
1000	70	10

*Based on 100,000 devices per system.

failure rate of 10 FIT is very desirable, a failure rate of 100 FIT is probably acceptable, and a failure rate of 1000 FIT is unacceptable, considering that such a rate would require repairs approximately twice a day and, depending on the number of devices per circuit board, the replacement of close to the entire complement of system circuit boards during the life of the system.

The previous example was based on an original semiconductor device system using 100,000 discrete transistors. Table 2 summarizes similar consideration for two modern systems: a large data set containing 150 to 225 ICs, and a private telephone system containing 10,000 or more ICs. The acceptable device failure rates for both of these systems are the same as for the first example (i.e., a device failure rate of less than 10 FIT is desirable, and a failure rate of 1000 FIT is unacceptable).

Table 2 Effect of device failure on modern system performance

Data set: 150 to 225 ICs		
Failure rate (FIT)	Mean time to failure (year)	% of sets failing per month
10	51	0.16
100	5	1.6
1000	0.5	16

Private telephone system: bulk of installations 5000 to 10,000 ICs; large installations 50,000 ICs		
Failure rate (FIT)	System failure/month	% of circuit packs failing in 10 years
10	0.07	1
100	0.7	10
1000	7	65

Table 3 Reliability prove-in

For 100 devices on test	
Failure rate (FIT)	Time for 1 failure (year)
10	114
100	11
1000	1

If we run 500 devices for 6 months with no failures	
Confidence level (%)	Failure rate (FIT)
Best estimate	325
60	430
90	1100
95	1400
99	2100

The next consideration is how to demonstrate that the devices actually have the desired failure rate of less than 10 FIT. Table 3 shows the expected time for one failure to occur in 100 devices being tested. Also tabulated are the estimated device failure rates if we test 500 devices for 6 months and observe no failures. In both cases we see that if we are to prove our desired device failure rate of 10 FIT, we must test large numbers of devices for totally unreasonable lengths of time, in terms of the development schedule of a device or system. Proof of adequate device reliability is time consuming and expensive; however, the cost of inadequate device reliability is disastrously expensive.

The following sections discuss methods to quantitatively measure and predict device failure rates, and to identify and eliminate the failure mechanisms.

14.5 MATHEMATICS OF FAILURE DISTRIBUTIONS, RELIABILITY, AND FAILURE RATES

The term "reliability" has many popular connotations. For a useful, mathematical description of reliability, however, we must first precisely define the term. A generally accepted definition of reliability is the probability that an item will perform a required function under stated conditions for a stated period of time.

The "required function" must include a definition of satisfactory operation and unsatisfactory operation, or failure. For an IC the required function is generally defined by a test program for an automatic test set. The program's output simply states "good" or "bad." In many cases, however, initial test programs are not complete; they do not test the circuit under all the required conditions. As new device failure modes are identified, the appropriate tests are included in later generations of the test program.

The "stated conditions" in the definition are the total physical environment, including the mechanical, thermal, and electrical conditions of expected use (and also of periods of disuse, such as storage, etc.)

The "stated period of time" is the time during which satisfactory operation is required. This will vary depending upon the usage of the system. In some cases the time can be relatively short, as in the case of an emergency beacon transmitter on an aircraft. Often the period of use is preceded by a long period of disuse. Other systems are in continuous usage. For such continuously operating systems another concept, "availability," is defined. Availability is the probability that an item will operate when needed, or, the average fraction of time that a system is expected to be in an operating condition.

The next sections present the basic reliability concept and related concepts in terms of mathematical functions. A more rigorous treatment can be found in Ref. 5.

14.5.1 Cumulative Distribution Function

Assume that a device or system is operating at time $t = 0$. The probability that the device will fail at or before time t is given by the function $F(t)$. This is a cumulative distribution function (or cdf) with the properties

$$
\begin{array}{ll}
F(t) = 0 & t < 0 \\
0 \le F(t) \le F(t') & 0 \le t \le t' \\
F(t) \to 1 & t \to \infty
\end{array} \tag{30}
$$

14.5.2 Reliability Function

The reliability function $R(t)$ is the probability that the device will survive to time t without failure. The function is related to $F(t)$ and is given by

$$ R(t) = 1 - F(t) \tag{31} $$

14.5.3 Probability Density Function

The derivative of $F(t)$ with respect to time is known as the probability density function (or pdf) and is represented by $f(t)$. The pdf is related to the cdf by

$$ f(t) = \frac{d}{dt} F(t) \tag{32} $$

or

$$ F(t) = \int_0^t f(x)\, dx \tag{33} $$

Similarly

$$R(t) = \int_t^\infty f(x)\, dx \qquad (34)$$

and

$$f(t) = - \frac{d}{dt} R(t) \qquad (35)$$

14.5.4 Failure Rate

In most applications the quantity which is of most concern is the instantaneous failure rate, often referred to as the hazard rate. By the term "failure rate" we will always mean the instantaneous failure rate, not the average failure rate.

The fraction of devices that were good at time t and that fail by time $t + \Delta$ is given by

$$F(t + \Delta) - F(t) = R(t) - R(t + \Delta) \qquad (36)$$

The average failure rate during the time interval Δ is

$$\lambda(t) = \frac{1}{\Delta} \frac{R(t) - R(t + \Delta)}{R(t)} \qquad (37)$$

In the limit as $\Delta \to 0$, the instantaneous failure rate is given by

$$\lambda(t) = - \frac{1}{R(t)} \frac{dR(t)}{dt} \qquad (38a)$$

$$= \frac{f(t)}{R(t)} \qquad (38b)$$

$$= \frac{f(t)}{1 - F(t)} \qquad (38c)$$

$$= - \frac{d}{dt} \ln R(t) \qquad (38d)$$

Using Eq. 38d, we can express the reliability function as

$$R(t) = \exp\left[- \int_0^t \lambda(x)\, dx \right] \qquad (39)$$

14.5.5 Mean Time to Failure

A common measure of reliability is the mean time to failure (MTTF) of the device or system, which can be expressed as

$$\text{MTTF} = \int_0^\infty t f(t)\, dt \qquad (40)$$

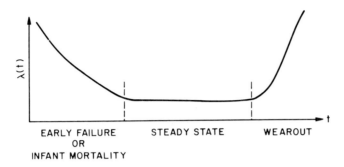

Fig. 6 Failure rate versus time for typical ICs.

MTTF is the device's average age at failure for a population whose reliability is represented by the function $R(t)$ with a pdf of $f(t)$.

14.6 COMMON DISTRIBUTION FUNCTIONS

It is desirable to have a single mathematical model that represents the failure rate of devices over their entire life. The failure rate of an IC, and of most common items as well, generally varies as a function of time in the manner illustrated in Fig. 6. During the early life of the device the failure rate is high, but it decreases as a function of time. The failures during this period are termed "early failures" or "infant mortality." The causes of early failure are generally manufacturing defects. In many devices the early failure period is the most important because of the very high failure rates that can be observed during this period. Section 14.7.4 discusses methods of eliminating or shortening this early failure period.

During the midlife or steady-state period, the failure rate is generally low and fairly constant. Device failures are a result of a large number of unrelated causes.

Eventually the device enters the final, or wearout, period of its life. This period is more commonly observed in other items such as electric light bulbs than in ICs. In most ICs no wearout mechanisms are observed during the useful life of the device. However, some failure mechanisms (e.g., mobile ion shift, corrosion, and electromigration) can be observed in low quality ICs.

Of the common distribution functions we will be discussing, none adequately represent all three periods of a device's life. At most, one distribution function can be used to represent two of the periods. Note that the distribution functions discussed in the next sections are treated in more detail in Refs. 5 and 6.

14.6.1 Exponential Distribution Function

The simplest distribution function, the exponential function, is characterized by a constant failure rate over the lifetime of the device. This function is useful for representing a device in which all early failure and wearout mechanisms have been eliminated. The exponential distribution is characterized by the following functions:

$$\lambda(t) = \lambda_0 \equiv \text{constant}$$

$$R(t) = e^{-\lambda_0 t}$$

$$F(t) = 1 - e^{-\lambda_0 t} \tag{41}$$

$$f(t) = \lambda_0 e^{-\lambda_0 t}$$

$$\text{MTTF} = \int_0^\infty t \lambda_0 e^{-\lambda_0 t} \, dt = \frac{1}{\lambda_0}$$

14.6.2 Weibull Distribution Function

In the Weibull distribution function the failure rate varies as a power of the age of the device. The failure rate is represented as

$$\lambda(t) = \frac{\beta}{\alpha} t^{\beta-1} \tag{42}$$

where α and β are constants. For $\beta < 1$ the failure rate decreases with time, and the Weibull distribution may be used to represent the early failure period of a device. For $\beta > 1$ the failure rate increases with time, and the Weibull distribution may be used to represent the wearout period of a device. If $\beta = 1$ the failure rate is constant, and the exponential distribution is a special case of the Weibull distribution with $\beta = 1$. From Eq. 42 we can also calculate that

$$R(t) = \exp\left[-\frac{1}{\alpha} t^\beta \right]$$

$$F(t) = 1 - \exp\left[-\frac{1}{\alpha} t^\beta \right] \tag{43}$$

$$f(t) = \frac{\beta}{\alpha} t^{\beta-1} \exp\left[-\frac{1}{\alpha} t^\beta \right]$$

In some applications of the Weibull distribution function we can obtain a better fit to the experimental failure rate by introducing a third parameter. In Eqs. 42 and 43 the time t is replaced by

$$X = t - \gamma \tag{44}$$

The parameter γ corresponds to some portion of the life of the device that has expired during device manufacture, device burn-in (Section 14.7.4), or device testing.

In fitting experimental data to an assumed distribution function to determine the quality of fit and the distribution parameters, we need to design plotting paper on which the experimental data lies on a straight line if the data can be represented by the assumed distribution. Such plotting paper is available for use with the Weibull distribution. For the Weibull distribution

$$1 - F(t) = \exp \left[- \frac{1}{\alpha} t^\beta \right] \qquad (45)$$

and

$$\ln \left\{ \ln \left[\frac{1}{1 - F(t)} \right] \right\} = \beta \ln t - \ln \alpha \qquad (46)$$

which is linear in the form

$$y = mx + b$$

Figure 7 shows an example of Weibull plotting paper; the ordinate is marked in cumulative percent failed devices $F(t)$, and the abscissa is $\ln t$. If the experimental data is represented by the two-parameter Weibull distribution, a straight line is obtained when the data is plotted (Fig. 7). The slope of the line is the parameter β.

The early failure rate of many ICs can be represented by a Weibull distribution, and Weibull plotting paper is commonly used in IC reliability studies.

14.6.3 Duane Plotting

In some areas, especially those relating to system design, we can analyze failure data by using a technique known as a Duane plot.[14] This graphical analysis technique allows a quick prediction of device failure rates when the number of device failures is small, and is most often used for evaluating prototype systems.

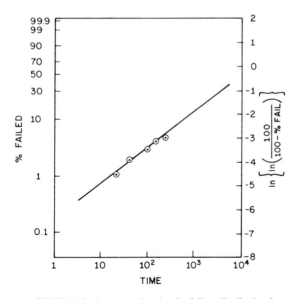

Fig. 7 An example of Weibull plotting paper showing the failure distribution for a typical device.

The technique consists of plotting the log of the average failure rate (AFR) versus the log of time. The average failure rate is defined as the fraction of failed devices at time t divided by the time:

$$\text{Average failure rate} = \text{AFR} = \frac{F(t)}{t} \tag{47}$$

Often on such a plot the data plots as a straight line with negative slope S, that is,

$$\ln \text{AFR} = -S \ln t + \ln K$$

or

$$\text{AFR} = \frac{F(t)}{t} = Kt^{-S} \tag{48}$$

where K is a constant. If that is the case, then

$$F(t) = Kt^{1-S}$$
$$f(t) = K(1-S)t^{1-S}$$

and

$$\lambda(t) = \frac{f(t)}{1-F(t)} = \frac{K(1-S)t^{-S}}{1 - Kt^{-S}} \tag{49}$$

If only a small fraction of devices have failed, then $F(t) \ll 1$ and

$$\lambda(t) \cong K(1-S)t^{-S} \tag{50}$$

or

$$\lambda(t) = (1-S)\text{AFR} \tag{51}$$

Equation 50 corresponds to a Weibull distribution with $\beta = (1-S)$ and $\alpha = 1/K$. Thus, a Duane plot is an alternative to a Weibull plot to analyze data in cases where $F(t) \ll 1$.

The danger in using a Duane plot is that the limitation $F(t) \ll 1$ is not always recognized, and Eq. 50 or 51 rather than Eq. 49 is used to extrapolate $\lambda(t)$ into the future. Also, extrapolating $\lambda(t)$ during the early failure period into the steady-state period may result in gross underestimation of the long-term failure rate.

14.6.4 Log-Normal Distribution Function

The log-normal distribution function has been used quite successfully to describe the failure statistics of semiconductor devices over wide spans of time. Depending on the values of the distribution parameters this function can represent any one of the three periods in the life of a device.

The probability density function (pdf) of the log-normal distribution is given by

$$f(t) = \frac{1}{\sigma t \sqrt{2\pi}} \exp\left[-\frac{1}{2}\left[\frac{\ln t - \mu}{\sigma}\right]^2\right]$$

(52)

$$= \frac{1}{\sigma t \sqrt{2\pi}} \exp\left[-\frac{1}{2}\left[\frac{1}{\sigma}\ln\frac{t}{t_{50}}\right]^2\right]$$

where the median time to failure (the time when 50% of the devices have failed) is given by

$$t_{50} = e^{\mu}$$

(53)

The average time to failure is

$$\bar{t} = \exp\left[\mu + \frac{\sigma^2}{2}\right]$$

(54)

and the scale parameter σ is approximately

$$\sigma \approx \ln\frac{t_{50}}{t_{16}}$$

(55)

where t_{16} is the time when 16% of the devices have failed (more precisely, 15.866%).

The cdf is given by

$$F(t) = \frac{1}{\sigma\sqrt{2\pi}} \int_0^t \frac{dx}{x} \exp\left[-\frac{1}{2}\left[\frac{\ln t - \mu}{\sigma}\right]^2\right]$$

(56)

and the failure rate is

$$\lambda(t) = \frac{f(t)}{1 - F(t)}$$

(57)

The log-normal distribution does not lend itself to easy numerical calculation. Most data analyses using the log-normal distribution are done either graphically or on a digital computer. Figure 8 shows the log-normal failure rate as a function of time and the cdf for different values of the scale parameter σ. The log-normal distribution may represent the early failure, the steady state, or the wearout period of a device.

Figure 9 shows a normalized plot of log-normal failure rate versus time. The normalized failure rate $\lambda(t)/t_{50}$ is a function only of t/t_{50} and the scale parameter σ. This figure also plots the locus of constant failure rate at a given time. If the constant failure rate is λ_0 and the given time is t_0, then

$$\lambda t_{50} = (\lambda_0 t_{50})(t_0/t_0) = (\lambda_0 t_0)/(t_0/t_{50})$$

This equation plots as a straight line with slope of -1 in the log-log plot of Fig. 9.

Consider, for example, the case of a failure distribution with $\sigma = 2$. For short times the failure rate is low, but increasing with time. When the normalized time

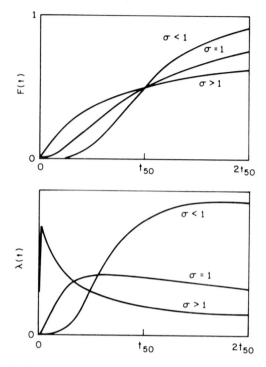

Fig. 8 Cumulative distribution function $F(t)$ and failure rate $\lambda(t)$ for the log-normal failure distribution.

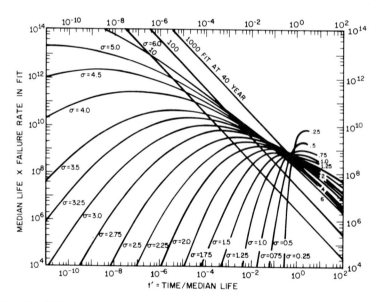

Fig. 9 Normalized failure rate versus normalized time for the log-normal failure distribution. In this plot, a line with slope -1 is the locus of a given failure rate at a given time with σ as the parameter. Loci are shown for three failure rate values at 40 years. *(After Goldthwaite, Ref. 15.)*

Table 4 Log-normal distribution

Minimum median life t_{50} required to meet a maximum failure rate specification at 40 years

σ	Minimum Median Life t_{50} (h)		
	10 FIT	100 FIT	1000 FET
0	3.5×10^5	3.5×10^5	3.5×10^5
0.5	2×10^6	1×10^6	7×10^5
1	7×10^6	3×10^6	8×10^5
2	1×10^8	1.3×10^7	1.3×10^6
3	1×10^9	$\mathbf{1 \times 10^{8}}$*	1×10^7
4	$\mathbf{3 \times 10^{10}}$	3×10^9	3×10^8
5	$\mathbf{2 \times 10^{12}}$	$\mathbf{2 \times 10^{11}}$	$\mathbf{2 \times 10^{10}}$

*Boldface values correspond to those cases where the maximum failure rate occurs in less than 40 years.

$(t' = t / t_{50})$ is approximately 3×10^{-3}, the failure rate equals 10 FIT, and the time equals 40 years. The median life of the failure distribution required for this to happen is 1×10^8 h. If the median life is greater than 1×10^8 h, the failure rate for $\sigma = 2$ will be less than 10 FIT at 40 years. If the median life is less than 1×10^8 h, the failure rate will be greater than 10 FIT at 40 years.

For the case $\sigma = 4.5$, the normalized failure rate has a maximum of 2×10^{12} at $t' \cong 2 \times 10^{-9}$. If this maximum failure rate is 10 FIT, then the median life is 2×10^{11} h, and the maximum failure rate of 10 FIT occurs at $t = 400$ h. If the median life, for $\sigma = 4.5$, is less than 2×10^{11} h, then the maximum failure rate is greater than 10 FIT.

Table 4 summarizes the minimum median life required to meet a maximum failure rate at 40 years. The boldface values of minimum median life correspond to those cases where the maximum failure rate occurs in less than 40 years.

Experimental data can be fitted to the log-normal distribution function by plotting the data on standard log-normal plotting paper (Fig. 10). The σ of the distribution is estimated by the intersection of a line, parallel to the experimental data and passing through the index point $\oplus$, with the σ scale on the right-hand side of the plot.

14.6.5 Two Failure Populations

The study of a device's failure rate is often complicated by the existence of more than one failure mechanism. The failure statistics of each individual mechanism may be governed by a log-normal distribution, but the parameters of the distribution, σ and μ, are different for each mechanism. Quite often one of the failure modes is characterized by a short median life and represents a small percentage of the total population, usually referred to as the sport population. This mode represents an early failure mechanism. The remainder of the population has a considerably longer median life and represents the steady-state failure mechanism. In actual use, the sport population may be the predominant failure mechanism seen during use of devices in the field.

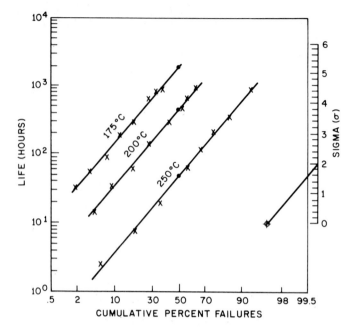

Fig. 10 An example of log-normal plotting paper showing device failure distribution at three temperatures.

Figure 11 shows bimodal failure data plotted on log-normal paper. Using trial and error techniques, the data is resolved into the two component distributions. By obtaining each distribution's parameters the failure rate can be expressed as

$$\lambda_T(t) = \lambda_S(t) \times \text{fraction of sports}$$

$$+ \lambda_M(t) \times \text{fraction in main population} \tag{58}$$

14.7 ACCELERATED TESTING

In the previous sections we have shown that the required failure rate of an IC is on the order of 100 FIT or less. From Tables 3 and 4 we see that, under normal operating conditions, with a failure rate of 100 FIT, the time required to observe one failure in 100 devices is about 100,000 h. The required median life is on the order of 10^5 to 10^{11} h, depending on the scale parameter σ. Clearly, it is impractical to study the failure characteristics of devices with the requisite reliability under the normal operating conditions. If we are to study failure characteristics, some means must be found to accelerate the mechanisms that cause devices to fail.

There are five common stresses used to accelerate device failure mechanisms: temperature, voltage, current, humidity, and temperature cycling. Temperature cycling is used to accelerate mechanical failure of the chip or assembled package (see Chapter 13).

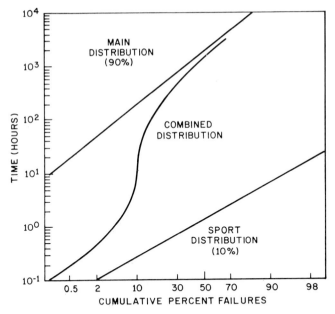

Fig. 11 A plot of a bimodal failure distribution with the two constituent log-normal distributions. *(After Peck, Ref. 29.)*

In any study performed under accelerated aging conditions, different failure mechanisms may be accelerated by different amounts for the same applied stress. A device may fail at normal operating conditions because of two completely different failure mechanisms. Under the applied stress, one of these failure modes may be accelerated much more than the other. Thus in these accelerated aging studies, we would see only one failure mode, and we may successfully eliminate that one mode. However, under normal operating conditions, we would have reduced the device failure rate by only a factor of 2.

Accelerated aging is a useful tool only if we know the device failure mechanisms and the acceleration of these mechanisms as a function of the applied stress. Adequate studies must be done under normal operating conditions to satisfy ourselves that no failure mechanisms remain that were not accelerated by the applied stress. The ultimate test of device reliability is long hours of operation under expected conditions of normal use.

14.7.1 Temperature Acceleration

Many of the mechanisms that cause failure are chemical or physical processes that can be accelerated by temperature. The reaction rate R at which these processes proceed is governed, in many cases, by the Arrhenius equation

$$R = R_0 \exp \left[- E_a / kT \right] \qquad (59)$$

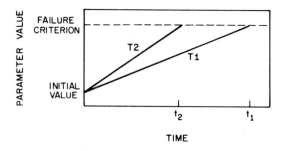

Fig. 12 Change of parameter value with time at different temperatures.

where E_a is the activation energy (in electron volts) of the process, k is the Boltzman constant equal to 8.6×10^{-5} eV/K, and T is the temperature in degrees absolute.

We can visualize some parameter of the IC that changes as a function of time. This parameter has some initial value, and IC failure takes place when the parameter exceeds some value which we call the failure criterion. Figure 12 shows the parameter increasing at different rates at the two temperatures T_1 and T_2, where $T_2 > T_1$. For the two temperatures, failure takes place at time t_1 and t_2, respectively, where $t_2 < t_1$. We have arbitrarily shown the parameter value to change linearly with time.

Assuming that the reaction causing the parameter value to change is governed by the Arrhenius equation, the ratio of the two times to failure is given by

$$\frac{t_2}{t_1} = \frac{R_2}{R_1} = \exp\left[\frac{E_a}{k}\left[\frac{1}{T_1} - \frac{1}{T_2}\right]\right] \tag{60}$$

The acceleration factor due to the increased temperature of operation is the ratio of the two times to failure

$$\text{Acceleration} = \frac{t_2}{t_1} = \exp\left[\frac{E_a}{k}\left[\frac{1}{T_1} - \frac{1}{T_2}\right]\right] \tag{61}$$

IC failure occurs when the destructive reaction has proceeded to some value equal to the failure criterion. The product of the reaction rate R and the time to failure t_F is a constant. Thus the time to failure at different temperatures is

$$t_F = \frac{\text{constant}}{R} = \text{constant} \times \exp\left(E_a/kT\right) \tag{62}$$

or

$$\ln t_F = \text{constant} + E_a/kT \tag{63}$$

The validity of the Arrhenius assumption and the value of the activation energy for a given failure mechanism can be determined by plotting the natural logarithm of the time to failure versus the inverse of the absolute temperature. If a linear plot over a reasonable range of temperature is observed, then extrapolating the failure time back to the operating temperature is valid. Note that the silicon chip temperature is

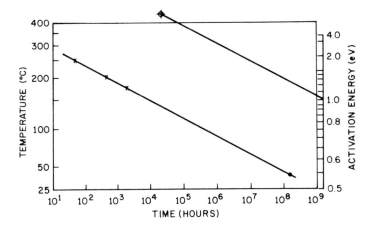

Fig. 13 Arrhenius plot of the failure data from Fig. 10. The median life at each temperature is plotted versus $1/T$.

higher than the package temperature; the difference is determined by the power dissipated by the chip and the thermal resistance of the package.

In this discussion we have assumed that the failure reaction is linear in time. Some reactions may have a different functional dependence. Generally the same treatment applies.[16]

To illustrate the acceleration of failure by temperature we show in Fig. 10 the failure distribution of a group of discrete transistors operated at three different temperatures. At all three temperatures the failure distribution is log-normal. The scale parameter at all three temperatures has a value of 2.0, supporting the assumption that the same failure mechanism is dominant at all three temperatures. This assumption should be verified by detailed analysis of the failed devices.

Figure 13 shows a plot of the median time to failure at the three temperatures, versus the inverse absolute temperature. The three data points fall on a straight line; the slope of the line corresponds to an activation energy of 1.0 eV. The median life extrapolated to an operating temperature of 40°C is approximately 2×10^8 h. The extrapolation in median time to failure from the data point at 175°C to the operating temperature is a factor of 10^5! Referring back to Table 4, we see that a median time to failure of 2×10^8 h, with a scale factor $\sigma = 2.0$, corresponds to a failure rate at the operating temperature which reaches a value of 10 FIT after 40 years of use.

Table 5 summarizes the acceleration factor due to operating at an increased temperature for two values of the activation energy. The table also shows the time at the increased temperature which is equivalent to 40 years of operation at a 60°C ambient temperature. For a failure mechanism with a 1.0-eV activation energy, 11 h of operation at 200°C is equivalent to 40 years of operation at 60°C.

The above discussion assumes that the activation energy is positive, that is, devices fail faster at higher temperatures. For several known failure mechanisms, however, the failure rate decreases at elevated temperatures. In one of these mechanisms[17] hot electrons generated in the silicon are captured in the gate oxide, causing a

Table 5 Acceleration factor and time equivalent to 40 years

T_{high}* (°C)	Acceleration factor		Time equivalent to 40 years (h)	
	$E_a = 1.0$ eV	0.5 eV	1.0 eV	0.5 eV
300	2.2×10^6	1500	0.2	233
250	3.2×10^5	570	1.1	616
200	3.1×10^4	176	11	2,000
150	1700	41	200	8,526
125	300	17	1,200	20,200
85	11.5	3.4	30,000	103,000

$*T_{amb} = 60°C$.

shift in the threshold voltage with time. These captured electrons are gradually released back to the silicon; the higher the temperature, the greater the release rate. At high temperatures no shift in threshold voltage is seen, because the release rate is equal to or greater than the capture rate. At low temperatures very large shifts in threshold voltage are seen, because the release rate becomes much smaller than the capture rate. It is imperative that the device failure rate be studied at *all* extremes of temperature.

14.7.2 Voltage and Current Acceleration

Voltage and current are effective accelerating stresses for many of the common failure mechanisms observed in ICs. Voltage (or in some cases electric field) causes acceleration of failure caused by dielectric breakdown, interface charge accumulation, charge injection, and corrosion. With most ICs, varying the applied voltage over a very large range is not possible. A device designed to operate at an applied voltage of 5 V usually will not function properly if the applied voltage is outside the 4- to 7-V range. Of course, some devices operate over a much wider range of applied voltage.

Most studies indicate that the reaction rate R of the failure mechanism is proportional to a power of the applied voltage. The power is usually a function of temperature, that is,

$$R(T,V) = R_0(T)V^{\gamma(T)} \tag{64}$$

The coefficient $R_0(T)$ is usually an Arrhenius function of T, and the parameter $\gamma(T)$ varies between about 1 to 4.5. Thus, if we operate a given device at 7 V rather than the nominal 5 V, we can accelerate the failure anywhere from 40% to as much as a factor of 4, not a very large acceleration.

In the case of dielectric breakdown, a different type of acceleration occurs.[18] For a given applied field a certain fraction of the devices fail in a very short time (on the order of seconds). Very few additional failures occur as the field is maintained. If the applied field is then increased, an additional fraction of failures occurs, again in a relatively short time. In a case such as this, operation of the device at an increased voltage is more in the nature of a screening or burn-in rather than a method of accelerating the failure mechanism. (Section 14.7.4 discusses this subject in more detail.)

Device operation at increased current levels is used primarily as a method of accelerating failures caused by electromigration in the metallic conductors. Studies show that the reaction rate R of electromigration is a function of temperature and current density of the form

$$R(T, J) = R_0(T)J^{\gamma(T)} \tag{65}$$

Again the coefficient $R_0(T)$ is an Arrhenius function whereas the parameter $\gamma(T)$ varies between about 1 and 4.

It is usually not possible to independently vary the current in an IC, as it is determined by the circuit design. Most studies of the acceleration of electromigration failures use special test structures.[19] These structures are used to determine the maximum allowable current density for which the failure rate of conductors will be acceptable. These values of maximum current density are then used as a design guide in IC layout.

14.7.3 Humidity-Temperature Acceleration

Usually, state-of-the-art ICs are packaged in hermetic packages, allowing evaluation and initial reliability studies to be carried out under optimum conditions. As the devices mature, competitive pressure to reduce costs results in the use of lower cost, nonhermetic packages. The presence of water vapor in the chip environment introduces a new variety of possible failure mechanisms.

Studies show that water vapor quickly permeates plastic packaging material.[20] In the first step of the permeation process, water vapor transports contaminants from the surface of the package through the plastic and leaches impurities from the plastic packaging material itself. The surface of the chip is very quickly exposed to the water vapor and various contaminants.

The second step is the diffusion of the contaminated water vapor through the passivation layer of the chip, usually a relatively slow process. However, if the passivation layer contains defects or cracks the water vapor penetrates through the passivation layer much more quickly. In these studies the penetration of water vapor through the passivation layer determined the reaction rate of the failure mechanisms.[20]

Once the water vapor reaches the metallization level of the chip, electrochemical corrosion can occur. The ions needed for this corrosion process can arise from two sources. As described above, ions can diffuse as a contaminant through the passivation layer along with the water vapor. However, if the intermediate dielectric of the chip is a phosphorus-doped glass, the water vapor can leach phosphorus from the intermediate dielectric. In either case, electrochemical corrosion is a very rapid process, which results in metallization failure. This type of failure mechanism can be accelerated by increasing the partial pressure of water vapor in the environment.

Table 6 Humidity induced failures[20]

Device	Bipolar	NMOS
Passivation layer	Sputtered SiO_2	SiN
Intermediate dielectric	Sputtered SiO_2	P-glass
Mechanism	Cl corrosion (contamination)	P corrosion (from P-glass)
σ	0.43	1.34
E_a	1.1 eV	0.3 eV
Accelerating factor (125−60°C)	300	5.5
40 yr/acceleration	1200 h	7.3 y

Increasing the ambient temperature also increases the rate-limiting diffusion of the water vapor through the passivation layer. Table 6 summarizes the analysis for the acceleration of the failure of two types of devices.[20] The acceleration factor obtained for the two types of devices is significantly different.

As with other types of stress, the application of an acceleration factor obtained for one type of failure mechanism in a given device to any other device is extremely dangerous. For each device, and each failure mechanism in that device, the accelerating effect of temperature, humidity, and voltage must be redetermined.

14.7.4 Burn-In

For mature products in high volume manufacture, a considerable effort is made in determining and eliminating the predominant failure mechanisms identified in the initial reliability studies. The steady-state failure rate of the device meets or exceeds the design goal. Possible wearout mechanisms will have been identified and eliminated, either through modifications in the process or in the design. However, the manufactured devices will normally show the existence of continuing early failure or infant mortality type of failures.

Generally, manufacturing defects cause the infant mortality type of failures. Examples of such defects include: oxide pinholes, photoresist or etching defects resulting in near opens or shorts, contamination on the chip or in the package, scratches, weak chip or wire bonds, and partially cracked chips or packages.

The contribution of the infant mortality failures to the overall failure rate can be modeled as a subpopulation of sport devices, using either the log-normal or some other distribution function. If the steady-state failure rate is extremely low, then the infant mortality failure rate is most easily modeled using the Weibull model, without making reference to any subpopulation. For this case the overall failure rate is expressed as

$$\lambda(t) = \frac{\beta}{\alpha} t^{\beta-1} + \lambda_{SS} \qquad (66)$$

where $\lambda(t)$ is the device failure rate, β and α are the parameters of the infant mortality failures, and λ_{SS} is the steady-state failure rate.

The purpose of a burn-in procedure is to operate the devices for some period of time during which most of the devices that are subject to infant mortality failure actually fail. Failing the burn-in procedure is preferable to having devices fail after they are installed in a system and delivered to the customer. Hopefully the conditions during burn-in accelerate the failure mechanisms that contribute to the infant mortality failures. The burn-in then may possibly consist of operating the devices under conditions of increased temperature, increased voltage, and high current (if possible) load conditions.

Studies of the acceleration of infant mortality failures under conditions of increased operating temperature[21] show that infant mortality failure mechanisms have an activation energy E_a of 0.37 to 0.42 eV. The acceleration factors for infant mortality failures by increased operating voltage have not been reported.

If devices are operated under burn-in conditions for some time t_{BI} where the acceleration of the infant mortality failures is A_{BI}, then the failure rate of the devices when they are placed in service is

$$\lambda_{EFF}(t) = \lambda(t + t_0)$$
$$= \lambda(t + t_{BI}A_{BI}) \qquad (67)$$

Figure 14 shows the effect of burn-in on the device failure rate. If the activation energy of the infant mortality failure is 0.4 eV, then a 168-h burn-in at 150°C is equivalent to over 4 months of operation at 60°C. This can result in a significant improvement in the reliability of the installed system.

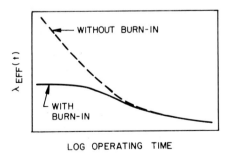

LOG OPERATING TIME

Fig. 14 Failure rate versus time in use for devices with and without burn-in. The broken curve is the failure rate $\lambda(t)$ without burn-in. The solid curve is the failure rate $\lambda_{EFF}(t) = \lambda(t + t_0)$ with burn-in.

Other methods of eliminating potential infant mortality failures from the devices shipped to customers are also in use, such as cycling the packaged devices over an extended temperature range to eliminate weak wire or chip bonds and partially cracked chips or packages. As mentioned in Section 14.7.2, device operation at an applied voltage above the nominal operating voltage screens out devices with oxide defects that may fail when the device is subjected to an overvoltage condition in actual service.

14.8 FAILURE MECHANISMS

Previous sections have examined the techniques for evaluating and predicting IC reliability and failure rates. Equally as important as the evaluation of reliability is the improvement of reliability. The viability of VLSI circuits is critically dependent upon the identification and elimination of those failure mechanisms that are introduced as technology makes its inexorable push toward higher density circuits of ever-increasing performance.

Table 7 summarizes the failure mechanisms that are commonly identified as degrading IC reliability. Many of these processes have been covered in previous chapters. More detailed information on specific failure processes is available in several review articles on the reliability of silicon ICs.[22, 23]

Table 7 lists those factors that contribute to each failure process and which factors may accelerate device degradation. In integrated circuit reliability studies the accelerating factors are varied, or applied at an increased value during a burn-in procedure. Table 7 includes the activation energy E_a for the processes that are accelerated by temperature and the coefficient for the processes that are accelerated as a power γ of the applied field or current density. Of prime importance in Table 7 is not the particular value of E_a or γ for a particular process, but the wide range over which these parameters vary. The exact value of E_a or γ must be determined for each process and each device made in that process.

In addition to the failure processes listed in Table 7, other failure mechanisms, two of which we will discuss next, are observed in silicon ICs.

14.8.1 Electrostatic Discharge Damage

The gate oxide in modern VLSI circuits is presently on the order of 250 Å thick and will be considerably thinner in future processes. The dielectric breakdown strength of SiO_2 is approximately 8×10^6 V/cm. Thus a 250-Å gate oxide will not sustain voltages in excess of 20 V. This is well in excess of the normal operating voltage of VLSI circuits. However, during handling of the devices, voltages higher than the breakdown voltage of the gate oxide may be placed upon the circuits.

A major source of excessive voltage arises from triboelectricity (electricity produced by rubbing two materials together). A person walking across a room or simply removing an IC from its plastic packaging material can generate from 15,000 to 20,000 V. The discharge of the triboelectrically generated charge into the IC can

Table 7 Time dependent failure mechanisms in silicon semiconductor devices[30]

Device association	Failure mechanism	Relevant factors	Accelerating factors	Acceleration (E_a = apparent activation energy for temp.)
Silicon oxide and silicon-silicon oxide interface	Surface charge accumulation	Mobile ions, V, T	T	$E_a = 1.0{-}1.05$ eV (depends upon ion density)
	Dielectric breakdown	$\mathcal{E}, T$	$\mathcal{E}, T$	$E_a = 0.2{-}1.0$ eV $\mathcal{E}^\gamma, \gamma(T) = 1{-}4.4$
	Charge injection	$\mathcal{E}, T, Q_f$	$\mathcal{E}, T$	$E_a = 1.3$ eV (slow trapping) $E_a \approx -1$ eV (hot electron injection)
Metallization	Electromigration	T, J, A, gradients of T and J, grain size	T, J	$E_a = 0.5{-}1.2$ eV $J^\gamma, \gamma(T) = 1{-}4$
	Corrosion (chemical, galvanic, electrolytic)	Contamination, H, V, T	H, V, T	Strong H effect $E_a \approx 0.3{-}1.1$ eV (for electrolysis) V may have thresholds
	Contact degradation	T, metals, impurities	Varied	
Bonds and other mechanical interfaces	Intermetallic growth	T, impurities, bond strength	T	Al-Au: $E_a = 1.0{-}1.05$ eV
	Fatigue	Bond strength, temperature cycling	Temp. extremes in cycling	
Hermeticity	Seal leaks	Pressure differential, atmosphere	Pressure	

Note: V—voltage; T—temperature; $\mathcal{E}$—electric field; J—current density; A—area; H—humidity.

either damage the circuit sufficiently to cause immediate failure or damage the device slightly, but enough to cause subsequent failure early in the operating life of the device.

Burn-in will not reduce device failure caused by electrostatic discharge (ESD). In fact, additional handling of the devices during the burn-in procedure may actually increase the early device failure rate if adequate safeguards are not taken to prevent the buildup of static electricity. Electrostatic charging theory, ESD failure models, and methods for preventing ESD failures are available.[24]

Even with proper precautions the discharge of possibly several hundred volts of static electricity occurs during IC handling. The input and output circuit terminals are designed with protection networks on the chip that provide a path for the discharge of the current and prevent the generation of excessive voltage across the gate oxide of

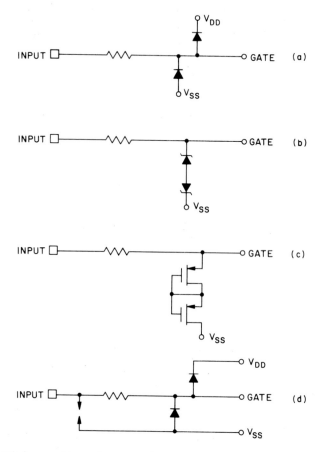

Fig. 15 Schematic for several types of gate protection networks. (a) Diode. (b) Zener diode. (c) Transistor. (d) Spark gap. *(After Wood, Ref. 23.)*

the devices. Figure 15 illustrates several types of gate protection networks. For the circuit in Fig. 15a, if a positive voltage is discharged into the input terminal, the upper diode is forward biased, and the discharge current will flow through the diode to the V_{DD} terminal of the device. No excess voltage is generated across the gate oxide. Similarly, for the discharge of a negative voltage into the input terminal, the current will forward bias the lower diode and flow to the V_{SS} terminal. For the circuit in Fig. 15b, when a discharge into the input terminal occurs, one of the zener diodes will be forward biased and the other will be biased to its reverse breakdown potential, allowing the discharge of the current to the V_{SS} terminal. For the circuit in Fig. 15c, the discharge into the input terminal either flows through the source diode to the V_{SS} terminal, or the potential turns on the high threshold voltage transistors, allowing the current to flow to V_{SS}. The spark gap in Fig. 15d provides an additional path for the discharge of current to V_{SS}. In all the circuits the resistor limits the peak current flow. The diodes and other current paths of these protection networks must be designed to handle the high currents and powers produced in a typical ESD event. A human body, initially charged to a potential of 2000 V, will cause a peak current flow of several amperes and a peak power dissipation of several kilowatts when the body is discharged into an IC.

14.8.2 α-Particle-Induced Soft Errors

Radioactive elements, such as uranium or thorium, are naturally occurring impurities in IC packaging materials. The α particles emitted by these materials can cause soft errors in the ICs.[25, 26] The term "soft error" refers to a random failure not related to a physically defective device. The penetration of an α particle into the silicon causes the generation of an electron-hole plasma along the path of the particle. The generated carriers can cause the loss of information stored in the memory cells of a dynamic memory or stored in the depletion region of the drains of devices making up circuits such as dynamic shift registers or other logic units. The adsorption of a 4-MeV α particle can generate 10^6 electron-hole pairs,[27] which is equal to or greater than the charge stored in a dynamic memory cell. To illustrate the seriousness of the problem, a typical soft error rate in a memory system[22] containing 1000 16K memory devices may be on the order of one soft error per 1000 h, corresponding to a device failure rate of 1000 FIT.

The incidence of soft errors can be reduced by surrounding or coating the IC chip with a material having a very low density of radioactive contamination.[28] For example, α particles with an energy up to 8 MeV are completely absorbed in 50 μm of silicone rubber, and this material does not emit any significant amount of α particles.

14.9 SUMMARY AND FUTURE TRENDS

The trend in modern VLSI technology, as exemplified in previous chapters, has the tendency to exacerbate the problems of maintaining high yield and reliability. Progress in VLSI technology is achieved at the cost of finer dimensions, larger chip area, more complex processes, and the introduction of new and more complex material systems. All of these inevitably lead to initially higher defect density. Identifying, characterizing, and ultimately eliminating the causes of circuit failure and low yield is a necessary part of the development of a new technology.

Similarly, the cost of progress will be devices operating with higher electric fields, higher current densities, and higher power dissipation. All of these factors will accelerate the failure of devices due to those failure mechanisms previously identified and characterized. In addition, the introduction of new material systems introduces new types of IC failure. The successful development of new technologies must include the identification and characterization of these failure processes. The combined efforts of technology development along with yield and failure mode analysis will provide the technology needed to produce VLSI circuits usable in economically viable systems.

REFERENCES

[1] W. Feller, *An Introduction to Probability Theory and Its Applications*, 3d ed. Wiley, New York, 1968.

[2] S. E. Schuster, "Multiple Word/Bit Line Redundancy for Semiconductor Memories," *IEEE J. Solid State Circuits*, **SC13**, 698 (1978).

[3] B. T. Murphy, "Cost-Size Optima of Monolithic Integrated Circuits," *Proc. IEEE*, **52**, 1537 (1964).

[4] C. H. Stapper, Jr., "On a Composite Model to the I.C. Yield Problem," *IEEE J. Solid State Circuits*, **SC10**, 537 (1975).

[5] W. W. Hines and D. C. Montgomery, *Probability and Statistics in Engineering and Management Science*, Wiley, New York, 1972.

[6] K. V. Bury, *Statistical Models in Applied Science*, Wiley, New York, 1975.

[7] H. Murrman and D. Kranzer, "Yield Modeling of Integrated Circuits," *Siemens Forsch. Entwicklungs Ber.*, **9**, 38 (1981).

[8] C. H. Stapper, "LSI Yield Modeling and Process Monitoring," *IBM J. Res. Dev.*, **20**, 228 (1976).

[9] A. Gupta, W. A. Porter, and J. W. Lathrop, "Defect Analysis and Yield Degradation of Integrated Circuits," *IEEE J. Solid State Circuits*, **SC9**, 96 (1974).

[10] C. H. Stapper, "Yield Model for 256K RAMs and Beyond", *IEEE Int. S.S.C.C. Digest of Tech. Papers*, 12 (1982).

[11] W. G. Ansly, "Computation of Integrated Circuit Yields from the Distribution of Slice Yields for the Individual Devices," *IEEE Trans. Electron Devices*, **ED15**, 405 (1968).

[12] A. Gupta and J. W. Lathrop, "Yield Analysis of Large Integrated Circuit Chips," *IEEE J. Solid State Circuits*, **SC7**, 389 (1972).

[13] T. Yanagawa, "Yield Degradation of Integrated Circuits Due to Spot Defects," *IEEE Trans. Electron Devices*, **ED19**, 190 (1972).

[14] J. T. Duane, "Learning Curve Approach to Reliability Monitoring," *IEEE Trans. Aerosp.*, **2**, 563 (1964).

[15] L. R. Goldthwaite, "Failure Rate Study for the Lognormal Lifetime Model," *IEEE Proc. 7th Symposium on Reliability and Quality Control*, 1961, p. 208.

[16] R. E. Weston and H. A. Schwartz, *Chemical Kinetics*, Prentice-Hall, Englewood Cliffs, N.J., 1972.

[17] R. C. Sun, J. T. Clemens, and J. T. Nelson, "Effects of Silicon Nitride Encapsulation on MOS Device Stability," *Reliability Physics, 18th Annual Proceedings*, 1980, pp. 244–251.

[18] E. S. Anolick and L-Y. Chen, "Application of Step Stress to Time Dependent Breakdown," *Reliability Physics, 19th Annual Proceedings*, 1981, pp. 23–27.

[19] S. Vaidya, D. B. Fraser, and A. K. Sinha, "Electromigration Resistance of Fine-Line Al for VLSI Applications," *Reliability Physics, 18th Annual Proceedings*, 1980, pp. 165–170.

[20] J. E. Gunn and S. K. Malik, "Highly Accelerated Temperature and Humidity Stress Test Technique (HAST)," *Reliability Physics, 19th Annual Proceedings*, 1981, pp. 48–51.

[21] D. S. Peck, "New Concerns About Integrated Circuit Reliability," *Reliability Physics, 16th Annual Proceedings*, 1978, pp. 1–6.

[22] C. M. Bailey, "Basic Integrated Circuit Failure Mechanisms," in L. Esaki, Ed., *Large Scale Integrated Circuits: State of the Art and Prospects*, Lange Voorkout, The Hague, 1982.

[23] J. Wood, "Reliability and Degradation of Silicon Devices and Integrated Circuits," in M. J. Howes and D. V. Morgan, Eds., *Reliability and Degradation*, Wiley, New York, 1981.

[24] B. A. Unger, "Electrostatic Discharge Failures of Semiconductor Devices," *Reliability Physics, 19th Annual Proceedings*, 1981, pp. 193–199.

[25] T. C. May and M. H. Woods, "A New Physical Mechanism for Soft Errors in Dynamic Memories," *Reliability Physics, 16th Annual Proceedings*, 1978, pp. 33–40.

[26] T. C. May and M. H. Woods, "Alpha-Particle-Induced Soft Errors in Dynamic Memories," *IEEE Trans. Electron Devices*, **ED26**, 2 (1979).

[27] C. M. Hsieh, P. C. Murley, and R. R. O'Brien, "Dynamics of Charge Collection from Alpha Particle Tracks in Integrated Circuits," *Reliability Physics, 19th Annual Proceedings*, 1981, pp. 38–42.

[28] M. L. White, et al., "The Use of Silicone RTV Rubber for Alpha Particle Protection of Silicon Integrated Circuits," *Reliability Physics, 19th Annual Proceedings*, 1981, pp. 43−47.

[29] D. S. Peck, "The Analysis of Data from Accelerated Stress Tests," *Reliability Physics, 9th Annual Proceedings*, 1971, pp. 69−77.

[30] D. S. Peck, "Practical Applications of Accelerated Testing-Introduction," *Reliability Physics, 13th Annual Proceedings*, 1975, pp. 253−254.

PROBLEMS

1 The wafer in Fig. 1 has a total of 266 chip sites and 99 good chips. If the wafer grid is divided into new grids containing double, triple, and quadruple chip sites, there are 126, 83, and 58 double, triple, and quadruple chip sites, respectively. The number of good double, triple, and quadruple chips is 29, 14, and 7, respectively. Examination of the wafer shows that no good chips are found over approximately 46.6% of the area of the wafer, corresponding to a Y_0 of 53.4%.

   (a) Estimate from the yield of good chips the D_0A based upon Poisson, triangular, and rectangular distribution functions of D.

   (b) Using these D_0A estimates, predict the yield of a chip with four times the chip area in Fig. 1. Compare your answer to the observed yield of good quadruple chips.

2 Using the data from Problem 1 for the observed yield of multiple chips, does the data give a good fit to Eq. 26, or would a better fit be obtained using the gamma distribution yield of Eq. 17?

3 A group of 10 devices is placed on an accelerated aging test at 150°C and operated for a total time of 3000 h. The devices are removed from the test and tested for electrical characteristics every 100 h. The devices fail at the following test times: 200, 800, 1000, 1700, 2800, and 2900 h. At 3000 h four devices are still operating. Fit the data to both log-normal and Weibull distribution functions. [Note: When plotting data of this type it is accepted practice to use (number failed)/(number on test + 1) for the fraction of devices failed.]

   (a) Does the data fit one distribution function better than the other?

   (b) What are the median life and σ for the log-normal distribution?

   (c) What is the median life for the Weibull distribution?

   (d) At what time do the two distributions predict that 99% of the devices will fail?

   (e) A second group of 10 devices from the same manufacturing lot are aged under identical conditions. Failures were found at 200, 1600, 1700, and 2400 h. At 3000 h six devices were still good. Analyze this data in the same manner as above.

   (f) Combine the results from the two experiments, and estimate the median life and σ for these devices when aged at 150°C.

4 A third group of 20 devices from the same production lot as in Problem 3 is aged at 200°C. The devices are tested after every 10 h of aging. Devices fail at the following times: 10, 30, 40, 50 (two failures), 80, 100 (two failures), and 110 h. Testing was discontinued at this time.

   (a) What is the best estimate of the log-normal median life and σ?

   (b) Assuming that failure mode analysis indicates that the devices are failing by the same mechanism at both 150 and 200°C and that the Arrhenius relation thus applies, calculate the activation energy and the estimated mean life at operating temperatures of 70 and 50°C.

   (c) Making reasonable estimates for the error in the determination of the median life at 200°C and 150°C, what is a reasonable estimate for the lower limit of the median life at 70°C?

5 Using the results of Problems 3 and 4:

   (a) What is the best estimate of the failure rate of the device when operated at a temperature of 70°C, at an operating time of 4 months, 1 year, and 10 years?

   (b) Using the lower limit for the value of the mean life at 70°C (from Problem 4), what is the worst-case prediction of the failure rate at 4 months, 1 year, and 10 years?

   (c) What are the best estimates of the failure rate if the operating temperature is reduced to 50°C?

6 For a median life of 4.0×10^6 h and a σ of 1.3;

(a) What fraction of devices will have failed after 10 years of use at 70°C?

(b) If 100 of these devices are used in a system, what fraction of the systems will have failed in 10 years? (Hint: Use the Poisson approximation to the binomial probability, Eq. 4, that no failed devices are found in a given system.)

(c) What is the expected fraction of failed devices after 10 years of operation using the lower limit for the median life of 1.8×10^6 h?

(d) What is the expected fraction of failed systems?

(e) What is the expected fraction of failed devices and systems after 10 years of operation if the system is operated at 50°C where the expected mean life is 4.0×10^7 h?

7 Twenty sample devices from another production lot of the same devices as in the previous problems were evaluated at an aging temperature of 150°C. Some of the devices in this lot were believed to have been improperly processed and contaminated with sodium. These 20 devices were tested after 2, 4, 8, ..., and 4096 h of aging. Failures were found at the following times: 32, 64 (three failures), 128, 512, 1024 (two failures), 2048 (three failures), and 4096 (two failures) h. Plot the failure distributions on log-normal plotting paper.

(a) Estimate from the inflection point of the failure distribution the fraction of contaminated or sport devices in the population.

(b) Separate the devices into sport and normal devices. Replot the failure data to determine the median life and σ of each population.

8 Assume that the sodium-contaminated sport devices comprise 25% of the population, that the median life at 150°C is 60 h, the σ is 0.7, and the activation energy is 1.0 eV. Assume that the main population has a median life at 150°C of 4800 h, a σ of 1.3, and an activation energy of 1.10 eV.

(a) Plot the failure rate of the devices as a function of time in use at a temperature of 70°C for devices used as produced.

(b) Would a 150-h burn-in at 150°C provide an effective method of reducing the failure rate of these devices during normal operation at 70°C?

PROPERTIES OF SILICON (at 300 K)

Properties	Si
Atoms/cm^3	5.0×10^{22}
Atomic weight	28.09
Breakdown field (V/cm)	$\sim 3 \times 10^5$
Crystal structure	Diamond
Density (g/cm^3): Solid Liquid (1412°C)	2.33 2.53
Dielectric constant	11.9
Effective density of states in conduction band, N_C (cm^{-3})	2.8×10^{19}
Effective density of states in valence band, N_V (cm^{-3})	1.04×10^{19}
Effective mass, m^*/m_0: Electrons Holes	$m_l^* = 0.98,\ m_t^* = 0.19$ $m_{lh}^* = 0.16,\ m_{hh}^* = 0.49$
Electron affinity (V)	4.05
Energy gap (eV)	1.12
Heat capacity (cal/g-mol-°C): Solid Liquid (1412°C)	4.78 6.76
Index of refraction	3.42
Intrinsic carrier concentration (cm^{-3})	1.45×10^{10}
Intrinsic Debye Length (μm)	24
Intrinsic resistivity (Ω-cm)	2.3×10^5
Lattice constant (Å)	5.43095

Properties	Si
Linear coefficient of thermal expansion, $\Delta L / L \Delta T (°C^{-1})$	2.6×10^{-6}
Melting point (°C)	1412
Minority-carrier lifetime (s)	2.5×10^{-3}
Mobility (drift) $(cm^2/V\text{-}s)$:	
μ_n (electrons)	1500
μ_p (holes)	475
Optical phonon energy (eV)	0.063
Phonon mean free path λ_0 (Å):	
Electrons	76
Holes	55
Poisson's ratio	0.42
Specific heat (J/g-°C)	0.7
Thermal conductivity (W/cm-°C):	
Solid	1.5
Liquid (1412°C)	4.3
Thermal diffusivity (cm^2/s)	0.9
Torsion modulus (kg/mm^2)	4050
Vapor pressure (Pa)	1 at 1650°C 10^{-6} at 900°C
Young's modulus (kg/mm^2)	10,890

LIST OF SYMBOLS

Symbol	Description	Unit
a	Lattice constant	Å
$\mathcal{B}$	Magnetic induction	Wb/m^2
c	Speed of light in vacuum	cm/s
C	Capacitance	F
D	Electric displacement	C/cm^2
D	Diffusion coefficient	cm^2/s
E	Energy	eV
E_F	Fermi energy level	eV
E_g	Energy bandgap	eV
$\mathcal{E}$	Electric field	V/cm
$\mathcal{E}_m$	Maximum field	V/cm
f	Frequency	Hz
h	Planck's constant	J-s
$h\nu$	Photon energy	eV
I	Current	A
J	Current density	A/cm^2
k	Boltzmann constant	J/K
kT	Thermal energy	eV
L	Length	cm or μm
m_0	Electron rest mass	kg
m^*	Effective mass	kg
$\bar{n}$	Refractive index	
n	Density of free electrons	cm^{-3}
n_i	Intrinsic density	cm^{-3}
N	Doping concentration	cm^{-3}
N_A	Acceptor impurity density	cm^{-3}
N_D	Donor impurity density	cm^{-3}

Symbol	Description	Unit
p	Density of free holes	cm^{-3}
P	Pressure	N/m^2
q	Magnitude of electronic charge	C
Q_{it}	Interface-trap density	charges/cm^2
R	Resistance	Ω
t	Time	s
T	Absolute temperature	K
v	Carrier velocity	cm/s
v_s	Saturation velocity	cm/s
V	Voltage	V
V_{bi}	Built-in potential	V
V_B	Breakdown voltage	V
W	Thickness	cm or μm
x	x direction	
∇T	Temperature gradient	K/cm
ϵ_0	Permittivity in vacuum	F/cm
ϵ_s	Semiconductor permittivity	F/cm
ϵ_i	Insulator permittivity	F/cm
ϵ_s / ϵ_0 or ϵ_i / ϵ_0	Dielectric constant	
τ	Lifetin e or decay time	s
θ	Angle	rad
λ	Wavelength	μm or Å
ν	Frequency of light	Hz
μ_0	Permeability in vacuum	H/cm
μ_n	Electron mobility	cm^2/V-s
μ_p	Hole mobility	cm^2/V-s
ρ	Resistivity	Ω-cm
ϕ	Barrier height or imref	V
ϕ_m	Metal work function	V
ω	Angular frequency ($2\pi f$ or $2\pi\nu$)	Hz
Ω	Ohm	Ω

INTERNATIONAL SYSTEM OF UNITS

Quantity	Unit	Symbol	Units
Length	meter	m	
Mass	kilogram	kg	
Time	second	s	
Temperature	kelvin	K	
Current	ampere	A	
Frequency	hertz	Hz	$1/s$
Force	newton	N	$kg\text{-}m/s^2$
Pressure	pascal	Pa	N/m^2
Energy	joule	J	N-m
Power	watt	W	J/s
Electric charge	coulomb	C	A-s
Potential	volt	V	J/C
Conductance	siemens	S	A/V
Resistance	ohm	Ω	V/A
Capacitance	farad	F	C/V
Magnetic flux	weber	Wb	V-s
Magnetic induction	tesla	T	Wb/m^2
Inductance	henry	H	Wb/A

PHYSICAL CONSTANTS

Quantity	Symbol	Value
Angstrom unit	$\overset{\circ}{A}$	$1\overset{\circ}{A} = 10^{-1}\,\text{nm} = 10^{-4}\,\mu\text{m}$ $= 10^{-8}\,\text{cm} = 10^{-10}\,\text{m}$
Avogadro constant	N_{AVO}	$6.02204 \times 10^{23}\,\text{mol}^{-1}$
Bohr radius	a_B	$0.52917\,\overset{\circ}{A}$
Boltzmann constant	k	$1.38066 \times 10^{-23}\,\text{J/K}\ (R/N_{AVO})$
Elementary charge	q	$1.60218 \times 10^{-19}\,\text{C}$
Electron rest mass	m_0	$0.91095 \times 10^{-30}\,\text{kg}$
Electron volt	eV	$1\,\text{eV} = 1.60218 \times 10^{-19}\,\text{J}$ $= 23.053\,\text{kcal/mol}$
Gas constant	R	$1.98719\,\text{cal mol}^{-1}\,\text{K}^{-1}$
Permeability in vacuum	μ_0	$1.25663 \times 10^{-8}\,\text{H/cm}\ (4\pi \times 10^{-9})$
Permittivity in vacuum	ϵ_0	$8.85418 \times 10^{-14}\,\text{F/cm}\ (1/\mu_0 c^2)$
Planck constant	h	$6.62617 \times 10^{-34}\,\text{J-s}$
Reduced Planck constant	$\hbar$	$1.05458 \times 10^{-34}\,\text{J-s}\ (h/2\pi)$
Proton rest mass	M_p	$1.67264 \times 10^{-27}\,\text{kg}$
Speed of light in vacuum	c	$2.99792 \times 10^{10}\,\text{cm/s}$
Standard atmosphere		$1.01325 \times 10^5\,\text{N/m}^2$
Thermal voltage at 300 K	kT/q	$0.0259\,\text{V}$
Wavelength of 1-eV quantum	λ	$1.23977\,\mu\text{m}$

INDEX